FROM GENES TO CELLS

FROM GENES TO CELLS

Stephen R. Bolsover
Jeremy S. Hyams
Steve Jones
Elizabeth A. Shephard
Hugh A. White

University College London

A John Wiley & Sons, Inc., Publication

New York • Chichester • Weinheim • Brisbane • Singapore • Toronto

Address All Inquiries to the Publisher
Wiley-Liss, Inc., 605 Third Avenue, New York, NY 10158-0012

This text is printed on acid-free paper.

Library of Congress Cataloging-in-Publication Data

From genes to cells / Stephen R. Bolsover . . . [et al.].
p. cm
Includes index.
ISBN 0-471-59792-9 (pbk. : alk. paper)
1. Molecular biology. 2. Cytology. I. Bolsover, Stephen R., 1954- .
QH506.F753 1997
574.87—dc20 96-28603

Printed in the United States of America

10 9 8 7 6 5 4 3 2 1

CONTRIBUTORS

Stephen R. Bolsover, Department of Physiology, Rockefeller Building, University College London, Gower Street, London, WC1E 6BT

Jeremy S. Hyams, Department of Biology, Darwin Building, University College London, Gower Street, London, WC1E 6BT

Steve Jones, Department of Biology, Galton Laboratory, Wolfson House, University College London, Gower Street, London, WC1E 6BT

Elizabeth A. Shephard, Department of Biochemistry and Molecular Biology, Darwin Building, University College London, Gower Street, London, WC1E 6BT

Hugh A. White, Department of Biochemistry and Molecular Biology, Darwin Building, University College London, Gower Street, London, WC1E 6BT

CONTENTS

PREFACE xvii

CHAPTER 1: CELLS AND TISSUES 1

Principles of Microscopy, 2
  The Light Microscope, 3
  The Electron Microscope, 6
  The Scanning Electron Microscope, 7
Only Two Types of Cell, 8
  Special Properties of Plant Cells, 8
  Origin of Eukaryotic Cells, 10
  Tissues: Cell Collectives, 10
Summary, 14

CHAPTER 2: MEMBRANES AND ORGANELLES 21

Basic Properties of Cell Membranes, 21
  Lipids are Esters of Glycerol and Fatty Acid, 22
  Membranes Are Made of Phospholipids, 23
  Straight Through the Membrane: Simple Diffusion, 24
  Beyond the Plasmalemma: The Extracellular Matrix, 24
  Cell Junctions, 29
Organelles Bounded by Double Membrane Envelopes, 32
  The Nucleus, 32

Mitochondria and Chloroplasts, 34
Mitochondria, 34
Chloroplasts, 34
Organelles Bounded by a Single Membrane, 36
Peroxisomes, 36
Endoplasmic Reticulum, 36
Golgi Apparatus, 37
Endocytotic Vesicles and Endosomes, 37
Lysosomes, 38
Summary, 38

CHAPTER 3: THE CYTOSKELETON AND CELL MOVEMENT 41
The Cytoskeleton is Composed of Three Classes of Filaments, 42
Microtubules, 42
Microtubule-Based Motility, 46
Cilia and Flagella, 46
Intracellular Transport, 51
Microfilaments, 53
Muscle Contraction, 54
Cell Locomotion, 54
Cytoplasmic Streaming, 56
Intermediate Filaments, 56
Summary, 57

CHAPTER 4: CELL DIVISION 59
Control of the Cell Division Cycle, 60
Molecular Regulation of the G2 (Interphase/Mitosis) Cell Cycle Control Point, 62
The G1 Control Point, 65
Cell Division: Mitosis and Meiosis, 66
Mitosis, 66
Prophase, 66
Prometaphase, 66
Metaphase, 68
Anaphase, 68
Telophase, 68
Meiosis, 68
Summary, 69

CHAPTER 5: DNA STRUCTURE AND THE GENETIC CODE 73
The Structure of DNA, 73
The DNA Molecule is a Double Helix, 76

Complementarity of DNA Chains, 78
Different Forms of DNA, 78
DNA as the Genetic Material, 78
The Packaging of DNA into Chromosomes, 79
Eukaryotic Chromosomes and Chromatin Structure, 79
Prokaryotic Chromosomes, 80
Plasmids, 80
Viruses, 82
The Genetic Code, 82
The Code is Degenerate but Unambiguous, 83
The Start and Stop Codes for Protein Synthesis, 85
The Code is Nearly Universal, 86
Missense Mutations, 86
Summary, 86

CHAPTER 6: PATTERNS OF INHERITANCE 89
Mendel and the Foundations of Genetics, 90
The Law of Segregation, 90
The Law of Independent Assortment, 92
Mendelian Inheritance in Humans, 94
Modifying Mendel: Changes in Ratios, 97
Lethal Alleles, 97
Gene Interaction, 99
Cytoplasmic Inheritance, 100
Multifactorial Inheritance, 100
Nature, Nurture, or Both? 101
Summary, 102

CHAPTER 7: MAPPING THE GENES 105
Genes and Chromosomes, 105
Linkage Mapping and the Structure of the Genome, 107
Making the Physical Map, 111
Introns and Exons: The Complexity of Eukaryotic Genes, 114
The Major Classes of Eukaryotic DNA, 115
Summary, 116

CHAPTER 8: GENE MUTATION 119
Measuring the Mutation Rate, 119
The "Natural" Rate of Mutation, 123
Mutagenesis, 123
Repairing Mutations, 125

The Nature of Mutation, 126
Mutation is Not Random, 128
Summary, 130

CHAPTER 9: DNA REPLICATION AND DNA REPAIR 133
DNA Replication, 133
DNA Replication is Semiconservative, 133
The DNA Replication Fork, 134
Proteins Open Up the DNA Double Helix During Replication, 134
DNA Helicases, 136
Single-Stranded Binding Proteins, 136
Topoisomerase I, 136
The Biochemistry of DNA Replication, 137
DNA Polymerase III, 137
DNA Synthesis Requires an RNA Primer, 138
The Self-Correcting DNA Polymerase, 139
DNA Repair, 140
Spontaneous and Chemically Induced Base Changes, 140
Repair Processes, 140
Summary, 143

CHAPTER 10: TRANSCRIPTION AND THE CONTROL OF GENE EXPRESSION 145
The Structure of RNA, 145
RNA Polymerase, 146
Gene Notation, 146
Bacterial RNA Synthesis, 148
Control of Bacterial Gene Expression, 149
lac, an Inducible Operon, 152
trp, a Repressible Operon, 156
Eukaryotic RNA Synthesis, 157
Messenger RNA Processing, 158
Control of Eukaryotic Gene Expression, 160
Summary, 163

CHAPTER 11: RECOMBINANT DNA AND GENETIC ENGINEERING 167
DNA Cloning, 167
Creating the Clone, 168
Introduction of Foreign DNA Molecules into Bacteria, 169
Cloning Vectors, 169

Joining Foreign DNAs to a Cloning Vector, 170
Introduction of Recombinant Plasmids Into Bacteria, 171
Selection of cDNA Clones, 172
Genomic DNA Clones, 175
Uses of DNA Clones, 178
DNA Sequencing, 178
Southern Blotting, 181
In Situ Hybridization, 181
Production of Mammalian Proteins in Bacteria, 183
Protein Engineering, 184
Polymerase Chain Reaction, 186
Summary, 186

CHAPTER 12: TRANSLATION AND PROTEIN TARGETING 189
The Attachment of an Amino Acid to its tRNA, 190
Transfer RNA, the Anticodon, and the Wobble, 190
The Ribosome, 192
Bacterial Protein Synthesis, 193
Chain Initiation, 193
Ribosome-Binding Site, 193
The 70S Initiation Complex, 194
Elongation of the Protein Chain, 194
The Polyribosome, 197
Termination of Protein Synthesis, 197
The Ribosome is Recycled, 198
Eukaryotic Protein Synthesis is a Little More Complex, 199
Antibiotics and Protein Synthesis, 200
Protein Targeting, 200
Golgi Apparatus, 204
Trans Golgi Network, 204
Targeting Proteins to the Lysosome, 205
Summary, 207

CHAPTER 13: PROTEIN STRUCTURE 209
Polymers of Amino Acids, 210
The Amino Acid Building Blocks, 210
The Unique Properties of Each Amino Acid, 216
Charge, 216
UV Absorbance, 217
Disulfide Bridging, 217
Phosphorylation, 217
Other Amino Acids Are Found in Nature, 218

The Three Dimensional Structures of Proteins, 219
Prosthetic Groups, 230
The Primary Structure Contains All of the Information Necessary to Specify Higher-Level Structures, 231
Summary, 232

CHAPTER 14: HOW PROTEINS WORK 235
Dynamic Protein Structures, 235
Allosteric Effects, 236
Chemical Changes that Shift the Preferred Shape, 238
Enzymes Are Protein Catalysts, 240
The Initial Velocity of an Enzyme Reaction, 241
The Effect of Substrate Concentration, 242
The Effect of Enzyme Concentration, 244
The Specificity Constant, 244
Enzyme Catalysis, 245
Cofactors and Prosthetic Groups, 246
Enzymes Can Be Regulated, 246
Summary, 248

CHAPTER 15: ENERGY TRADING WITHIN THE CELL 251
Cellular Energy Currencies, 251
Nicotinamide Adenine Dinucleotide (NADH), 253
Nucleoside Triphosphates (ATP plus GTP, CTP, TTP, and UTP), 253
The Hydrogen Ion Gradient across the Mitochondrial Membrane, 255
The Sodium Gradient across the Plasmalemma, 256
The Energy Currencies are Interconvertible, 256
Exchange Mechanisms that Convert between the Four Energy Currencies, 256
The Electron Transport Chain, 258
ATP Synthase, 258
Sodium/Potassium ATPase, 259
ADP/ATP Exchanger, 260
All Carriers Can Change Direction, 260
Summary, 261

CHAPTER 16: METABOLISM 263
The Krebs Cycle: The Central Switching Yard of Metabolism, 264
From Glucose to Pyruvate: Glycolysis, 266
Glycolysis without Oxygen, 268
Glycogen Can Feed the Glycolytic Pathway, 269
From Fats to Acetyl-CoA: Lipolysis and β Oxidation, 269

Amino Acids As Another Source of Metabolic Energy, 269
Making Glucose: Gluconeogenesis, 272
Making Fatty Acids, 274
Synthesis of Amino Acids, 278
Carbon Fixation in Plants, 278
Control of Energy Production, 279
  Feedforward and Feedback, 279
  Negative Feedback Control of Glycolysis, 280
  Feedforward Control in Muscle Cells, 281
Summary, 282

CHAPTER 17: IONS AND VOLTAGES 285
The Potassium Gradient and the Resting Voltage, 285
  Potassium Channels Make the Plasmalemma Permeable to Potassium Ions, 286
  Concentration Gradients and Electrical Voltage Can Balance, 287
The Chloride Gradient, 291
General Properties of Channels, 291
General Properties of Carriers, 292
  The Glucose Carrier, 293
  The Sodium/Calcium Exchanger, 294
  Carriers with an Enzymic Action: The Calcium ATPase, 295
Summary, 297

CHAPTER 18: THE ACTION POTENTIAL 303
The Calcium Action Potential in Sea Urchin Eggs, 303
  Effect of Egg Membrane Voltage on Sperm Fusion, 303
  The Voltage-Gated Calcium Channel, 305
  The Calcium Action Potential, 306
The Voltage-Gated Sodium Channel in Nerve Cells, 309
  Myelination and Rapid Action Potential Transmission, 313
Summary, 314

CHAPTER 19: INTRACELLULAR MESSENGERS 317
Calcium, 317
  Calcium Can Enter From the Extracellular Medium: Exocytosis at the Axon Terminal, 318
  Calcium Can Be Released From the Endoplasmic Reticulum: Platelet Activation, 320
  The Processes Activated by Cytosolic Calcium are Extremely Diverse, 323
  Return of Cytosolic Calcium to Resting Levels, 323
Cyclic Adenosine Monophosphate (cAMP), 325

Cyclic Guanosine Monophosphate (cGMP), 327
Multiple Messengers, 328
Summary, 328

CHAPTER 20: INTERCELLULAR COMMUNICATION 331
Classifying Transmitters and Receptors, 331
Ionotropic Cell Surface Receptors, 332
Metabotropic Cell Surface Receptors, 332
Intracellular Receptors, 333
Intercellular Communication in Action: The Gastrocnemius Muscle, 334
Telling the Muscle to Contract: The Action of Motoneurones, 336
Controlling the Blood Supply: Paracrine Transmitters, 336
New Blood Vessels in Growing Muscles: Growth Factors, 339
Summary, 340

CHAPTER 21: CYSTIC FIBROSIS 345
Cystic Fibrosis is a Severe Genetic Disease, 346
The Fundamental Lesion Lies in Chloride Transport, 347
Homing in on the Gene Using Classical Genetics, 348
Molecular Genetics Carried the Search Forward, 350
A Cure for CF—or a Premature Hope? 351
Summary, 353

SUGGESTED FURTHER READING 355

GLOSSARY 361

INDEX 403

PREFACE

This book gives an introduction to a substantial part of modern cell biology in 353 pages. It is intended to provide the student entering this field for the first time with a guide that embraces areas as diverse as genetics and neuroscience. This is a task that is especially necessary now, when advances in molecular genetics are changing almost every aspect of our lives. It is also, paradoxically, a task that is possible now in a way that would not have been possible even ten years ago. The explosion of knowledge within biology that has occurred in the last 30 years has revealed that phenomena that seemed widely disparate are part of a simple pattern.

Hidden beneath the appearance of a simple, introductory text to cell biology is therefore a radical message—that the division of biology into the traditional historical subjects such as genetics, biochemistry, and physiology is no longer valid. We hope that this book will help our readers see the whole of cell biology—*From Genes to Cells*—as a single, and ever more exciting, science.

CHAPTER

1

CELLS AND TISSUES

The cell is the basic unit of all living organisms. Microorganisms such as bacteria, yeast, and amoebae exist as single cells. By contrast, the adult human is made up of about thirty trillion cells (1 trillion = 10^{12}). These are mostly organized into collectives or *tissues*. Cells are, with a few notable exceptions, small (Box 1.1). Their discovery stemmed from the conviction of a small group of seventeenth-century microscope makers that a new and undiscovered world lay beyond the limits of the human eye. These pioneers set in motion a science and an industry that continues to the present day.

The first person to observe and record cells was Robert Hooke (1665) who described the *cella* (open spaces) of cork and elder pith. But the colossus of this era of discovery was Anton van Leeuwenhoek, a man with no university education but with unrivaled talents as both a microscope maker and as an observer and recorder of the microscopic living world. Despite van Leeuwenhoek's herculean efforts it was another 150 years before, in 1838, a botanist, Matthias Schleiden, and a zoologist, Theodor Schwann, formally proposed that all living organisms are composed of cells. Their "cell theory," which nowadays seems so obvious, was a milestone in the development of modern biology.

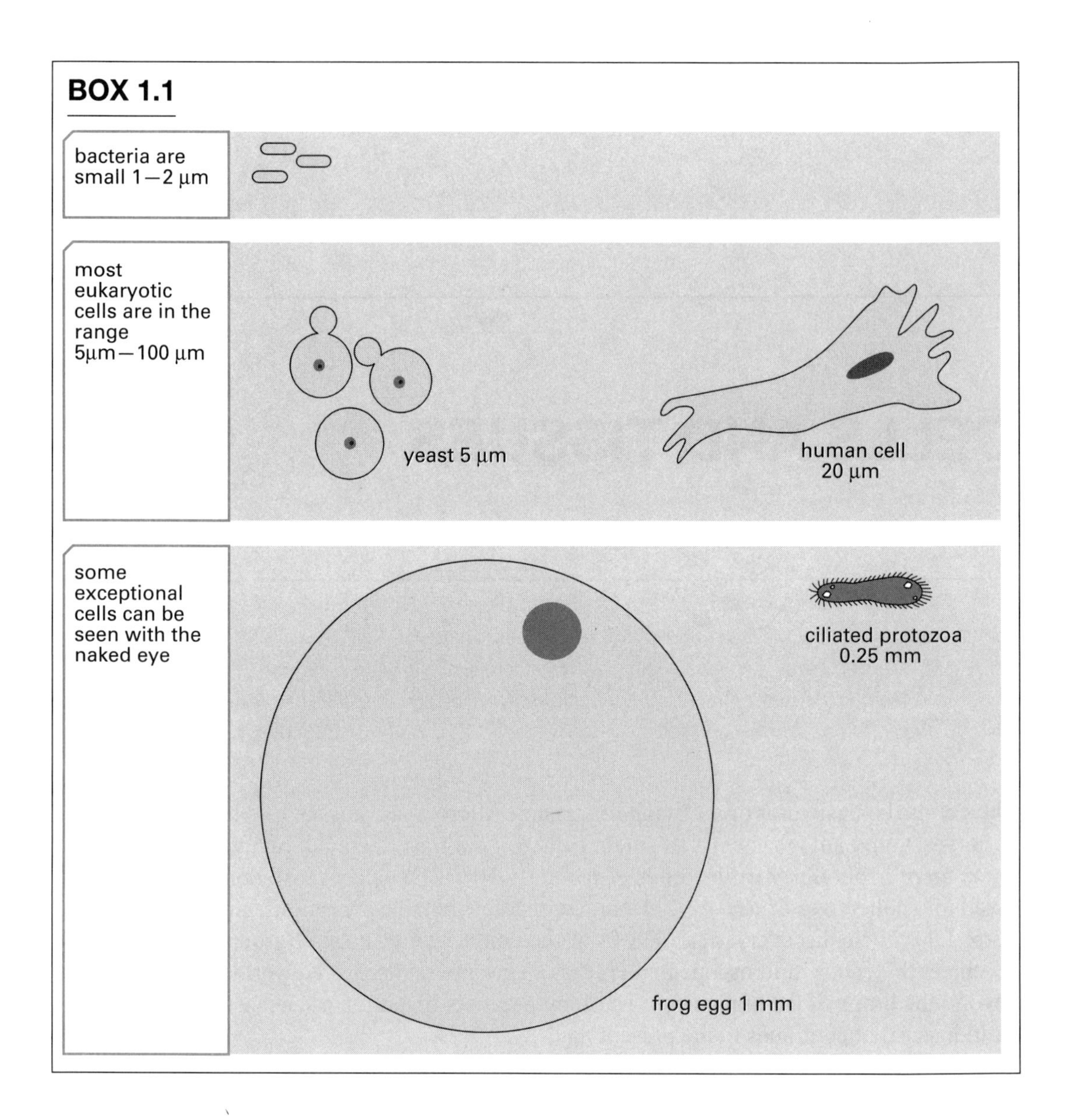

PRINCIPLES OF MICROSCOPY

Microscopes make small objects appear bigger. A *light microscope* will magnify an image up to 2,000 times its original size. *Electron microscopes* can achieve magnifications up to 1,000,000 times. However, bigger is only better when more details are revealed. The fineness of detail that a microscope can reveal is its resolving power. This is defined as the smallest distance that two objects can approach one another yet still be recognized as separate. The resolution that a microscope achieves is mainly a function of the wavelength of the illumination source it employs. The smaller the wavelength, the smaller the object that can still deflect the rays and thus be detected. Thus short wavelengths mean better resolving power. The light microscope can distinguish objects as small as 0.25 μm, roughly half the wavelength of visible light, 500 nm (Box 1.2). It can be used to visualize the smallest cells and the major intracellular structures or *organelles*. The wavelength of an electron beam is about 100,000 times less than that of white light. By using electrons instead of light the electron microscope reveals the ultrastructure of the cell and its organelles (Fig. 1.1). In practice, the electron microscope can distinguish structures about one thousand times smaller than the light microscope. The microscope study of cell structure organization is known as cytology.

The Light Microscope

A light microscope consists of a light source, which can be the sun or artificial light, plus three glass lenses: a condenser lens to focus light on the specimen, an objective lens to form the magnified image, and a projector lens, usually called the eyepiece, to convey the image to the eye (Fig. 1.2). The magnification achieved depends on the focal length of the various lenses and their arrangement but is typically between 100 and 2,000 times. In bright field microscopy the image that reaches the eye consists of all the colors originally present in white light minus those absorbed by the cell. Most living cells have little color (plant cells are an obvious exception) and are therefore largely transparent to transmitted light. Structures within the cell may be seen more clearly by cytochemistry, the use of colored stains to highlight particular organelles selectively. However, many of these compounds are highly toxic, and to be effec-

BOX 1.2

A question of scale:

1 m = 1000 mm (1 mm = 10^{-3} m)
1 mm = 1000 μm (1 μm = 10^{-6} m)
1 μm = 1000 nm (1 nm = 10^{-9} m)

In the light microscope, measurements are ususally expressed in μm; in the electron microscope, measurements are usually expressed in nm.

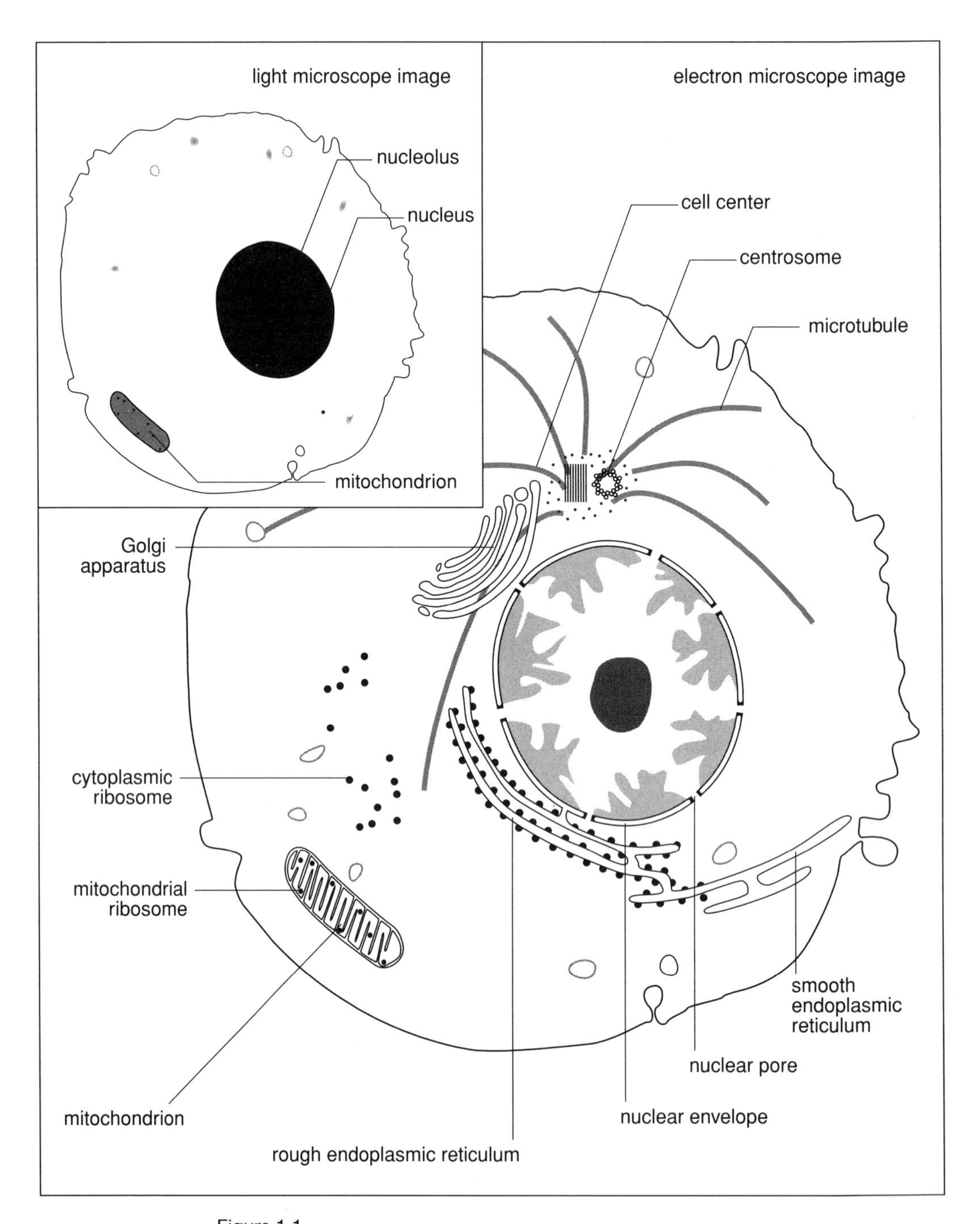

Figure 1.1
Cell structure as seen through the light and electron microscopes

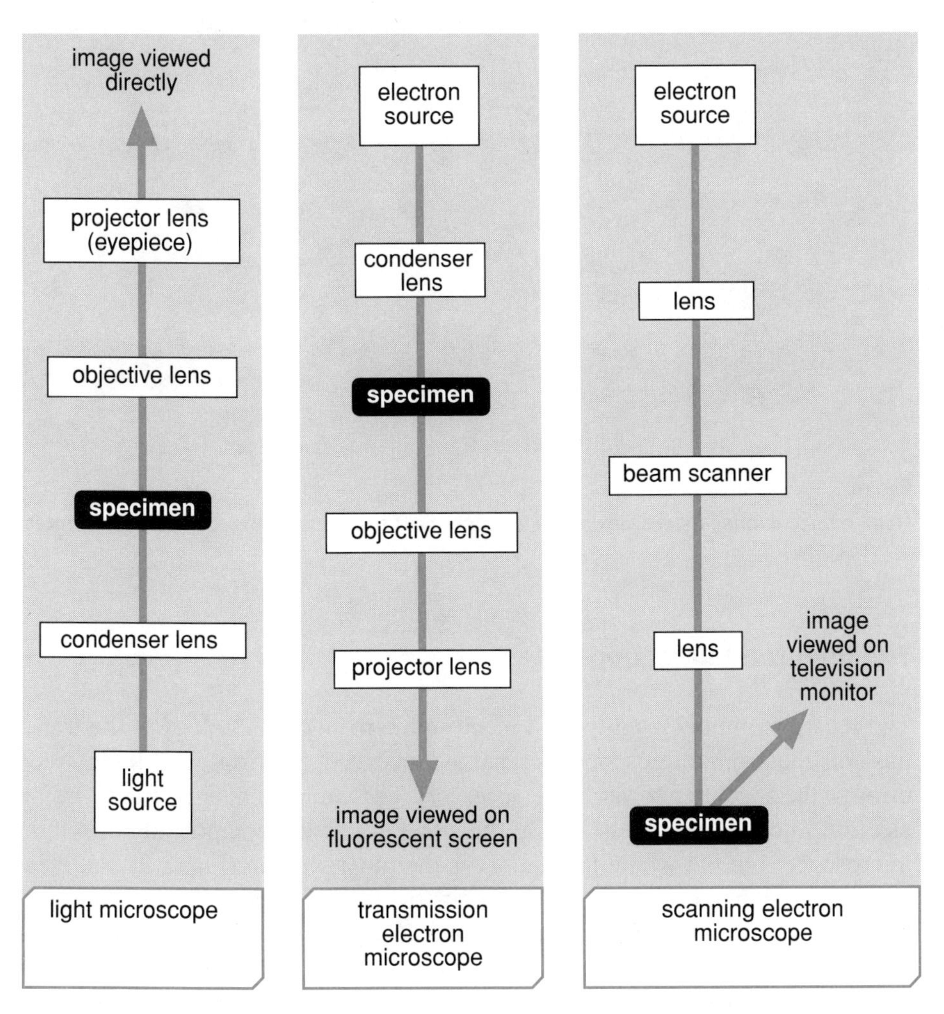

Figure 1.2
The basic design of light and electron microscopes

tive, they often require that the cell or tissue is first subjected to a series of harsh chemical treatments.

A different approach, and one that can be applied to living cells, is the use of phase contrast microscopy. This relies on the fact that light travels at different speed through regions of the cell that differ in composition. The phase contrast microscope converts these differences in refractive index into differences in brightness so that considerably more detail is revealed (Fig. 1.3).

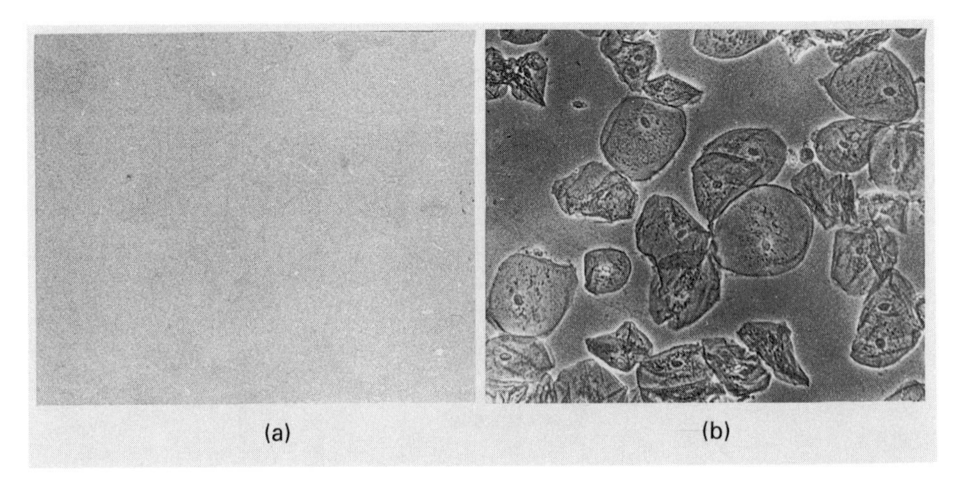

Figure 1.3

Human buccal (cheek) cells viewed (a) by bright field microscopy and (b) by phase contrast microscopy.

The Electron Microscope

The most commonly used type of electron microscope in biology is the transmission electron microscope (so named because electrons are transmitted through the specimen towards the observer). The components of a transmission electron microscope do essentially the same job as the components of a light microscope, but the whole instrument is the other way up (Fig. 1.2). An electron source generates a beam of electrons by heating a thin V-shaped piece of tungsten wire to 3,000°C. A large voltage accelerates the beam down the microscope column, which is under vacuum because otherwise the electrons would be slowed by collision with air molecules. An image is obtained by a similar arrangement of lenses to that used in the light microscope, although electromagnets are used for focussing rather than glass. While the electron microscope offers great improvements in resolution, electron beams are potentially highly destructive, and biological material must be extensively processed before it can be examined. The preparation of cells for electron microscopy is summarized in Box 1.3.

The image produced by the transmission electron microscope is rich in detail. However, the image is static, two-dimensional, and artificial. Often only a small region of what was once a dynamic, living, three-dimensional cell is revealed. The image is essentially a snapshot taken at the particular instant that the cell was killed. Such images must be interpreted with great care; nevertheless, the transmission electron microscope is the main source of our information on cell ultrastructure.

BOX 1.3

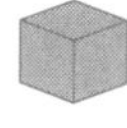

A small piece of tissue (~1 mm^3) is immersed in glutaraldehyde and osmium tetroxide. These chemicals bind all the component parts of the cells together; the tissue is said to be **fixed.** It is then washed thoroughly.

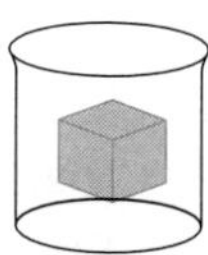

The tissue is **dehydrated** by soaking in acetone or ethanol.

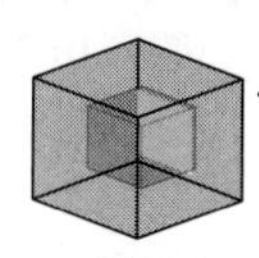

The tissue is **embedded** in resin which is then baked hard.

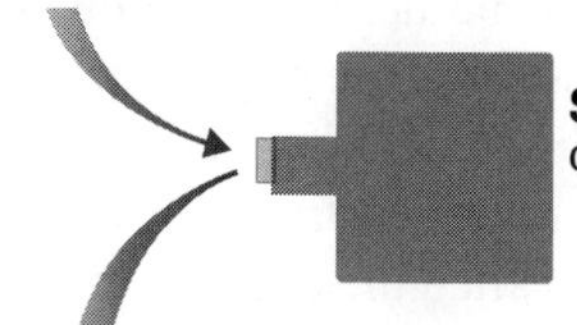

Sections (thin slices less than 100 nm thick) are cut with a machine called an ultramicrotome.

The sections are placed on a small copper grid and **stained** with uranyl acetate and lead citrate. When viewed in the electron microscope, regions that have bound lots of uranium and lead will appear dark because they are a barrier to the electron beam.

The Scanning Electron Microscope

Whereas the image in a transmission electron microscope is formed by electrons transmitted through the specimen, in the scanning electron microscope it is formed from electrons that are reflected back from the surface of a specimen as the electron beam scans rapidly back and forth over it. These reflected electrons are processed to generate a picture on a television monitor. The scanning electron microscope operates over a wide magnification range, from 10× to 100,000×. Its greatest advantage is a large depth of focus, which gives a three-dimensional image. The scanning electron microscope therefore provides useful topographical information about the surfaces of cells or tissues. Modern instruments have a resolution of about 1 nm.

ONLY TWO TYPES OF CELL

At first sight cells exhibit a staggering diversity. Some lead a solitary existence, others live in communities; some have defined geometric shapes, others have flexible boundaries; some swim, some crawl, and some are sedentary. Many are green; others are red, blue, or purple or have no obvious coloration. Given these differences, it is perhaps surprising that there are only two types of cell.

Those of most organisms, from protozoa to mammals and from fungi to plants, are *eukaryotic*. They are generally 5–100 μm across, although a few are large enough to be seen with the naked eye (Box 1.1). Eukaryotic cells contain a variety of specialized structures including membrane-bound *organelles* (page 32), the largest of which, the nucleus, contains the genetic material. Bacteria are said to be *prokaryotic*. They are small (1-2 μm across) and have very little internal organization so that, for instance, the genetic material is free within the cell.

The structure and function of organelles will be described in detail in subsequent chapters. Table 1.1 provides a brief glossary of the major intracellular structures and organelles and summarizes the differences between prokaryotic and eukaryotic cells. These differences are further illustrated in Fig. 1.4. The entire insides of the cell, except the nucleus, is called the cytoplasm. Cytoplasm consists of all the bits—ribosomes, centrosomes, organelles—suspended in a watery medium called the cytosol. In fact despite all their complexity, organisms are mostly water and Box 1.4 describes the chemical properties of this unique liquid.

Special Properties of Plant Cells

Perhaps the most striking difference among eukaryotic cells is between those of animal and plants (Fig. 1.4). Higher plants have evolved a sedentary life-

Table 1.1 Differences between prokaryotic and eukaryotic cells

	Prokaryotes	Eukaryotes
Size	1–2 μm	5–100 μm
Nucleus	Absent	Present: bounded by nuclear envelope
DNA	Single, circular	Multiple, linear, associated with protein (chromatin)
Cell division	Simple fission	Mitosis or meiosis
Internal membranes	Rare	Complex (Golgi apparatus, ER, etc.)
Ribosomes	70S	80S (70S in organelles)
Cytoskeleton	Absent	Microtubules, microfilaments, intermediate filaments
Motility	Flagella; rotary motor	9 + 2 cilia and flagella, sliding motor
First appeared	3.5×10^9 years ago	1.5×10^9 years ago

prokaryotic

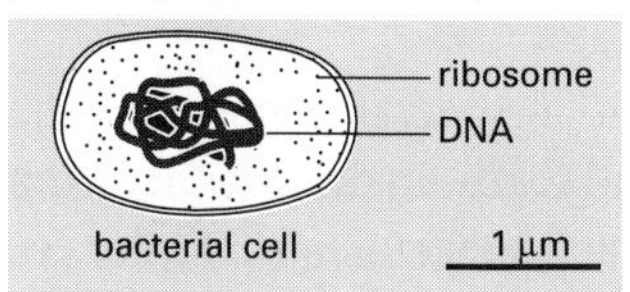

eukaryotic

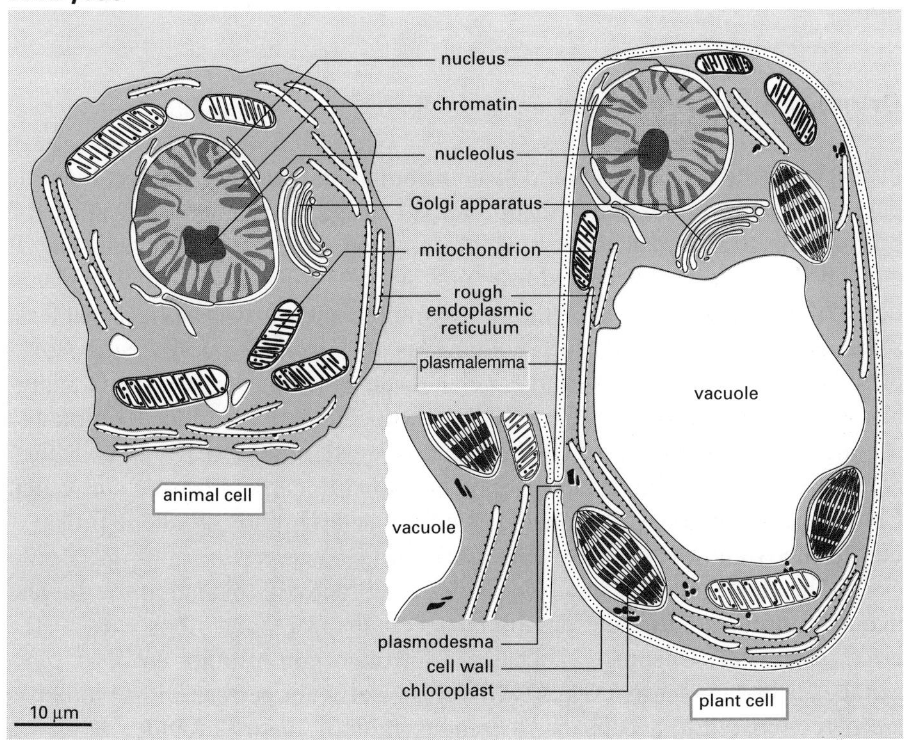

Figure 1.4
The organization of prokaryotic and eukaryotic cells

style and a mode of nutrition that demands that they hold up a leaf canopy. Unlike the flexible boundaries of animal cells, plant cells are enclosed within a rigid cell wall that gives shape to the cell and structural rigidity to the organism. Plant cells frequently contain one or more large bags called vacuoles that contain a simple, watery solution with a high concentration of sugars and other soluble compounds. Water enters the vacuole to dilute these sugars, generating hydrostatic pressure that is counterbalanced by the rigid wall. In this way the cells of the plant become stiff or turgid, in the same way that a balloon inflated inside a stocking becomes stiff. Vacuoles are frequently pigmented and the spectacular colors of petals and fruit reflect the presence of compounds such as anthocyanins (purple pigments) in the vacuole.

Cells of photosynthetic plant tissue contain a special organelle, the chloroplast, which houses the light-harvesting and carbohydrate-generating systems (Chapters 15 and 16). Higher plant cells lack centrioles (Chapter 3), which are present in most animal cells and also in some simple plants such as algae.

Origin of Eukaryotic Cells

Prokaryotic cells are simpler and more primitive in their organization than the eukaryotic cells. Fossils show that prokaryotic organisms antedate by at least 2 billion years the first eukaryotes, which appeared some 1.5 billion years ago. It is likely that eukaryotes evolved from prokaryotes. The most plausible explanation of this process is known as the endosymbiotic theory. Its basis is that at least some, and perhaps all, eukaryotic organelles originated as free-living prokaryotes. These were engulfed by larger cells inside which they established a mutually beneficial relationship. For example, mitochondria would have originated as free-living aerobic bacteria—'aerobic' because, unlike the host cell, they could use oxygen in the air to oxidize their food to carbon dioxide and water. Chloroplasts would have originated as cyanobacteria (photosynthetic prokaryotes formerly known as blue-green algae).

The endosymbiotic theory provides an attractive explanation for the fact that both mitochondria and chloroplasts contain DNA and ribosomes of the prokaryotic type (Chapter 12). The case for the origin of other eukaryotic organelles is less persuasive. Thus, while it is clearly not perfect, most biologists are now prepared to accept that the endosymbiotic theory provides at least a partial explanation for the evolution of the eukaryotic cell from a prokaryotic ancestor. On the other hand, living forms having a cellular organization intermediate between prokaryotes and eukaryotes are rare. Some intracellular parasites do seem to represent the "missing link." These primitive protozoa possess a bona fide nucleus but lack mitochondria and other typical eukaryotic organelles. They also have the prokaryotic type of ribosome. Microspora, an organism that infects AIDS patients, is a member of this group.

Tissues: Cell Collectives

Multicellular organisms are composed of tissues, groups of cells that share a common function. In animals there are four main tissue types: epithelium, connective tissue, nerve and muscle.

EPITHELIA are sheets of cells that cover the surface of animal's bodies and line their internal cavities such as the lungs and intestine. The cells may be columnar (taller than they are broad), or squamous (flat). To increase the area available for the absorption of nutrients, the intestine wall is folded to form fingers called villi (singular, villus). Focusing in on the columnar epithelial cells lining the intestine shows that their surface, too, is folded into much smaller fingers called microvilli (singular, microvillus) (Figure 1.5). The bottom or basal surface of the epithelial cells sits on a supporting layer called the basement membrane. Many of the epithelial cells of the airways, such as those lining the trachea and bronchioles, have cilia on their surfaces. Cilia are hairlike appendages that actively beat back and forth, moving a layer of mucus away from the lungs (Chapter 3). Particles and bacteria are trapped in the mucus layer, preventing them from reaching the delicate air-exchange membranes in the lung. The epithelium of the skin is referred to as stratified because it is composed of several layers.

CONNECTIVE TISSUES provide essential support for the other tissues of the body. They include blood, bone, cartilage, and adipose tissue in which fat is stored. In general, connective tissue tends to contain relatively few cells within a large volume of extracellular matrix consisting of different types of fibers embedded in amorphous ground substance. The most abundant of the fibers is collagen. Collagen forms extremely long, flexible fibers and accounts for about a third of the protein of the human body. Other fibers are elastic so that the supported tissues will return to their original position after they are bent out of shape. The amorphous ground substance absorbs large quantities of water, through which chemicals, oxygen and carbon dioxide can easily move to and from the cells in the other tissues and organs. Of the many cell types found in connective tissue, the two most important are fibroblasts, which secrete the ground substance and fibers, and macrophages, which remove foreign, dead, and defective material from it. A number of inherited diseases are associated with defects in connective tissue. Marfan's syndrome, for example, is characterized by long arms, legs, and torso and by a weakness of the cardiovascular system and eyes. These various symptoms have a single underlying cause, a defect in the organization of the collagen fibers.

NERVOUS TISSUE forms the brain and spinal cord as well as the nerves that run throughout our bodies. It is a highly modified epithelium that is composed of several cell types. Principal among these are the nerve cells, or neurons, and the glial cells. The job of neurons is to send and process electrical information. Neurons extend extremely long wire-like processes called axons which, in the leg, for instance, can be over a meter in length. Like wires, axons transmit electrical signals rapidly over long distances (Chapters 18–20). One of the many jobs of glial cells is to make myelin, the electrical insulation around axons.

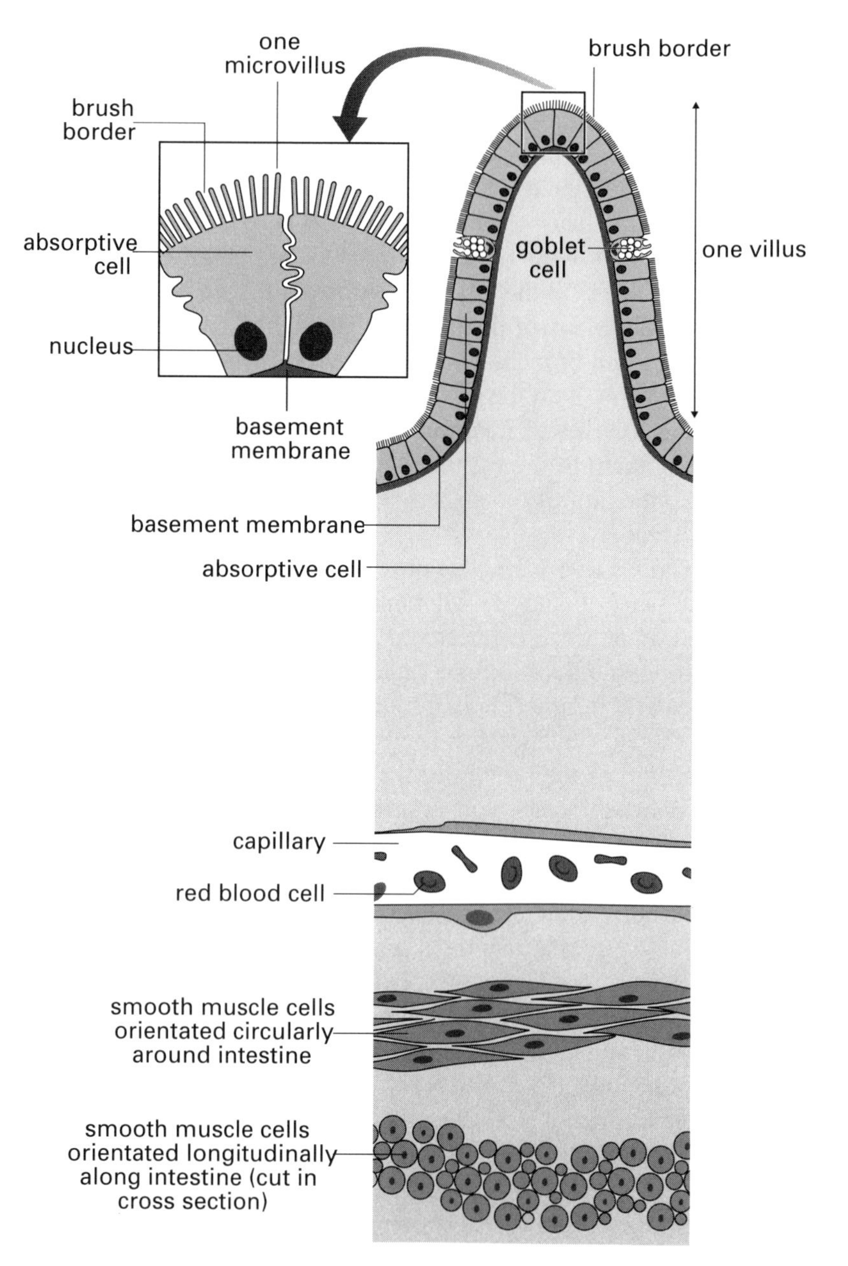

Figure 1.5

The intestine is folded at large and small scales

MUSCLE TISSUE can be of two types, smooth and striated. Smooth muscle cells are long and slender; they are usually found in the walls of tubular organs like the uterus, stomach and many blood vessels. In general, smooth muscle cells contract more slowly and can maintain the contracted state for a long period of time. There are two classes of striated muscle: cardiac and skeletal, and these make up the big blocks of tissue that we recognize as muscle. The heart is made of cardiac muscle cells while the muscles we use to move our limbs and bodies are composed of skeletal muscle cells. Skeletal muscles are

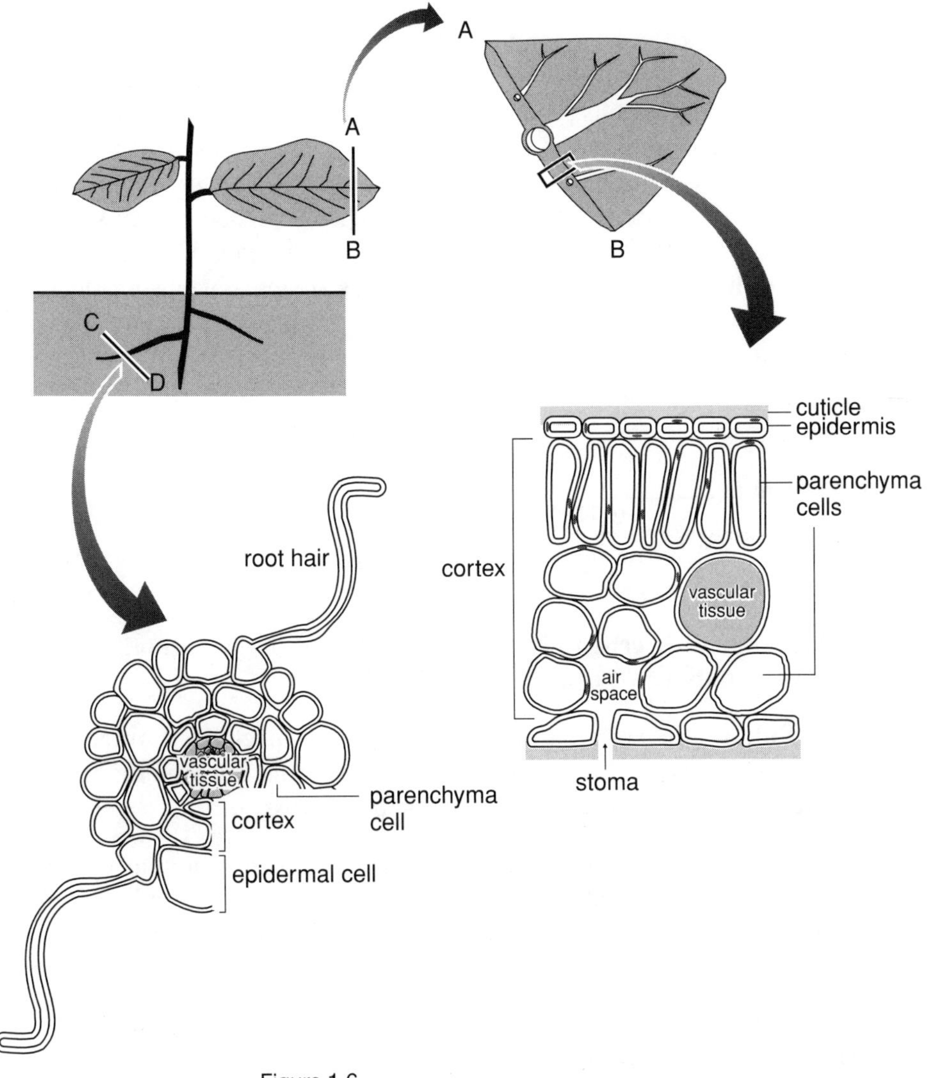

Figure 1.6
Plants are composed of several tissues

made of hundreds or even thousands of muscle fibers, each fiber being a giant single cell with many nuclei. This rather unusual situation is the result of an event that occurs in the embryo when the cells that give rise to the fibers fuse together, pooling their nuclei in a common cytoplasm. How muscles contract will be described in Chapter 4.

Plant cells are also organized into tissues. The basic organization of a shoot or root is into an outer protective layer or epidermis, a central vascular tissue that provides support and transport and a cortex that fills the space between the other two (Fig. 1.6). The epidermis consists of one or more layers of closely packed cells. Above the ground, these cells secrete a waxy layer, the cuticle, that helps the plant retain water. Holes in the cuticle called stomata allow gas exchange between the air and the photosynthetic cells and are also the major route for water loss from the plant by the process of transpiration. Below ground, the epidermal cells produce root hairs that are important in the absorption of water and minerals. The vascular tissue is composed of xylem and phloem. The xylem transports water and its dissolved solutes from the roots; the phloem conveys the products of photosynthesis, predominantly sugars, to their site of use or storage. Photosynthesis occurs primarily in the parenchyma cells that form the bulk of the cortex. These cells are less obviously specialized and are the major site of metabolic activity in the plant. They generally lack thick cell walls.

SUMMARY

1. The cell is the basic organizational unit of all living organisms. There are only two types of cells, prokaryotic cells and eukaryotic cells.

2. Prokaryotic cells have very little internal organization. They usually measure only 1 to 2 μm across.

3. Eukaryotic cells contain a variety of specialized internal organelles. The largest of these, the nucleus, contains the genetic material. Eukaryotic cells usually measure 5 to 100 μm across.

4. In eukaryotic cells the major cellular organelles such as the nucleus and mitochondria can be seen with the light microscope. It takes an electron microscope to reveal the detailed structure of these organelles and to resolve smaller cellular structures.

5. In multicellular organisms cells are organized into tissues. In animals there are four tissue types: epithelium, connective tissue, nervous tissue, and muscle tissue.

BOX 1.4

WATER IS A POLAR MOLECULE

One molecule of water contains one oxygen and two hydrogen atoms. The three atoms are not in a straight line but form an open V:

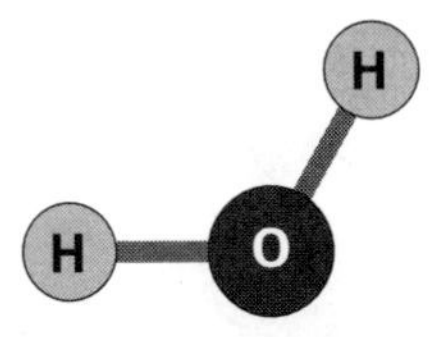

Oxygen tends to grab more than its fair share of the electrons that form the bonds. Since these negatively charged electrons spend more time close to the oxygen than to hydrogen, the atoms within water have on average a small net charge. The δ in the figure is used to indicate that this charge is less than the charge on a single electron or proton:

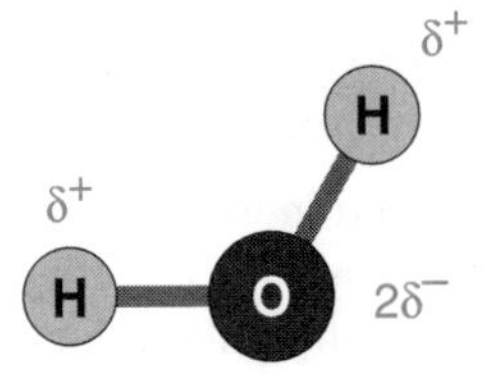

Molecules that, like water, have distinct positive and negative regions are called polar.

Octane, the main constituent of gasoline, is an example of a nonpolar solvent. The electrons that form the bonds are shared equally between carbon and hydrogen, and the component atoms do not bear a net charge.

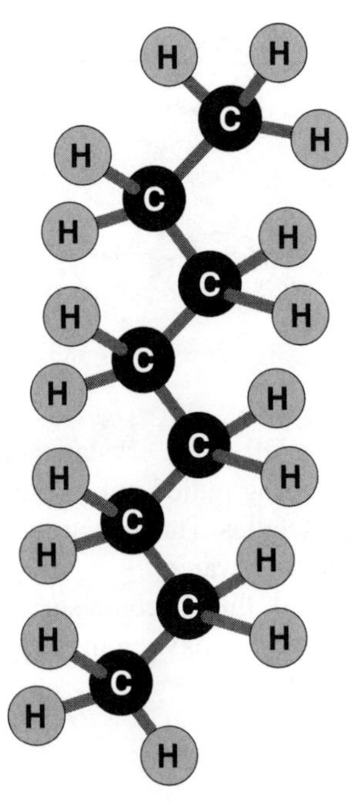

Ionic Compounds Will Dissolve Only in Polar Solvents

At the edge of a crystal of sodium chloride, positively charged sodium ions are being pulled in toward the center of the crystal by the negative charge on chloride ions, and negatively charged chloride ions are being pulled in toward the center of the crystal by the positive charge on sodium ions. When a crystal of sodium chloride is placed in a nonpolar solvent like octane, there is nothing to offset this inward force. The sodium and chloride ions do not leave the crystal. Sodium chloride is therefore insoluble in octane:

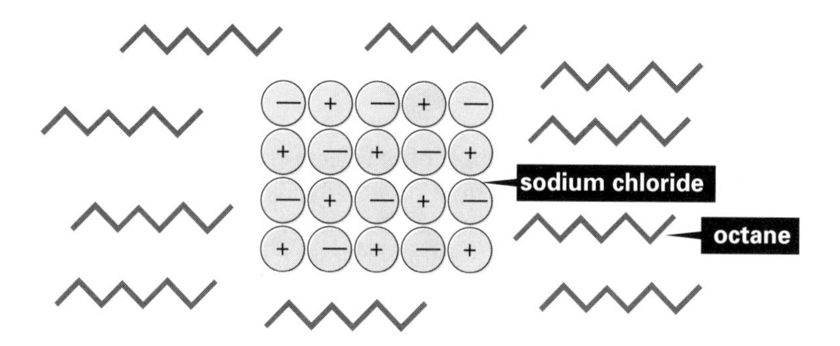

When a crystal of sodium chloride is placed in water, the chloride ion, shown at the top left of the crystal in the diagram below, is pulled into the crystal by the positive charge on its sodium ion neighbors, but at the same time it is being pulled out of the crystal by the positive charge on the hydrogen atoms of nearby water molecules:

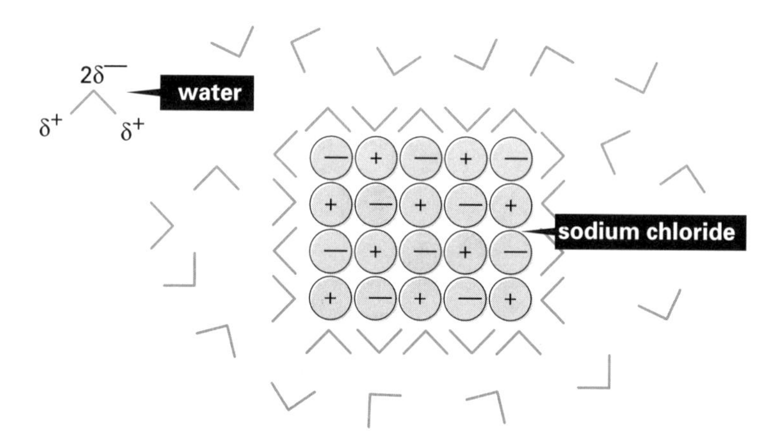

Similarly the sodium ion at the bottom left of the crystal is pulled in by the negative charge on its chloride ion neighbors while it is being pulled out of the crystal by the negative charge on the oxygen atoms of nearby water molecules. The ions are not held in the crystal so tightly and can leave, so sodium chloride is soluble in water.

Once the ions have left the crystal, they become surrounded by a hydration shell of water, all oriented in the appropriate direction—oxygen inward for a positive ion like sodium; hydrogen inward for a negative ion like chloride:

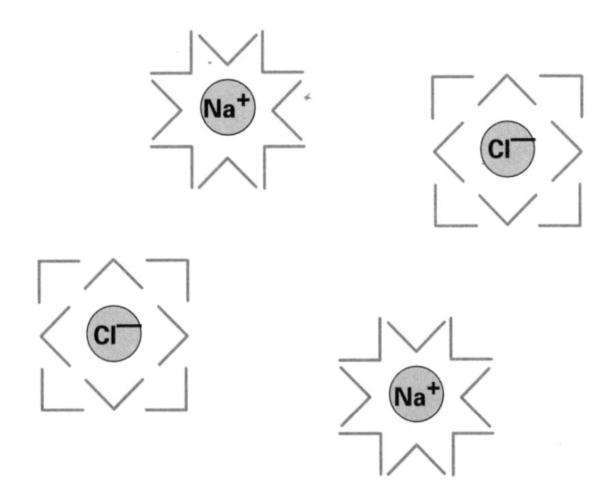

Acids Are Molecules That Give H⁺ to Water

We will meet lactic acid in Chapter 16. Pure lactic acid has the structure:

```
        OH  OH
        |   |
H3C —C —C═O
        |
        H
```

lactic acid

The –COOH on the right is called a carboxyl group. Both oxygen atoms in the carboxyl group tend to grab the electron on the hydrogen atom. In the presence of water the hydrogen gives up its electron completely and leaves to attach to a water molecule. The electron is left behind, giving a negative charge to what is now a lactate ion:

$$CH_3COOH + H_2O \rightleftharpoons CH_3COO^- + H_3O^+$$

For convenience we often write this as:

$$CH_3COOH \rightleftharpoons CH_3COO^- + H^+$$

Remember that H^+ is a bare proton and immediately attaches to water to form H_3O^+.

Dissolving lots of lactic acid in water produces an acid solution with a high concentration of H^+ (really H_3O^+) ions. For historical reasons the acidity of a solution is given as the pH, which is defined:

$$\mathrm{pH} = -\log_{10}[H^+]$$

where $[H^+]$ is measured in moles per liter. Pure water has a pH of 7. This is said to be *neutral* as regards pH. When the pH is lower than 7, there is more H^+ present, so the solution is acidic.

The pH of a solution of lactic acid determines the position of equilibrium between lactic acid and lactate:

$$CH_3COOH \rightleftharpoons CH_3COO^- + H^+$$

As the pH falls, the concentration of H^+ ions increases. The frequency of collisions between lactate ions and H^+ ions increases, and more and more recombine to make lactic acid. At pH 3.9, the concentration of lactic acid and lactate are the same. In general, the pH at which half the molecules of an acid have the H still attached (we call this the protonated state) and half have given H^+ to water (we call this the deprotonated state) is called the pK_a. Thus the pK_a of lactic acid is 3.9. The lower the pK_a, the stronger the acid.

Because many biologically important molecules are acids and so are mainly ionized at neutral pH we usually speak of them as ions—"glutamate" and "lactate" instead of glutamic and lactic acids.

Bases Are Molecules That Take H^+ from Water

We will meet nicotine, the drug in tobacco, in Chapter 20. Pure nicotine has the structure:

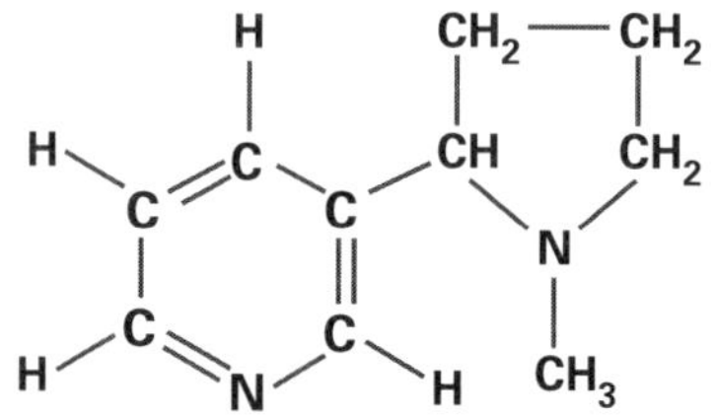

nicotine

When nicotine is dissolved in water, it accepts a H^+ from water to become the positively charged, protonated nicotine ion:

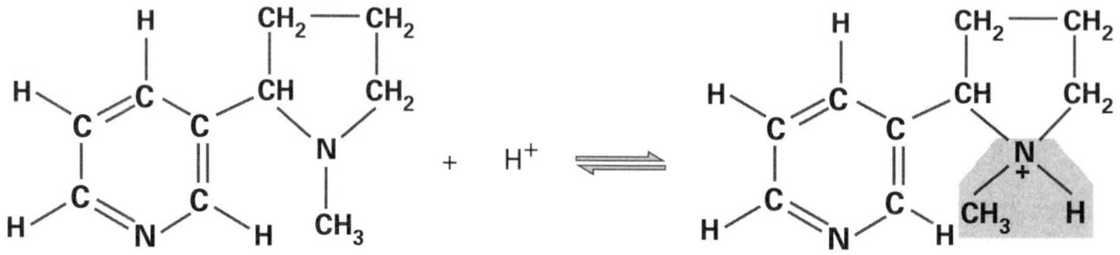

nicotine + H^+ $\rightleftharpoons$ nicotine ion

Dissolving lots of nicotine in water produces an alkaline solution, that is, one with a low concentration of H^+ ions and hence a pH greater than 7. The solution never runs out of H^+ completely because new H^+ are formed from water:

$$H_2O \rightleftharpoons OH^- + H^+$$

As more and more nicotine is added to water the H^+s are used up, producing a low concentration of H^+ but lots of OH^-.

The pH of this solution determines the position of equilibrium between protonated and deprotonated nicotine. As before, the pK_a is defined as the pH at which the concentration of the protonated and deprotonated form are the same. The pK_a of nicotine is 7.9, meaning that the concentration of H^+ must fall to the low level of $10^{-7.9}$ moles/liter or 13 nm, before half of the nicotines will give up their H^+s. Cytoplasm has a pH that lies very slightly on the alkaline side of neutrality, at about 7.2.

A Hydrogen Bond Forms When a Hydrogen Atom Is Shared

We have seen how oxygen tends to grab electrons from hydrogen, forming a polar bond. Nitrogen and sulfur have similar electron-grabbing properties. If a hydrogen attached to an oxygen, nitrogen, or sulfur by a covalent bond gets close to a second electron-grabbing atom, then that second atom also grabs a small share of the electrons to form what is known as the hydrogen bond. The atom to which the hydrogen is covalently bonded is called the donor because it is losing some of its share of electrons; the other electron-grabbing atom is the acceptor. For a hydrogen bond to form, the donor and acceptor must be within a fixed distance of one another (typically 0.3 nm) with the hydrogen on a straight line between them. Liquid water is so stable because the individual molecules can hydrogen bond to each other:

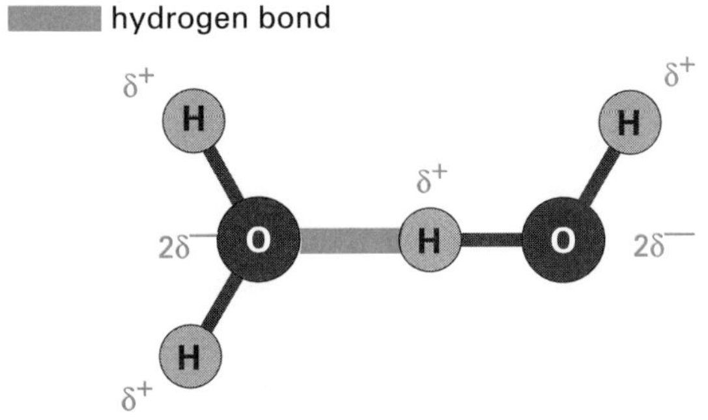

hydrogen bond

Hydrogen bonding between the base pairs of DNA (page 74) involves sharing of hydrogen between nitrogen and oxygen and between nitrogen and nitrogen:

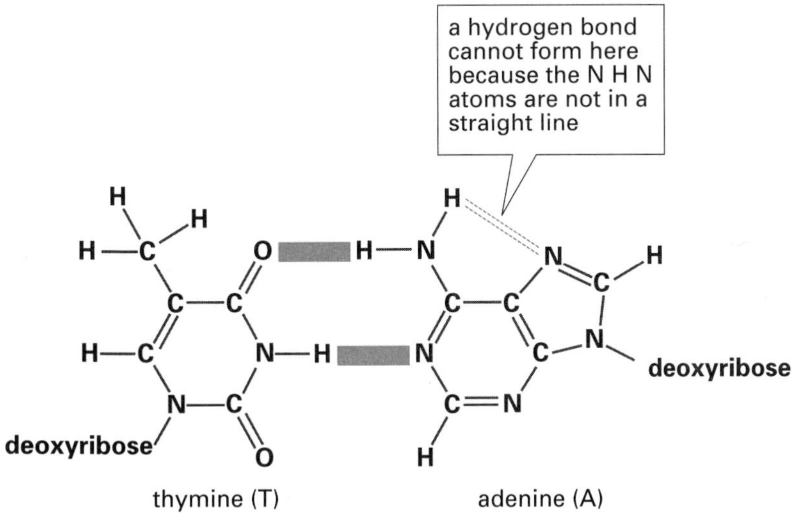

a hydrogen bond cannot form here because the N H N atoms are not in a straight line

deoxyribose

deoxyribose

thymine (T)

adenine (A)

CHAPTER

2

MEMBRANES AND ORGANELLES

In much the same way that our homes are divided into rooms adapted for particular activities, so eukaryotic cells contain distinct compartments or organelles to house specific functions. The term organelle is used rather loosely. At one extreme it embraces any distinct cellular structure; at the other it includes only those cellular compartments containing their own DNA and having some limited genetic autonomy. In this book we define organelles as those cellular components whose limits, like those of the cell itself, are defined by membranes. It is first necessary to consider some of the fundamental properties of cell membranes.

BASIC PROPERTIES OF CELL MEMBRANES

It is impossible to overstate the importance of membranes to living cells; without them life cannot exist. The plasmalemma, also known as the plasma membrane or cell membrane, defines the boundary of the cell. It regulates the movement of materials into and out of the cell and facilitates electrical signaling. Other membranes define the boundaries of cell organelles and provide a matrix upon which complex chemical reactions can occur. Some of these themes will be developed in subsequent chapters. In the following section the basic structure of the cell membrane will be outlined.

Lipids Are Esters of Glycerol and Fatty Acid

Before proceeding to the membrane structure, the properties of lipids should be examined. Figure 2.1 shows two lipids. On the left is trioleoylglycerol, the main component of olive oil. It is a triacylglycerol (or triglyceride), and it has four components: three long-chain organic acids, called fatty acids, joined by ester bonds to the small alcohol glycerol. Olive oil does not mix with water. It

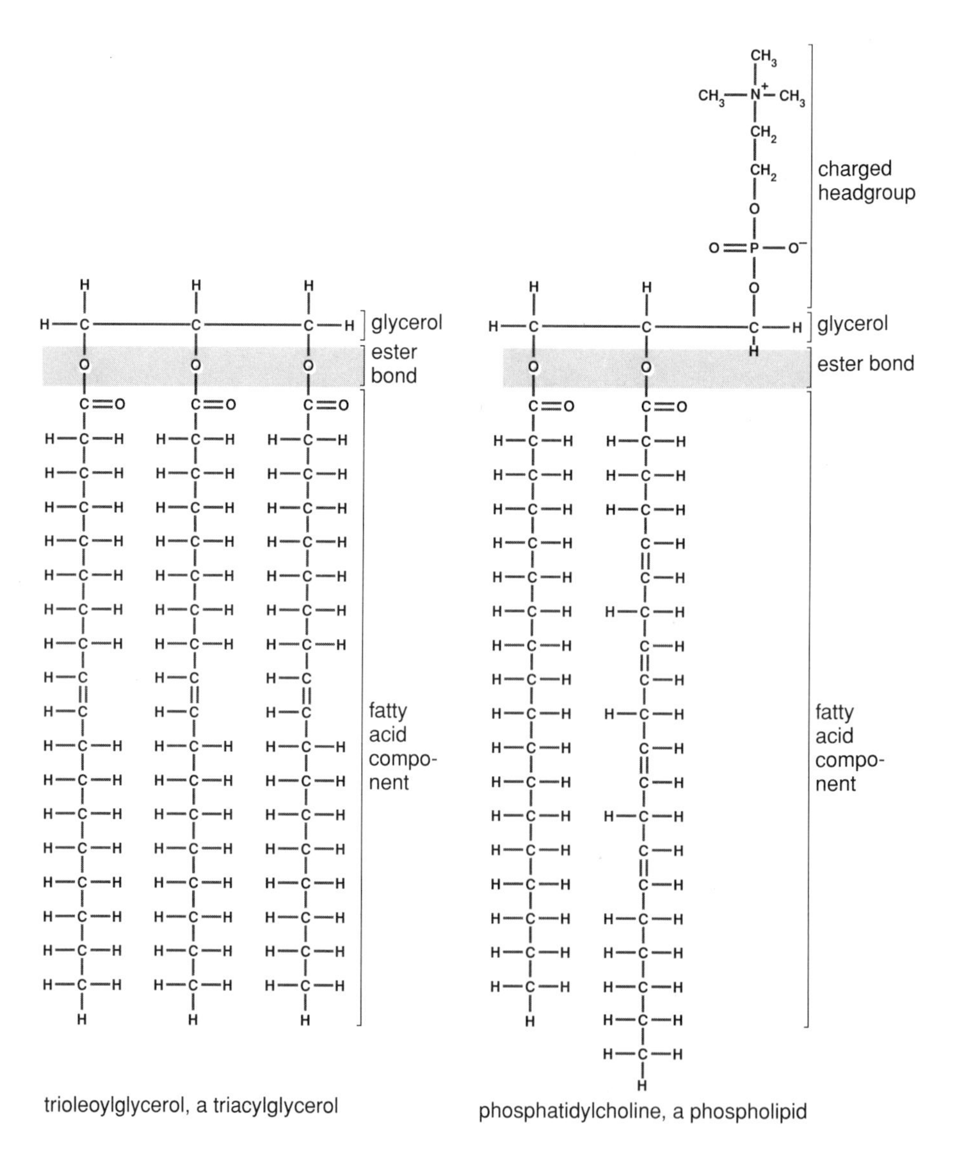

Figure 2.1

The chemical structure of lipids.

is therefore said to be *hydrophobic*. Since the majority of the triacylglycerol molecule, which is just carbon and hydrogen, cannot hydrogen bond with water molecules (Box 1.4), when we shake up olive oil and vinegar to make salad dressing, they do not mix but rather the olive oil breaks into small droplets that float in the vinegar. In the same way cells contain droplets of triacylglycerol within the cytosol. These are usually small, but specialized fat cells can be full of fat droplets (fats and oils are similar molecules; they differ only in their melting point).

Membranes Are Made of Phospholipids

At the right of Fig. 2.1 is phosphatidylcholine, an example of the phospholipids that make membranes. Like triacylglycerols, *phospholipids* have fatty acids attached to glycerol but only two of them. In place of the third fatty acid is a highly electrically charged head group. Since it is charged, the head group tends to associate with water (Box 1.4). It is said to be *hydrophilic*. The two fatty acids, on the other hand, form a tail which, like olive oil, is hydrophobic. Phospholipids can therefore neither dissolve in water (because of their hydrophobic tails) nor remain completely separate, like olive oil (because then the head group could not associate with water). These lipid molecules therefore spontaneously form lipid bilayers about 7 nm thick (Fig. 2.2) in which each part of the molecule is in its preferred environment.

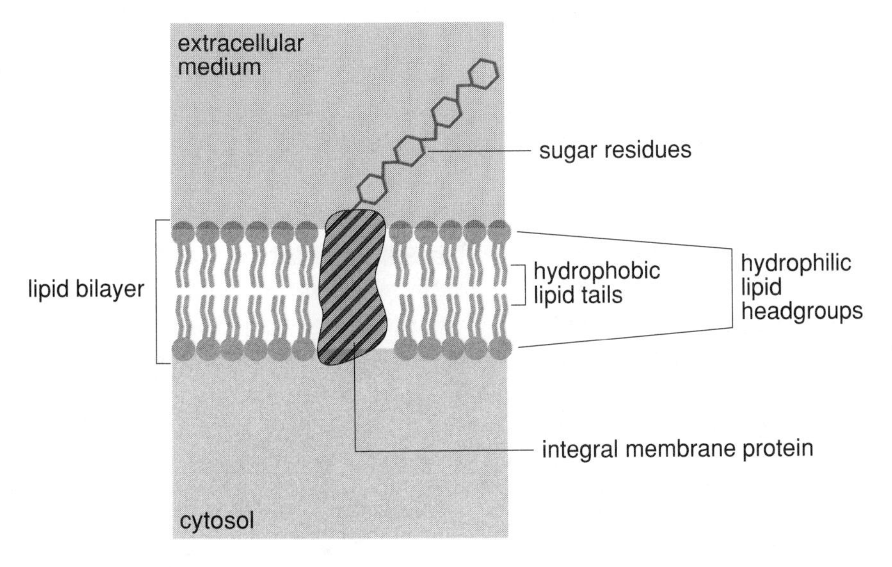

Figure 2.2
Membranes have a fluid mosaic structure.

Proteins, polymers of amino acids, are found everywhere in the cell and do most of the cell's work. Their structure and operation will be described in Chapters 13 and 14. All of the membranes of the cell, including the plasmalemma, contain proteins. These may be tightly associated with the membrane so that they can only be extracted from it with great difficulty, the so-called integral proteins; or they may be only loosely attached so that they can be separated with relative ease, in which case they are termed peripheral proteins. Some integral proteins are receptors for specific extracellular molecules (Box 2.2 and page 37). Membrane proteins are free to move within the plane of the membrane, and membranes are often described as having a fluid mosaic structure (Fig. 2.2).

Straight through the Membrane: Simple Diffusion

Molecules of oxygen are uncharged. Although they dissolve readily enough in water, they are not so hydrophilic that they cannot also dissolve in the hydrophobic interior of lipid bilayers. Oxygen molecules can therefore pass from the extracellular fluid into the interior of the plasmalemma, and from there pass into the cytoplasm. This process by which a solute gets into or out of cells (or into and out of membrane-bound organelles), passing through the lipid component of the membrane, is called *simple diffusion*. Three other small molecules with important roles in biology—carbon dioxide, nitric oxide, and water itself—also pass across the plasmalemma by simple diffusion, as do the uncharged hormones of the steroid family. In contrast, charged ions associate strongly with water molecules and are therefore strongly hydrophilic. They cannot dissolve in hydrophobic regions, as anyone who has tried to dissolve salt in oil will know. Thus ions cannot cross membranes by simple diffusion but only through channels or via carriers (Chapter 17).

Beyond the Plasmalemma: The Extracellular Matrix

In connective tissue (Chapter 1) the spaces between cells are filled by an extracellular matrix consisting of polysaccharides (Box 2.1) and proteins together with combinations of the two called proteoglycans. The proteins provide tensile strength and elasticity while the polysaccharides form a hydrated gel that expands to fill the extracellular space. The sugar-based extracellular matrix reaches its highest form of expression in the cell walls of plants. The plant cell wall consists primarily of tiny strands called microfibrils that are made of cellulose, a long unbranched polymer of the monosaccharide (or sugar) glucose (Box 2.1). The fibrils are linked together by other polysaccharide molecules

BOX 2.1

Sugars and Polysaccharides

Carbohydrates are a very large group of molecules which are built up from simple sugars—monosaccharides. The distinguishing feature of a monosaccharide is the presence of a central skeleton of carbon atoms, to which are attached many hydroxyl groups. Classically, a monosaccharide had to have the general formula $C_n(H_2O)_n$, hence the name carbohydrate. Glucose, $C_6H_{12}O_6$, fits this rule, but many compounds that it is convenient to call sugars do not.

Glucose easily switches between three forms, or isomers:

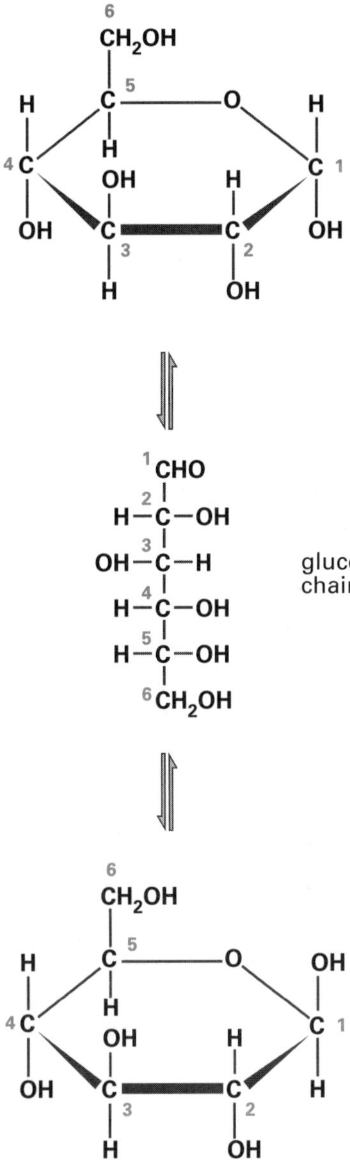

α-glucose

glucose in open chain form

β-glucose

The six-membered ring structure of a monosaccharide is called a pyranose ring. The two pyranose ring structures are optical isomers because each can be formed from the other by reflection in a mirror. (Optical isomers are discussed again in Box 13.1 on page 224.) The two optical isomers, named α and β, continually interconvert in solution via the open-chain form which is only present in very low concentration.

Five other monosaccharides are described in this book. They are shown below, each as a common isomer.

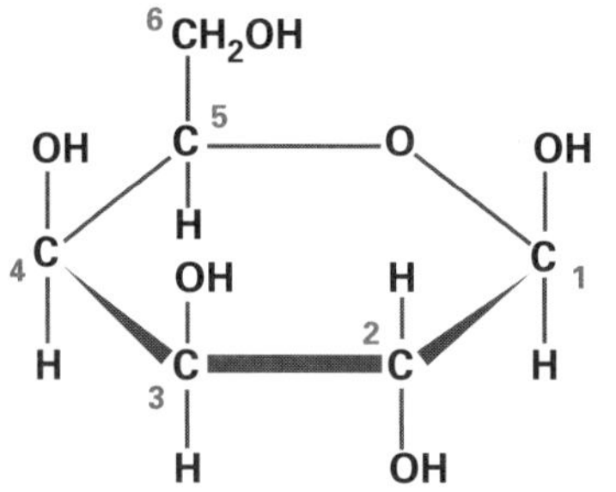

β-galactose

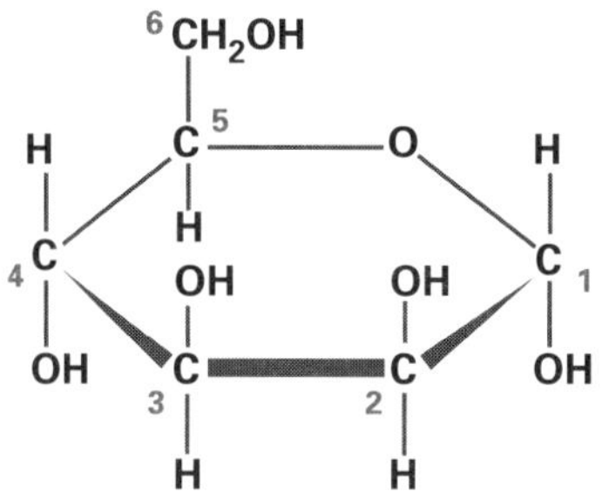

α-mannose

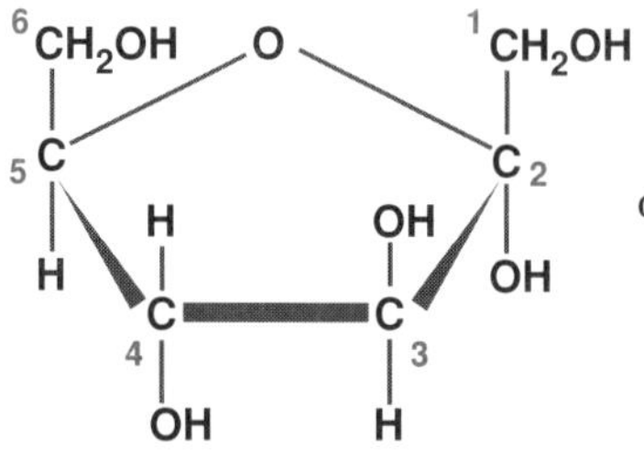

α-fructose

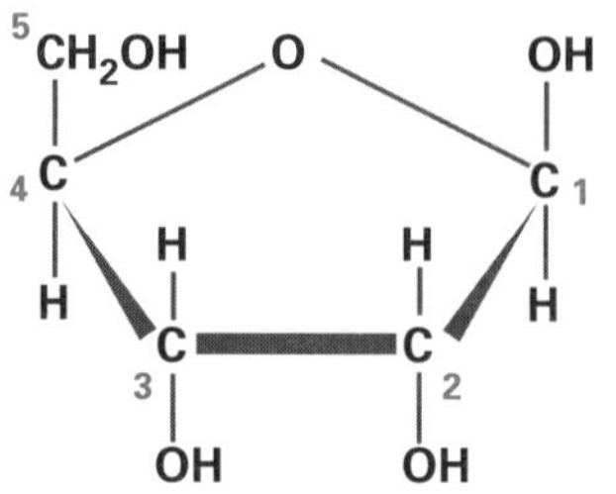

β-ribose

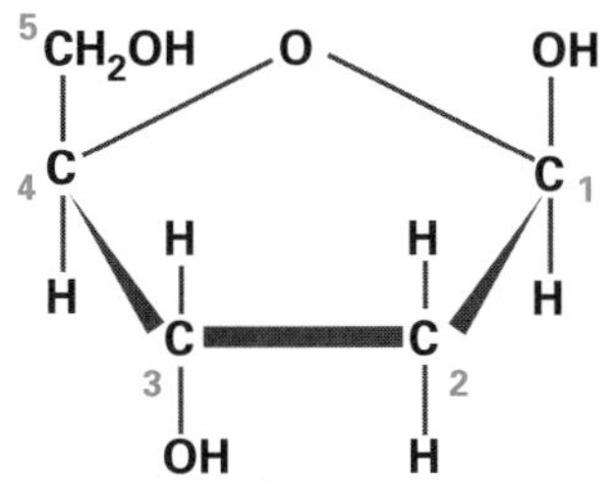

β-deoxyribose

Monosaccharides can easily be joined together by glycosidic bonds in which the carbon backbones are linked through oxygen. A water molecule is lost. The bond is identified by the carbons linked. For instance, lactose, the sugar that is found in milk, is formed when glucose and galactose are linked by a (1→4) glycosidic bond.

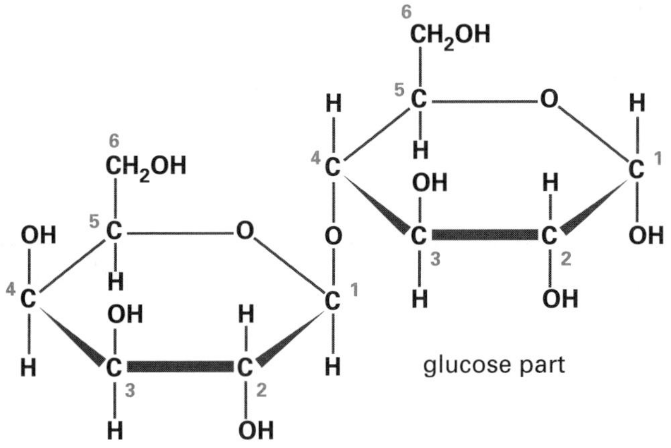

glucose part

galactose part

A complication arises here. Although free monosaccharides can easily switch between the α and β forms, formation of the glycosidic bond locks the shape. Thus, although galactose in solution spends equal time in the α and β forms, the galactose residue in lactose is locked into the β form, so the bond in lactose is specified as β(1→4). The deoxyribose and ribose sugars in DNA and RNA (Chapters 5, 10) are similarly locked into the β form.

Formation of glycosidic bonds can continue almost indefinitely. A short chain of linked sugars is called an oligosaccharide. Glycogen is a polymer made exclusively of glucose monomers with α(1→4) links. It also has branches at intervals, each branch being an α(1→4) linked chain of glucose residues linked to the main chain with an α(1→6) bond.

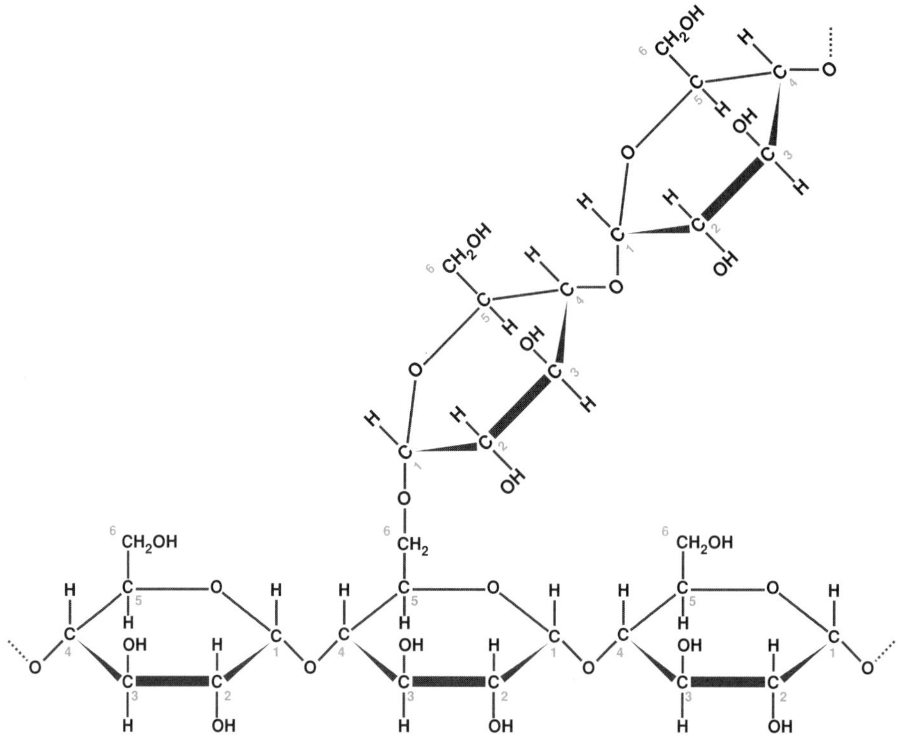

Solid lumps of glycogen are found in the cytosol of muscle, liver, and some other cells. These glycogen granules are 10 to 40 nm in diameter with up to 120,000 glucose residues. They are not membrane bound and so do not qualify as organelles. Glycogen is broken down to release glucose when the cell needs energy (Chapter 16).

Cellulose, which makes up the cell wall of plants is, like glycogen, a polymer of glucose, but this time the links are $\beta(1 \rightarrow 4)$:

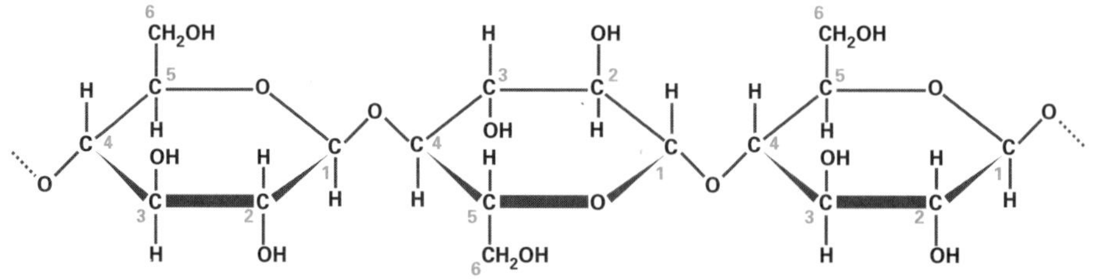

On the page, cellulose looks very like glycogen, but the difference in bond type is critical. Glucoses linked by $\alpha(1 \rightarrow 4)$ links arrange themselves in a floppy helix, while glucoses linked by $\beta(1 \rightarrow 4)$ links form extended chains, ideal for building the rigid plant cell wall. Animals have enzymes that can break down the $\alpha(1 \rightarrow 4)$ bond in glycogen, but only certain bacteria and fungi can break the $\beta(1 \rightarrow 4)$ link in cellulose. All animals that eat plants rely on bacteria in their intestines to provide the enzymes to digest cellulose.

such as hemicellulose and pectin. The thickness of the wall is determined largely by its pectin content. The plant cell wall is laid down in stages; in growing cells the orientation of cellulose fibrils in the primary cell wall determines the direction in which the cell expands as the result of its internal turgor pressure. Once the growth of the cell is complete, other compounds, most notably lignin, are added to form the secondary cell wall.

Cell Junctions

In multicellular organisms, and particularly in epithelia, it is often necessary for neighboring cells within a tissue to be connected together. This function is provided by cell junctions. In animal cells there are three types of junctions: those that form a tight seal between adjacent cells are known as *tight junctions*; those that anchor cells together, allowing the tissue to be stretched without tearing, are called *adhering junctions*; while a third class of cell junctions that allows communication between cells are known as *gap junctions*. Plant cells contain a unique class of communicating junction known as *plasmodesmata* (Fig. 1.4).

TIGHT JUNCTIONS are generally associated with epithelial cells such as those lining the small intestine. The plasma membranes of adjacent cells are pressed together so tightly that no intercellular space exists between them. This arrangement ensures that the only way that molecules can get from the lumen of the intestine to the blood supply that lies beneath is by passing through the cells, a route that can be selective.

ADHERING JUNCTIONS are of two types: spot desmosomes are like rivets between cells. They are found in epithelial tissues such as skin that are subjected to mechanical stress. Belt desmosomes have a different structure from spot desmosomes and are usually associated with actin filaments (Chapter 3); they often form a distinct belt, linking sheets of cells in register. Contraction of the filaments leads to changes in the shape of the individual cells and thus the epithelium in which it is located.

GAP JUNCTIONS are concerned with cell to cell communication. When two cells form a gap junction, ions and small molecules can pass directly from the cytosol of one cell to the cytosol of the other cell without going into the extracellular fluid. The structure that makes this possible is the gap junction channel. Channels, as we will see in Chapter 17, are water-filled tubes through membranes. Each gap junction channel runs all the way through the plasmalemma of the first cell, across the small gap between the cells, and through the plasmalemma of the second cell. In the middle of the channel is a continuous hole about 1.5 nm in diameter. This is large enough to allow small ions through (and therefore pass electrical current) together with amino acids and nucleotides, but it is too small for proteins or nucleic acids (Chapters 5, 13). Gap junctions are especially important in the heart, where they allow an electrical signal to pass

BOX 2.2

How It All Fits Together: Protein to Protein Binding

Proteins, described in detail in Chapters 13 and 14, are the main stuff of which cells are made. This box illustrates the fact that many of the jobs that proteins do depend on recognition and binding.

Binding That Is Strong and Permanent—Cadherin at Adhering Junctions

An adhering junction will form if neighboring cells produce the integral membrane protein cadherin. As the figure shows, the extracellular region (or domain) of cadherin has just the right shape and properties to bind to a second cadherin, locking the two cells together. Even though there are many other proteins around, none has the same shape and properties, so cadherin only binds to cadherin: The binding is specific. Since one cadherin binds a second, identical molecule, this is called homophilic binding.

Recognition and binding need not be homophilic. The cytoplasmic domain of cadherin binds specifically to a cytoplasmic protein called catenin, which in turn binds specifically to strands of actin molecules called microfilaments (page 53).

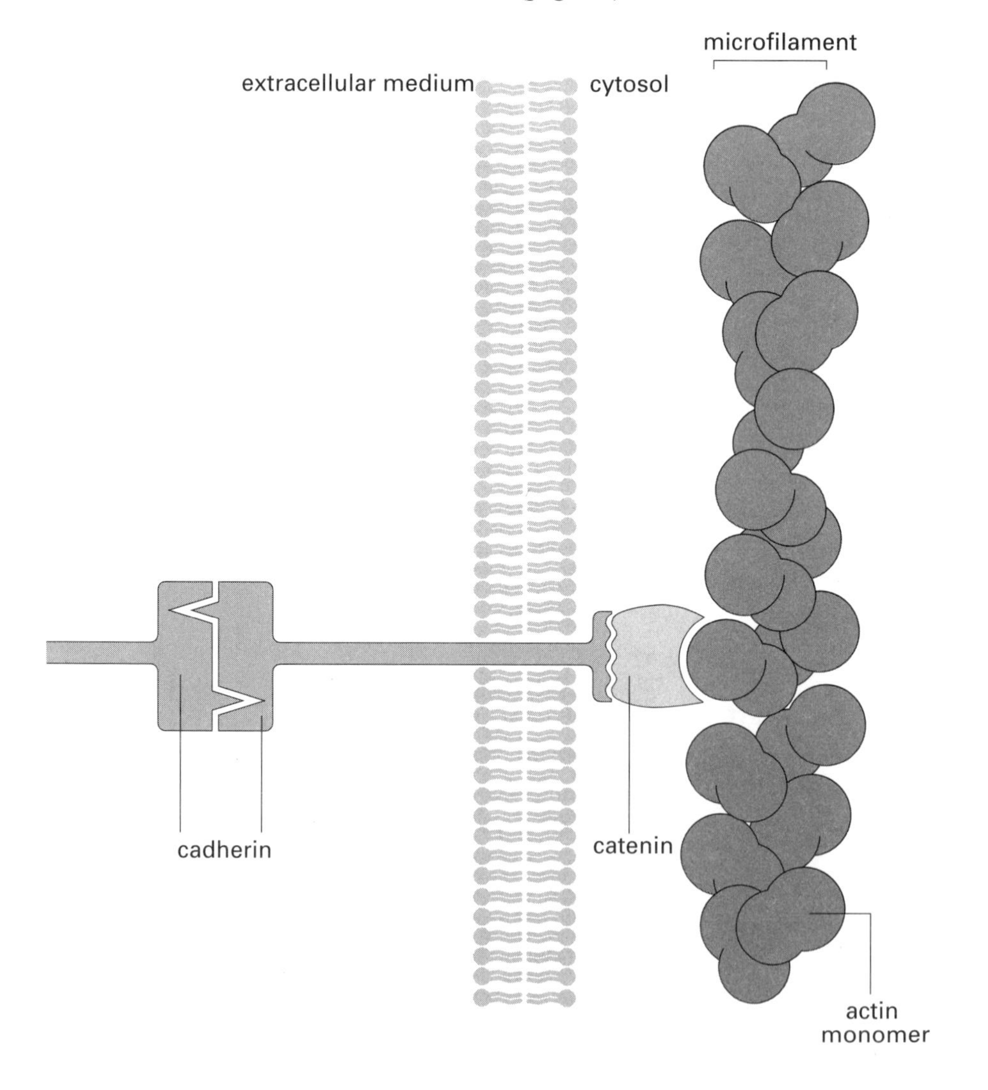

Phosphorylation Can Turn Binding On or Off—Signaling through SH2 Domains

Growth factors such as fibroblast growth factor (FGF) are small proteins released from one population of cells that act to cause cell growth and division in another population (page 339). The receptor for fibroblast growth factor, the FGFR, is typical and is shown below. It is an integral membrane protein with an extracellular domain that binds FGF. One FGF can bind to two FGFR molecules, so FGF causes FGFRs to associate in pairs—to dimerize.

The cytoplasmic domain of the FGFR has tyrosine kinase activity. This means that the protein is an enzyme that catalyses the attachment of a negatively charged phosphoryl group to tyrosine, one of the amino acid building blocks of a protein (page 214). This process is called phosphorylation (page 218). (The phosphoryl group comes from adenosine triphosphate, ATP; page 253.) In the absence of FGF, the FGFR does not encounter any tyrosine, so the enzyme does nothing. However, when FGF causes FGFR dimerization, the kinase activity on each FGFR phosphorylates a tyrosine on the other FGFR.

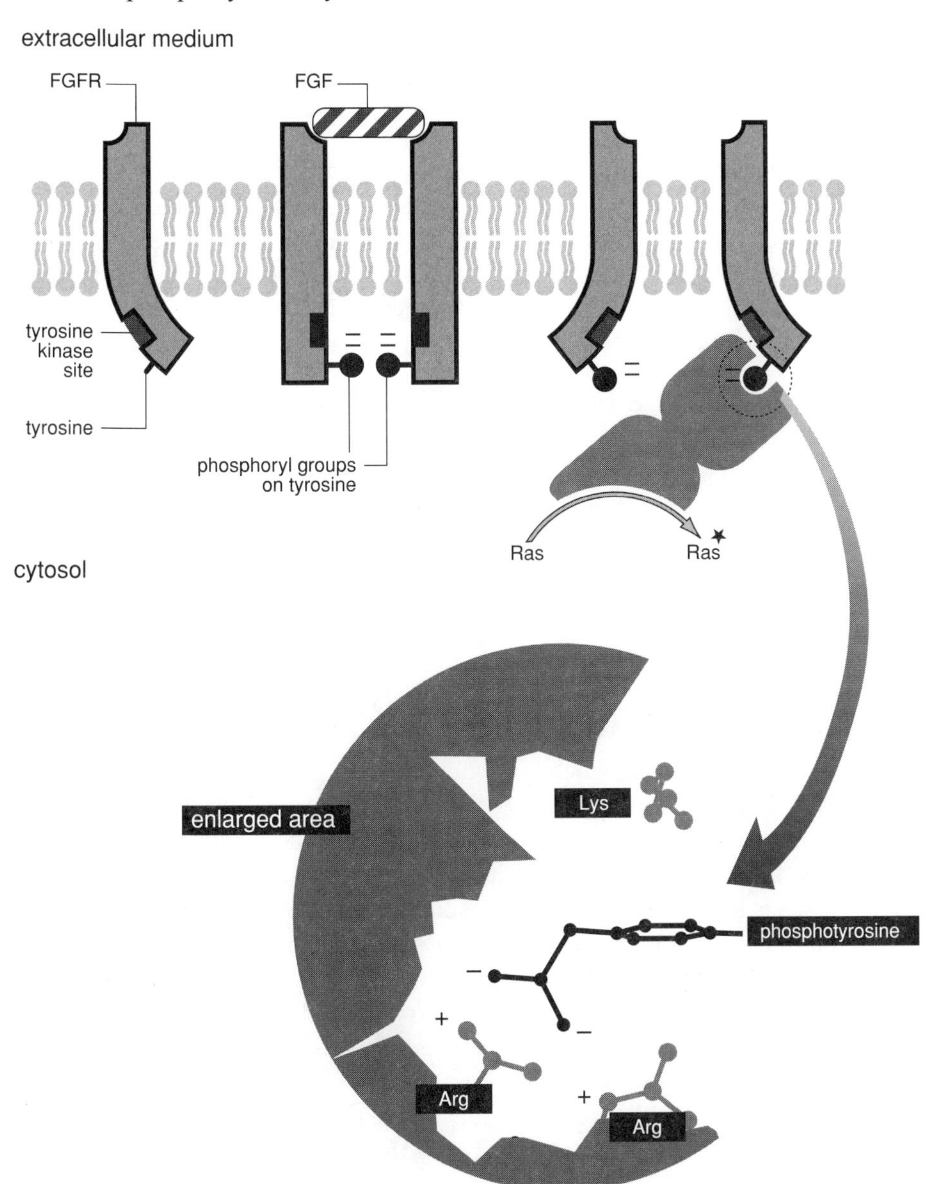

A number of cytoplasmic proteins have a domain called SH2 that is just the right shape to stick to phosphorylated tyrosine. At the bottom of a deep pocket in the protein surface is the positively charged amino acid arginine (Arg) (page 212). Although proteins can be phosphorylated on amino acids other than tyrosine (page 219), only tyrosine is long enough to insert down into the pocket so that the negative phosphoryl can stick to the positive arginine. Proteins with SH2 domains therefore stick to dimerized FGFRs. Of course SH2 domains do not stick to solitary (monomeric) FGFRs whose tyrosines do not carry the negatively charged phosphoryl group.

A chain of protein–protein binding now leads to the activation of Ras, a protein that plays a central role in activating cell division (Box 4.2, page 70). The end result is that if a cell has FGFR in its plasmalemma, exposing it to FGF will cause it to divide.

rapidly between all the cardiac muscle cells, ensuring that they all contract at the same time. Gap junction channels can open and close in a process called gating (page 292). The plasmodesmata that perforate the cell walls of many plant tissues serve much the same purpose, though they are much bigger and cannot shut quickly. Some plant viruses use plasmodesmata to spread from cell to cell.

ORGANELLES BOUNDED BY DOUBLE MEMBRANE ENVELOPES

Three of the major cell organelles—the nucleus, mitochondrion and, in plant cells, the chloroplast—share the distinctive feature of being enclosed within an envelope consisting of two parallel membranes.

The Nucleus

The nucleus is often the most prominent cell organelle. It contains the genome, the cell's database, which is encoded in molecules of the nucleic acid, DNA. In later chapters we will see how blocks of information within the DNA molecules, called genes, are copied onto another nucleic acid, RNA (Chapter 10). The information on RNA is then used to create proteins (Chapter 12). The nucleus is bounded by a nuclear envelope composed of two membranes enclosing an intermembrane space (Fig. 2.3). The inner membrane of the nuclear envelope is lined by a meshwork of proteins called the nuclear lamina, which provides rigidity to the nucleus. The two-way traffic of proteins and nucleic acids between the nucleus and the cytoplasm passes through holes in the nuclear envelope called *nuclear pores*. The nucleus of a cell that is synthesizing proteins at a low level will have a few nuclear pores. In cells that are undergoing active protein synthesis, however, virtually the whole nuclear surface is perforated.

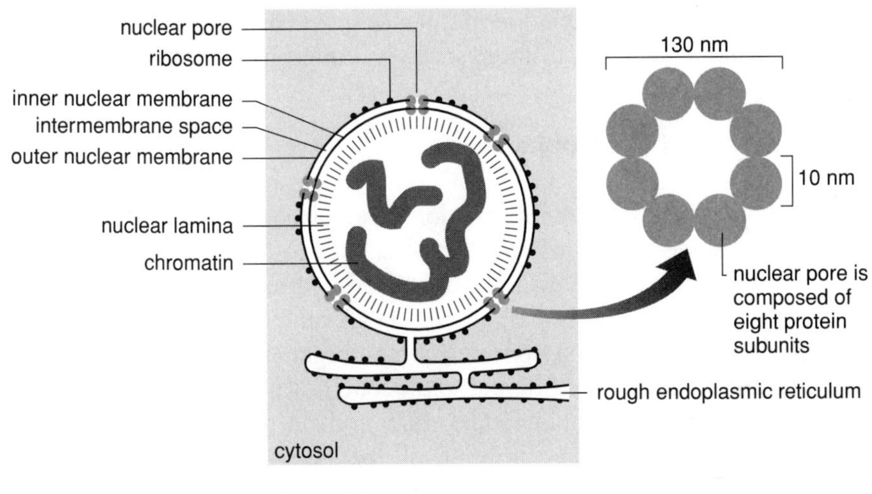

Figure 2.3
Structure of the nuclear envelope.

Each pore is about 0.13 μm across and about 0.07 μm deep. It consists of eight protein subunits arranged around a large central channel (Fig. 2.3).

Within the nucleus it is usually possible to recognize discrete areas. Most of it is occupied by chromatin, a complex of DNA and certain DNA-binding proteins such as histones (Chapter 5). In most cells it is possible to recognize two types of chromatin: Lightly staining euchromatin is that portion of the cell's DNA that is being actively transcribed into RNA (and hence into proteins); heterochromatin, on the other hand, represents the inactive portion of the genome where no RNA synthesis is occurring. The DNA in heterochromatin is tightly coiled then supercoiled (see Fig. 5.6), leading to its dense appearance in both the light and electron microscopes.

RNA, unlike DNA, is not confined within the nucleus and other organelles but is found within the cytoplasm. Here it is associated with particles called ribosomes whose function is to make proteins. Ribosomes are made in the nucleus, in specialized regions called nucleoli that form at specific sites on particular chromosomes called the nucleolar organizer regions. These contain blocks of genes that code for ribosomal RNA (rRNA) (Chapter 21). Once made, the ribosomes are exported through the nuclear pores to the cytoplasm.

It should be stressed that the appearance of the nucleus we have described thus far relates to the cell in interphase, the period between successive rounds of cell division. As the cell enters mitosis (Chapter 4), the organization of the nucleus changes dramatically. The DNA becomes more and more tightly packed, and is revealed as a number of separate rods called *chromosomes*—46 in human cells. The nucleolus disperses, and the nuclear envelope fragments into tiny vesicles. Upon completion of mitosis, these structural rearrangements are reversed, and the nucleus resumes its typical interphase organization.

Mitochondria and Chloroplasts

Although most of the genetic information of a eukaryotic cell is encoded by nuclear genes, some of the information necessary to make both chloroplasts and mitochondria is encoded by genes contained within these organelles themselves. These genes reside in small circular DNA molecules that are similar to the chromosomes of bacteria (from which both organelles are thought to have evolved; Chapter 1) but very different to the long linear DNA molecules in the nucleus. Chloroplasts and mitochondria both contain ribosomes (again, more like those of bacteria than the ribosomes in the cytoplasm of their own cell), and synthesize a small subset of their own proteins, although the great majority of proteins that form chloroplasts and mitochondria are encoded by nuclear genes and synthesized in the cytoplasm. Perhaps the most distinctive feature of these organelles, however, is that the inner of their two membranes is markedly elaborated and folded to increase its surface area.

MITOCHONDRIA. Mitochondria are among the most easily recognizable organelles due to the extensive folding of their inner membrane to form shelflike projections named cristae (Fig. 2.4). The number of cristae, like the number of mitochondria, is adjusted to the energy requirements of the cell type in which they are found. In striated muscle cells which must contract and relax repeatedly over long periods of time, there are many mitochondria that contain numerous cristae: in fat cells which need little energy, there are few mitochondria and their cristae are less well developed. This gives a clue as to the function of mitochondria: They are the cell's power stations. Mitochondria produce the molecule adenosine triphosphate (ATP). ATP is one of the cell's energy currencies that provide the energy to drive a host of cellular reactions and mechanisms (Chapter 15). The double membrane structure provides four distinct domains: the outer membrane, the inner membrane, the intermembrane space, and the matrix. Each domain houses a distinct set of enzyme functions. Details are given in Chapters 15 and 16.

CHLOROPLASTS. In the chloroplast the inner membrane is folded into thylakoids which at least in higher plants, form stacks called grana (Fig. 2.4). The thylakoids contain the proteins and other molecules responsible for the light reaction of photosynthesis. The dark reaction of photosynthesis, on the other hand, takes place in the matrix, called the stroma, which also contains the DNA and ribosomes. These two reactions of photosynthesis are described in Chapters 15 and 16 respectively.

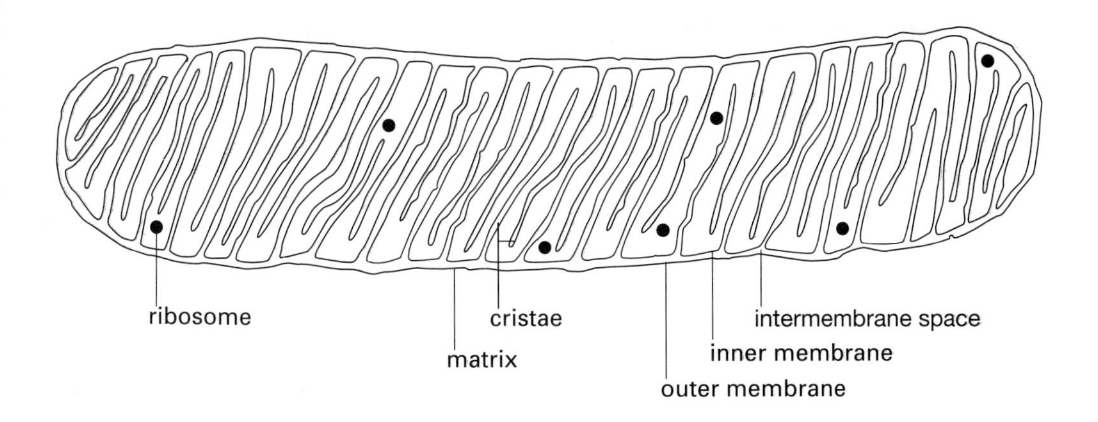

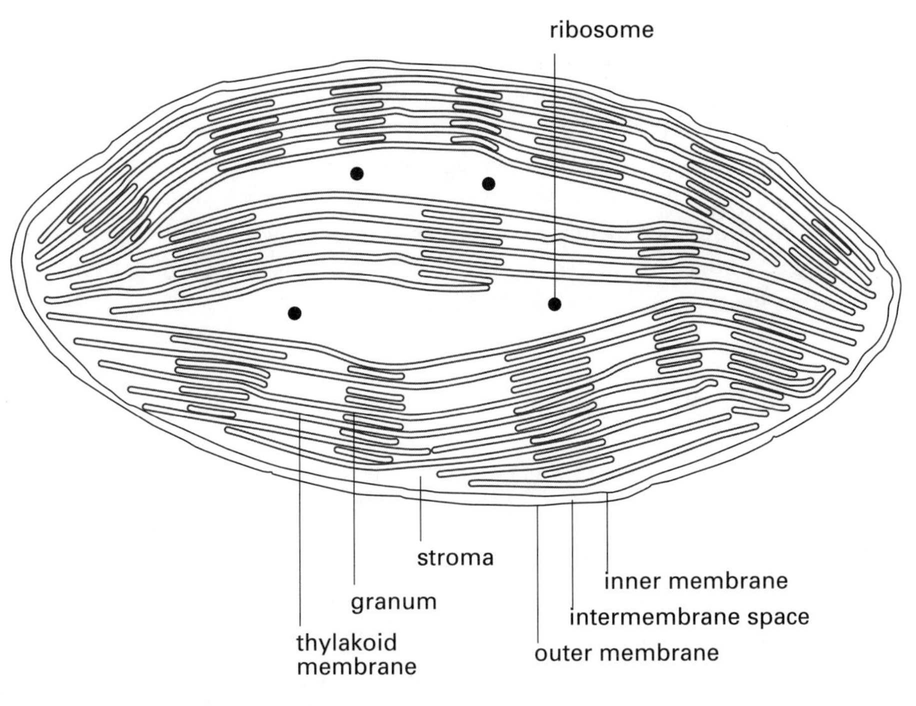

Figure 2.4

Mitochondrion and chloroplast.

ORGANELLES BOUNDED BY A SINGLE MEMBRANE

Eukaryotic cells contain many sacs and tubes bounded by a single membrane. Although these are often rather similar in appearance, they can be subdivided into different types specialized to carry out distinct functions.

Peroxisomes

Mitochondria and chloroplasts are frequently found close to another membrane-bound organelle, the peroxisome. In human cells peroxisomes have a diameter of about 1 μm, and their dense matrix contains a heterogeneous collection of proteins concerned with a variety of metabolic functions, some of which are only now being begun to be understood. Peroxisomes are so named because they are frequently responsible for the conversion of the highly reactive molecule hydrogen peroxide (formed as a by product of oxidation) into water. This reaction is carried out by the protein catalase which sometimes forms an obvious crystal within the peroxisome. Catalase is an enzyme—a protein catalyst that carries out chemical reactions (Chapter 14). In fact, it was one of the first enzymes to be discovered. In humans, peroxisomes are primarily associated with lipid metabolism. Understanding peroxisome function is important for a number of inherited human diseases such as Zellweger's syndrome where an inability to metabolize lipid properly as a result of peroxisome malfunction leads to death in early childhood.

Endoplasmic Reticulum

The endoplasmic reticulum (ER) is a system of membrane-bounded chemicals that run throughout the cell, forming a continuous network whose lumen is at all points separated from the cytosol by a single membrane (Fig. 1.1). Two regions can be recognized in most eukaryotic cells. These are known as smooth endoplasmic reticulum and rough endoplasmic reticulum.

The function of the smooth endoplasmic reticulum varies from tissue to tissue. In ovaries, testes and the adrenal gland it is where steroid hormones are made; in the liver it is the site of detoxification of foreign chemicals. Probably the most universal role of the smooth endoplasmic reticulum is the storage and sudden release of calcium ions. Calcium ions are pumped from the cytosol into the lumen of the smooth endoplasmic reticulum to more than 100 times the concentration found in the cytosol. Many stimuli can cause this calcium to be released back into the cytosol, where it activates many cell processes (Chapter 19).

The rough endoplasmic reticulum is where the cells make the proteins that will end up as integral membrane proteins in the plasmalemma, and proteins that the cell will export to the extracellular medium. This process is described in detail in Chapter 12, so we will only describe it briefly here. Ribosomes making these proteins attach to the surface of the endoplasmic reticulum, and as the protein is made, it is fed through the membrane, either all the way through if it is to be exported or part way through if it is an integral membrane protein.

Golgi Apparatus

The Golgi apparatus, named for its discoverer Camillo Golgi, is a distinctive stack of flattened sacks (cisternae) situated close to the centrosome (Chapter 3) at the so-called cell center immediately adjacent to the nucleus (Fig. 1.1). From the Golgi, proteins pass to the extracellular medium by the process of exocytosis (page 205) or to other intracellular organelles.

Endocytotic Vesicles and Endosomes

Useful molecules in the extracellular medium that are too large to pass through the plasmalemma are taken up into cells by the process of endocytosis. This begins when the plasmalemma invaginates to form an inward-curving pocket. The pocket then closes off so that the molecules are trapped within a membrane-bounded vesicle called an endocytotic vesicle.

Some of the endocytotic vesicles that form at the cell surface are relatively nonspecific. They carry materials into the cell that have been attracted to the cell surface by the charged surface of the membrane or materials that were simply abundant in the aqueous environment at the cell surface. The endocytotic vesicle may fuse with a primary endosome (see below). Membrane brought into the cell in this way must, at some point, be recycled back to the surface.

Receptor-mediated endocytosis is a highly specific cellular mechanism for selecting molecules at the cell surface, even those present at very low concentration. Receptors in the plasmalemma stick out into the extracellular medium and bind the desired molecule (the ligand) with high specificity. At the same time, a three-legged protein called clathrin associates with the cytosolic side of the plasmalemma and assembles into a basket or cage-like network on the surface of the plasmalemma. This forces the plasmalemma to invaginate to form a coated pit, so-called because it is surrounded by a coat of clathrin. Eventually the pit pinches off from the plasmalemma to form a coated vesicle surrounded by a complete clathrin cage. Coated vesicles have a lifetime of only a few minutes before an enzyme strips off the clathrin cage. The molecular components of the cage are then recycled to the cell membrane, leaving the naked vesicle to fuse with a membrane compartment called the primary endosome. The pH (see Box 1.4) within the endosome drops rapidly to between 5.0 and 5.5. In this acid environment the receptor-ligand complex is unstable, and the ligand is released to float free in the lumen. Now two vesicles pinch off from this compartment. One carries the ligand and goes to fuse with a primary lysosome. The other vesicle carries the receptor back to the plasmalemma.

Lysosomes

Lysosomes are a kind of intracellular digestive vesicle, degrading unwanted materials. Lysosomes that are waiting to be used bear a coat of clathrin. When digestion is required the clathrin coat is lost, allowing the inactive or primary lysosome to fuse with a vesicle containing the material to be digested to make a secondary lysosome. The vesicles with which the primary lysosomes fuse may be bringing materials in from outside the cell, or they may be vesicles made by condensing a membrane around worn out or unneeded organelles in the cells own cytoplasm. The latter are sometimes called autophagic vacuoles. Much of the material hydrolyzed in secondary lysosomes is rendered small enough to diffuse through the membrane and is used in the cytoplasm in the cell's own metabolism. Some materials may not be digestible, even by such an array of enzymes, and they remain in the lysosome for the lifetime of the cell. These small dense remnant lysosomes are called residual bodies.

SUMMARY

1. Membranes define the boundary of the cell and of the cell organelles. They are composed primarily of phospholipids and protein.

2. Phospholipids have two fatty acids that form a *hydrophobic* tail, plus a *hydrophilic* head group. The three subunits are attached to a glycerol backbone.

3. The plasmalemma is the membrane that surrounds the whole cell. Outside the plasmalemma lies an extracellular matrix of polysaccharides and proteins.

4. A number of uncharged, relatively small solutes, can dissolve in the hydrophobic interior of membranes and hence pass through in a process called simple diffusion.

5. Cells are joined together by three types of cell junction. Tight junctions occlude the extracellular space. Adhering junctions anchor cells

together. Gap junctions, together with plasmodesmata in plants, allow direct passage of solutes from cytosol to cytosol of adjacent cells.

6. The nucleus, mitochondria, and, in plant cells, chloroplasts are bounded by double membrane envelopes. The nuclear envelope contains numerous nuclear pores.

7. Endocytosis is the process by which cells bring molecules in from outside. Receptor-mediated endocytosis allows them to take up specific molecules that are essential for growth.

BOX 2.3

The Kinks Have It: Double Bonds, Membrane Fluidity, and Evening Primroses

One of the fatty acids commonly found as a component of animal fat is stearic acid, shown below. Its eighteen carbon atoms are joined by single bonds, making a long straight molecule. In contrast, oleic acid has a double bond between the ninth and tenth carbons. This introduces a kink in the chain. A fatty acid with kinks in is less ready to solidify because it is less able to pack in a regular fashion. The more double bonds in a fatty acid, the lower its melting point. Fatty acids containing double bonds between the carbon atoms are said to be unsaturated.

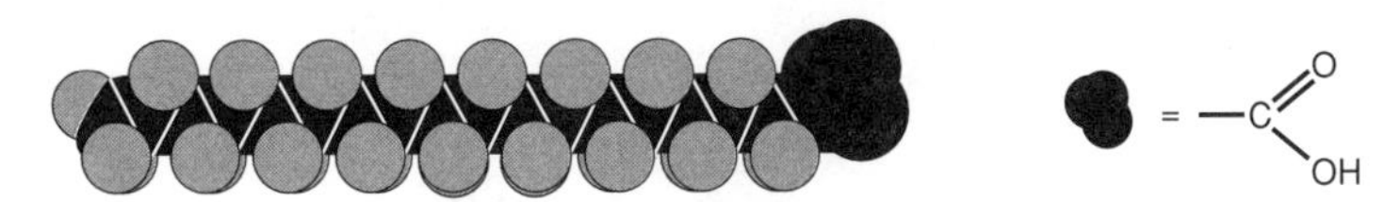

Stearic acid, C18 no double bonds, melts at 69.6°

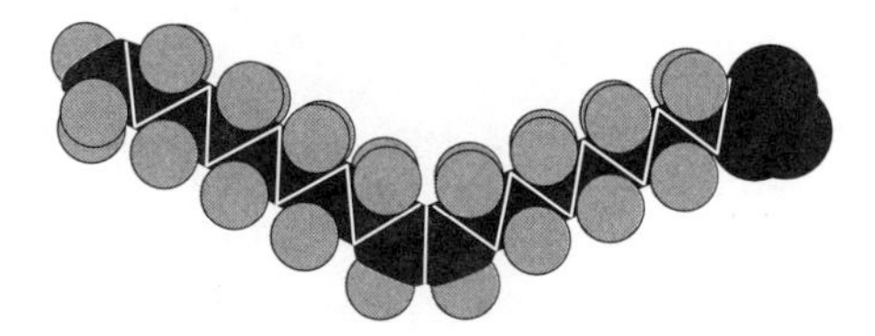

Oleic acid, C18 one double bond, melts at 13.4°

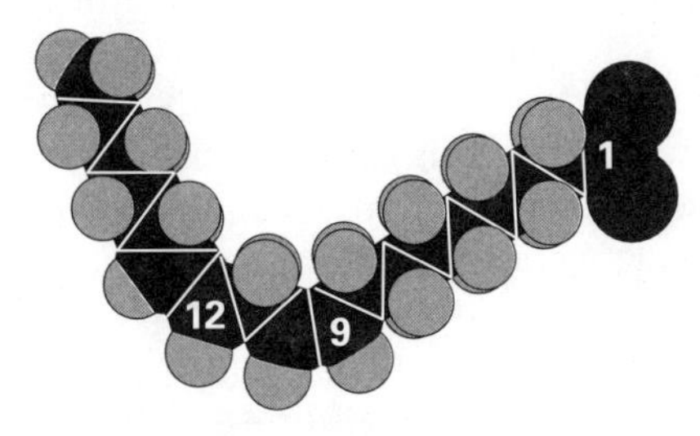

Linoleic acid, C18 two double bonds, melts at –9°

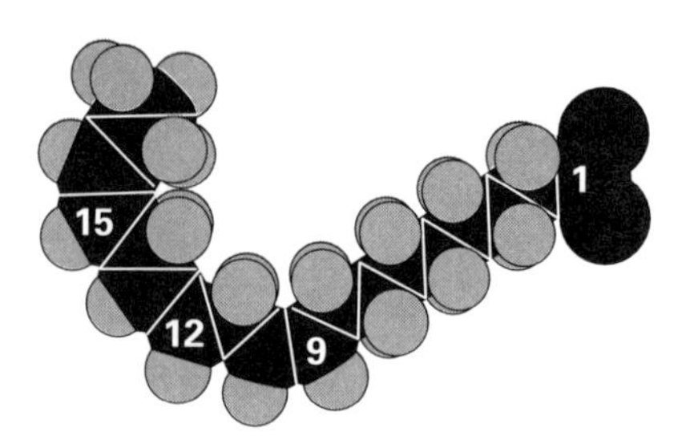

Linolenic acid, C18 three double bonds, melts at –17°

For the triacylglycerols and the diacylglycerols within membranes the same rule applies—the more double bonds in the fatty acid, the lower the melting point. Membranes must not solidify—if they did then they would crack and the cell contents would leak out each time the cell was bent. Unsaturated fatty acids play an essential role in maintaining membrane liquidity.

Mammals are unable to introduce double bonds beyond carbon 9 in the fatty acid chain. This means that linoleic and linolenic acids must be present in the diet. They are known as essential fatty acids. Fortunately the biochemical abilities of plants are not so restricted and plant oils form a valuable source of unsaturated fatty acids.

The normal form of linolenic acid is α-linolenic acid, which has its double bonds between carbons 9 and 10, 12 and 13, and 15 and 16. Some plant seed oils contain an isomer of linoleic acid with double bonds between carbons 6 and 7, 9 and 10, and 12 and 13. This is called γ-linolenic acid and is shown below.

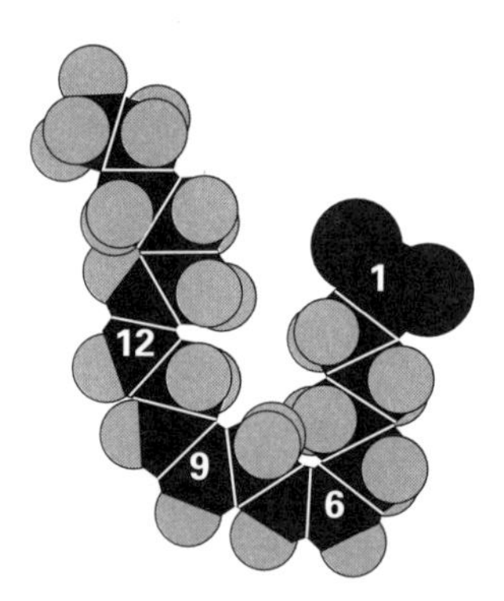

The attractive, yellow-flowered garden plant the evening primrose (*Oenethera perennis*) has seeds that contain an oil with γ-linolenic in its triacylglycerols. The 6 to 7 double bond introduces a kink closer to the glycerol. This is thought to increase fluidity when incorporated into membrane lipids. No one really knows if it really has the marvelous health effects that some claim for it or, if it does, how it works. This ignorance has not stopped some people from making lots of money from evening primrose oil. It is included in cosmetics and in alternative medicines for internal and external application. γ-linolenic acid occurs in other plants—borage (*Borago officinalis*) seeds are one of the richest sources. It is even found in some fungi. However, these lack the romance of the evening primrose!

CHAPTER

3

THE CYTOSKELETON AND CELL MOVEMENT

In a series of letters to the Royal Society in London in the late seventeenth century, Anton van Leeuwenhoek described his observations of a hitherto unimagined world of exotic microscopic creatures. In particular, he marveled at the nimble movements of some of his newly discovered "little animalcules" (which we now know to be bacteria and protozoa) and speculated at length on the nature of their "little legs" and "little feet." The man we now acknowledge as the father of microscopy also began the study of cell motility, currently one of the most exciting areas of cell biology.

Cells move for a variety of reasons. Human spermatozoa in their millions swim frantically toward an ovum; the soil amoeba, *Acanthamoeba* (said to be the most abundant organism on earth), crawls peripatetically over and between soil particles, engulfing bacteria and small organic particles as it does so; cells in the early human embryo show similar crawling movement as they reorganize to form tissues and organs and the invasive properties of some cancer cells is due their reverting to this highly motile embryonic state. Of course not all cells show these overt forms of motility, but careful observation of even the most sedentary cells often reveals a remarkable repertoire of intracellular movements. Cell motility in its many forms is based on a complex of intracellular filaments known as the cytoskeleton. As the name implies, the cytoskeleton also has a structural role and is largely responsible for determining the shape of the cell and its intracellular organization.

THE CYTOSKELETON IS COMPOSED OF THREE CLASSES OF FILAMENTS

Eukaryotic cells contain three cytoplasmic filament networks. Two of them, *microtubules* and *microfilaments*, are concerned with the repertoire of movements of and within cells. The third, *intermediate filaments*, have a purely structural role. The organization of the three filament systems of the cytoskeleton is most easily studied by techniques such as indirect immunofluorescence microscopy (Box 3.1).

MICROTUBULES

The physical properties of microtubules allow them to do many things. They can form bundles of rigid fibers that make structural scaffolds that serve to determine cell shape; they have an inherent structural polarity that determines the direction in which the cell performs actions such as growth and exocytosis; they provide a system of intracellular highways that supports a two-way traffic of organelles and small vesicles powered by protein motors that interact with the microtubule surface. What is more they can be rapidly formed and broken down, allowing the cell to respond to subtle environmental changes. This property is central to the most exquisite of all movements

BOX 3.1

Indirect Immunofluorescence Microscopy of the Cytoskeleton

Antibodies are proteins made by the immune system that bind to, and help remove, foreign molecules. Because they must not bind to the body's own molecules, antibodies are very specific to their targets, called antigens. Indirect immunofluorescence microscopy exploits this specificity to reveal in the light microscope structures that may be well below the instrument's theoretical limit of resolution. Two antibodies are used: a primary antibody, raised in an animal such as a rabbit against a purified cytoskeletal protein, and a secondary antibody, prepared against rabbit antibodies in a different host such as a goat. The secondary antibody is conjugated to a compound that fluoresces when illuminated with light of the appropriate wavelength, such as rhodamine isothiocyanate (which gives off red light) or fluorescein isothiocyanate (which glows green). Antibody molecules are large and do not cross the membrane of living cells, so the cell must first be fixed and its plasma membrane gently stripped away with a dilute detergent solution. The detergent is washed away and the two antibodies applied in sequence. The first antibody binds specifically to structures containing the antigen against which it was raised. The second antibody binds to the first, coating each filament in a double-decker antibody sandwich. When the cells are illuminated with the appropriate light the complete cellular array of microtubules, microfilaments, and intermediate filaments lights up spectacularly as colored threads.

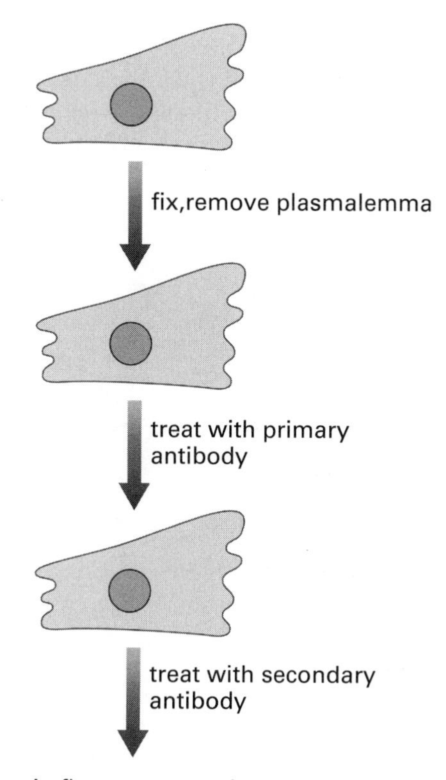

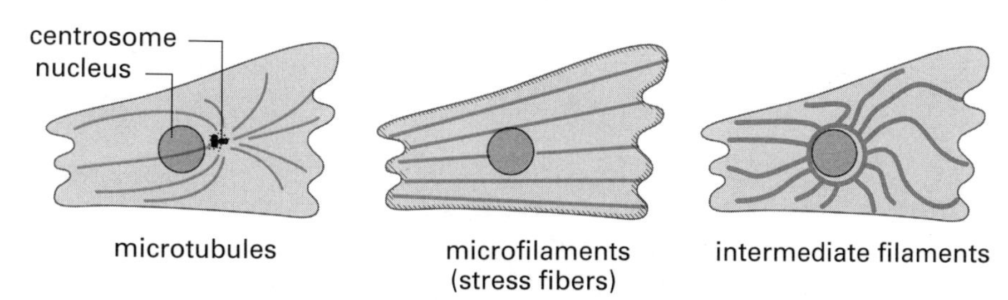

Antibodies provide highly specific and versatile reagents for cytology. Used in combination and with other fluorescent probes such as DAPI (which specifically stains DNA), several cellular structures can be revealed simultaneously. A similar technique can also be used to identify structures in the electron microscope, although in this situation the secondary antibody is conjugated to electron-dense particles such as 5 or 10 nm gold particles. Antibodies are also widely used in biochemistry for detecting small quantities of protein.

within the cell, the segregation of chromosomes at mitosis and meiosis (Chapter 4).

Animal cells contain a network of several thousand microtubules, all of which can be traced back to a single structure tightly attached to the surface of the nucleus. This is the *centrosome* which is said to be the *microtubule-organizing* center of the cell (Fig. 3.1). The centrosome is located at the cell center, which also contains the Golgi apparatus (Chapter 2), and consists of amorphous pericentriolar material enclosing pair of centrioles. These have the same structure as the basal bodies of cilia and flagella (Fig. 3.1; see below). Micro-

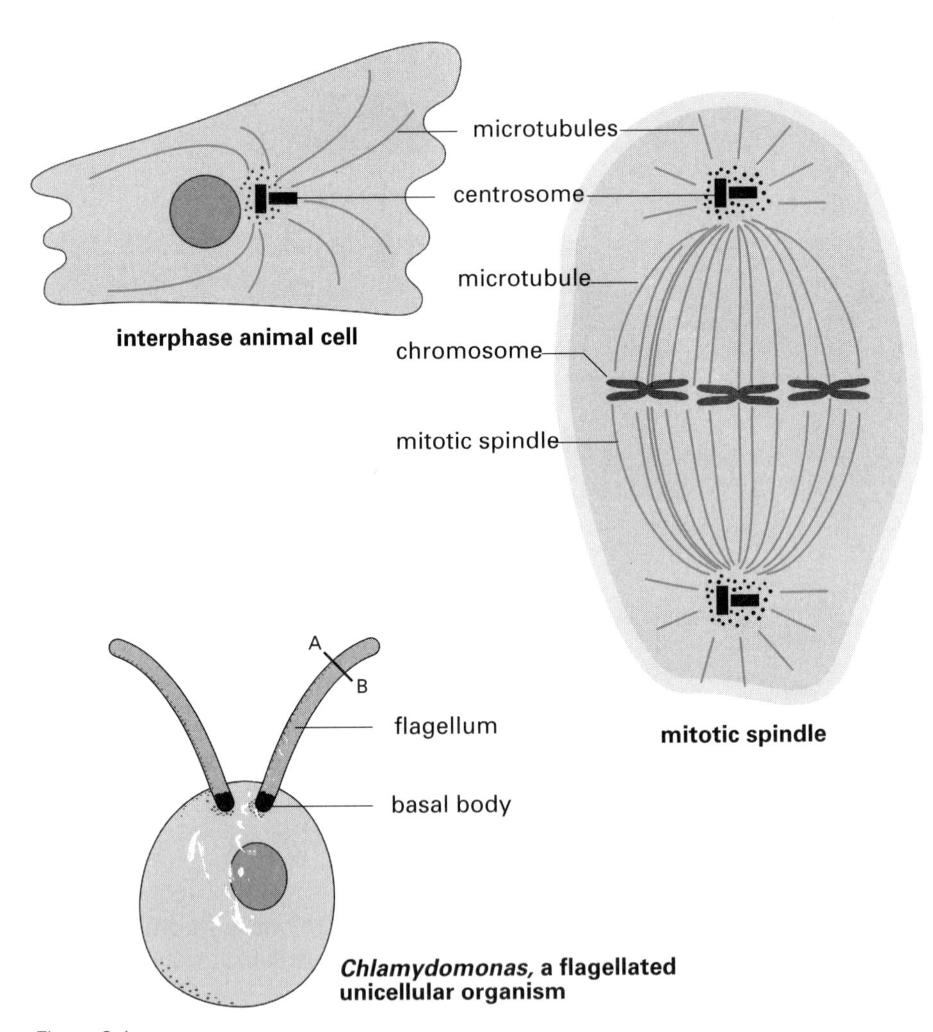

Figure 3.1

Centrosomes and basal bodies organize the microtubule cytoskeleton. A section along the line AB would appear as shown in Fig. 3.5 (page 50).

tubules in plant cells have a quite different organization. Here they lie immediately beneath the cell membrane, orientated at right angles to direction of cell expansion. The role of microtubules in plants is to direct the deposition of cellulose fibers on the outside of the cell membrane. As a consequence of the position of the microtubules beneath the membrane, cellulose is laid down in hoops that encase the plant cell in a rigid corset and allow it to expand only in one direction. Plant cells do not contain centrioles. It is not yet clear whether their microtubules arise from a defined organizing center.

Microtubules are composed of a subunit protein called tubulin. *Tubulin* is a dimeric protein; it is composed of two similar but distinct monomers designated α and β. There is a third type of tubulin, γ-tubulin, which is not a part of the microtubule as such but is found at the centrosome where it plays a role in initiating microtubule assembly. α-tubulin/β-tubulin dimers assemble into chains called protofilaments, 13 of which make up the microtubule wall. Within each protofilament the tubulin dimers are arranged head to tail, α to β, many times. This molecular polarity is reflected in the pattern of growth. Tubulin subunits are added to, and lost from, one end more rapidly than the other. By convention, the fast growing end is referred to as the (+) end and the slow growing end as the (–) end.

In cells the minus end of each microtubule is embedded in the centrosome so that only the plus ends are free to grow or shrink. This process is surprisingly complex. Individual microtubules undergo periods of slow growth followed by rapid shrinkage, sometimes disappearing completely. This phenomenon is referred to as dynamic instability. By chance, the growing ends of certain microtubules may be "captured" by sites at the cell membrane and stabilized. Further growth influences the shape of the cell (Fig. 3.2). Groups of microtubules having a common orientation make a stable structural framework which can, because microtubules are dynamic, be remodeled as the needs of the cell change.

One of the most important tools in establishing microtubule function in cells has been the plant alkaloid colchicine. Extracted from the corms of the autumn crocus, *Colchicum autumnale,* colchicine has been used since Roman times as a treatment for gout (although, curiously, for reasons that have little to do with microtubules). Cells exposed to colchicine lose their shape and the movement of organelles within the cytoplasm ceases. When the drug is washed away, microtubules reassemble from the centrosome and normal functions are resumed (Fig. 3.3). Another drug, taxol, obtained from the bark of the Pacific yew, *Taxus brevifolia,* has the opposite effect, causing large numbers of very stable microtubules to form in the cell, an effect that is difficult to reverse. Since microtubules must break down and reform as spindles at cell division (Chapter 4), cells treated with taxol cannot divide, and for this reason it is now used as an anticancer drug.

BOX 3.2

Bacteria Use a Unique Type of Motility

Although by no means all bacteria are motile, some such as *Escherichia coli* and *Salmonella typhimurium* have developed tactic responses, that is, the ability to perceive environmental signals and to move towards or away from them. Response to chemicals such as amino acids and sugars is referred to as chemotaxis, to light as phototaxis, to oxygen as aerotaxis, and so on. Bacteria do not swim in the way that we generally understand that term. If you, the reader, dive into a swimming pool, your momentum will carry you some distance because your mass is great relative to the viscosity of the water. For a tiny bacterium, water is a highly viscous medium. A bacterium swimming in water experiences roughly what a human swimmer in treacle might feel! Even so, a bacterium can travel at 100 $\mu m\ s^{-1}$, albeit in an erratic way, a brief period of smooth progression (a run) being followed by a rapid and random change of direction (a tumble). In the presence of an attractant, the frequency of tumbling is reduced as long as the bacterium is heading up the concentration gradient toward its source. As a result the bacterium swims smoothly for longer periods (Fig. *a*). If it starts to move down the gradient away from its goal, the frequency of tumbles increases until, by chance, the bacterium becomes correctly reoriented. In a gradient of a repellent, the situation is reversed.

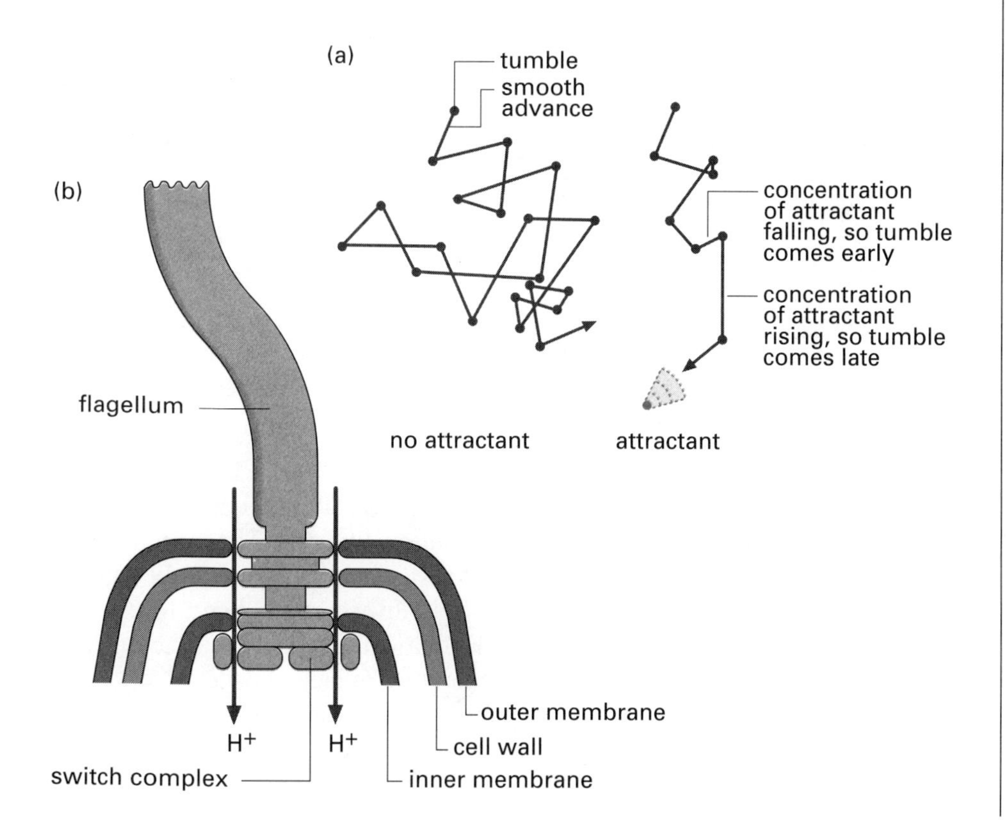

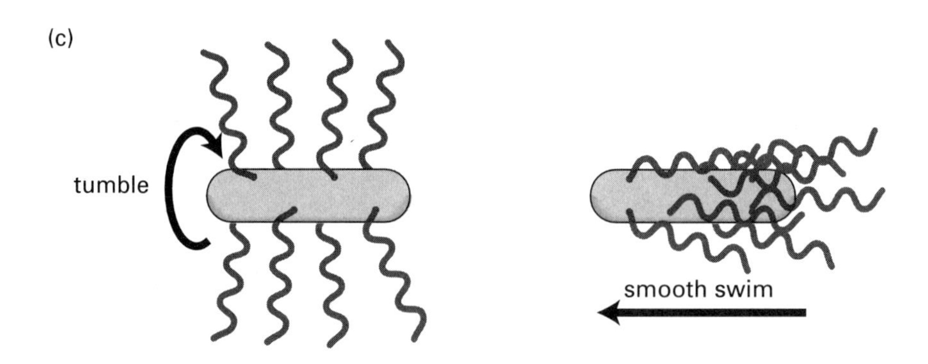

These observations pose two questions: What is the bacterium's sensory mechanism, and how does it execute such abrupt changes of direction? The answer to the first is that spanning the cell membrane are four classes of proteins, one for each of the categories of attractants to which the cell responds. Binding of attractant activates a sequence of chemotaxis proteins, one of which diffuses rapidly across the cytoplasm to interact with the flagellum. At the same time the ability of the membrane proteins to bind more attractant is reduced. The concentration of attractant must hence rise constantly to give the same message "OK, keep going this way."

The bacterial flagellum is a remarkable example of a rotary biological "motor." Each flagellum (*E. coli* has 8–10) consists of a rigid helical filament, 20–40 nm in diameter and extending up to 10 μm from the cell surface. It is composed of a single protein, flagellin. At its base, embedded in the cell membrane, is a minute motor consisting of a series of rings that rotate within the complex layers of membranes and cell wall that makes up the bacterial cell surface to generate force (Fig. *b*). A flow of H^+ across the membrane causes rotation between two of the rings at the base. The motor turns at up to 100 Hz (= 100 times per second), driving the flagellum like a propeller. The direction of rotation determines the pattern of behavior. During smooth swimming (runs) the motor operates counterclockwise. In the case of *E. coli,* the flagella rotate together as a bundle. A switch to clockwise rotation causes the bundle to fly apart and a tumble is initiated (Fig. *c*). It is the message from the chemotaxis proteins to a switch complex at the base of the motor that determines how often its direction is reversed.

numbers—ciliated cells often have hundreds of cilia, but flagellated cells usually have only one or two flagellae (Fig. 3.1). The real difference, however, lies in the nature of their movement. Cilia row like oars. The movement is biphasic, consisting in an effective stroke in which the cilium is held rigid and bends only at its base and a recovery stroke in which the bend formed at the base passes out to the tip (Fig. 3.4). Flagella wriggle like eels. They generate waves that pass from base to tip (and sometimes in the opposite direction) at constant amplitude (Fig. 3.4). Thus the movement of water by a flagellum is parallel to its axis, while a cilium moves water perpendicular to its axis and, hence parallel to the surface of the cell.

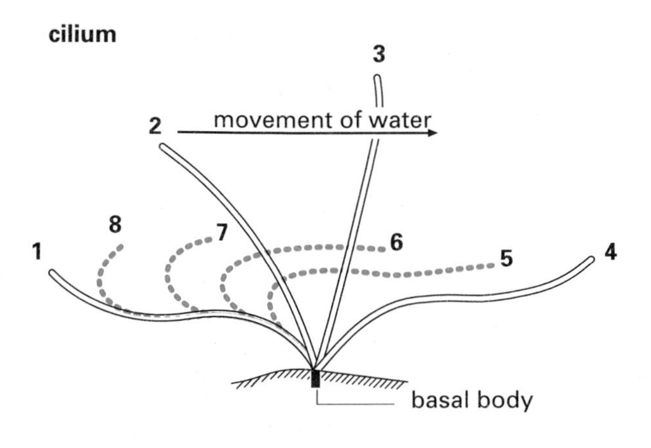

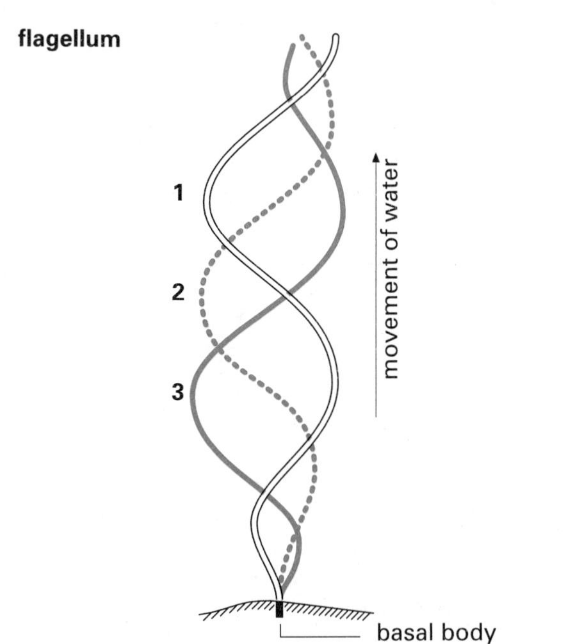

Figure 3.4
How cilia and flagella bend.

One group of protozoa has such conspicuous and numerous cilia that they are called ciliates. A paramecium swims as the result of the coordinated motion of several thousand cilia on the cell surface, producing waves of ciliary motion. Cilia are important in many ways in the human body. The respiratory tract is lined with about 0.5 m^2 of ciliated epithelium, bearing something like 10^{12} cilia. The beating of these cilia moves a belt of mucus containing inhaled particles and microorganisms away from the lungs. This activity is paralyzed by cigarette smoke, causing mucus to accumulate in the smoker's lung and causing the typical cough.

Despite their different pattern of beating, cilia and flagella are indistinguishable structurally. A cross section through either reveals two central microtubules surrounded by a ring of nine microtubule doublets (Fig. 3.5). The entire microtubule skeleton, called the 9 + 2 axoneme, is enclosed by an extension of the plasmalemma. Attached to the nine outer doublet microtubules are projections or arms composed of the protein *dynein*. Dynein is the motor that powers ciliary and flagellar beating. Using ATP produced by mitochondria near the

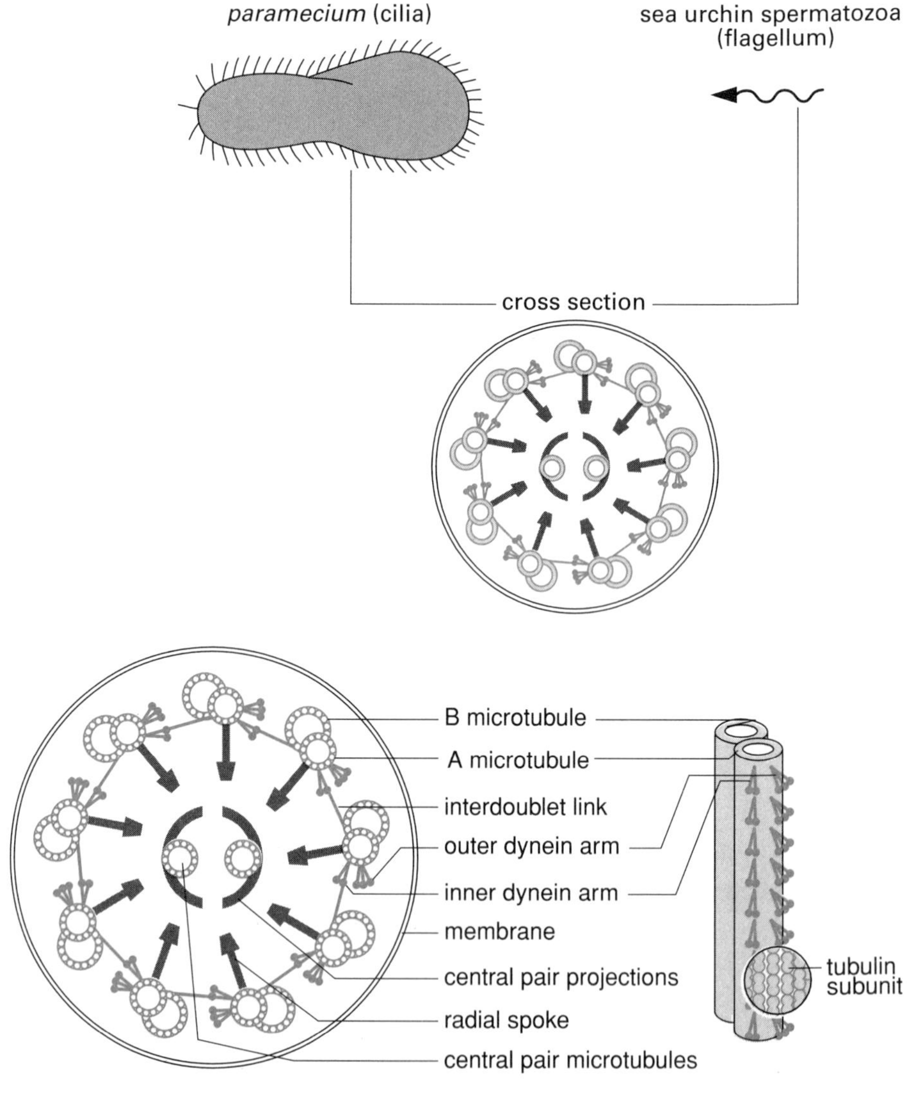

Figure 3.5
Cilia and flagella have identical structure.

base of the cilium or flagellum as fuel, the dynein arms cause a sliding movement to occur between adjacent outer doublets. Because the arms are activated in a strict sequence both around and along the axoneme, and because the amount of sliding is restricted by the radial spokes and interdoublet links, the sliding movement of the adjacent doublets causes a bending of the entire cilium or flagellum.

There are many variations on the basic 9 + 2 microtubule pattern, especially in spermatozoa. The sperm of mosquitoes and scorpions has a 9 + 1 axoneme, while the caddis fly has 9 + 7. Spider sperm has a 9 + 3 axoneme while the sperm of the eel has no central microtubules at all—a 9 + 0 pattern.

Some changes in the structure of flagella take place as the result of mutation. These are often detected during investigations of male infertility. Males producing nonmotile sperm may also suffer from chronic respiratory problems (bronchitis, sinusitis), because the cilia lining their airways are also defective. Most surprising is the fact that half of these individuals also exhibit Kartegener's syndrome. This is characterized by a mirror-image arrangement of the internal organs, where the heart is on the right, the appendix on the left, and so on. Ciliary beating must hence play a role in establishing the body plan during development. In the event of malfunction, the organs are positioned by chance and not design.

Intracellular Transport

The beating of cilia and flagella is an obvious manifestation of movement generated by a microtubule-based structure having a defined geometric shape. However, motility is a general property of microtubules within cells. It is best seen in specialized cell types such as the pigment cells, chromatophores, of fish and amphibia. The inward and outward movement of pigment granules along radial arrays of microtubules underlie their remarkable ability to change color (Fig. 3.6). But this is just an exaggerated example of processes that occur less spectacularly in all cells. For instance, neurons extend axons for up to 1 m from the cell body. Organelles and small vesicles are transported in both directions at speeds of up to 4 $\mu m\ sec^{-1}$. This phenomenon is referred to as neuronal transport. This is subdivided into outward or anterograde transport and inward or retrograde transport. Both are dependent upon microtubules that are abundant in neurons (Fig. 3.7).

The two forms of neuronal transport are dependent upon different microtubular molecular motors. Dynein, sometimes referred to as cytoplasmic dynein to distinguish it from its relative in cilia and flagella, moves vesicles and organelles in the retrograde direction, while another protein, *kinesin*, is the motor for movement along microtubules in the anterograde direction. The di-

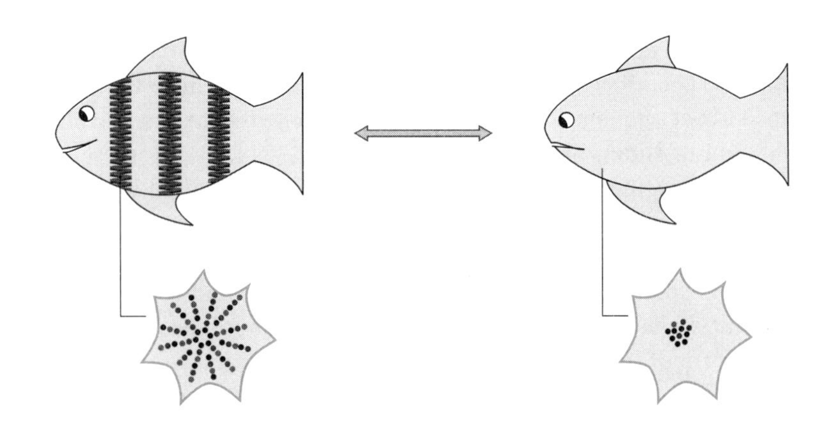

Figure 3.6
Pigment migration in fish chromatophores.

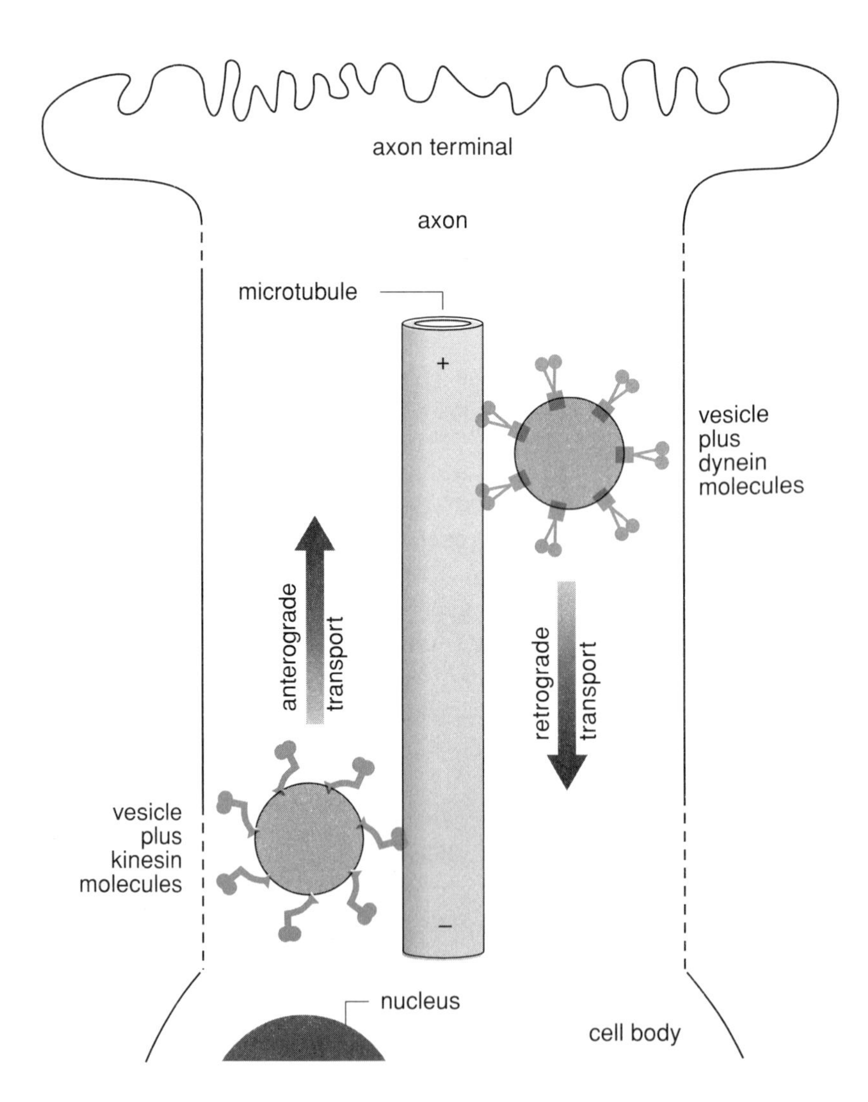

Figure 3.7
Axonal transport.

rection of organelle movement is specified by the polarity of the microtubules. Thus dynein moves along a microtubule in a plus to minus direction, while kinesin works in the opposite direction. Both proteins consist of a tail that binds to the cargo to be transported and two (kinesin) or three (dynein) globular heads that interact with the surface of the microtubule to generate movement. A single cell has more than one type of dynein and kinesin, the different motors being responsible for transporting different specific cargoes.

MICROFILAMENTS

Microfilaments are fine fibers, about 7 nm in diameter that are made up of subunits of the protein *actin*. Because it is a globular protein, the actin monomer is called G-actin, while the filament that forms from it is referred to as F-actin. Each actin filament is composed of two chains of actin monomers twisted around one another like two strands of pearls (Box 2.2). In animal cells actin is particularly associated with the cell periphery. Nonmotile cells can have bundles of actin filaments called stress fibers that help to anchor the cell to its substrate (Box 3.1), while a loose meshwork of filaments underlies the cell membrane. In actively moving cells the stress fibers disappear and actin becomes concentrated at the leading edge. Projections from the cell surface such as microvilli (page 000) are maintained by rigid bundles of actin filaments. Actin bundles are also associated with the acrosome of some spermatozoa. The acrosome is a harpoon-like projectile that helps the sperm to penetrate the egg. In the sea cucumber, *Thyone,* a spear-like process 0.5 μm in diameter and 90 μm in length and packed with microfilaments shoots forward when the sperm contacts the egg jelly.

The *Thyone* spermatozoan is a good example of the role of actin-binding proteins, a large family of proteins that associate with, and regulate the function of, both G- and F-actin. Prior to activation, a vesicle at the anterior of the *Thyone* sperm head contains G-actin at a concentrations way above that which would normally result in spontaneous polymerization into F-actin filaments. The reason this does not happen is that actin is stored as a complex with the protein profilin which maintains the actin in the monomeric form. When the sperm contacts the egg, the two proteins dissociate, and the actin rapidly assembles to form the acrosomal process.

Actin binding proteins may also cross-link F-actin filaments to form bundles. In the microvillus this role is served by the protein villin. Yet other members of this family cross-link actin to form a viscous, three-dimensional cytoplasmic gel. Many of the fundamental properties of cells such as cell locomotion and cytoplasmic streaming are based on the regulation of cytoplasmic viscosity in this way (see below).

Muscle Contraction

The study of microfilaments owes much to a long history of structural and biochemical research on the mechanism of muscle contraction. Skeletal muscle is made up of giant cells called muscle fibers (page 13). The cytoskeleton inside the fiber is composed of repeating units called sarcomeres (Fig. 3.8). The boundaries of each sarcomere are delineated in adjacent myofibrils by Z-discs that lie in register giving the muscle its characteristic striped (striated) appearance. Each sarcomere consists of overlapping thin and thick filaments. The thin filaments are composed of actin and are identical in structure to microfilaments. The actin filaments extend from the discs toward the center of the sarcomere where they are interspersed with thick filaments composed of the protein myosin. The myosin molecule has a distinctive structure, consisting of a tail and two globular heads. The thick filament is formed from a large number of myosin molecules arranged in a tail-to-tail manner. This design means that the thick filament is a bipolar structure with myosin heads at both ends. Powered by ATP, the myosin heads crawl along the actin (thin) filaments. Because of the geometry of the system, the two Z-discs are pulled toward each other and the myofibril shortens.

In the cytoplasm of nonmuscle cells, a superfamily of at least ten different types of myosin exists. One of them, called myosin II because it is two-headed, is indistinguishable from muscle myosin but does not assemble into filaments to the same extent, probably because the levels of force required within non-muscle cells is a fraction of that generated by muscles. Myosin I has only a single head and is sometimes referred to as "mini-myosin" because it also has only a short tail. These nonmuscle myosins generate movement by interacting with actin filaments in much the same way that dynein and kinesin interact with the surface of microtubules.

Cell Locomotion

For a cell such as an amoeba to crawl across a substrate, it must generate traction. Less than 1% of the ventral surface of a crawling cell is in contact with the surface over which it is moving. These transient points of attachment, rather like footprints, are called focal contacts. A cell does not slither along on its belly but rather "tiptoes" delicately across its substrate. An amoeba extends projections or pseudopodia in the direction of progress (Fig. 3.9). Contraction at the rear of the cell drives a fluid stream of liquid endoplasm forward into the pseudopodium where it is converted into more solid ectoplasm at the tip. The new ectoplasm is then displaced toward the rear where it is reconverted back into endoplasm. Ectoplasm is semi-solid because it contains long actin fila-

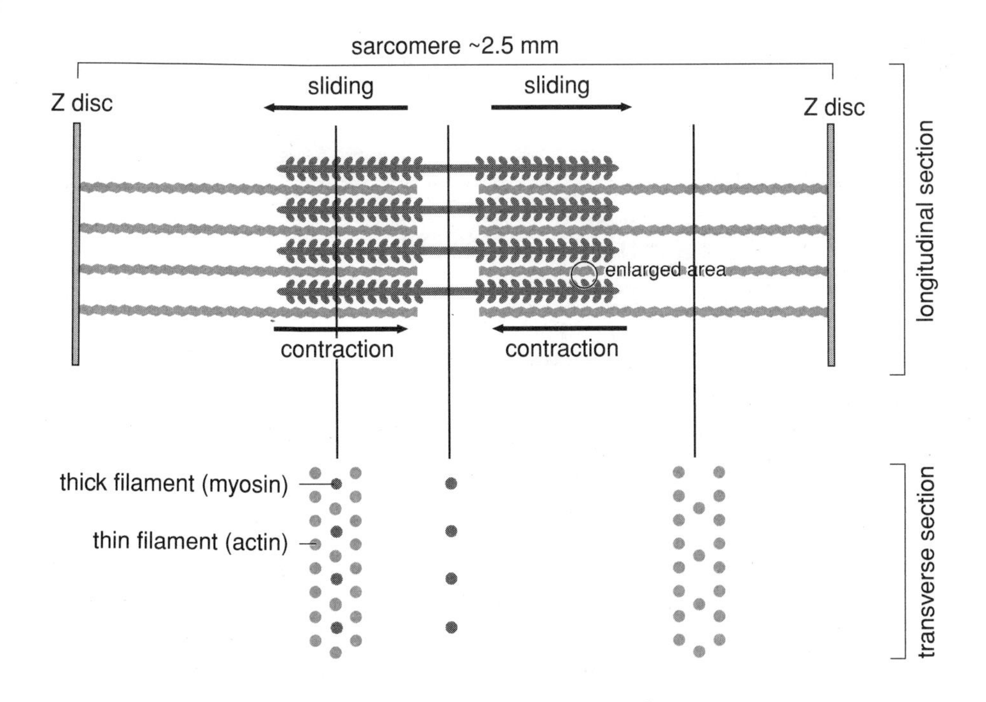

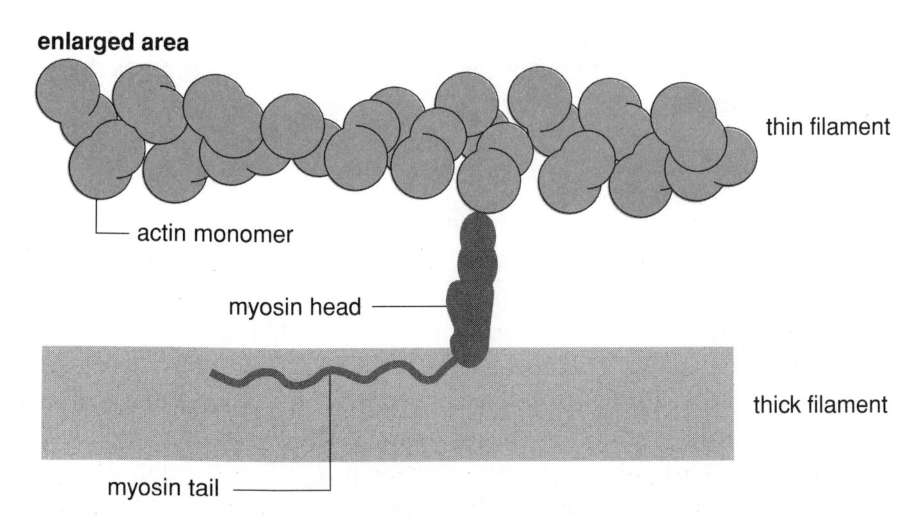

Figure 3.8
Muscle contraction

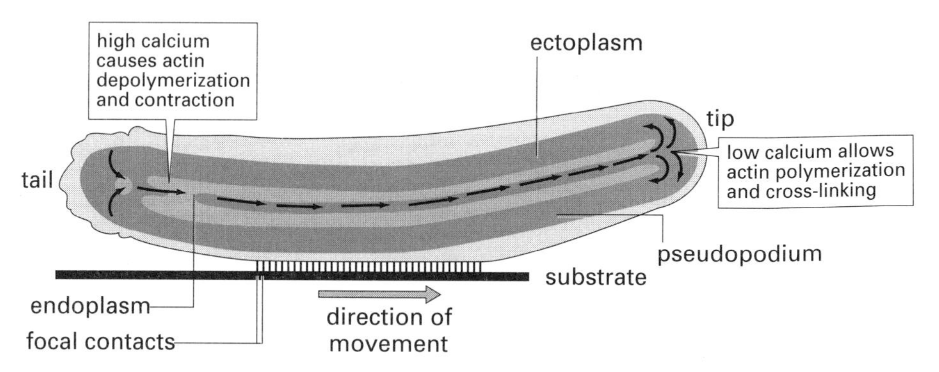

Figure 3.9

Amoeboid movement. Alternating contraction (C) and relaxation (R) results in the forward extension of a pseudopodium.

ments that are cross-linked by actin binding proteins to form a three-dimensional gel that is resistant to distortion. In regions of higher calcium concentrations, for example, at the tail, both the viscosity and the resistance of the gel is reduced because proteins such as gelsolin bind to and fragment the actin filaments. Calcium also activates myosin, the resulting contraction driving the now fluid endoplasm forward. Contraction at the tail is due to the action of myosin II, while extension of the pseudopod relies on myosin I which is localized at the tip. This cycle of extension, traction, and retraction is the basis of crawling movements.

Cytoplasmic Streaming

Even though plant cells themselves do not move, their cytoplasm streams as actively as that of animal cells. In the epidermal cells of the leaves of the water plant *Elodea,* a belt of moving cytoplasm carries the chloroplasts and other organelles in a continuous, unidirectional stream around the central vacuole. In the giant cells of the alga *Nitella,* the chloroplasts are embedded in a layer of viscous ectoplasm, while other organelles are carried along in the following endoplasm. The basis of cytoplasmic streaming is the interaction of myosin molecules attached to the moving organelles with cables of actin filaments in the ectoplasm.

INTERMEDIATE FILAMENTS

Intermediate filaments are so-named because their diameter, 10 nm, lies between that of the thin and thick muscle filaments. Although they look the same, intermediate filaments from different mammalian tissues are in fact composed of different subunit proteins. Nerve cells contain neurofilaments, muscle cells contain desmin filaments, connective tissue cells such as fibroblasts contain vimentin filaments, and epithelial cells contain filaments composed of keratin, the protein that makes up hair, skin, and fingernails and forms the horns and hooves of domestic animals. Little wonder then the intermediate filaments are the most stable of the cytoplasmic filament systems. Different proteins can generate a common structure because all share the same basic design. This consists of a central α-helical rod (page 223) and a non helical head and tail. Most of the variation between intermediate filament proteins lies in the head and tail, and these regions probably confer subtly different properties on different intermediate filament classes.

In the cell, intermediate filaments tend to form wavy bundles that extend from the nucleus to the cell surface (Box 3.1). In epithelial cells keratin filaments are typically associated with spot desmosomes, the membrane anchorage points by which cells attach to each other (page 29). The interaction of intermediate filaments with the nucleus became clearer with the surprising discovery that intermediate filaments share a high degree of amino acid sequence similarity with the nuclear lamins, proteins that make up the nuclear lamina (page 32). The nucleus is suspended by intermediate filaments stretching to the cell membrane, rather like a sailor in a hammock. Other than this, there is still very little known about the precise function of intermediate filaments.

SUMMARY

1. The cytoskeleton is composed of microtubule, microfilaments, and intermediate filaments. It is responsible for determining both cell shape and cell movement. The organization of the three filament classes can be determined by indirect immunofluorescence microscopy.

2. Microtubules are composed of tubulin. They form the 9 + 2 backbone of cilia and flagella, and they provide a transport system through the cell cytoplasm. All microtubules can be traced back to the centrosome or microtubule-organizing center. Organelles move along microtubules using motor proteins such as dynein and kinesin.

3. Microfilaments are composed of actin. Actin filaments are involved in muscle contraction and cell locomotion.

4. Intermediate filaments are largely structural.

CHAPTER

4

CELL DIVISION

An adult human is made up of about 30 trillion cells, all of which originate from a single fertilized egg. This simple statistic, however, considerably underestimates the true number of cells that each fertilized egg gives rise to because many of them will die by a process called apoptosis or programmed cell death. For example, hands form initially as paddle-shaped buds; linear columns of cells then die, leaving the five digits. In humans the cells of some tissues, such as the skin, the lining of the gut, and the bone marrow continue to divide throughout life, but others, the eye, nervous system, and skeletal muscle, show almost no replacement. These cells, laid down in infancy, must last a lifetime. Clearly cell division is precisely controlled, and the breakdown of this control system is one of the events that leads to a cell becoming cancerous.

The life of a cell from the time it is "born" by the division of its parent to the time it in turn divides is called the *cell cycle*. Its duration varies from 2–3 hours in a simple microbe like the budding yeast *Saccharomyces cerevisiae* to around 24 hours in a human cell grown in a culture dish. Under the microscope it is possible to distinguish only two elements of the cell cycle: interphase, an extended period of synthesis and growth during which the cell roughly doubles in mass but without displaying obvious morphological changes, and *mitosis*, a brief period of profound structural rearrangements resulting in the creation of two daughter cells.

About 40 years ago our view of interphase was revolutionized with the discovery that DNA replication was confined to a discrete period of interphase called *S phase* (for DNA synthesis). This did not immediately follow the previ-

S phase: DNA synthesis
M phase: cell division (mitosis and cytokinesis)

G1 phase: gap 1, the period between M phase and S phase
G2 phase: gap 2, the period between S phase and M phase

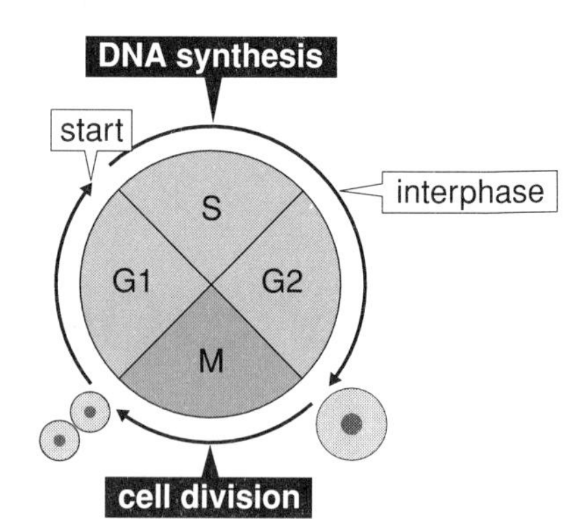

Figure 4.1
The cell division cycle.

ous mitosis but was separated from it by an interval called *gap1* or *G1*. A second gap, *G2*, separates the completion of S phase and the start of cell division or *M phase* (for mitosis). These four phases comprise the classic cell cycle "clock" (Fig. 4.1). The only exception is in the rapid early divisions of some embryos during which cells divide without growth and there are no gaps.

CONTROL OF THE CELL DIVISION CYCLE

A cell will not embark on its division cycle until the conditions are such that it will be able to complete the exercise successfully. If they are not favorable, the cell will instead enter a period of quiescence, sometimes called the G0 phase, where it can remain viable for months or even years. Cells also leave the cell cycle in order to differentiate into a specific cell type. Cells such as neurons, the cells of the nervous system, are said to be terminally differentiated because they are unable to return to the cell division cycle. A cell cannot leave the cell cycle wherever it likes but only at two specific points. The first is late in G1, just prior to the start of S phase, and the second, late in G2 at the point of entry into mitosis. These are the two crucial control points of the cell cycle, and passage across them is carefully regulated.

All cells possess both cell cycle control points, although one is called upon only in unusual circumstances. Figure 4.2 shows the results of an experiment to determine the point in the cycle at which human cells arrest. The upper graph shows the pattern found in well-fed and actively growing cells. The

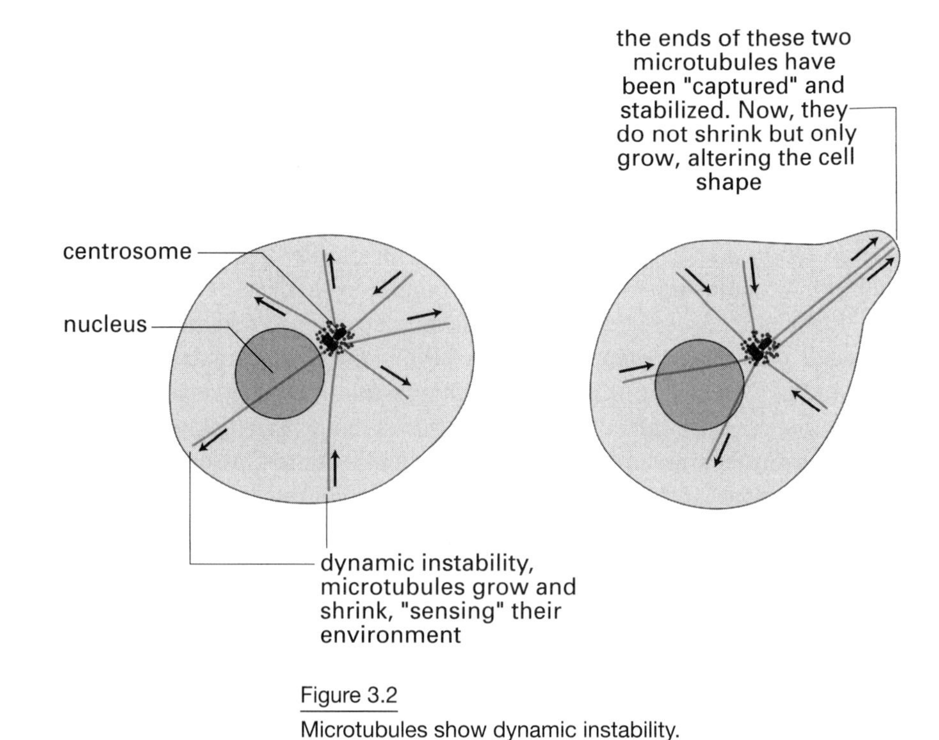

Figure 3.2
Microtubules show dynamic instability.

MICROTUBULE-BASED MOTILITY

Cilia and Flagella

Cilia and flagella appeared very early in the evolution of eukaryotic cells and have remained essentially unchanged to the present day (some bacteria are also propelled by structures called flagella, but these are quite different; see Box 3.2). The terms *cilium* (meaning an eyelash) and *flagellum* (meaning a whip) are often used arbitrarily. Generally, cilia are shorter than flagella (<10 μm compared to >40 μm) and are present on the surface of the cell in much greater

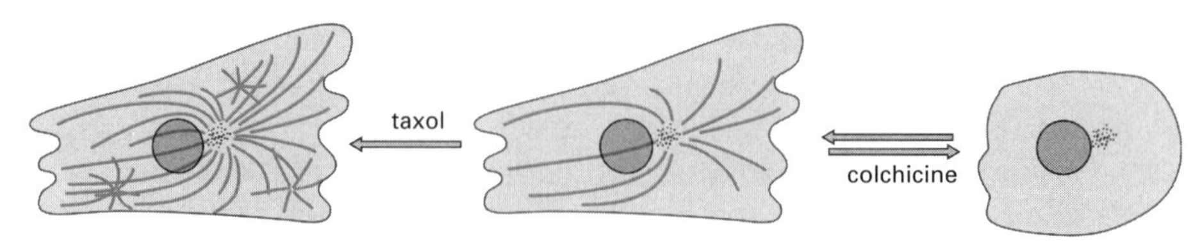

Figure 3.3
Taxol and colchicine have opposite effects on microtubules.

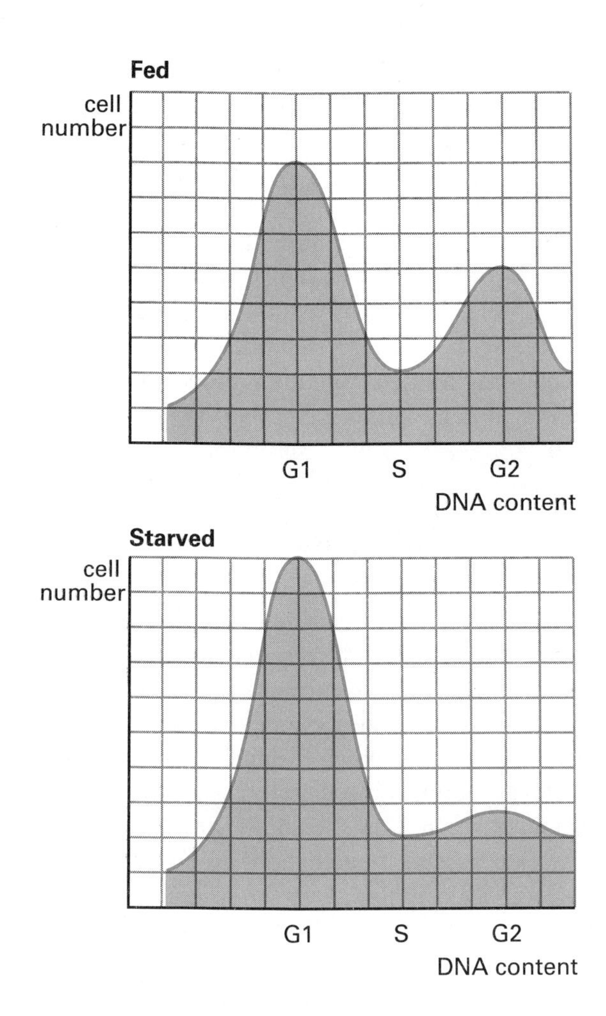

Figure 4.2
Removal of growth factors arrests human cells in G1.

amount of DNA in each cell is determined, and a frequency graph plotted, so that the height of the line at a particular place on the graph represents the relative number of cells with that amount of DNA. The fed cells give two distinct peaks: one peak at twice the DNA content of the other peak. This means that most of the cells are either in G1 or G2, with G2 being the right-hand peak that has twice the DNA content. The lower graph shows the result obtained for cells that have stopped dividing because growth factors have been removed from the culture medium. The peak of G1 has grown, and that at G2 has shrunk, indicating that the cells have arrested in G1. The same is true in the body: most of the nondividing cells left the cell cycle prior to S phase.

Many microorganisms such as the budding yeast also leave the cell cycle in G1. Their simple cells do not differentiate, but they do have an alternative fate that occurs only at this point in the cell cycle. This is to conjugate (that is, to mate) with a suitable partner and enter the sexual, or meiotic, pathway. The G1 control point in yeast is called START because it is the "point of no return," beyond which the cell is committed to completing the division cycle irrespective of conditions.

The two control points allow the cell to monitor both intrinsic and extrinsic factors that are relevant to the successful completion of cell division. In G1 the cell must decide whether it is big enough and whether nutritional conditions are appropriate to begin the crucial process of replicating its genome. In G2 the primary concern is that its DNA is in perfect condition before entering mitosis. There are sensitive mechanisms for detecting the presence of unreplicated or damaged DNA, and cells will not commit themselves to mitosis until any defects have been attended to.

Molecular Regulation of the G2 (Interphase/Mitosis) Cell Cycle Control Point

As cells enter mitosis, they undergo a remarkable sequence of structural changes: the chromosomes condense, the nuclear envelope breaks down, the nucleolus disperses, the membranes of the Golgi apparatus and endoplasmic reticulum fragment, and the cytoskeleton undergoes remodeling to form a completely new cellular structure, the mitotic spindle. How are so many and diverse structural changes coordinated in a narrow "window" of the cell cycle, often lasting only a few minutes? One possibility is that all are under the control of a cytoplasmic "master switch." During the 1960s some remarkable experiments started the hunt that has led to the identification of just what this switch might be.

When cultured mammalian cells in interphase and mitosis were fused together allowing the two cytoplasms to mix, the interphase nucleus entered mitosis well before it was expected to. The same thing happened when cytoplasm from M-phase *Xenopus* eggs was injected into interphase oocytes (Fig. 4.3). These experiments imply that the cytoplasm of dividing cells must contain some factor capable of controlling entry into mitosis. The factor was called *M-phase promoting factor*, or *MPF*.

Defining MPF in biochemical terms took almost 20 years. Some of the neatest experiments were carried out in the yeasts, *Saccharomyces cerevisiae* (the budding yeast) and *Schizosaccharomyces pombe* (the fission yeast). These organisms hold center stage in modern studies of the cell cycle because they can be used to isolate cell division cycle (cdc) mutants (Box 4.1). One such mutation, in the gene cdc2, was of particular interest, since it appeared to arrest the cell cycle at each of the two control points.

Modern molecular techniques made it possible to isolate the normal

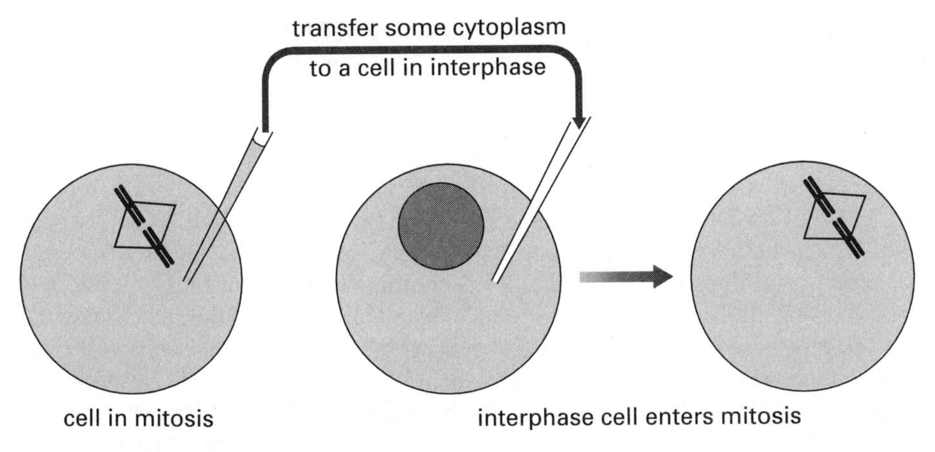

Figure 4.3
An interphase cell is caused to enter mitosis by injection of cytosol from a mitotic cell.

cdc2 gene and to determine that it was present in all eukaryotic organisms from yeast to humans. The protein encoded by the cdc2 gene has a molecular weight of 34 kD and is now called *p34*^{cdc2}. We now know that p34^{cdc2} functions as a cell cycle "master switch" by its action as a protein kinase (page 238). It is an enzyme that modifies the function of proteins associated with the chromosomes, nuclear envelope, nucleolus centrosomes, etc., initiating mitosis. Although p34^{cdc2} is present throughout the cell cycle, it is only active as an enzyme for a brief period at the interphase-mitosis boundary. This activation is achieved in two steps (Fig. 4.4). The first step is the association of p34^{cdc2} with another important cell cycle protein, *cyclin B*. This is one of a family of pro-

BOX 4.1

Yeast Cell Division Mutants as Tools to Study Cell Cycle Controls

Biologists often use simple organisms to provide new ways of studying complex problems. Multicellular organisms such as the nematode worm *Caenorhabditis elegans* and the fruit fly *Drosophila melanogaster* are particularly valuable for asking questions about the development of the body plan and the formation of complex tissues such as the nervous system. Unicellular organisms such as the yeasts *Saccharomyces cerevisiae* and *Schizosaccharomyces pombe*, on the other hand, have provided important insights into more fundamental biological problems such as the control of cell division. What all these model systems have in common is the ability to generate mutants that affect particular cellular or developmental pathways.

Since cell division is so essential to the growth and development of an organism, mutants affecting this process are likely to be lethal. In yeast this problem can be overcome by isolating temperature-sensitive mutants, strains that can divide and form colonies at one temperature (25°C, called the permissive temperature). This is an example of conditional mutation (page 120). Such mutants are called *cell division cycle* or *cdc* mutants. Budding yeasts, shown at the top of the figure, grow by enlargement of a bud, the mother cell remaining constant in size. When the bud is almost as large as its mother it detaches and forms a bud of its own. At 25°C a culture of a budding yeast cdc mutant will contain cells with a range of bud sizes, indicating that the culture is growing normally. Raising the temperature inactivates the cdc protein and imposes a blockage: Cells continue to grow until they hit the block and then stop. The result is that all the cells in the culture will eventually arrest at the same stage in the cycle. If the mutant cdc gene is required for a step early in the cell cycle, then all the cells will arrest with no buds. If the mutant cdc gene is required for a step late in the cell cycle, then all the cells will arrest with large buds, and so on. Fission yeasts, shown on the bottom of the figure, divide by medial fission, which produces two small sausage-shaped cells that subsequently elongate before they too divide. Cdc mutations have a very different effect on the cell's appearance. At the restrictive temperature the cells become extremely long due to the fact that, although cell cycle progress is blocked, the cells are still able to synthesize proteins and hence get bigger.

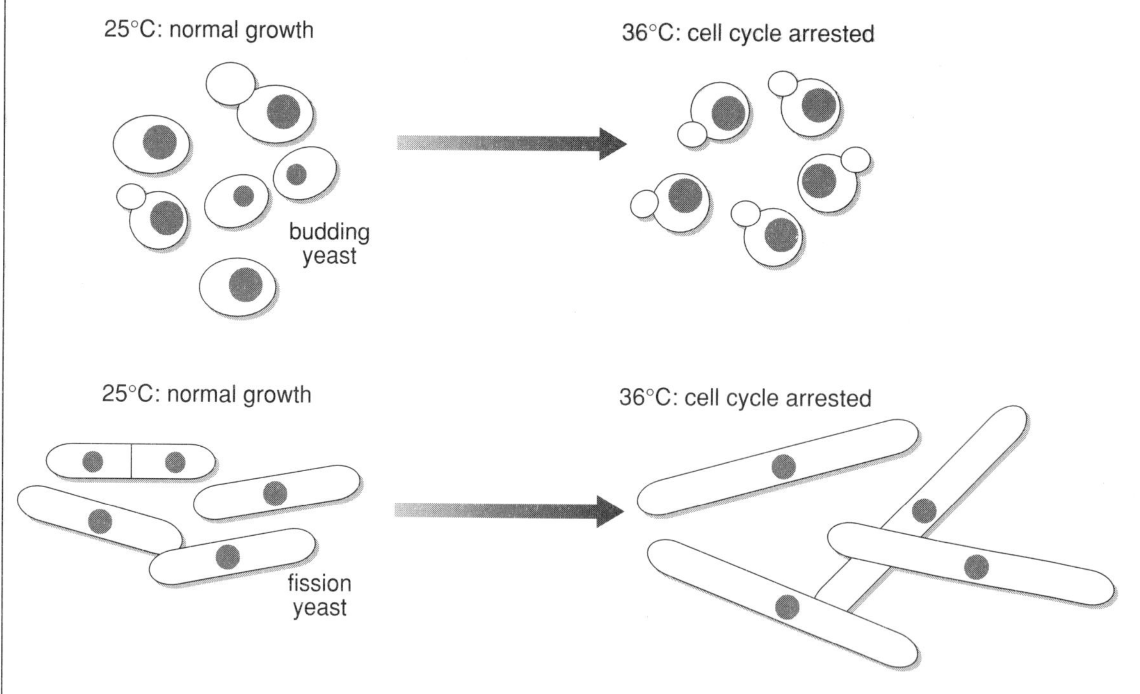

The collection of cdc mutants in both yeast species has turned out to be an inspired move. The cdc mutants were first isolated more than 20 years ago, and although it was not fully appreciated at the time, cell cycle control in the two yeasts is quite different. In the budding yeast, as in human cells, cell cycle regulation occurs in the G1 phase, while in fission yeast it occurs primarily in G2. Analysis of both groups of cdc mutants has led directly to the identification of human cell cycle genes, work that has been very important in cancer research.

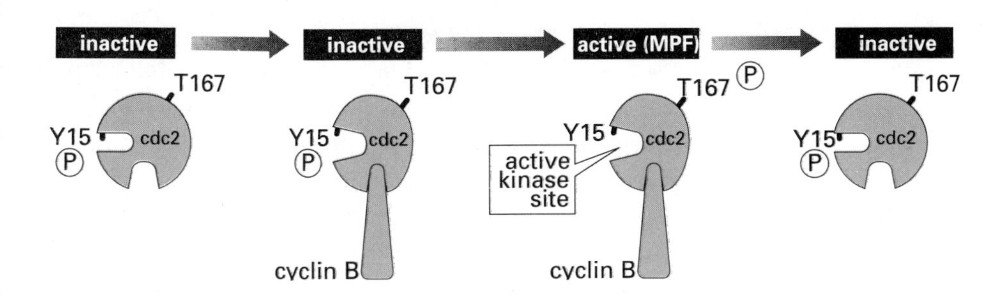

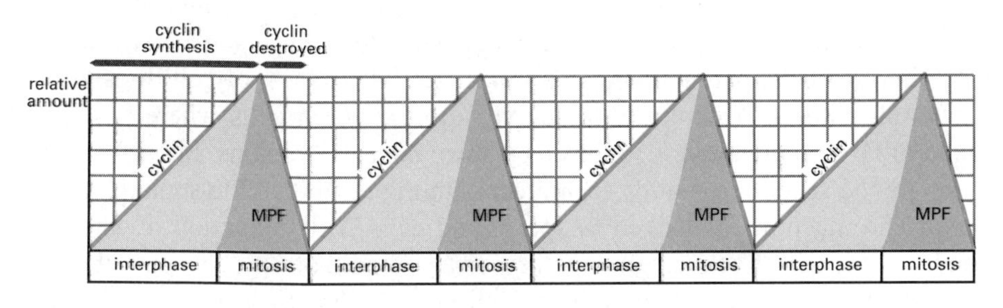

Figure 4.4
The control of cdc2 by cyclin B and phosphorylation.

teins whose concentration rises and falls (cycles) through the cell cycle. Cyclin B and $p34^{cdc2}$ make up MPF, but the complex is only activated when $p34^{cdc2}$ is further modified by the addition of a phosphoryl group, represented as P in the figure, to one of its amino acids (phosphorylation) and the removal of a phosphoryl group from another (dephosphorylation). Once mitosis is initiated, MPF must be deactivated so that the cell can return to its interphase organization. This is achieved by the destruction of cyclin B and the reversal of the modifications to $p34^{cdc2}$. This rather complex sequence of events is summarized in Fig. 4.4.

The G1 Control Point

Entry into S phase is also regulated by protein kinases. It is not $p34^{cdc2}$ that plays the important role in G1 but a close relative called $p33^{cdk2}$ (cdk stands for cyclin-dependent kinase). This enzyme has at least two cyclin partners, cyclins D and E. The complex of $p33^{cdk2}$ and a cyclin is the G1 equivalent of MPF, that is, S-phase promoting factor or SPF. This is activated in a similar way to MPF, but acts on a different set of proteins to trigger DNA synthesis.

CELL DIVISION: MITOSIS AND MEIOSIS

In sexually reproducing organisms, the germ cells that give rise to the eggs and sperm arise by a different type of cell division from the somatic cells that make up most of the body. Somatic human cells are diploid; that is, their nuclei contain two almost identical sets of 23 chromosomes. One set is inherited from the male parent, the other from the female parent. During DNA replication (S phase of the cell cycle) each chromosome is duplicated and in G2 consists of two sister chromatids that remain joined at their centromeres (Fig. 4.5). In somatic cells there is a single division of the nucleus (in which the sister chromatids split apart) to produce two diploid daughter cells, each inheriting an identical set of the parental cell's chromosomes. This form of cell division is known as mitosis. It contrasts to the situation in germ cells where a single round of DNA replication is followed by a two nuclear divisions. In only one of these are the sister chromatids separated, with the result that the somatic number of chromosomes is halved in sperm and eggs. This reduction division is known as meiosis. Each daughter cell receives a single or *haploid* copy of each parental chromosome. Following fertilization of the egg by a sperm, male and female nuclei fuse to restore the diploid chromosome number.

Mitosis

The fact that the vast majority of humans have identical sets of chromosomes in every one of their 30 trillion cells is testament to the remarkable accuracy of mitosis. Based on light microscopic observations made initially in the nineteenth century, mitosis is classically divided into five stages. Each is characterized by changes in both chromosome position and appearance and in the organization of the mitotic spindle (Fig. 4.5).

PROPHASE. The first visible sign of mitosis in most cells is the compaction of the chromosomes to form structures visible in the light microscope. Chromosome condensation allows long DNA molecules to be separated without tangling. Each chromosome has a constriction called the kinetochore, a structure that forms around the centromere and is the point of attachment of the chromosome to the spindle. At the same time as the chromosomes are condensing within the nucleus, the centrosomes begin to separate to establish the mitotic spindle (Fig. 3.1).

PROMETAPHASE. The nuclear envelope breaks down, allowing the chromosomes to attach to the forming spindle. Microtubule assembly from the centrosomes is random and dynamic. The growing ends of individual mi-

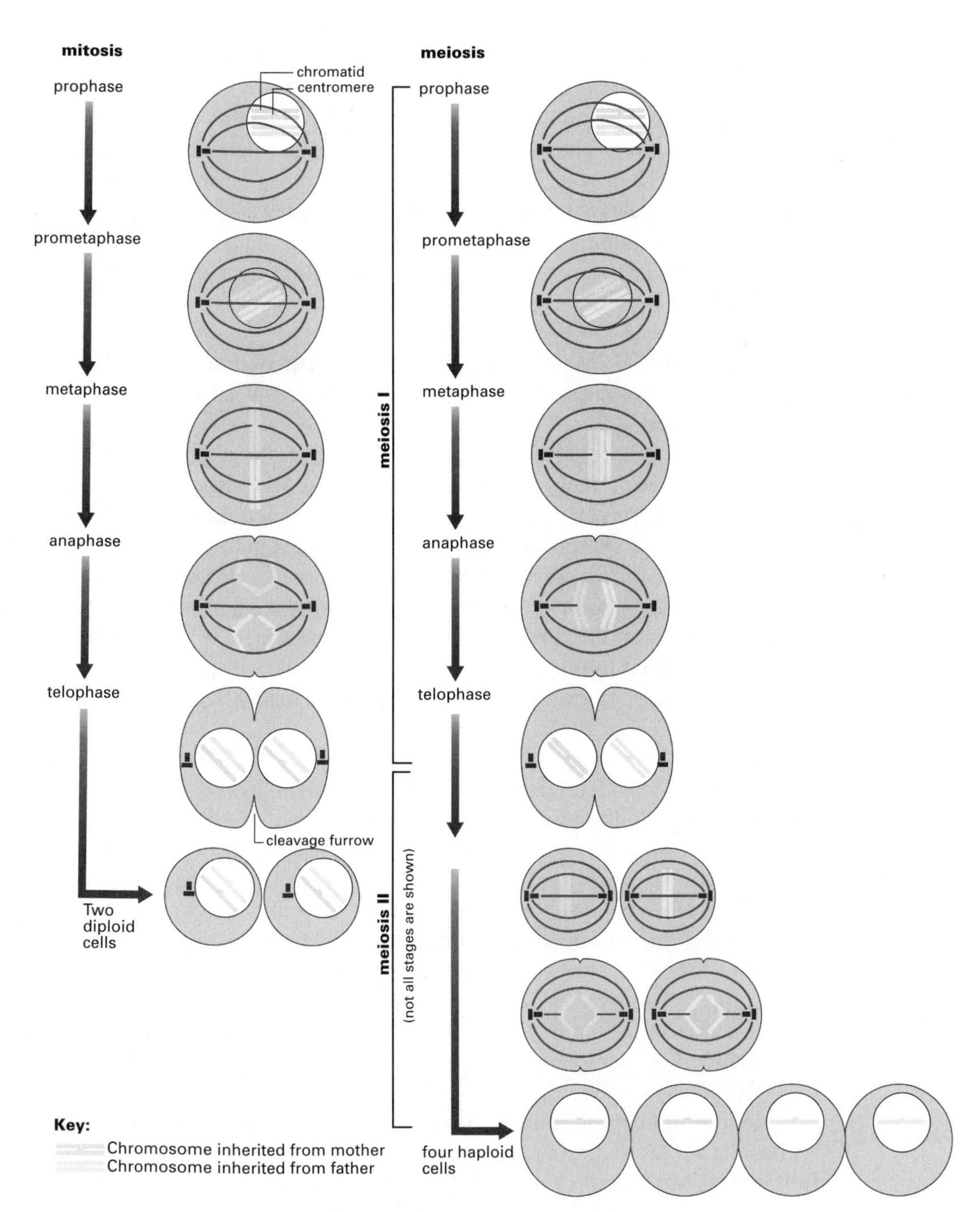

Figure 4.5

Stages of mitosis and meiosis.

crotubules make chance contact with and are "captured" by the kinetochores. Because of the random nature of these events, the two kinetochores of a sister chromatid are initially associated with different numbers of microtubules, and the forces acting upon each chromosome are unbalanced. Initially therefore the spindle is highly unstable, and chromosomes make frequent excursions toward and away from the poles. Gradually a balance of forces is established, and the chromosomes become aligned at the equator in a metaphase plate, with the kinetochores of each sister chromatid oriented toward opposite poles.

METAPHASE. Metaphase is the most stable period of mitosis. The system is in a steady state with the chromosomes lined up rather like athletes at the start of a race.

ANAPHASE. The trigger for the separation of the sister chromatids and the start of their journey to the spindle poles is unknown. In anaphase A there is movement of the chromosomes to the poles of the spindle, while anaphase B is an elongation of the spindle that further extends the distance between the poles. During anaphase A, the chromosomes move as a V with the kinetochores, upon which the force for chromosome movement is applied, leading the way. Compared to other forms of cell motility, the movement of chromosomes at anaphase is extremely slow, less than 1 μm per minute (page 51, Chapter 3).

TELOPHASE. This stage sees the reversal of many of the events of prophase: The chromosomes decondense, the spindle disassembles, the nuclear envelope reforms, the Golgi apparatus and endoplasmic reticulum reform, and the nucleolus reappears. Simultaneously *cytokinesis,* the physical division of the cell into two daughters, begins. In animal cells a cleavage furrow constricts the middle of the cell; in plants a structure called a phragmoplast forms at the equator of the spindles where it directs the formation of a new cell wall.

Meiosis

The distinction between mitosis and meiosis is quite simple: in mitosis DNA replication is followed by a single division of the nucleus, while in meiosis it is followed by two rounds of nuclear division so that the chromosome number is halved (Fig. 4.5). Each meiotic division (meiosis I and II) involves the formation of a meiotic spindle and the same sequence of prophase, prometaphase, metaphase, anaphase, and telophase. Mitosis is usually a rather brief process, but meiosis is often extended and, in different organisms, can last for months or even years. Most of this is occupied by a lengthy prophase of meiosis I called prophase I. It is here that new genetic arrangements are generated by crossing over (Chapter 7).

Prophase I is subdivided into five stages defined by chromosomal events: condensation, leptotene; pairing, zygotene; crossing over, pachytene; synthesis, diplotene; and recondensation, diakinesis. Leptotene resembles mitotic prophase. The chromosomes become visible as fine threads, although the two chromatids are so close that their double nature cannot be distinguished. At zygotene the two copies of each chromosome become aligned, a process called synapsis, to form bivalents (pairs of paired sister chromatids and their two centromeres). This arrangement continues through pachytene in which the chromosomes shorten and fatten, being held together by the synaptonemal complex. At diplotene the chromosomes of each bivalent separate, although they still touch at a number of chiasmata. The chromosomes become "fuzzy" in appearance as they decondense. At diakinesis the chromosomes recondense, the nucleoli disappear, and the events of prometaphase (nuclear envelope breakdown and the interaction of the chromosomes with the spindle) begin.

At metaphase I the bivalents align at the metaphase plate with both kinetochores of each pair of sister chromatids oriented toward the same pole. At anaphase I the homologous pairs separate, but unlike mitosis, the sister chromatids remain attached and journey together to the pole. The two daughter nuclei formed during telophase I almost immediately enter meiosis II. Prophase II is often so brief as to be undetectable. Metaphase II and anaphase II resemble their mitotic counterparts, the sister chromatids finally separating to give haploid cells.

Although, meiosis in male and female animals follows roughly the same lines, there are some important distinctions. In males the four haploid products of meiosis, the spermatids that subsequently give rise to spermatozoa, all survive. In females, both meiotic divisions are asymmetric, resulting in one large cell that survives, the oocyte, and three small cells, the polar bodies, that are discarded.

SUMMARY

1. The cell cycle is the period from the birth of a cell by the division of its parent to the time that it too divides.

2. The cell cycle consists of four phases: S phase, the period of DNA synthesis; M phase, the period of cell division called mitosis; and the gaps that separate them, called gap 1 (G1) and gap 2 (G2).

3. Entry into M phase is controlled by a protein kinase called $p34^{cdc2}$. $p34^{cdc2}$ is only active when it binds another protein, cyclin B. The active complex is called M-phase promoting factor or MPF.

4. Mitosis consists of five stages: prophase, prometaphase, metaphase, anaphase, and telophase.

5. Meiosis is a special type of cell division that gives rise to haploid reproductive cells called gametes. It comprises the same stages as meiosis, but prophase is much more complex.

BOX 4.2

The Cell Division Cycle Out of Control: The Cell Cycle and Cancer

The growth of an embryo from a fertilized egg is a period of frantic cell proliferation. Once adulthood is reached, however, cell division is limited to the repair of damaged cells and to the replacement of those (like the cells of the skin and blood) that exist for only a few days or weeks. Occasionally, however, a cell will ignore the signals telling it not to divide and begin to proliferate. The repeated division of a single rogue cell gives rise to an undifferentiated cell mass or tumor that may become malignant and spread.

The uncontrolled proliferation of cancer cells results from changes in the function of either of two classes of genes, oncogenes and tumor suppressor genes. Oncogenes are part of the normal pathways by which cells are stimulated to divide in response to external signals (Box 2.2). Cells become cancerous when the normal pattern of expression of an oncogene is disturbed through mutation. The identification of oncogenes came from studies of small RNA-containing retroviruses. These viruses cause cancer because they contain a mutant form of an oncogene.

The most common change to an oncogene that is associated with human cancers is in one that is known as ras. c-Ras encodes the small Ras protein, one element in the pathway by which extracellular signals from growth factors are transmitted to the nucleus to control gene expression (Box 2.2). Mutations in ras cause the cell division machinery to receive a constant instruction to "go," irrespective of whether division is appropriate or not.

Tumor suppressor genes normally inhibit cell division. More than half of the 6.5 million people who die from cancer each year carry mutations in the tumor suppressor gene p53. In a normal cell division cycle, p53 has no known function. However, in response to DNA damage, the level of p53 protein is increased and the cell cycle delayed until the damage is repaired. If the extent of the damage is too great to be repaired, the cell commits suicide by apoptosis. We can see this operating if we get too big a dose of ultraviolet light from the sum. The DNA in the cells of the outer skin is so damaged that they apoptose and a layer of dead skin falls off—we call this sunburn. If this mechanism is impaired through mutation of the p53 gene, then cells with DNA damage are able to survive and even divide, leading to chromosome instability and cancer. Another tumor suppressor gene is associated with childhood tumors of the developing retina called retinoblastomas. Children with this disease have mutations in the retinoblastoma gene, Rb. The Rb protein is part of a mechanism that ensures that the G1 phase of the cell cycle is completed correctly before a cell enters the crucial process of DNA replication. Loss of Rb function destroys this mechanism and uncontrolled cell proliferation follows. The same effect is observed in cells infected with certain DNA tumor viruses that produce a protein that binds to and inactivates Rb.

CHAPTER

5

DNA STRUCTURE AND THE GENETIC CODE

Chromosomes are made of deoxyribonucleic acid (DNA). This remarkable molecule contains all the information necessary to make a cell and is able to pass on this information when a cell divides. This chapter describes the structure and properties of DNA molecules, the way in which our DNA is packaged into chromosomes, and how the genetic information stored within them is retrieved via the genetic code.

THE STRUCTURE OF DNA

The building blocks (or monomers) of DNA are units called deoxyribonucleotides or more simply *nucleotides*. Fig. 5.1 shows one of these, deoxyadenosine triphosphate. On the left is a chain of three phosphoryl groups, then comes the sugar deoxyribose (Box 2.1). Lastly comes adenine, one of the family of nitrogen-rich compounds called *bases*. The three phosphoryl groups are denoted by the Greek symbols α, β, and γ. The numbers on the sugar—$1'$, $2'$, etc.—form the same numbering system we have seen before (Box 2.1), the $'$ is pronounced "prime" and is there to indicate that we are identifying the atoms of the sugar, not the atoms of the base. Four bases are found in DNA, they are the two *purines* called *adenine* (A) and *guanine* (G) and the two *pyrimidines* called *cytosine* (C) and *thymine* (T) (Fig. 5.2). The bases are attached to the $1'$-carbon

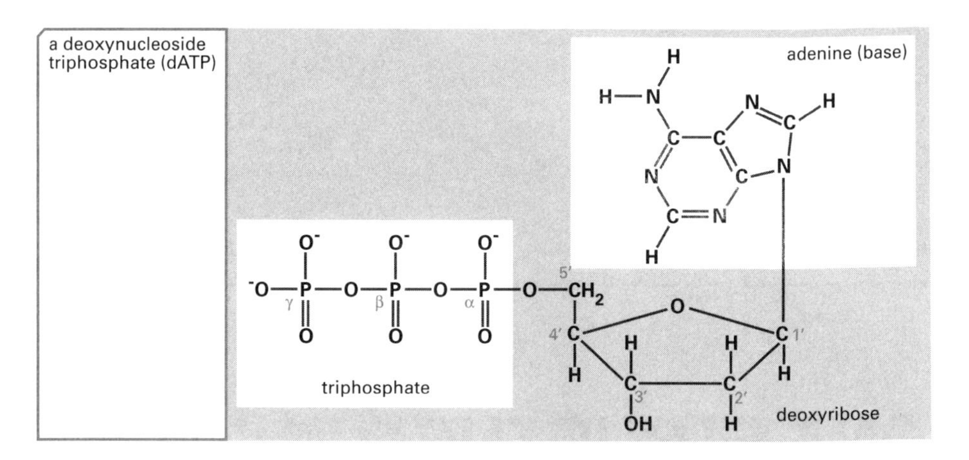

Figure 5.1
2′-deoxyadenosine 5′-triphosphate.

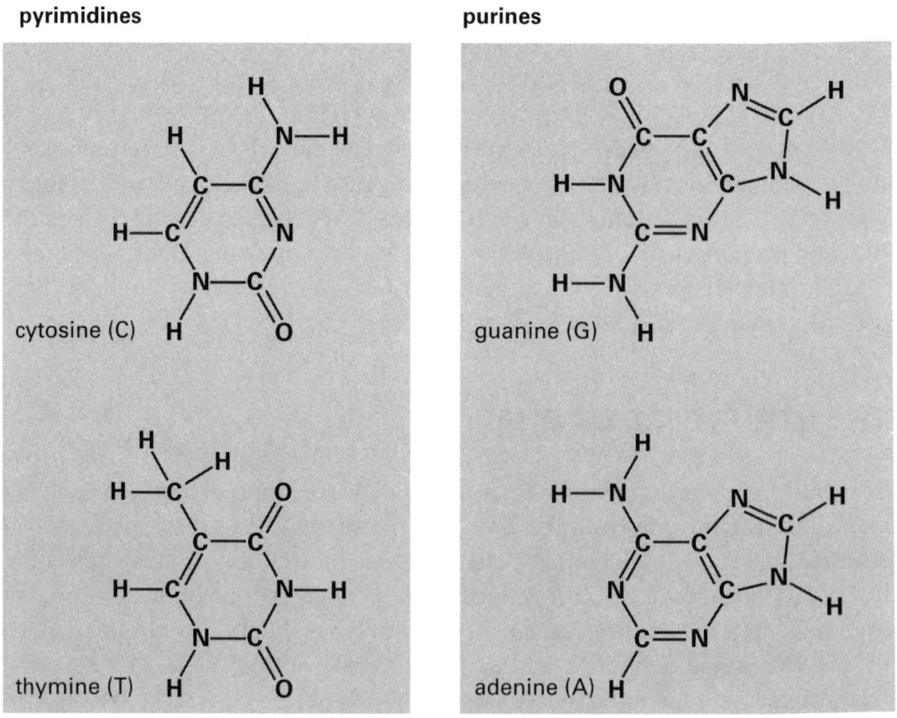

pyrimidines

H
H N—H
C—C
H—C N
N—C
cytosine (C) H O

H
H
H—C O
C—C
H—C N—H
N—C
thymine (T) H O

purines

O N H
C—C C
H—N C—N
C=N H
H—N
guanine (G) H

H
H—N N H
C—C C
N C—N
C=N H
adenine (A) H

Figure 5.2
The four bases found in DNA.

atom of deoxyribose. The combined base and sugar, without any phosphoryl groups, is known as a nucleoside. The phosphoryl groups are attached to the 5′ carbon of deoxyribose. The resulting phosphorylated nucleoside is called a nucleotide.

Four different nucleotides join to make DNA. They are the deoxyribonucleoside triphosphates: 2′-deoxyadenosine 5′-triphosphate (dATP), 2′-deoxyguanosine 5′-triphosphate (dGTP), 2′-deoxycytidine 5′-triphosphate (dCTP), and 2′-deoxythymidine 5′-triphosphate (dTTP).

DNA molecules are very large. The chromosome of the bacterium *E. coli* is made up of two strands of DNA, each containing 4.5 million nucleotides. In humans the genome comprises 46 chromosomes, which together contain about 6×10^9 nucleotides. Figure 5.3 illustrates the structure of the DNA chain. As nucleotides are added to the chain by the enzyme DNA poly-

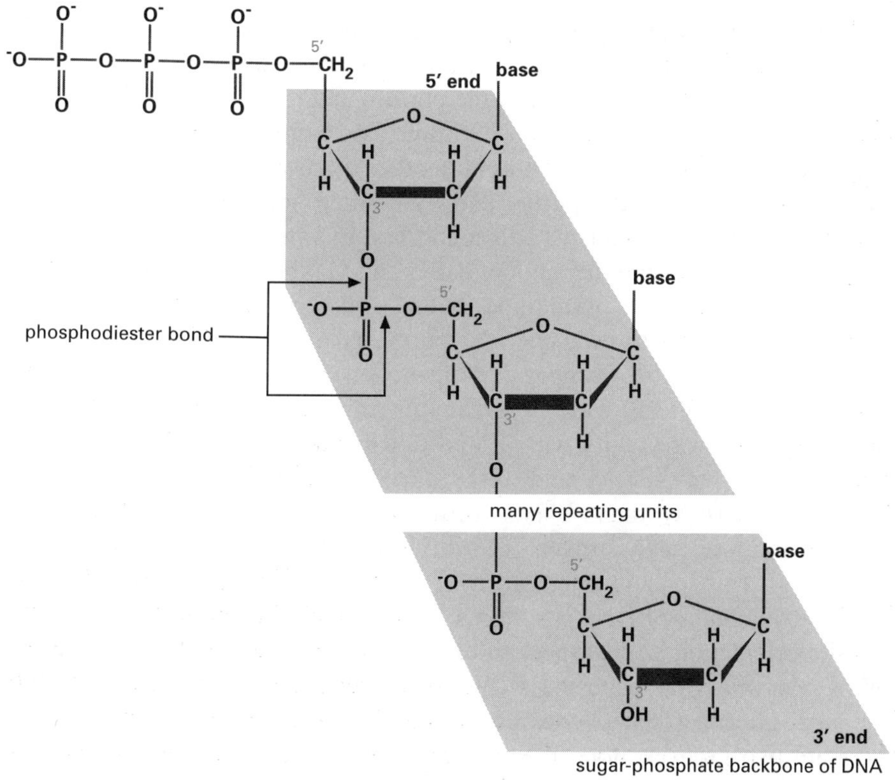

5′ end
base
phosphodiester bond
base
many repeating units
base
3′ end
sugar-phosphate backbone of DNA

Figure 5.3

The phosphodiester bond and the sugar-phosphate backbone of DNA.

merase (Chapter 9), they lose two phosphoryl groups. The last remains, and forms a *phosphodiester bond* between successive sugars. The bond forms between the hydroxyl group on the 3′ carbon of the sugar group of one nucleotide and the phosphoryl group attached to the 5′ carbon of the next nucleotide. Adjacent nucleotides are hence joined by a 3′-5′ phosphodiester bond. The linkage gives rise to the *sugar-phosphate backbone* of a DNA molecule (Fig. 5.3). A DNA chain has polarity because its two ends are different. The first nucleotide in the chain has a free phosphoryl group attached to the 5′ carbon of its deoxyribose, whereas the last nucleotide has a free hydroxyl group on the 3′ carbon of its sugar.

The DNA Molecule is a Double Helix

In 1953 Rosalind Franklin and Maurice Wilkins used X-ray diffraction to show that DNA was a helical (twisted) polymer. James Watson and Francis Crick demonstrated by building three-dimensional models that the molecule is a double helix (Fig. 5.4). Two hydrophilic sugar-phosphate backbones lie on the outside of the molecule, and the purines and pyrimidines lie on the inside of the molecule. In addition to fitting the X-ray data, the *double helix model* also provided an explanation of another discovery of the 1950s. Erwin Chargaff had analyzed the purine and pyrimidine content of DNA isolated from many different organisms and found that the amounts of A and T were always the same, as were the amounts of G and C. There is just enough space for one purine and one pyrimidine in the center of the double helix. The Watson–Crick model showed that the purine guanine would fit nicely with the pyrimidine cytosine, forming three hydrogen bonds (Box 1.4). The purine adenine would fit nicely with the pyrimidine thymine, forming two hydrogen bonds. Thus A always pairs with T, and G pairs with C. The three hydrogen bonds formed between G and C produce a relatively strong base pair. Because only two hydrogen bonds are formed between A and T, this weaker base pair is more easily broken. The difference in strengths between a G–C and an A–T base pair is important in certain biological processes (Chapter 10). The two chains of DNA are said to be antiparallel because they lie in the opposite orientation with respect to one another with the 3′-hydroxyl terminus of one strand opposite the 5′-phosphate terminus of the second strand. The sugar phosphate backbones do not completely conceal the bases inside. There are two grooves along the surface of the DNA molecule. One is wide and deep—the major groove—and the other is narrow and shallow—the minor groove (Fig. 5.4). Proteins can use the grooves to gain access to the bases (page 163).

cytosine-guanine base pair (C≡G)

hydrogen bond

deoxyribose

deoxyribose

cytosine (C) guanine (G)

thymine-adenine base pair (T=A)

deoxyribose

deoxyribose

thymine (T) adenine (A)

schematic representation of the double-helical structure of DNA

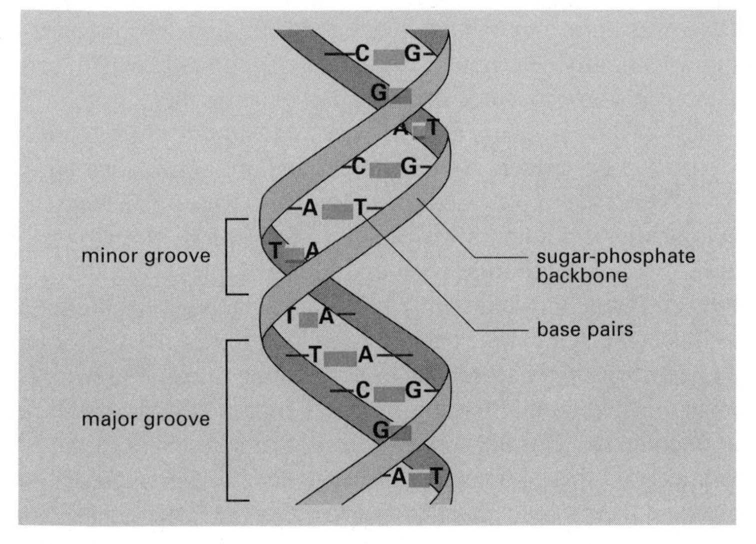

Figure 5.4

The DNA double helix held together by hydrogen bonds.

Complementarity of DNA Chains

A consequence of the base pairs formed between the two strands of DNA is that if the base sequence of one strand is known, then that of its partner can be inferred. A G in one strand will always be paired with a C in the other. Similarly an A will always pair with a T. The two strands are therefore said to be *complementary.*

Different Forms of DNA

The original Watson–Crick model of DNA corresponds to the B form in which the two strands of DNA form a right-handed helix. If viewed from either end it turns in a clockwise direction (Fig. 5.4). This, called B-DNA, is the predominant form in which DNA is found. There are, however, several variations of the B-form double helix. One of these, Z-DNA, so-called because its backbone has a zigzag shape, forms a left-handed helix and occurs when the DNA sequence is made of alternating purines and pyrimidines. The structure adopted by DNA is a function of its base sequence.

DNA AS THE GENETIC MATERIAL

DNA was known to contain genetic information several years before its structure was finally determined. As early as 1928 Fred Griffith carried out the now famous pneumococcus transformation experiment (Fig. 5.5). There are several strains of the bacterium *Diplococcus pneumoniae.* Some cause pneumonia and are said to be virulent; others do not and are nonvirulent strains. The virulent bacteria possess a polysaccharide coat, and when grown on an agar plate, their colonies have a smooth appearance—S bacteria. Nonvirulent bacteria do not have this polysaccharide coat, and their colonies are rough in appearance—R bacteria. Mice injected with S bacteria developed pneumonia and died, whereas those injected with R bacteria were unaffected. When S bacteria were killed by heat treatment before injection, the mice remained healthy. However, those mice injected with a mixture of live R bacteria and heat-killed S bacteria died of pneumonia. This observation meant that something in the heat-killed S bacteria carried the information that enabled bacteria to make the polysaccharide coat and therefore to change—transform—the R bacteria into a virulent strain. Later experiments in the mid-1940s by Oswald Avery, Maclyn McCarty, and Colin MacLeod clearly demonstrated that the transforming factor was DNA. They made extracts of S bacteria and treated them with enzymes that destroy either DNA, RNA (ribonucleic acid), or protein. These extracts were then

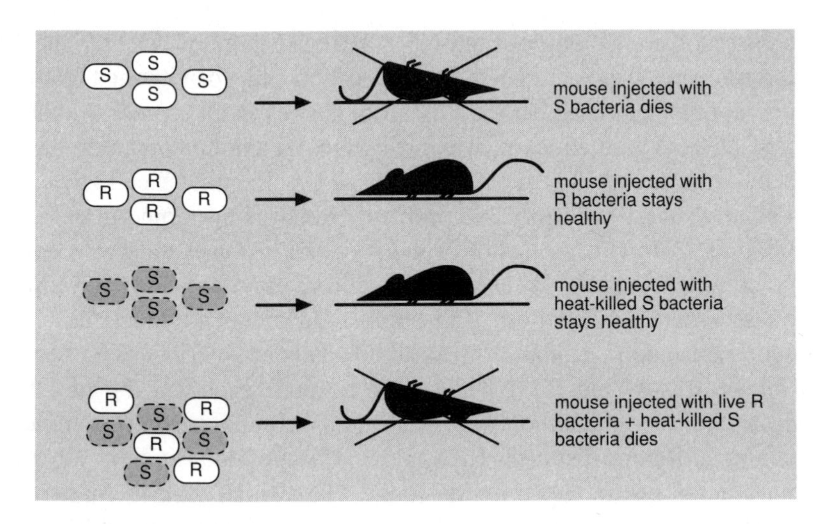

Figure 5.5
The *Pneumococcus* transformation experiment: R = rough, S = smooth.

mixed with R bacteria. Only the extract rich in DNA was able to transform R bacteria into S bacteria.

We now know that DNA molecules are indeed the cell's database and that DNA carries the genetic information encoded in the sequence of the four bases—adenine, guanine, cytosine, and thymine. The information in DNA is transferred to its daughter molecules through replication (the duplication of DNA molecules) and subsequent cell division (Chapter 9). DNA directs the synthesis of proteins through an intermediary molecule—RNA. The DNA code is transferred to RNA by a process known as transcription (Chapter 10). The RNA code is then translated into a sequence of amino acids during protein synthesis (Chapter 12). This is the central dogma of molecular biology: "DNA makes RNA makes protein."

Retroviruses are an exception to this rule. Inside the virus coat is a molecule of RNA plus an enzyme that can make DNA from an RNA template by the process known as reverse transcription.

THE PACKAGING OF DNA MOLECULES INTO CHROMOSOMES

Eukaryotic Chromosomes and Chromatin Structure

A human cell contains 46 chromosomes, each of which contains a single DNA molecule bundled up with various kinds of proteins. On average, each human chromosome contains about 6.5×10^7 base pairs of DNA. If the DNA in a chromosome were stretched as far as it would go without breaking, it would be about 5 cm long, so the 46 chromosomes in all represent about two meters of DNA. The nucleus in which this DNA must be contained has a diameter of only about 10 μm (Box 1.2, page 3). To pack a cell's DNA into such a small space is a formidable problem, which is overcome in eukaryotes by binding DNA to two groups of proteins called *histones* and *nonhistones* to form a complex known as *chromatin.* Chromatin contains approximately equal masses of DNA, histones, and nonhistone proteins. Under the electron microscope extended chromatin looks like "beads on a string." The beads are called *nucleosomes* (Fig. 5.6). They arise because each 146 bp of DNA is wound around an octamer of histone proteins composed of two molecules each of four different histones, H2A, H2B, H3 and H4. Histones are rich in the amino acids lysine and arginine (page 212) which give them an overall positive charge so that they bind tightly to the negatively charged phosphates of the DNA molecule. Each nucleosome is separated from its neighbor by about 50 base pairs of linker DNA which is bound by a fifth type of histone molecule, histone H1. This is only the first stage of DNA packaging. Higher-order packaging folds and refolds the chromosome still further (Fig. 5.6). This is accomplished with the help of various nonhistone proteins. Chromosomes reach their most condensed state at cell division. Metaphase chromosomes are 1400 nm wide compared to the 2 nm diameter of the DNA double helix (Fig. 5.6).

Prokaryotic Chromosomes

The chromosome of *E. coli* is a single circular DNA molecule of about 4.5×10^6 base pairs. Bacterial cells do not contain histones but have other basic proteins that do the same job of complexing the DNA so that it fits inside the cell. Prokaryotes do not have nuclear envelopes so the condensed chromosome lies free in the cytoplasm (Fig. 1.4).

Plasmids

Plasmids are small circular "mini-chromosomes" found in bacteria and some eukaryotes. They are several thousand base pairs long and are probably tightly coiled (supercoiled) inside the cell. Plasmids often code for proteins that confer resistance to a particular antibiotic (Chapter 11).

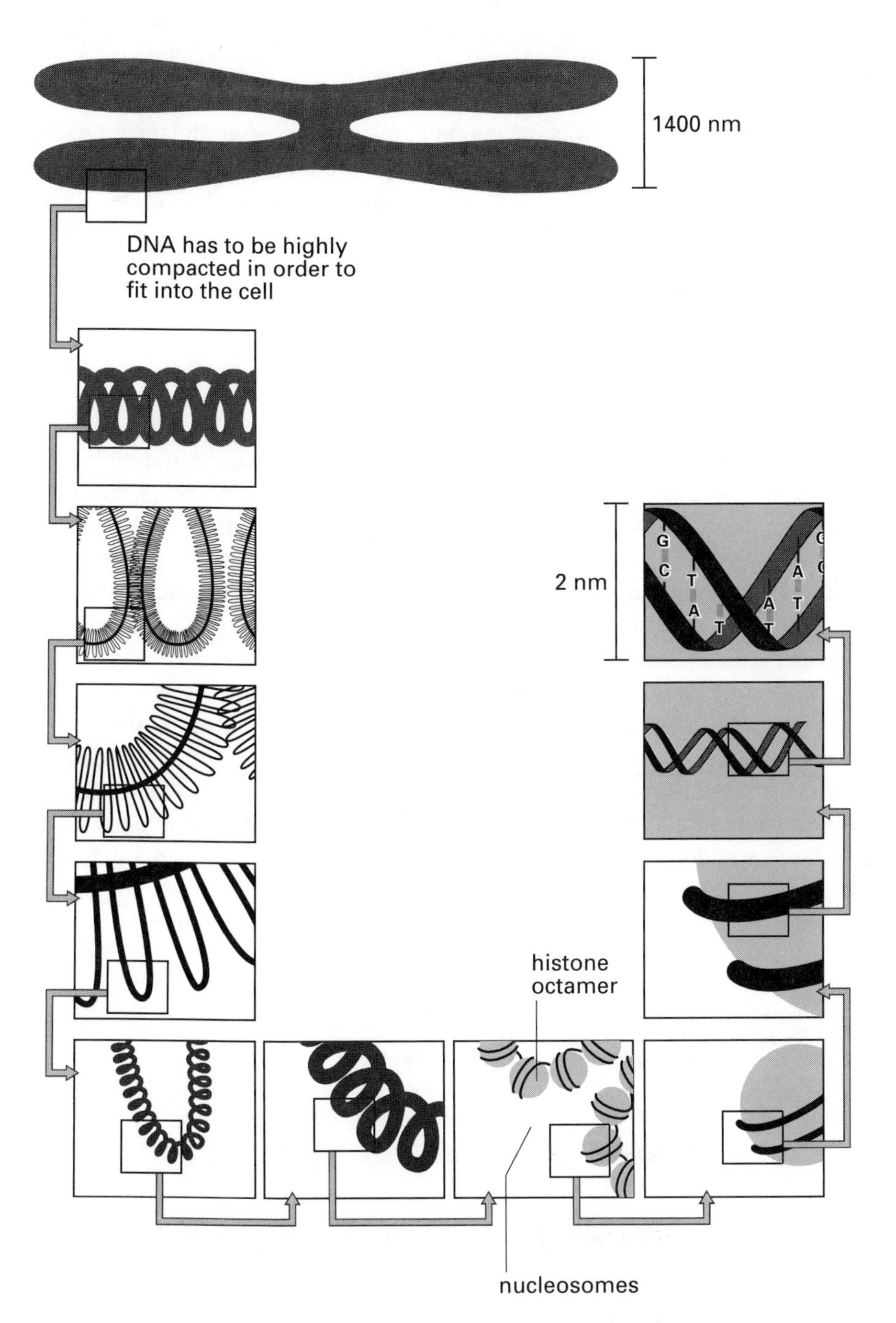

Figure 5.6
DNA packaged into chromosomes.

Viruses

The genetic material of a virus, its genome, may be single- or double-stranded DNA or even RNA. The genome is packaged within a protective protein coat. Viruses that infect bacteria are called bacteriophages. One of these called lambda has a fixed-size DNA molecule of 4.5×10^4 base pairs. However, the protein coat of the bacteriophage M13 will adjust to fit any size of chromosome.

THE GENETIC CODE

The sequence of bases along the DNA strand determines the amino acid sequence of proteins. Proteins are linear polymers of individual building blocks called amino acids (page 210). There are 20 different amino acids, but only four different DNA bases. The sequence is expressed on the three-letter *genetic code*. A single letter (base) could specify only four different amino acids, and a two-letter code only 16 (4×4) amino acids. The three-letter code, however, gives 64 ($4 \times 4 \times 4$) possible combinations, more than enough for all twenty amino acids. Each group of three bases is known as a *codon*.

This code is read in sequential groups of three, codon by codon. Adjacent codons do not overlap, and each triplet specifies one particular amino acid. This discovery was made by Francis Crick and his colleagues by studying the effect of various mutations on the bacteriophage T4 (a virus that infects a common bacteria called *E. coli*). If a mutation caused either one or two nucleotides to be added or deleted from one end of the T4 DNA, then a defective polypeptide was produced, with a completely different sequence of amino acids (Fig. 5.7). However, if three bases were added or deleted, then the protein

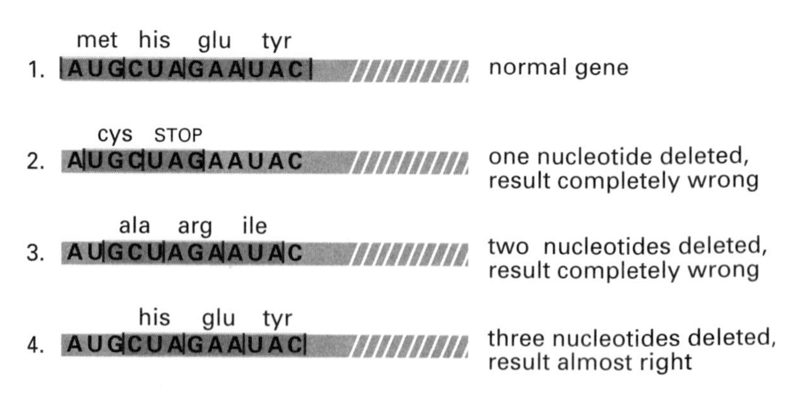

Figure 5.7

Reading frames of the genetic code, read in blocks of three.

produced often retained its normal function. These proteins were found to be identical to the original protein, except for the addition or loss of a single amino acid.

Despite the fact that the sequence of codons on DNA determines the sequence of amino acids in proteins, the DNA helix does not itself play a role in protein synthesis. The translation of the sequence from codons into amino acids occurs through the intervention of members of a third class of molecule—messenger RNAs (mRNA). These molecules act as templates, guiding the assembly of amino acids into a polypeptide chain. mRNA uses a triplet code similar to DNA; however, in mRNA the base uracil is used in place of thymine.

The identification of the triplets encoding each amino acid began in 1961. This was made possible by using a cell-free protein synthesis system prepared by breaking open *E. coli* cells. Synthetic RNA polymers, of known sequence, were added to the cell-free system together with the 20 amino acids. When the RNA template contained only U residues (poly U) the polypeptide produced contained only phenylalanine. The codon specifying this amino acid must be UUU. A poly A template produced a polypeptide of lysine, and poly C one of proline. AAA and CCC therefore specify lysine and proline, respectively. Synthetic RNA polymers containing all possible combinations of the bases A, C, G, and U, were added to the cell-free system to determine the codons for the other amino acids. A template made of the repeating unit CU gave a polypeptide with the alternating sequence leu–ser. Because the first amino acid in the chain was found to be leu its codon must be CUC and that for ser UCU. Although much of the genetic code was read in this way, the amino acids defined by some codons were particularly hard to determine. Only when specific tRNA molecules (page 190) were used was it possible to demonstrate that GUU codes for valine. The genetic code was finally solved by the combined efforts of several research teams. The leaders of two of these, Marshall Nirenberg and Gobind Khorana received the Nobel prize in 1968 for their part in cracking the code.

The Code Is Degenerate but Unambiguous

The 64 codons of the genetic code are shown in Table 5.1 together with the side chains of the amino acids that each codes for (Chapter 13). Sixty-one codons specify an amino acid and the remaining three act as stop signals for protein synthesis. Methionine and tryptophan are the only amino acids coded for by single codons. The other 18 are encoded by 2, 3, 4, or 6 codons, and so the code is said to be degenerate. No triplet codes for more than one amino acid and so the code is unambiguous. Notice that when two or more codons specify

TABLE 5.1 The genetic code and the corresponding amino acid side chains.

Amino acid	Side chain	Codons
alanine (ala)	CH_3	GCU, GCC, GCA, GCG
asparagine (asn)	O, NH_2, C, CH_2	AAU, AAC
aspartate (asp)	O, O^-, C, CH_2	GAU, GAC
arginine (arg)	NH_2, $C{=}\overset{+}{N}H_2$, NH, $(CH_2)_3$	CGU, CGC, CGA, CGG, AGA, AGG
cysteine (cys)	SH, CH_2	UGU, UGC
glutamine (gln)	O, NH_2, C, $(CH_2)_2$	CAA, CAG
glutamate acid (glu)	O, O^-, C, $(CH_2)_2$	GAA, GAG
glycine (gly)	H	GGU, GGC, GGA, GGG
histidine (his)	$H\overset{+}{N}$—CH, HC, N—C, H, CH_2	CAU, CAC
isoleucine (ile)	CH_3, CH_2, H_3C—CH	AUU, AUC, AUA
leucine (leu)	H_3C, CH_3, CH, CH_2	UUA, UUG, CUU, CUC, CUA, CUG
lysine (lys)	$\overset{+}{N}H_3$, $(CH_2)_4$	AAA, AAG
methionine (met)	CH_3, S, $(CH_2)_2$	AUG
phenylalanine (phe)	CH_2	UUU, UUC
proline (pro)	H_2C—CH_2, H_2N^+, CH_2, C, H, COO^-	CCU, CCC, CCA, CCG
serine (ser)	OH, CH_2	AGU, AGC, UCU, UCC, UCA, UCG
threonine (thr)	CH_3, HO—CH	ACU, ACC, ACG, ACA
tryptophan (trp)	HN, CH_2	UGG
tyrosine (tyr)	OH, CH_2	UAU, UAC
valine (val)	H_3C, CH_3, CH	GUU, GUC, GUG, GUA
STOP		UGA
STOP		UAA, UAG

the same amino acid, they only differ in the third base of the triplet (except for leucine and serine which are encoded by six codons). Thus mutations can arise in this position of the codon without altering amino acid sequences. Perhaps degeneracy evolved in the triplet system to avoid a situation in which 20 codons each specified a single amino acid, and 44 specified none. If this were the case, then most mutations would stop protein synthesis dead.

The Start and Stop Codes for Protein Synthesis

The order of the codons in DNA and the amino acid sequence of a protein are colinear. The start signal for protein synthesis is the codon AUG specifying the incorporation of methionine. Because the genetic code is read in blocks of three, there are three potential reading frames in any mRNA. Figure 5.7 shows that only one of these results in the synthesis of the correct protein. A mutation that inserts or deletes a nucleotide will change the normal reading frame and is called a frameshift mutation (Fig. 5.8).

The codons UAA, UAG, and UGA are *stop signals* for protein synthesis.

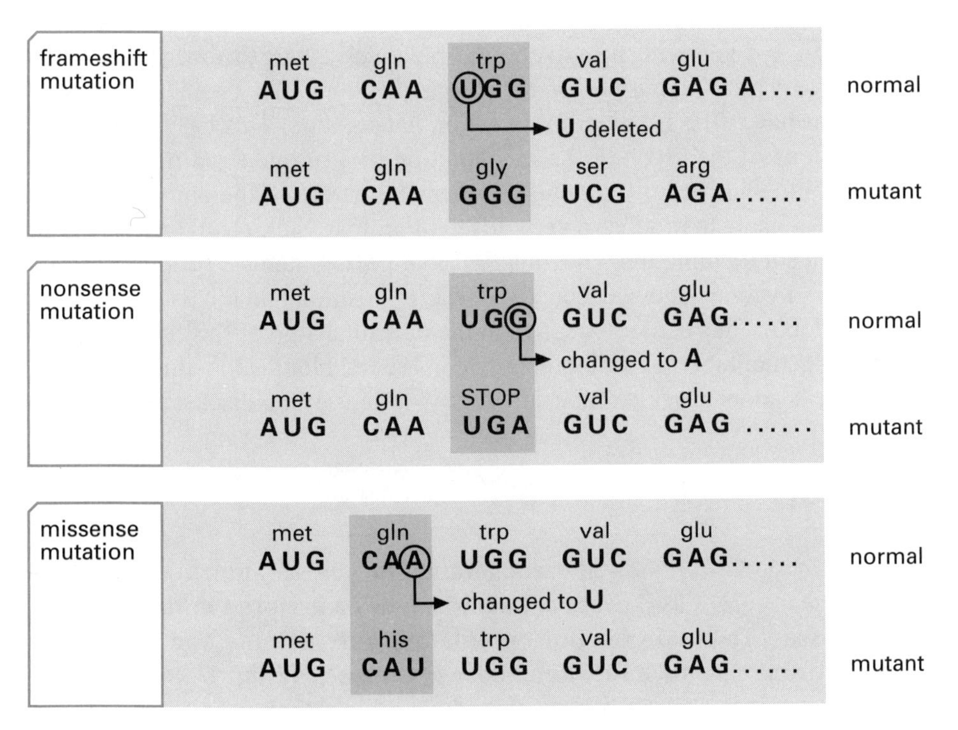

Figure 5.8
Mutations that alter the sequence of bases.

A base change that causes an amino acid codon to become a stop codon is known as a nonsense mutation (Fig. 5.8). If, for example, the codon for tryptophan UGG changes to UGA, then a premature stop signal will have been introduced into the messenger RNA template. A shortened protein, that will probably not work, is produced.

The Code Is Nearly Universal

Organisms as diverse as *E. coli* and mammals use the same genetic code. Therefore it was originally assumed that the code would be completely universal. However, the mammalian mitochondrial genome and a small number of organisms such as ciliated protozoa like *Paramecium* have some differences in their code.

Missense Mutations

A mutation that changes the codon from one amino acid to that for another is a missense mutation (Fig. 5.8). As Table 5.1 shows the second base of each codon shows the most consistency with the chemical nature of the amino acid it encodes. Amino acids with negatively and positively charged, hydrophilic side chains have A or G—a purine—in the second position. Those with hydrophobic side chains have C or T—a pyrimidine—in that position. This has implications for mutations of the second base. Substitution of a purine for a pyrimidine is very likely to significantly change the chemical nature of the amino acid side chain and is likely to produce a very different protein. Sickle-cell anaemia is an example of such a mutation. The mutation changes a glutamate residue encoded by GAG to a valine residue encoded by GTG at position 6 in the β-globin chain of hemoglobin. This causes a change in the overall charge of the chain, and the abnormal hemoglobin tends to precipitate in the red blood cells of those affected. The cells adopt a sickle shape, and the symptoms of the disease result.

SUMMARY

1. DNA, the cell's database, contains the genetic information necessary to encode RNA and protein. The information is stored in the sequence of four bases. These are the purines adenine and guanine and the pyrimidines cytosine and thymine. Each base is attached to the 1′-carbon atom of this sugar deoxyribose. A phosphoryl group is attached to the 5′ carbon of the sugar. The base + sugar + phosphoryl is called a nucleotide. The enzyme DNA polymerase joins nucleotides together by forming a phosphodi-

ester bond between the hydroxyl group on the 3′ carbon of deoxyribose of one nucleotide and the 5′-phosphoryl group of another. This gives rise to the sugar-phosphate backbone structure of DNA.

2. The two strands of DNA are held together in a double helical structure because guanine hydrogen bonds with cytosine and adenine hydrogen bonds with thymine. This means that if the sequence of one strand is known, that of the other can be inferred. The two strands are complementary in sequence.

3. DNA forms a complex with histone and nonhistone proteins to form chromatin. DNA is wrapped around histones to form a nucleosome structure. This is then folded again and again, such packaging compressing the DNA molecule to a size that fits into the cell.

4. The genetic code specifies the sequence of amino acid residues in proteins. The code is transferred from DNA to mRNA and is read in groups of three bases (a codon) during protein synthesis. There are 64 codons; 61 specify an amino acid and 3 are the stop signals for protein synthesis.

BOX 5.1

DNA—A Gordian Knot

At the start of his career Alexander the Great was shown the Gordian knot, a tangled ball of knotted rope, and told that whoever untied the knot would conquer Asia. Alexander cut through the knot with his sword. A similar problem occurs in the nucleus, where the forty-six chromosomes form two meters of tangled, knotted DNA. How does the DNA ever untangle at mitosis? The cell adopts Alexander's solution—it cuts the rope. At any place where the DNA helix is under strain, for instance, where two chromosomes press against each other, an enzyme called *topoisomerase II* cuts one chromosome so that the other can pass through the gap. Then, surpassing Alexander, the enzyme rejoins the cut ends. Topoisomerases are active all the time in the nucleus, relieving any strain that develops in the tangled mass of DNA.

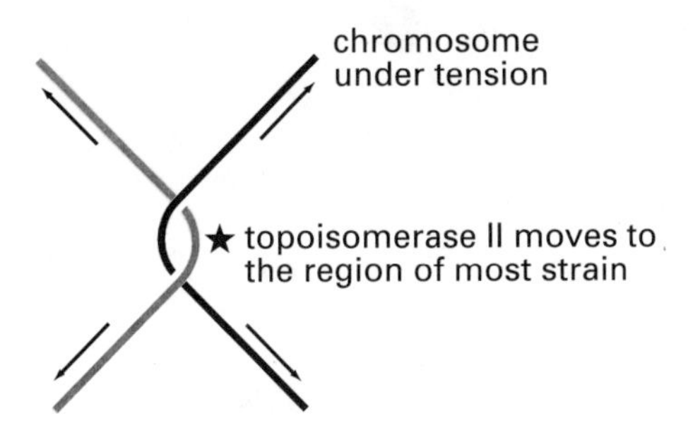

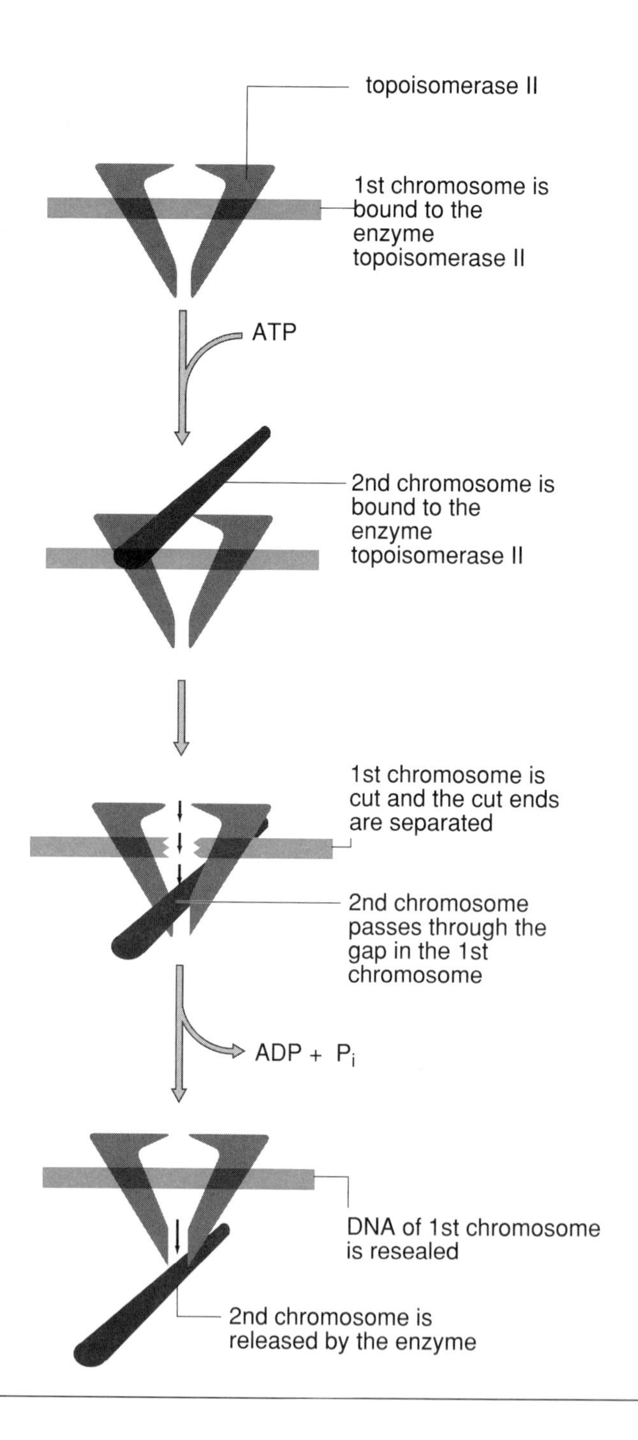
topoisomerase II
1st chromosome is bound to the enzyme topoisomerase II
ATP
2nd chromosome is bound to the enzyme topoisomerase II
1st chromosome is cut and the cut ends are separated
2nd chromosome passes through the gap in the 1st chromosome
ADP + P_i
DNA of 1st chromosome is resealed
2nd chromosome is released by the enzyme

CHAPTER 6

PATTERNS OF INHERITANCE

Genetics is, conventionally, about sex—which may help to explain why it is popular among biologists. Once it involved only mating males with females and observing the offspring. The most fundamental change in geneticists' view of their own subject, though, has been to realize that sex is not necessary. Sex in the familiar sense is just one of many genetic processes, most of which do not involve mating at all. They include patterns of inheritance in body cells other than those giving rise to sperm and egg. Somatic cell genetics, as this field is known, is central to the understanding of cancer and of ageing. Certain genes are passed on only through females or only through males, and some species have dispensed with males altogether. Modern genetic technology relies on our new ability to break through the sex barrier and to transfer genes between creatures as different as bacteria and humans. Nevertheless, sex is still important. The transfer of information between generations depends on the operation of simple rules that depend on the cooperation of males and females. They were first worked out in the middle of the nineteenth century by a monk, Gregor Mendel, who studied patterns of inheritance in peas. Mendel's laws apply, we now know, to most genes in most organisms.

MENDEL AND THE FOUNDATIONS OF GENETICS

Genetics differs from the rest of biology in that it depends largely on counting rather than measuring. Mendel was the first to see the importance of studying discrete characters that can be tallied up in each generation. He used peas because there existed many pure lines, kept separate for many years. Each line differed from the others in a number of distinct traits—pea shape or color, plant height, or flower color. Within a line all individuals were identical, and the offspring of two plants with, say, round yellow peas from within the same line would all have round peas, yellow in color, like those of their parents.

Mendel's breakthrough was to count the number of offspring of each type in crosses made between pure lines. He took advantage of the fact that peas (and many other plants) are hermaphrodite, with both male and female sex organs on the same plant, and self-compatible; that is, egg cells from a particular plant can be fertilized by pollen from the same individual. This is the process of self-fertilization or selfing.

THE LAW OF SEGREGATION

Mendel's first experiment is illustrated in Figure 6.1. He considered the inheritance of a single character—the color of the pea—in crosses made between two

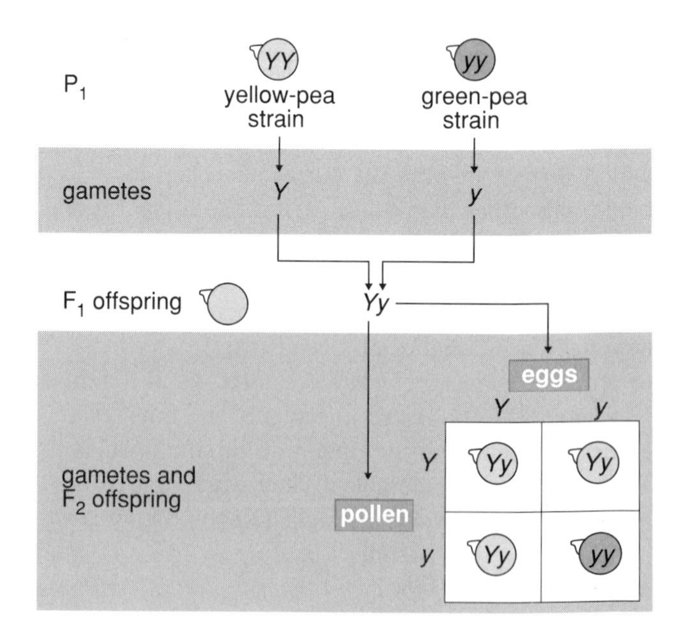

Figure 6.1
Mendel's interpretation of his first cross.

pure lines, one with yellow and one with green peas. When in the parental generation (the P_1 generation) pollen from a plant with yellow peas was used to fertilize the eggs from a plant whose peas were green, all of the offspring (the first filial or F_1) generation had yellow peas. This was an important result. Before Mendel, those few biologists who thought about heredity had vague ideas based on the blending of the qualities of the parents—the mixing of bloods. In Mendel's cross, though, the offspring resembled one parent but not the other. Heredity did not depend on the mingling of characteristics so that offspring were the average of their parents but on distinct and separate qualities in each one of them.

The next step was to self-fertilize the yellow F_1 offspring, that is, to fertilize an egg with pollen from the same plant. Again there was a surprise. In the succeeding generation (the second filial or F_2 generation) individuals with green peas reappeared. Although its effects had been masked in the F_1, whatever made peas look green surfaced unchanged after one generation of oblivion. Counting the peas said more. In this, the second generation, there was a ratio of three yellow peas to one green. From this simple pattern Mendel formulated the basis of inheritance: the separation of an organism's appearance, its *phenotype*, from the instructions, the *genotype*, that produce it. Inheritance was based on discrete "particles," or *genes*.

Mendel suggested that pollen and egg (the gametes) each carried one copy of the instruction to produce either a yellow or a green pea. A fertilized egg (the zygote) contained two copies, one from each parent. The particles came in different flavors, or *alleles*. For some alleles—*dominants*—only one copy was needed to produce an individual with the appropriate phenotype; what it was paired with (an allele the same as or different from itself) did not matter. For others, *recessives*, two copies of the same allele were required to produce a particular phenotype. The alleles for pea color were present at an unspecified site in the plant, the pea color *locus* (plural: loci).

The P_1 plants from the pure lines each had two copies of the same allele—they were *homozygotes*. This is indicated at the top of Figure 6.1: Plants with yellow peas are described as YY, and plants with green peas as yy. When they were crossed to produce the F_1, their offspring had one copy of each allele—they were Yy—and are known as *heterozygotes*. In Mendel's experiment, the allele for yellow pea color was dominant to that for green, so that the peas on the F_1 plants were yellow in appearance. A heterozygote for pea color looked just like an individual homozygous for the dominant allele, although its genotype was different. Homozygotes produce only one type of gamete, while heterozygotes make two, half with the dominant and half with the recessive allele. The box at the bottom of Figure 6.1 shows how the random joining of the gametes produced by the F_1 plants gives rise to the observed ratio of three yellow pea plants to one green pea plant.

To check his hypothesis, Mendel made a reciprocal cross. Instead of fertilizing the eggs of a green-pea plant with pollen from one whose peas were yellow, he did the experiment the other way round: "green" pollen with "yellow" eggs. It made no difference to the results. Even better, the same experiment carried out on pure lines differing in pea shape (round or wrinkled), flower color (purple or white), or plant height (tall or short) gave exactly the same ratios in the F_1 and F_2 as he had obtained with pea color. There seemed to be a pattern to heredity that applied whatever character was inherited.

From this, Mendel formulated his first law, the law of segregation: characters are controlled by pairs of alleles that segregate from each other during the formation of gametes and are restored at fertilization. The idea of blending inheritance was dead.

THE LAW OF INDEPENDENT ASSORTMENT

The pure lines differed from each other in several characters at once: pea color and shape, plant height, flower color, and the like. Mendel's second series of experiments considered the inheritance of characters considered in pairs: a dihybrid cross. Figure 6.2 shows the results of a cross between a plant with yellow and round peas with one having green and wrinkled peas. In the F_1, exactly as before, every plant resembled just one of its parents: all its peas were yellow and round. This pattern was easily explained. All these plants were heterozygous at both loci, that for loci for pea color and that for pea shape. Their phenotype was that of the dominant allele at each locus.

Selfing these F_1 plants gave what at first sight seemed an odd result in the F_2. A little thought showed Mendel that it could be interpreted in the same simple terms as those in his first experiment. The F_2 of the dihybrid cross contained four kinds of plant. Two were just like those in previous generations: They had round and yellow peas or, alternatively, wrinkled and green. Two, though, were new. The phenotypes were present in novel arrangements, round with green and wrinkled with yellow. These are recombinants, new combinations of color and shape not present in the original pure lines.

Mendel suggested that this was due to the patterns of inheritance at the loci for shape and for color being independent. It made no difference to, say, the allele for round peas (R is for round, r is for wrinkled) whether or not the plant in which it resided had the allele for green or for yellow peas at a separate locus.

Figure 6.2 shows the sixteen possible genotypes when two loci, each with two alleles, are inherited independently. Only one produces plants with wrinkled and green peas, that is, plants that are recessive homozygotes at both loci. The reason is easy to see. The chance of any fertilized egg receiving two copies of the allele for wrinkled peas is ¼, and two copies of that for green also

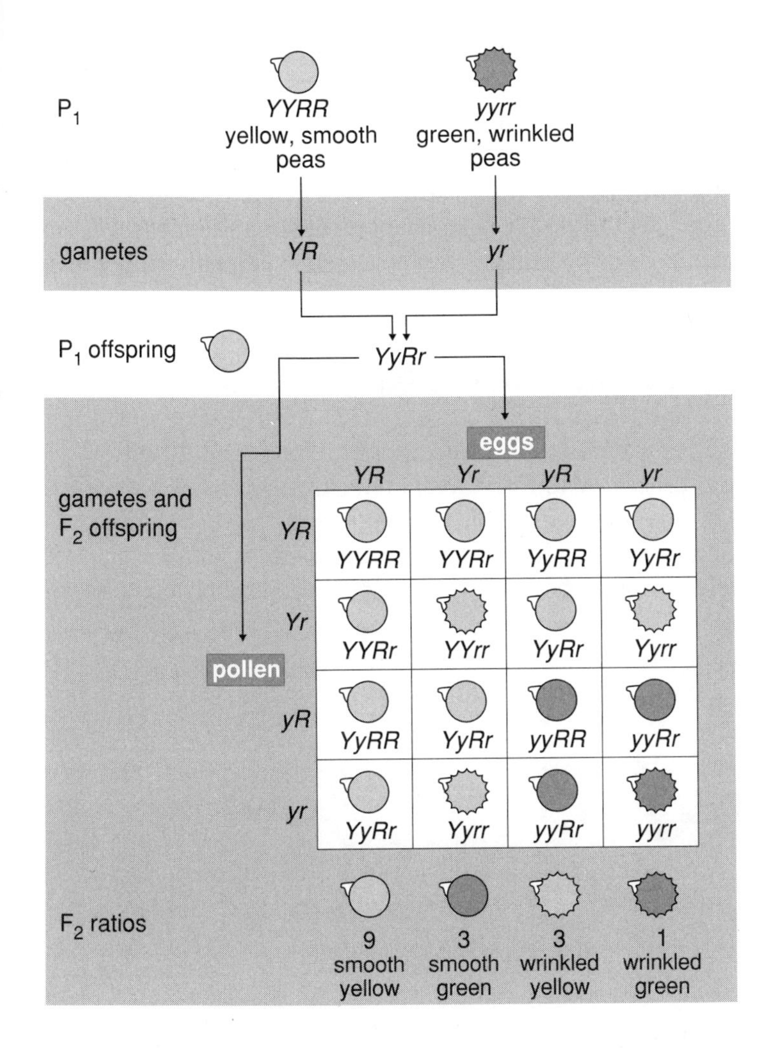

Figure 6.2
Mendel's interpretation of a dihybrid cross.

¼. The chance of it receiving two copies of the recessive allele at each of these two loci considered simultaneously is ¼ × ¼ or one in sixteen.

Nine sixteenths of the F_2 plants have yellow and round peas, but this phenotypic class conceals four genotypes. Six out of sixteen are, phenotypically, recombinants. Each represents new mixtures of alleles and phenotypes found in neither parent.

From this 9:3:3:1 ratio Mendel formulated his second law, the law of *independent assortment*: During the formation of gametes, patterns of segregation at one locus are independent of those at other loci. Inheritance was, it seemed, remarkably simple, depending only on the reshuffling of alternative alleles at different loci as one generation succeeds another.

Mendel was, fundamentally, right. His laws explain the patterns of inheritance of thousands of characters in hundreds of creatures. The fact that there are many exceptions to his simple rules does not diminish their importance. They form the basis against which breeding experiments can be tested. Exceptions often turn out to provide new insights into what genes are and how they work. Only in the past few years has genetics gone beyond Mendelism to an era in which the machinery of inheritance can be examined directly, rather than by inferring its workings from the nature of its products.

MENDELIAN INHERITANCE IN HUMANS

Mendel's work was disregarded by his contemporaries, but around 1900 it was uncovered by plant breeders. Within five years the first patterns of Mendelian inheritance had been found in humans. It is of course impossible to carry out planned breeding experiments with people. Instead, geneticists have to depend on the unplanned results of matings. They use pedigrees to establish patterns of inheritance.

Figure 6.3 shows the pedigree of a family transmitting *Huntington's disease*. This causes a gradual degeneration of the nervous system, leading to a loss

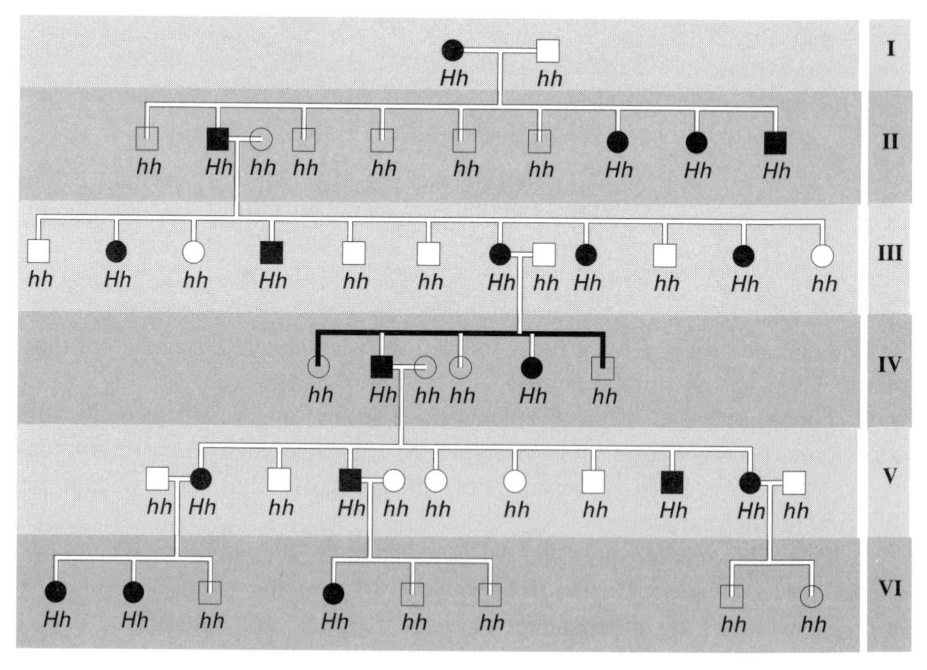

Figure 6.3
The pedigree of a family transmitting Huntington's disease.

of mobility and (usually) to an early death. The symbols are straightforward. Squares are males, circles females. Marriages are shown by a horizontal line joining two symbols, and succeeding generations are shown in series down the page. The black symbols indicate anyone showing the phenotype of interest.

Every affected person in this pedigree has at least one affected parent. The condition does not skip generations but can be traced back in an unbroken succession to the beginning of the pedigree. In large families there are hints of a predictable ratio; in generation V, for example, there is a family of eight children with an affected father. Four are affected, and four not. Even in smaller families about half of the children are affected, although the accidents of sampling mean that not every family shows exactly that ratio.

This is exactly the pattern expected for a phenotype controlled by a dominant allele. The probable genotypes for the individuals involved are marked in the diagram.

There are plenty of examples of dominant inheritance in humans. The Rhesus blood group is a familiar example. This, like all blood groups, is a system of diversity that is manifest when blood from different people is mixed. In certain combinations mixing blood of different types can be dangerous. Rhesus positive children always come from families in which one or both parents is Rhesus positive. Rhesus negative parents never have a Rhesus positive child—just the pattern expected if the positive allele is dominant over the negative.

Cystic fibrosis is an example of recessive inheritance in humans. It affects many organs including the lungs, and until recently sufferers died very young. To manifest the disease children must receive two copies of the appropriate allele, one from each parent. Figure 6.4 shows part of a family tree. In generation IV two affected individuals, a boy and a girl, were born. These were

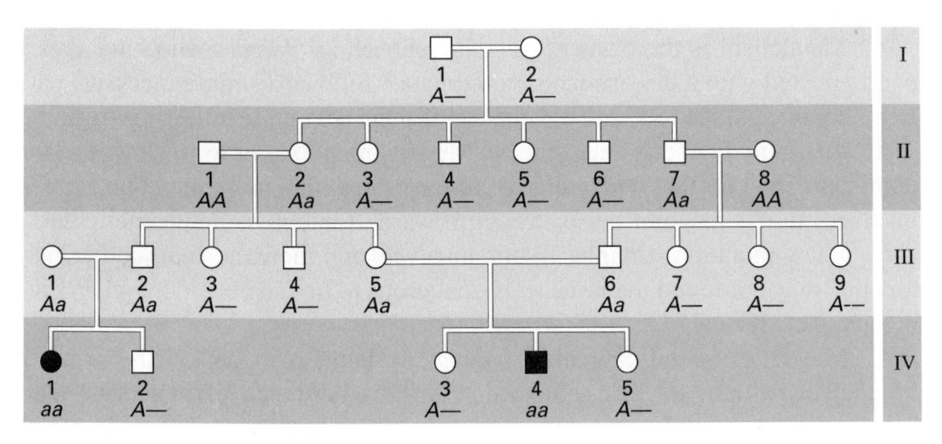

Figure 6.4.
The pedigree of a family transmitting cystic fibrosis.

therefore homozygous for the recessive cystic fibrosis gene—their genotype was aa. It is possible to work out what the genotypes of some of their siblings and ancestors must have been. These have been filled in in the figure. Where it is not possible to deduce the identity of an allele, it is marked with a dash –. Working back from the affected children, we can see that their parents must have been Aa, but since none of their uncles or aunts had children, we can only say that these individuals carried one A allele. The identity of the other allele cannot be deduced. It is worth noting that two of the children's grandparents (generation II) were a brother and sister. It is hence very likely that, as marked on the figure, these grandparents were both carriers of a cystic fibrosis allele—that is, they were heterozygotes, Aa—and that the history of cystic fibrosis in this family began because one of the great-grandparents in generation I was a carrier. We do not know, however, which of the great-grandparents it was. Both are therefore marked A–.

Occasionally a person affected with cystic fibrosis marries. If his or her spouse is not a relative (and is hence, for a rare recessive condition such as CF, likely to be a homozygote for the dominant allele), then all their children will be heterozygotes and phenotypically normal. The general pattern is quite different from that of dominant inheritance. The condition skips generations and often appears in people who have no obvious family history of the disease. What is more, an affected parent may have a large family, none of whom show symptoms of the disease as each one is a heterozygote.

When two heterozygotes marry, the chances of any child being affected is one in four. Since most families nowadays are small, it is often hard to see the ratio (three normal to one affected) that was so obvious in Mendel's peas. It is not of course true that if one affected child is born, then the next three will be normal. In heterozygotes each sperm and each egg has a one in two chance of bearing the recessive allele. Every zygote hence has one chance in four of carrying two copies of that gene.

Mendelism is the basis of genetic counseling. Parents who have had a child affected with a disease whose inheritance follows a simple recessive pattern may be anxious to know the chance of their next child suffering from the same condition. For such genes the answer is simple: one in four. However, for dominants (and for that wide range of illnesses in which more than one gene is involved) things may not be so straightforward. Sometimes new genetic damage—a new mutation (Chapter 8)—is involved, and there are many other factors that may confound the issue. It is dangerous to be too confident when giving genetic advice.

Mendel's second law also applies to humans. One of the earliest Mendelian patterns to be established was the inheritance of maleness: men have a dominant sex-determining allele not present in women. The pedigree of Huntington's disease shows that the allele for the condition is inherited quite

BOX 6.1

Multiple Alleles

Often there are more than two allelic forms at a particular locus. One of the first mutations to be discovered in the fruit fly *Drosophila* changed the color of the eye from red to white. The white allele was recessive, giving at first sight a simple two-allele system. Since then, though, many more alleles at the eye color locus have been found. They include eosin, apricot, blood, wine, and satsuma. Each allele gives rise to a distinct eye color, and on simple crossing experiments at least, each appears to be a different allele at the same locus, with a hierarchy of dominance among them.

Humans show a similar pattern of inheritance. The ABO blood group system is a familiar example. Furthermore molecular genetics often shows that what at first seemed to be a straightforward recessive allele for an inherited disease in fact conceals a number of quite distinct variants at a particular locus, each arising from a unique genetic lesion. Sometimes the multiple alleles at the relevant locus have subtly different phenotypic effects, but more often they can be detected only by looking directly at the DNA.

This is true, for example, of Tay–Sachs disease, a recessive condition leading to an inherited degeneration of the nervous system that is relatively common in some Jewish populations. Family studies suggested a straightforward pattern of Mendelian inheritance, with the normal allele dominant to a single simple recessive. Looking at the gene directly, though, has uncovered three distinct genetic variants at the locus responsible. Each involves a change at a different site in the DNA. In other conditions, such as the inherited failure of blood clotting known as hemophilia (which was once regarded as a classic of simple Mendelian inheritance), there may be hundreds of different allelic variants in different families, underlying what at first appeared to be a simple and unitary genetic disease.

independently from that of maleness; men and women are equally likely to inherit or to transmit the disease. Just like pea color and pea shape, there is independent assortment of the two attributes.

MODIFYING MENDEL: CHANGES IN RATIOS

Mendel was lucky in studying characters that are easy to identify and have simple patterns of inheritance. There are numerous exceptions to his rules, many of which illuminate the subtleties of how genes work.

Lethal Alleles

Mendel's results depended on plants of different phenotype having equal chances of survival. In fact that is not quite true. After all, one rarely sees peas with yellow or with wrinkled seeds in nature. However, the differences in fitness in his experiments were small enough not to matter. Often, though, particular genotypes are much less likely to survive than are others. Quite frequently one or the other is lethal. Those carrying certain genes die, distorting Mendelian ratios.

Manx cats have no tails. If two of them are mated, there is a ratio of two tail-less to one normal among their offspring. However, if the embryos inside a pregnant female are examined, a third class of offspring, small and deformed, is found. These do not survive until birth. Manx cats are all heterozygotes for the "Manx" allele. In the F_1, four genotypes are created: homozygous for the normal allele, heterozygotes, and homozygotes for the Manx allele. The Manx homozygotes die early in development while heterozygotes survive, but with no tail. Thus the Manx allele is dominant in its effects on tail length (heterozygotes have short tails) but recessive in its effects on viability (heterozygotes—just like cats with two copies of the normal allele—are born alive).

Human achondroplastic dwarfism works in the same way. If two people with this conditions (which interferes with the development of arms and legs) marry, about one child in four dies before, or soon after, birth from a severe skeletal abnormality. He or she has inherited two copies of the gene involved. Heterozygotes are viable but grow up short.

Lethal genes are common. As we will see, radiation damages genes, often producing dominant lethals. Individuals inheriting a dominant lethal die immediately (or, at best, do not survive for long enough to pass on their genes). This is the basis of the sterile male technique for controlling insect pests, such as the mosquitoes that carry malaria. Males are irradiated, and released in their millions. Their sperm carry dominant lethals, so that any offspring resulting from their mating with normal females die.

Dominant lethals contain within themselves the seeds of their own destruction, as no carrier can transmit the allele. Many—perhaps most—human pregnancies do not succeed, often without the woman knowing she was pregnant in the first place. Much of this wastage of fertilized eggs is due to dominant lethal alleles. Alleles at other loci are (or were) effectively dominant lethal in the sense that although their bearers might survive, they rarely have children. For instance, achondroplastic dwarfs rarely used to find a mate although they were generally healthy. Nowadays, though, many of them marry.

Many inborn diseases in humans are due to *recessive lethals*: Individuals with two copies of the gene die young. Around three thousand recessive lethals are known, and there may be many more. A crude calculation based on the overall incidence of such births suggests that every human being is likely, without knowing it, to carry a recessive lethal that would kill the carrier were it present as a homozygote.

Gene Interaction

Some phenotypes depend on the simultaneous action of alleles at two or more gene loci.

The police have, for many years, matched the blood groups of a bloodstain at the scene of a crime with those of a suspect. Even a simple ABO genotype (Box 6.2) may be powerful evidence in a court of law. Sometimes criminals leave other clues instead, such as saliva or semen. There is genetic variation in the extent to which blood group antigens (including those in the ABO system) are secreted into these body fluids. Some individuals, secretors, produce A, B, or O blood group substances in saliva and semen; others do not. The Secretor allele is dominant to the nonsecretor form. The variation involves a locus quite distinct from that of the ABO blood groups themselves. Traces of, say, A and B antigens in the saliva used to lick an envelope containing a death threat hence give clues about two distinct gene loci. The criminal must have carried both the Secretor allele at one locus, and those for A and B at another. The two loci interacted to produce the phenotype.

There are many instances in which the actions of one gene influence the phenotype produced by another. Sometimes the effect is dramatic. Children

BOX 6.2

Complications with Dominance

Dominance and recessivity are properties of phenotypes, not genotypes, and may not be absolute. Sometimes an allele is dominant for some of its attributes and recessive for others. Many people of African descent have the sickle-cell mutation (page 86). Individuals who carry either one or two sickle-cell alleles show a greatly increased resistance to malaria. Thus resistance to malaria is a dominant trait. However, individuals who carry two sickle-cell alleles (but not those with one) show the debilitating failure of red blood cells that is sickle-cell anemia. The sickle-cell allele is therefore dominant in endowing resistance to malaria but recessive in causing sickle-cell anemia.

The ABO blood system shows another complication of dominance. There are three alleles (A, B, and O) and hence six possible genotypes (AA, BB, OO, AB, AO, and BO). There is complete dominance between some alleles. Individuals heterozygous for AO have blood group A, as do AA homozygotes. Exactly the same pattern is true for blood group B. AB heterozygotes, though, have the attributes of both alleles, giving blood group AB.

Genetic counseling has been revolutionized by molecular biologists' new ability to get around the problem of dominance by looking directly at genes rather than their products. Heterozygotes can be distinguished from either homozygote, so that effectively all characteristics lose the attribute of dominance or recessivity altogether.

homozygous for an albino allele (which abolishes the ability to make dark skin pigment) may carry, at other loci, alleles for brown eyes or black hair, but these cannot show themselves because the albino allele interrupts the pathway of pigment synthesis before the other alleles have a chance to manifest their effects.

Cytoplasmic Inheritance

Some plants have attractive type of leaf coloration called variegation. A few leaves are green, those on other branches are white, and yet more have patches of white or green pigment. Since every type of branch on such a plant can flower, crosses can be made between the various patterns. When pollen from a white branch is used to fertilize the eggs of normal green leaved plants, the results are straightforward: all offspring are green. The reciprocal cross—pollen from green with eggs of white—produces white seedlings. However, any cross involving an egg from a variegated parent (whatever the genotype of the pollen) produces a mixture of green, white and variegated offspring. What is going on? Remember that organelles called chloroplasts give plants their green color—and chloroplasts contain their own DNA. Pollen grains are small and contain very few chloroplasts, eggs are large, with plenty of them. Almost all the chloroplasts in the seedlings are inherited from the egg. The number of white chloroplasts in a variegated plant differs at random from egg to egg. If the egg has only white chloroplasts that cannot make chlorophyll, so will the seedlings.

The same is true in humans—not for chloroplasts but for the other class of organelles with their own DNA, mitochondria. Every individual's mitochondrial DNA is inherited down the maternal line, from their mother, grandmother, and so on back into history.

Multifactorial Inheritance

Mendel was sensible in that he looked at characters—round or wrinkled, green or yellow—that involved alternative forms quite distinct from each other. However, most variation is not like this. There are tall people, short people, and those of intermediate height. The same is true of the clever, the stupid, and the broad mass of average intelligence. Continuous characters such as these are often—at least in part—under the control of genes.

Measuring how many genes are involved in characters such as size, or milk yield in cattle, is not easy. One approach is to ask how similar is the size, or milk yield, of parents, offspring, and more distant relatives. Another is to breed each generation from individuals with extreme values of a particular character and to see how quickly subsequent generations diverge from average value of their parents. Yet another is to take pairs of similar individuals from various parts of the size or milk-yield distribution and to breed them together for many generations. By so doing, a series of inbred lines is produced. If these differ consistently from each other in the character being studied, then there is a good case to be made that much of the variation in the original population was due to the segregation of genetic differences. In principle at least, it is possible to sort out how many loci affect the character by asking how distinct the lines of descendants are from each other.

All these are approaches to measuring heritability—the proportion of the total variation in a population which is due to genetic variation. This is an important endeavor in agriculture, as traits with high heritability are more easily improved by selective breeding. In humans too such information can be useful. The fact that relatives may share a tendency toward heart disease or cancer was the first step to uncovering some of the many genes now known to control these traits.

Nature, Nurture, or Both?

The Siamese cat shows, at first sight, a classic instance of a recessively inherited phenotype. Such cats have pigmented ears, noses, and tails, but the main part of their bodies is creamy white. Crossing a Siamese with a black cat gives all-black offspring, and mating these among themselves produces the standard 3:1 ratio of black to Siamese in the next generation. This simple pattern is complicated by the fact that the temperature at which Siamese cats are raised makes a large difference to the way it looks. The allele responsible reduces the stability of an enzyme in the pathway making fur pigment. It works normally at low temperatures (e.g., on the ears of the cat) but fails to make pigment on the warmer parts of the body, such as the trunk. A Siamese cat kept in the cold is darker than one kept in a warm room—sometimes, in fact, almost uniform in color. Therefore much of the variation in this character, with its classic pattern of recessive inheritance, is due to an interaction between the appropriate allele and the environment in which it is placed. To say that Siamese pattern is inborn, or that it is acquired, is not to say anything very useful.

Gene-environment interaction is almost universal in genetics. Often it has practical importance. Families with an allele interfering with the removal of cholesterol from the blood are in danger of heart disease when they eat a fatty diet, but a low cholesterol diet removes much of the risk. Many other genes are more or less able to respond to the environment in which they find themselves. p53 is a protein much involved in the control of cell division (Box 4.2).

It exists in a number of distinct forms. People with certain forms of the gene are much more susceptible to carcinogenic chemicals than are others. Their cancer is due both to their genes and to the environment they face. As is often said, if everyone smoked, lung cancer would be a genetic disease. Discussions about the heritability of attributes such as intelligence tend to sink into tedious, and usually sterile, arguments about the relative importance of gene and of environment. Taller people, or fatter pigs, can result from good food as well as good genes. Only too often the endless debate about nature versus nurture is vain. It misses a central point about inheritance, which is that phenotypes may be separated by many steps from the genotypes that produce them.

SUMMARY

1. Inheritance is based on "particles" or genes, passed more or less unchanged from generation to generation. Sometimes the attributes of an individual—its phenotype—can be used to infer what its genotype might be. Mendel, with his peas, was the first to use patterns of inheritance to deduce simple genetic laws that apply to humans as much as to other organisms.

2. In Mendel's simple model every attribute is under the control of a single gene locus, present in alternative forms or alleles. Body cells are diploid, receiving half their alleles from sperm and half from egg, each of which is haploid. Some alleles, dominants (a human example being the neurodegenerative disease Huntington's disease), need to be present only in a single copy. For others, recessives (e.g., cystic fibrosis), two copies are needed to show their effects. Individuals with two copies of the same allele are homozygotes, while those with one copy of each of two different alleles are heterozygotes. Occasionally new alleles are produced by genetic accidents or mutations as genes are passed from one generation to the next.

3. In general, alleles at different loci are inherited independently. They show independent assortment, generating new mixtures of alleles at each locus (recombinants) in their offspring. These simple laws underlie much of genetic counseling.

4. There are many exceptions to simple Mendelian rules. Sometimes homozygotes die—the alleles are recessive lethal, confusing the ratios. Many human genes are recessive lethals, killing sperm, eggs, or fetuses before they are born. A more striking exception is that a few genes, those in mitochondria, are passed down only through females and do not follow Mendel's laws.

5. Sometimes alleles at different loci interact to produce a phenotype, and sometimes so many alleles are involved that they give the appearance of a continuous distribution, for instance, as is true of human height or intelligence. The issue is further complicated by the intimate interaction of gene and environment, both of which affect nearly all characters to some extent.

CHAPTER

7

MAPPING THE GENES

Mendel saw genes as "particles", as independent units, each responsible for a different character. He knew nothing of chromosomes, let alone DNA. All his work—and that of most geneticists for most of the subject's history—depended on inference, working back from phenotypic ratios to deduce what must be going on at the genotypic level. However, soon after the rediscovery of Mendel's results, there came a hint that his particles could be identified with structures in the cell. This clue was the first in what became an enormous scientific effort: making a physical map of the genome and matching it to the linkage map, a map based not on the material composition of the DNA but on patterns of inheritance.

GENES AND CHROMOSOMES

Chromosomes were obvious candidates as bearers of genes. Like Mendel's units, they come in pairs, one inherited from each parent. Sperm and egg have half the number of chromosomes of body cells and are produced by a distinct form of cell division, meiosis rather than mitosis (Fig. 4.5). What is more, different chromosomes separate, as did Mendel's particles, independently into sperm and egg cells. Thus a gene is not only a unit of information but has a physical reality.

A gene can be defined as a length of DNA that performs a particular job, for instance, coding for a polypeptide chain. Generally speaking, the locus of a particular gene corresponds to the position of the DNA on the chromosome. Dominance and recessivity too can be described in physical terms. They are, in the end, functions of proteins. The cystic fibrosis gene codes for a channel protein required in many cells, including cells in the lungs (page 230). Heterozygotes have one working copy of the gene. They can make the protein and are healthy. Homozygotes for the cystic fibrosis allele have two damaged copies. Neither codes for a working protein, and they show the symptoms of the disease.

The proof of the theory that genes are physical entities that can be equated with structures in the cell came with the discovery of *sex-linked inheritance*. The American geneticist Thomas Hunt Morgan found that in some crosses using the fruit fly *Drosophila melanogaster* the results depended on which parent (male or female) had which phenotype. In one laboratory culture every individual had white eyes rather than the normal red. When white-eyed males were crossed with red-eyed females, and the offspring of that cross mated among themselves, the results were not very different from those obtained by Mendel. All the F_1 flies had red eyes, and there was a ratio of 3 red to 1 white in the F_2. The white eye allele (w) is therefore recessive to the red eye allele (W). There was, however, one oddity: all the F_2 white-eyed flies were male! The oddity became more marked in the reciprocal cross: white-eyed females with red-eyed males. This time white-eyed flies appeared in the F_1: half the flies had red eyes and half white. Once again, all the white-eyed flies were male. When F_1 flies were mated, an F_2 ratio of half white- and half red-eyed flies appeared. This time, though, there were equal numbers of male and female red-and white-eyed individuals.

It struck Morgan that the pattern of inheritance of the white-eye allele followed exactly that of the X chromosome. He deduced from this that the locus for eye color must actually be on the X chromosome, with no corresponding locus on the Y chromosome. Figure 7.1 shows his interpretation of what was going on. X^w represents an X chromosome that bears the white-eye allele, and X^W is an X chromosome with the dominant red-eye allele. In males, with only one X chromosome, any allele on the single X chromosome shows its effects because there is no second X to mask it. These males are hemizygous for this gene. In a cross between red-eyed males and the white-eyed females, the hemizygous F_1 males have white eyes, while the heterozygous females have the normal (or "wild-type") red eyes.

In these experiments Morgan saw the first real fit between a genetic model based on patterns of inheritance and a physical structure in the cell. The attempt to reconcile genetic maps based on breeding experiments with physical maps based on examining the structure of chromosomes and of DNA has continued ever since. The relationship between the two has turned out to be much more complex than could have been imagined and has involved technology be-

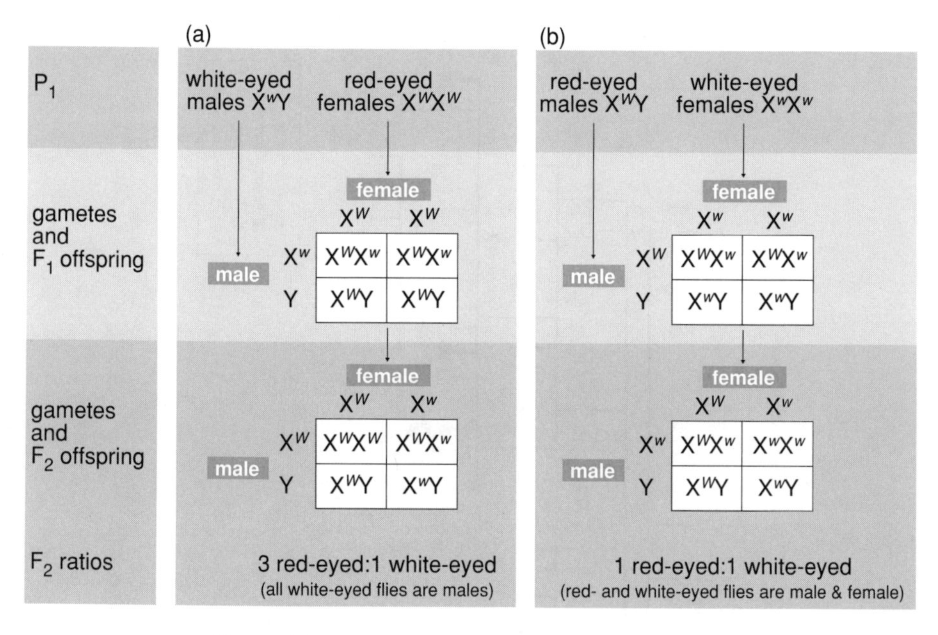

Figure 7.1
Sex linkage of the white-eye character in *Drosophila*.

yond Morgan's wildest dreams. Nevertheless, the fit between genes and chromosomes spotted by him is at the basis of the whole of modern genetics.

Many instances of sex linkage are known in humans and their pattern of inheritance is exactly the same as that for white eye in *Drosophila*. Figure 7.2 shows a pedigree of the blood-clotting disease hemophilia in one family, with the inferred genotype of selected individuals. It is obvious that males show the condition more often than do females and that most males receive their damaged allele from their heterozygous and phenotypically normal mother. Female children can inherit the damaged allele from a father, but he would of course have the disease, and until recently hemophiliacs rarely lived long enough to have children. Leopold, Duke of Albany, was an exception.

LINKAGE MAPPING AND THE STRUCTURE OF THE GENOME

Figure 7.3 shows the results of another of Morgan's experiments. It reveals what is at first sight a surprising pattern of inheritance of two abnormal characteristics: purple eye and vestigial wing in *Drosophila*. In the first generation a fly homozygous for two recessive alleles, purple eye and vestigial wing (which reduces the wing to a stump), is crossed to a wild-type fly with red eyes and

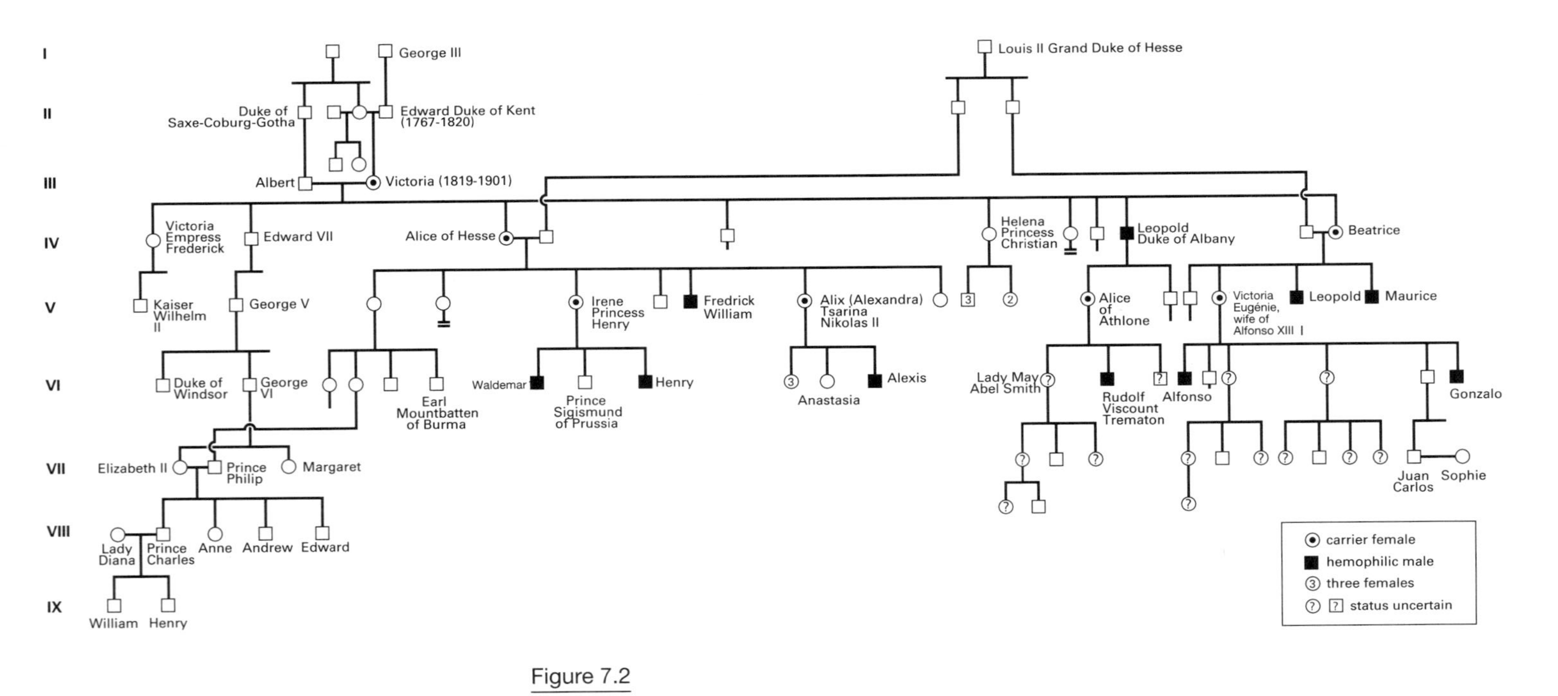

Figure 7.2
The pedigree of a family transmitting hemophilia.

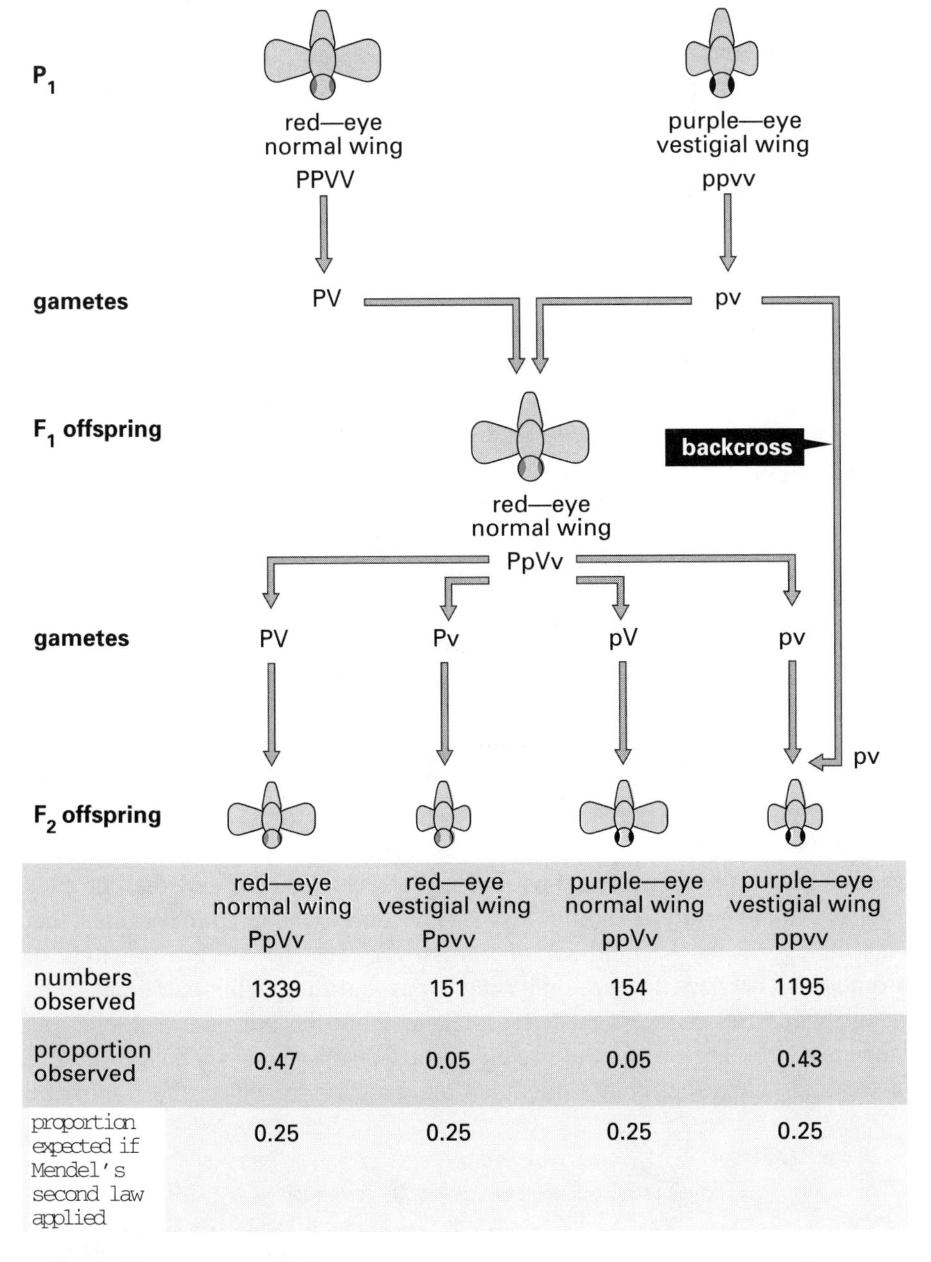

	red—eye normal wing PpVv	red—eye vestigial wing Ppvv	purple—eye normal wing ppVv	purple—eye vestigial wing ppvv
numbers observed	1339	151	154	1195
proportion observed	0.47	0.05	0.05	0.43
proportion expected if Mendel's second law applied	0.25	0.25	0.25	0.25

Figure 7.3

A dihybrid cross for the linked characters purple eye and vestigial wing in *Drosophila.*

normal wings. In the F_1 of course all these flies are phenotypically wild-type, since they are heterozygous at both loci. The female F_1 heterozygotes are then back-crossed with purple-eyed, vestigial-winged males. What would we expect, given Mendel's laws, in the F_2? The F_1 female is contributing equal numbers of red eye (P) and purple eye (p) alleles; offspring receiving P will have red eyes, and offspring receiving p will have purple eyes because the homozygous male must contribute p. We therefore expect equal numbers of red and purple eyes in the F_2. In the same way the F_1 female is contributing equal numbers of normal wing (V) and vestigial wing (v) alleles; offspring receiving V will have normal wings, and offspring receiving v will have vestigial wings because the homozygous male must contribute v. We therefore expect equal numbers of normal and vestigial wings in the F_2. Mendel's second law says that patterns of segregation at one locus are independent of those at other loci, so we expect to find equal numbers of all four possible phenotypes, the two parental phenotypes, red eye–normal wing and purple eye–vestigial wing, and the two recombinants, red eye–vestigial wing and purple eye–normal wing.

In fact recombinants appear, but there are far fewer than Mendel's second law of independent assortment predicts. The loci are in some way *linked.* Morgan suggested that this is because they lay on the same chromosome. But in that case, why are there any recombinants at all? If the loci are on the same chromosome and are simply inherited as a unit, then a gamete that carries the P allele should also carry the V allele, and vice versa. The solution lies in the behavior of chromosomes during meiosis. After homologous chromosomes have paired up in leptotene (Fig. 4.5), crossing over occurs, more or less at random, at a number of sites along each chromosome. At a crossover site the DNA strand is cut at the equivalent position in both chromatids, and the cut ends swapped around before rejoining (Fig. 7.4). If, during the meiotic divisions that make the egg in the F1 mothers, one crossover event occurs in the length of chromosome between the vestigal-wing locus and the purple-eye locus, then the egg will carry the vestigal-wing allele without the purple-eye allele, and vice versa. Crossing-over in the stretch of chromosome between the purple-eye and vestigial-wing loci is infrequent, so the gametes shown on the right-hand

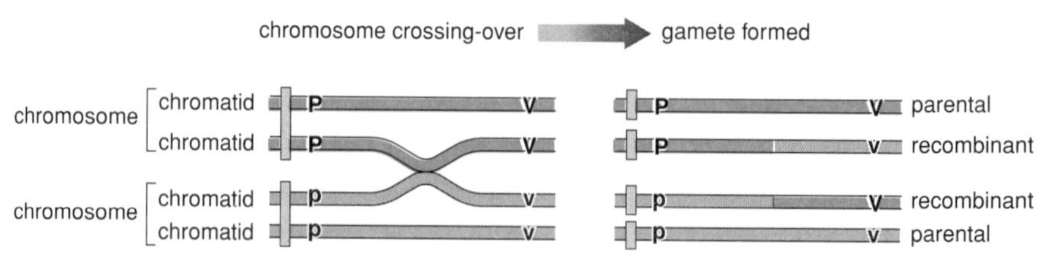

Figure 7.4
Crossing over between the purple-eye and vestigial-wing loci.

side of Fig. 7.4 that will give rise to recombinants are relatively few. Crossing over can be detected in the light microscope, because each point of crossover creates a chiasma (page 69).

The order of the genes along chromosomes can be determined by seeing just how much each pair (or group) of loci deviates from Mendel's second law—how much of a shortage of recombinants there is—and using the degree of deviation to make linkage maps. Pairs (or groups) of loci are followed through the generations. If recombinants appear very rarely, then the loci must be close together. If recombination is common, then the loci are separated by a long piece of DNA in which crossing-over can occur. The proportion of recombinants is a measure of how far apart the two loci are on the chromosome. In practice, most linkage maps are made by considering gene loci in threes rather than in pairs. This makes it easier to arrange adjoining sections of the chromosome map in order by using one of the three loci as a bridge between adjacent sections of the chart. By splicing together the results of thousands of "three-point crosses" in this way, a complete *linkage map* of the *Drosophila* genome has been made (Fig. 7.5).

Needless to say, linkage mapping is not nearly as easy in humans as it is in fruit flies. It has, though, been slow but steady. Linkage mapping is limited by the impossibility of setting up experimental crosses and by our reluctance to have families large enough to look for deviations from Mendel's second law. Many gene loci are known to be linked to the X chromosome; and a rather smaller number have been mapped to each of the autosomes.

MAKING THE PHYSICAL MAP

Linkage mapping depends on sex because crossing-over occurs during meiosis. *Physical mapping* circumvents sexual reproduction. It depends on natural or artificially induced genetic accidents and on deciphering the order of DNA bases.

As we have seen, the first examples of physical mapping happened early in the history of genetics, when a visible difference in the structure of the sex chromosomes correlated with the pattern of inheritance of a gene. Soon there were lots of other hints that the linkage map had its counterpart in the physical organization of chromosomes. Many came from looking at visible changes in the chromosomes themselves.

Sometimes a small *deletion* in a chromosome coincides with a change in patterns of inheritance. In heterozygotes such a deletion may expose the effects of recessive alleles on the homologous segment of the sister chromosome. A deletion of a small section of the X chromosome in *Drosophila,* for example, showed the exact location of several mutants affecting eye color and wing shape.

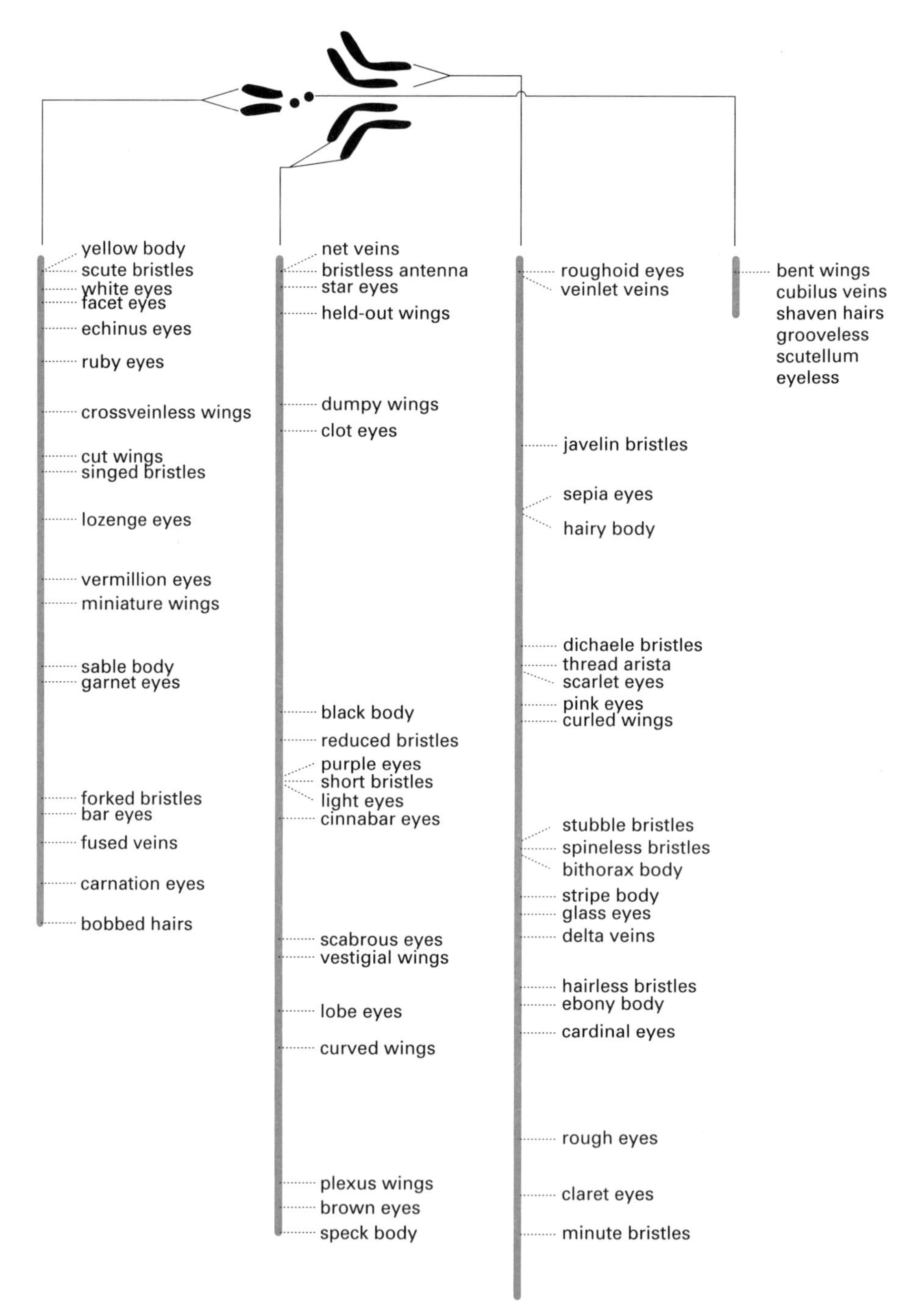

Figure 7.5

A very simplified version of the linkage map for the *Drosophila* genome.

BOX 7.1

Gene Mapping for Schizophrenia

Before molecular genetics, deletions could give only a vague idea of the location of a gene when a section of chromosome large enough to be seen down a microscope had to be missing before its absence could be noticed. Now very small deletions can be identified by sequencing. They can be used to track down genes.

Schizophrenia is a common disease, with one person in a hundred liable to show symptoms at some time. The name schizophrenia means "divided self." It was once thought to be due to possession by demons. It is a devastating disease involving visions, voices and failures of emotion. Sometimes the patient freezes almost solid; sometimes there are inescapable movements. All this can begin in adolescence and continue throughout life—a life, all too often, ended by suicide.

The disease runs in families. A tenth of children who have a schizophrenic parent will manifest symptoms. This might show of course only that members of a family share the same stresses. Psychiatrists once thought that schizophrenia was a sane response to an insane world. Recently studies using twins have indicated strongly that there is a genetic basis for the condition. When one twin has schizophrenia, a nonidentical twin has a 10% risk of developing the disease, but for an identical twin the risk is 50%. Studies using adopted children tell the same story—it is the genetic family, not the adoptive family, that is the best predictor of whether an individual is at risk.

The genetics of schizophrenia is not simple. Many genes are involved, and it may, like breast cancer, be influenced by genes in some families but not in others. Nevertheless, geneticists are beginning to home in on the genes that determine the appearance of schizophrenia. The first clue came from a condition called velocardiofacial syndrome, whose symptoms include heart disease and cleft palate. One in ten velocardiofacial syndrome patients develop severe schizophrenia. When geneticists mapped the same region of chromosome 22 in otherwise normal schizophrenics, they found that two in every hundred had a tiny deletion. This is evidence that one of the genes affecting schizophrenia has its locus on chromosome 22.

In humans physical changes in chromosomes lead to distinctive inborn abnormalities, suggesting where particular genes are. Children born with Cri—du—chat syndrome are mentally retarded and have a characteristic cat-like cry (which explains the odd name of the disease). Each affected child also has a missing section (a deletion) of one particular short segment of chromosome 5. The gene or genes involved can hence be located to this small part of the genome without having to make crosses at all.

Occasionally chromosome pairing during leptotene can go wrong so that non-homologous chromosomes pair. If this happens, crossing over will create chromosome *translocations* in which genetic material is transferred from one chromosome to another. In some creatures (such as mice) such events are common and may be an important force in evolution. Comparison of humans and chimps shows that nearly all the chromosomal differences between the two species involve movements of sections among (and sometimes within) chromosomes, rather than changes in the amount of genetic material.

Translocations can be helpful in gene mapping. They were important in tracking down the gene that pushes a very early embryo onto the path of manhood. The maleness gene (or testis-determining factor) must be somewhere on the Y chromosome. Normally only those with a Y chromosome develop testes and, further emphasizing the physical link between that chromosome and masculinity, babies born without a Y (the XO phenotype, or Turner's syndrome) are female (Box 7.2, page 117).

Very occasionally, though, children are born who are male (albeit sterile) but who have two X chromosomes. In fact a tiny portion of the Y has been translocated to the tip of one of the X chromosomes. The translocated section of Y chromosome was so small as to allow the search for the maleness gene to be narrowed down to one small segment of that chromosome—where it was duly found.

All these instances of physical mapping based on changes in chromosome structure do little more than make an approximate estimate as to where a gene might be. It once seemed that this would be its main role: as an extra tool in the genetic geographers' armory that might supplement linkage mapping. Now everything has changed. Physical mapping has triumphed, and genetics as a science has reversed its direction. In many creatures (notably in humans) it is easier to go straight to the DNA than to make a linkage map based on crossing experiments.

The goal of the mappers is to make the complete chart of the three thousand million base pairs in the human genome. Improvements in technology mean that this may soon be achieved. Already there is a more or less complete map of ordered segments of long pieces of DNA spanning all the human chromosomes. Since most inherited diseases arise from errors in genes that code for proteins, there is naturally more interest in mapping this part of the genome (which turns out to involve around sixty thousand different genes). Already the sequences of the genes which go wrong in most of the common single-gene defects has been established. When the DNA sequence of the first eukaryotic genes was determined, there was an enormous surprise.

INTRONS AND EXONS—THE COMPLEXITY OF EUKARYOTIC GENES

Genes should be simple things: DNA makes RNA makes protein, and a gene codes for the amino acids of a protein by the three-base genetic code. In prokaryotes, indeed, a gene is a continuous series of bases which, read in threes, code for the protein. This simple and apparently sensible system does not apply in higher eukaryotes. Instead, the protein-coding regions of almost all eukaryotic genes are separated by noncoding regions. The protein-coding regions of the "split" genes are called *exons*. The regions between are "intervening" sequences or *introns* for short. Introns are often very long compared to exons. As happens in prokaryotes, messenger RNA complementary to the DNA is synthesized but then the introns are spliced out before the mRNA leaves the nucleus (page 159). In fact there is an evolutionary rationale to this apparently perverse arrangement. As we will see (page 225), a single protein is often composed of a series of domains, with each domain performing a different role. The breaks between exons usually correspond to domain boundaries. During evolution re-ordering of exons has created new genes that have some of the exons of one gene, and some of the exons of another, and hence generates novel proteins composed of new arrangements of domains, each of which still do their job.

THE MAJOR CLASSES OF EUKARYOTIC DNA

Even taking introns into account, much of the DNA in eukaryotes does not represent genes. Interspersed throughout the human genome are sequences that occur many times. The are known as *repetitious DNA*. Some sequences are repeated more than a million times. These are called satellite DNA. The repeating unit is usually several hundred base pairs long and many copies are often lined up next to each other in tandem repeats. Shorter sequences of about 30 base pairs are also found as tandem repeats. These are called minisatellites, and occur much less frequently than does satellite DNA. Sometimes DNA that encodes RNA is also tandemly repeated. These include the genes that code for pre-ribosomal RNAs (about 250 copies/cell), transfer RNAs (50 copies/cell), and histone proteins (20–50 copies/cell). The products of these genes are required in large amounts. However, most protein-coding genes occur only once in the genome as single-copy genes.

Many genes have been duplicated at some time during their evolution. Mutation over the succeeding generations causes the initially identical copies to diverge in sequence and produce what is known as a *gene family*. Its members usually have a related function—the immunoglobulin gene family, for example, makes antibodies. Different members of a gene family sometimes encode proteins that carry out the same specialized function but at different times during development. Figure 7.6 shows the α and β globin gene families. The β-globin cluster is on human chromosome 11, and the α-globin gene cluster is on human chromosome 16. Hemoglobin is composed of two α-globins and two β-globins. The gene clusters encode proteins expressed at specific times dur-

α-globin gene cluster - chromosome 16

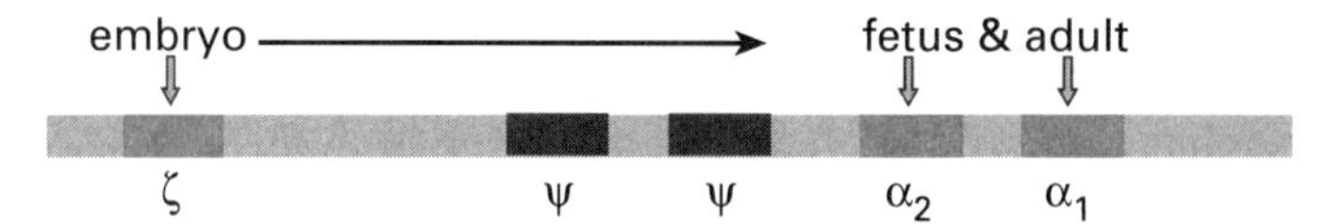

β-globin gene cluster - chromosome 11

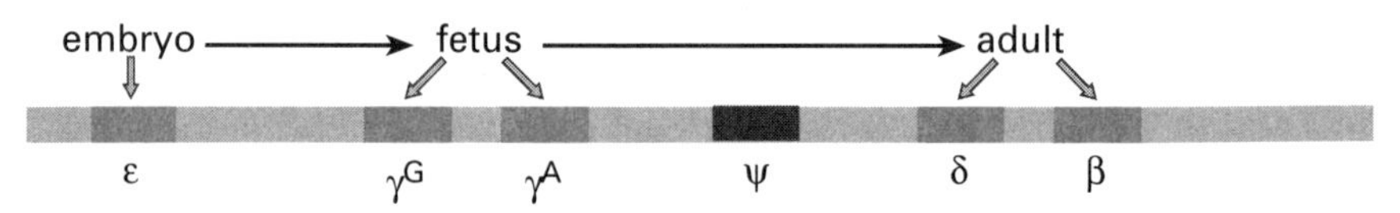

Figure 7.6

The human α- and β-globin gene family clusters. In an adult, the expression of δ is very much less than that of β.

ing development; from embryo, to fetus, to adult. The different globin proteins are produced at different stages of gestation to cope with the different oxygen transport requirements during development. The duplication of genes and their subsequent divergence has led to the expansion of the gene repertoire, the production of new protein molecules, and the elaboration of ever more specialized gene functions during evolution.

Some sections of DNA are very similar in sequence to other members of their gene family but do not produce mRNA. These are known as *pseudogenes*. There are two in the α-globin gene cluster (Ψ in Fig. 7.6). Pseudogenes may be former genes that have mutated to such an extent that they can no longer be transcribed into RNA. Some pseudogenes have arisen because an mRNA molecule has inadvertently been integrated into the genome. Such pseudogenes are immediately recognizable because some or all of their introns were spliced out before the integration occurred. Some have the poly A tail characteristic of intact mRNA (page 159). These are called processed pseudogenes.

SUMMARY

1. Mendel saw genes as independent "particles," with no association with each other or with a cellular structure. The discovery of sex-linked inheritance showed that genes were on chromosomes. Genes that are located on the same chromosome are linked.

2. The order of genes relative to each other and to the chromosome

can be mapped. Linkage mapping involves crosses between individuals differing at two or more loci. Any shortage of recombinants suggests that the loci are linked. Recombinants are only formed when crossing over occurs between the loci during meiosis. The greater the deficiency of recombinants, the closer the loci.

3. Physical mapping depends on reading along the DNA chain in order to associate genes with particular chromosomal locations. Sometimes a small deletion in a chromosome produces a particular phenotype, indicating the location of a gene. Sometimes one chromosome becomes attached to another (a translocation) changing the pattern of inheritance of a particular gene.

4. New methods of sequencing the DNA will soon allow the complete physical map, the order of the DNA bases, to be made. The physical and genetic maps are remarkably different. Large amounts of DNA are not directly copied into gene products. Sometimes introns are inserted into the expressed sequences (the exons) of the working genes. In addition there is a large amount of DNA, whose function is not obvious, in between the expressed genes. This includes much repetitious DNA, whose sequence is multiplied many times.

5. Functioning genes may be found in repeated groups of slightly diverging structure, gene families, either close together or scattered over the genome. Some of the family members have lost the ability to operate—they are pseudogenes.

BOX 7.2

On Being a Man: Genes and Maleness

In humans, being a male consists largely of having a Y chromosome. Those born without this crucial element (XX, and unusual karyotypes such as XO) are female; those born with one or more (XY, and unusual karyotypes such as XXY and XYY) are male. The Y chromosome has about 60 million DNA base pairs, but much of this consists of short repeated untranscribed sequences so that the segment conferring maleness is a fairly small one. There is much less variation on the Y—both between individuals and between people from different parts of the world—than on other chromosomes, suggesting that all of them are inherited from a male ancestor who lived in the relatively recent past, a genetic Adam.

The rule that Y means male is broken in a few men who have an XX karyotype. This very rare phenomenon is due to a translocation; a small segment of a Y chromosome becomes attached to an X and is inherited in concert with it. This was the clue that allowed the maleness gene itself, the so-called testis-determining factor, to be identified and cloned. Once discovered, it was named SRY (for sex-determining region, Y chromosome). It is a small gene, only a couple of hundred bases long, encoding a DNA-binding protein (page 163) that controls expression of a number of genes, some of which in turn control other genes. The presence of the SRY gene product therefore acts as a switch that diverts the very early embryo into the male, rather than the female, pattern of development.

Other species do things differently. In *Drosophila* males are XY and females XX, but it is the balance between the number of X chromosomes and the number of autosomes that is important rather than a single specific Y-borne gene. In alligators sex is determined by the temperature at which the egg is incubated.

CHAPTER

8

GENE MUTATION

Without diversity there could be no genetics. Mendel's work (Chapter 6) depended on inherited differences. Even physical mapping of the human genome relies largely on identifying families who have a genetic constitution distinct from that of the rest of the population. All this variation has arisen through errors in the copying process as genetic information passes from generation to generation. These mistakes—*mutations*—have taken place since the first life, and the first gene, appeared more three thousand million years ago.

However, inheritance is, in general, stable. If it were not, Mendel could not have worked out his ratios and molecular biologists would be confused by massive differences between parents and offspring. This balance between stability and change has molded the living world. Mutation is the raw material of evolution. What is it, how often does it happen, and why is it there? Before the development of modern genetic technology, these questions seemed easy to answer. Now everything has been transformed. Sequencing the DNA shows that mutation is a much more complicated and surprising business than once seemed reasonable to expect.

MEASURING THE MUTATION RATE

Sometimes mutations can be counted directly. Offspring show a phenotype that has never before been seen in their family. Charles Darwin recorded such an event in sheep. Suddenly one lamb in an otherwise normal flock was born with short stubby legs. The condition was passed on to its own offspring. Achondroplastic dwarfism in humans (page 98) can behave similarly—normal parents have a child with short legs. Since the allele is dominant, only one copy is needed to manifest its effects. Any new mutation can be seen at once. In principle, all that is needed to measure the mutation rate is to count the numbers of such children born to normal parents. For instance, in Fig. 8.1 one lamb in a group of 5,000 born to normal parents has achondroplastic dwarfism. Since the condition is dominant, neither parent can have possessed the allele, which must have arisen by mutation in the germ cells of one parent. If this frequency of dwarf births were repeated in a larger sample, one would be able to estimate the mutation rate as 2×10^{-4} per generation, or half this, 10^{-4}, per gamete.

There have been many attempts to measure the rate of mutation in this way. There are problems in so doing. For example, errors at distinct loci may be counted as if they were independent mutations at a single gene, leading to an overestimate of the mutation rate at that locus. Just counting new dominant mutations as they arise is also ineffective in measuring the overall mutation rate because it misses recessive mutations, and those alleles that kill their carriers before they are born.

The problem can be resolved in various ways. Haploid organisms such as bacteria are much used for the study of mutation because there is no difficulty with dominance: New variants are not masked by a copy of the unchanged allele at that locus. Bacteria also have many *conditional mutations* that can be used as a screen for genetic change. Such mutations alter their carriers' ability to cope with a certain environment. This allows the cunning biologist, by manipulating the circumstances under which they are grown, to identify individuals who have undergone a mutation. One producing, for example, resistance to the antibiotic streptomycin can be identified simply by sowing bacteria on a medium that contains that substance. Only those descended from a mutant individual are able to grow and form colonies. By comparing patterns of growth on normal medium (upon which both mutated and unmutated

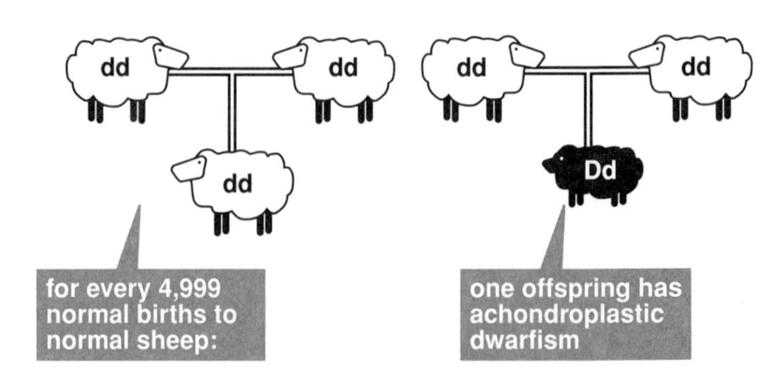

Figure 8.1
The rate at which a dominant allele arises by mutation (here, about 2×10^{-4} per generation) can in principle be estimated directly from observing the rate of appearance of the trait in a normal population.

colonies can flourish) with that on the antibiotic it is possible to work out the mutation rate at that locus directly.

The conditional mutant approach can also be used in *Drosophila.* High temperature, rather than antibiotics, is often used as the screen. Mutants fail to grow at the higher, restrictive temperature but develop normally in cooler, permissive conditions. Sometimes it is possible to search out mutations by looking directly at genes (or their immediate products) rather than at a phenotype whose attributes are a long way away from the locus itself (Box 8.1).

BOX 8.1

Hiroshima as a Genetic Experiment

On August 6, 1945, an atom bomb exploded over Hiroshima. Tens of thousands were killed, either immediately or as a result of radiation sickness. The latter was a surprise: one physicist who helped design the bomb said "I expected that anyone harmed by radiation would have been killed by a brick first."

Large doses of radiation cause mutations in both the germ line and in body cells. They will therefore tend to cause cancer. Experiments revealed the effect on germ cells on *Drosophila* (page 124), and many of the early workers on X rays died of cancer. But what is a safe radiation dose? At the time of the bomb, estimates of the amount of radiation needed to double the human mutation rate varied by more than a hundred times.

From 1947 onward, tens of thousands of children born to bomb-irradiated parents were examined for new mutations. In the first few years, only gross changes—new dominant alleles for disease, or chromosomal rearrangements—could be looked for. Now there is biochemical technology. Two-dimensional electrophoresis detects differences in the charge or size of protein molecules that might arise from mutation. Proteins in the blood are separated by applying a powerful electric field to a sample placed on a gel. Then the gel is rotated through 90 degrees, the proteins denatured (page 230) by chemical means, and the field applied again. Finally, the proteins are stained so that their position on the gel can be seen.

With some exceptions it is true to say that if a cell contains two working copies of a gene, one on each of two sister chromosomes, then both will be transcribed and translated into protein. If the genes differ enough so that the charge or size of the proteins are significantly different, then two distinct spots will appear on the gel. Any shift in the position of a particular protein in a child compared to its parents must be due to a new mutation. The mutation usually produces no change in the phenotype of the child, but two-dimensional gel electrophoresis will nevertheless detect it. The figure illustrates this approach. A typical gel would show a thousand or so spots, but for simplicity we show the products of only four genes that are identical in the two parents. Their child has the same four dots—plus one more. The child has inherited one unmutated allele at this locus from one parent producing a protein spot at the same place on the gel as for the parents, while the other allele has mutated producing a protein spot at a new location.

Well over a million genes were looked at. The results were simple and fairly reassuring.

Four mutations were found in a control group whose parents had not been exposed to the bomb, two in the exposed group. There was no evidence that the radiation from the bombs had caused a detectable increase in the human mutation rate. However, new work on DNA mutations in the children of those living around the Chernobyl nuclear power station at the time of the accident suggests that there was a marked increase in mutation rate, so that the news about radiation is not completely comforting.

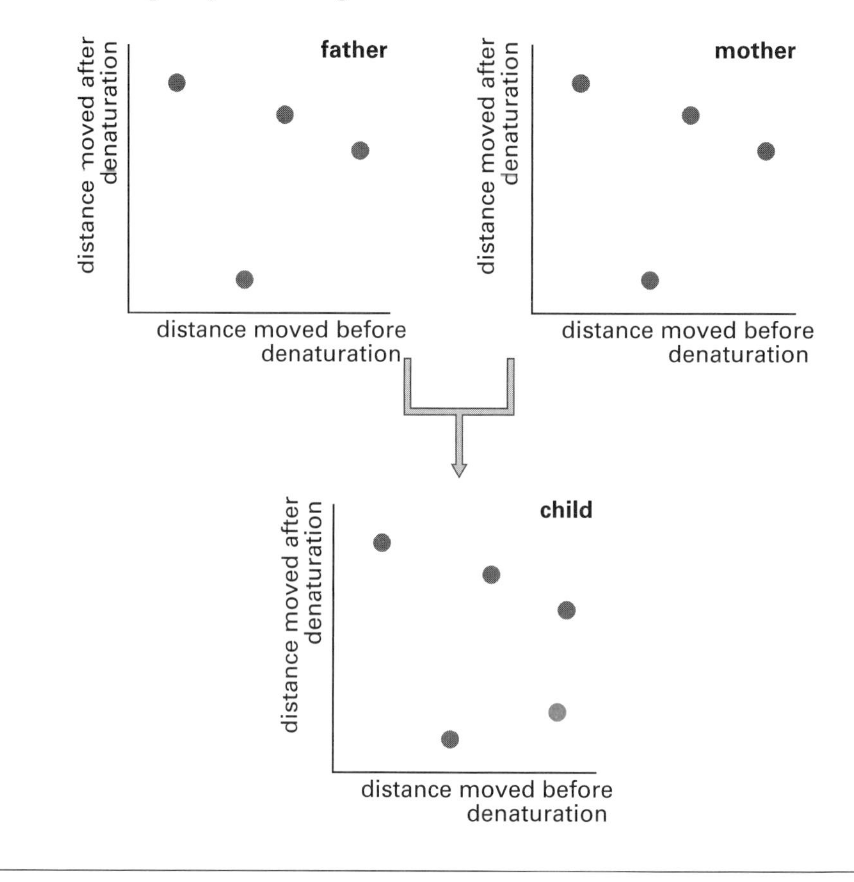

It is now possible to look for mutational changes in DNA itself. The most straightforward approach is to identify a family with an inherited disease and to track down and sequence the DNA change involved (page 178). The parents of a child with a particular abnormality can be tested to see whether they are themselves unknowing carriers of the altered gene or whether a new mutation is at fault.

There are also various indirect methods of looking for new mutations. Genetic technology is powerful, but it is expensive, slow, and, most important, can usually detect mutations only at the specific loci being tested. There are other ingenious ways of looking for recessive mutations, often for many at the same time. Most depend on attempts to bring together, in homozygous condition, whole lengths of chromosome that may reveal hidden genetic damage. New recessive mutations will show their effects either as changed in phenotype or (more often) by killing the individuals carrying them.

THE "NATURAL" RATE OF MUTATION

How often do mutations happen? Certainly the picture that emerges when waiting for new phenotypes to appear suggests that the answer is "not very often." There are many estimates of just what the mutation rate might be. They vary widely. Often there are large differences between particular genes. Sometimes the reason is obvious. For instance, large genes often have a higher mutation rate than small genes do. Muscular dystrophy is due to a mutation in a huge gene, and its mutation rate is relatively high. However, other differences in mutation rate are not so easy to explain. Table 8.1 gives some estimates of mutation rate in a variety of creatures.

MUTAGENESIS

Although mutations, generally speaking, are rare, early in the history of genetics it became clear that their rate of appearance could be greatly increased by

TABLE 8.1 The mutation rate measured in different organisms and at different loci

Organism	Mutation	Value
Bacteriophage T2 (a virus)	Lysis inhibition	1×10^{-8} per gene replication
	Host range	3×10^{-9} per gene replication
Escherichia coli (a bacterium)	Lactose fermentation	2×10^{-7} per cell division
	Histidine requirement	4×10^{-8} per cell division
Chlamydomonas reinhardtii (a simple plant)	Streptomycin sensitivity	1×10^{-6} per cell division
Neurospora crassa (a simple fungus)	Inositol requirement	8×10^{-8} per asexual spore
	Adenine requirement	4×10^{-8} per asexual spore
Drosophila melanogaster (insect)	White eye color	4×10^{-5} per gamete
Human (mammal)	Achondroplasia	1×10^{-4} per gamete
	Duchenne muscular dystrophy	6×10^{-5} per gamete
	Hemophilia A	3×10^{-5} per gamete
	Huntington's disease	1×10^{-6} per gamete

Source: Data from R. Sager and F. J. Ryan, *Heredity* (New York: Wiley, 1961).

external agents. Since most of the mutations produced in such experiments were harmful, this led to an era in which mutation was seen as a damaging process, interfering with an otherwise unblemished set of genes.

In the 1930s an ingenious series of experiments with *Drosophila* showed that X rays—previously thought to be innocuous—were powerful agents of mutation. Like other *ionizing radiations* their wavelength is short so that they can enter cells and ionize some of the molecules therein. The experiments depended on giving a dose of radiation to a fly and then, by controlled breeding, producing individuals homozygous for the chromosomes irradiated in their ancestors. Any new recessive alleles will then show their effect—usually damaging, and often lethal.

Figure 8.2 shows the relationship between the incidence of new lethal alleles and X-ray dose. There is a striking relationship between them. Even a small measure of radiation increases genetic damage: a finding that gave rise to concern that no increase in radiation over natural levels could safely be accepted. X rays are not the only forms of radiation that can cause injury. Even heat increases the mutation rate (which may partly explain why the testes—the site at which mutations may be particularly important—are held outside the body, where life is cool). Ultraviolet light is a potent *mutagen* as well. Although it cannot penetrate to the germ cells, it can damage the DNA of cells in the skin.

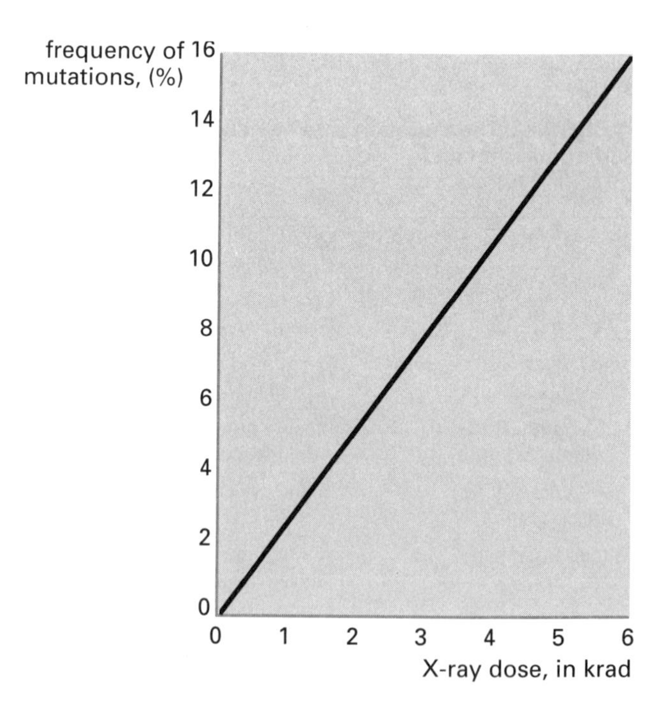

Figure 8.2

In *Drosophila* there is a linear relationship between X-ray dose and the appearance of new lethal alleles.

Most people can, if they wish, escape most radiation. However, the same is not true for chemicals. The first hint that chemicals could cause mutations came from work on the war gases, such as mustard gas. These were known to cause long-lasting burns similar to those produced by large doses of X rays. For most chemical mutagens, the relationship between dose and damage is not as simple as it is for X rays. Certainly, some chemicals are potent agents of genetic injury. One of the first—acridine mustard—can increase the mutation rate by more than five thousand times, and other chemicals are even more dangerous. Some chemicals actually damage DNA directly. The alkylating agent ethyl methyl sulphonate, for example, adds an ethyl group to one or more bases. Others, the base analogues such as the dangerous 5-bromo uracil, insinuate themselves into the DNA chain, interfering with the pairing mechanism. Certain substances such as acridine orange are not chemically related to purines or pyrimidines but have about the same shape and can creep into the replicating chain.

There is a close relationship between mutation and cancer. The biochemist Bruce Ames tested whether a series of chemicals (such as those found in tobacco smoke) that were known to cause cancer would cause mutations in a test stock of the bacterium *Salmonella typhimurium.* It was immediately obvious that there was a striking relationship between chemical *carcinogenesis* and mutagenesis. This same *Ames test* has been applied to thousands of chemicals to see if they can cause mutations in bacteria (and might hence be carcinogenic). Hundreds of potential carcinogens have been found.

Some carcinogens—DDT and the polychlorinated biphenyls (PCB)—are artificial, and they have caused much public alarm. However, it is less widely known that many of the most potent carcinogens are perfectly natural. Black pepper and other spices, coffee, celery, and even lettuce contain mutagens. Most of these poisons are important in the plant's defenses against insect pests; their presence in foods that have been eaten for thousands of years puts fears that civilization will end in a mutational nightmare into a more reasonable context. On an even more positive note, Ames tests show that some foods—particularly fruits and vegetables—contain substances that are *antimutagens*, reducing the damaging effects of other toxic chemicals.

Repairing Mutations

The chemical instability of the DNA bases is enough to destroy the genetic message quickly. Without correction, about one cytosine base in a thousand would mutate to uracil each generation (page 140)—which would rapidly lead to chaos. Evolution has come up with a series of repair mechanisms to rectify the damage as it happens.

Figure 8.3

The reactions catalyzed by superoxide dismutase and catalase.

$$2O_2^{\cdot -} + 2H^+ \xrightarrow{\text{superoxide dimutase}} O_2 + H_2O_2$$

$$2H_2O_2 \xrightarrow{\text{catalase}} 2H_2O + O_2$$

Some mechanisms involve enzymes that break down chemicals in the cell and are potent enough to attack DNA. The enzyme *superoxide dismutase* breaks down chemically active oxygen radicals to produce hydrogen peroxide, which is itself destroyed by the enzyme catalase in the peroxisomes (Fig. 8.3). As we will see in the next chapter, there is also a whole series of enzymes that repairs damage to the DNA.

The repair machinery is very effective. Mutant strains of bacteria lacking repair enzymes have a mutation rate that is a thousand times the normal. Just how important they are in humans is shown by the effects of the inherited disease *Xeroderma pigmentosum* (Box 9.1). Patients with this condition develop malignant skin cancers that can kill them. They lack part of the excision repair mechanism, which cuts out and replaces the inappropriate chemical bonds between thymines that form upon exposure to ultraviolet light (page 140). If this system is not working, then any thymine dimer that is formed remains as a mutation. Since so many cancers are now known to be closely related to damage to DNA, there is talk of screening people to check how active their repair machinery is before allowing them to be exposed to carcinogenic chemicals in their jobs.

THE NATURE OF MUTATION

All mutations are changes in DNA. Most were once assumed to change a single base at a time. Now, although such simple changes certainly exist, mutation is known to be a much more complex process, with many surprising attributes of its own. The first hint of just how complicated the process may be came from looking at mutations producing gross changes in the chromosomes.

Some mutations arise through rearrangement of whole sections of genetic material. Occasionally, there is *polyploidization*, the multiplication of whole chromosome sets. This has often happened in evolution, particularly in plants. The chromosomes of many familiar crops (such as wheat) show that they arose from the doubling of the number of chromosomes of an ancestor. Not surprisingly, such drastic changes are usually lethal in humans. Occasionally, though, a severely ill child is born and survives for long enough to show that some of its cells have twice the normal number of chromosome sets—they are tetraploid.

More often there is a mutational change in the number of copies of just one chromosome to give an aneuploid. In humans such mutational changes are usually lethal early in development. In fact the majority of fertilizations end in spontaneous abortion, usually before the woman knows she is pregnant. Many lost fetuses are *aneuploid*. Some, however, survive. Trisomy-21, or Down's syndrome, is due to a mutational error (nondisjunction: a failure of chromosome separation) in the formation of sperm or egg. The child has an extra copy of chromosome 21, one of the smaller chromosomes. Such children often survive to adulthood, although they do show signs of abnormality. About one birth in six hundred produces a Down's child.

Sometimes a piece of chromosome shifts to a new home—a *translocation* (page 113). This explains why occasionally children are born with all the symptoms of Down's syndrome but apparently with no extra chromosome 21. A segment of this chromosome (which must contain the crucial gene for the condition) has become attached to another because of a mutational change in an earlier generation. Other, less drastic chromosome changes also exist. As we have seen (page 111), *deletions* are the loss of a section of a single chromosome. Sometimes a deletion has an immediate phenotypic effect such as in the *Drosophila* wing mutation notch which is the visible manifestation of a small heterozygous deletion. Deletions in human chromosomes that are large enough to be seen by the microscope are usually damaging. Cri du chat syndrome (page 113) results from a tiny deletion of a piece of chromosome 5. *Duplications* involve the gain of a short section of material. The *Drosophila* mutant Bar (which leads to a change in the shape of the eye) can be seen, on the chromosomal level, to result from the doubling-up of a short piece of the X chromosome. *Inversions* are more subtle mutational events. They involve two simultaneous chromosome breaks. The section between the breaks swivels round and is re-inserted in reverse order. Some translocations, deletions, duplications, and inversions involve sections of chromosome long enough to be seen with a microscope. Others occur on a smaller scale and can only be detected by DNA sequencing (page 178). This technology, as well as revolutionizing our view of the genetic map, has revealed a gamut of new and surprising mechanisms of change at the molecular level that are changing our views of what mutation actually is.

Some DNA mutations are simple: they involve single-base *substitutions*. Diseases such as sickle cell anemia are of this kind. This is the commonest inherited illness in the world and is due to the mutation of an adenine to a thymine in β-globin. As we have described (page 86) this leads to a change in the protein sequence, with the replacement of a glutamic acid by a valine. In turn, the solubility of the hemoglobin is affected, leading to severe disease in sickle-cell homozygotes.

Single-base mutations have happened again and again. Two people—or two fruit flies—are very likely to differ in the details of DNA sequence because of mutations that have taken place recently or in the distant past. Such polymorphisms are most frequent in the less functional parts of the DNA. The third position of codons, for example, often differs from person to person but codes for the same amino acid (page 84). The same is true for the DNA sequence within introns (page 114). Polymorphisms must, once again, be due to mutations, perhaps in the distant past. No doubt such events also take place at other more crucial sites, but if the changed base is less effective in doing its job than what went before, then the individual carrying it will have a reduced chance of surviving and of passing on the altered gene.

Sometimes a single-base change in the DNA can have a drastic effect. If a coding triplet is converted to a "stop" codon (page 85), the growing protein chain will come to an abrupt halt. Often this is lethal early in development, but there are some inherited diseases that arise because of this form of mutation. The thalassemias—which are common around the Mediterranean—are mutations of this kind. They stop the synthesis of the α or β chain of hemoglobin at one or other point along its length.

Mutation Is Not Random

Remarkably it seems that third parties (such as viruses or parasitic and unstable DNA within the genome itself) play a crucial part in causing mutation. Mutation is as much an inherent property of the genome as it is a response to external damage. This revolution in our understanding began with some work on maize. Maize is useful to geneticists. Each cob is effectively a family of individual seeds held together. In the 1930s a stock was found in which the mutation rate (in this case from albino back to colored seeds) was so enormously high that separate seeds on the same cob could look quite different because of mutations during early development. The effect is attractive, and these cobs are sold as ornaments. Crosses with a normal stock showed that this tendency toward genetic instability was itself inherited. A mutator gene (a gene which increases the mutation rate at one or more other loci) was involved.

Many such genes—some of which increase the mutation rate at several loci simultaneously—are now known. They have an additional peculiarity: Mendelian crosses show that they can hop around the genome, sometimes appearing in one place and sometimes in another. They are known as transposable elements. Remarkably enough, some show "infectious heredity." When a *Drosophila* that lacks a particular transposable element is crossed with one that carries it, the element can jump into the chromosomes that originated in the non-carrier parent. As it does, there may be an increase in the mutation rate at other loci. All this suggests that a new and bizarre form of mutation is at work.

The study of change at the molecular level shows that such curiosities are, in fact, almost the rule. Many genetic errors (in humans as much as in maize) result from the interspersion of autonomous genetic elements into working genes. Artificial mutagens frequently act by mobilizing them. The maize transposable element is fairly typical. It is about 4,500 base-pairs long and, in its complete form, contains within itself the code for an enzyme called a *transposase* that copies the DNA element to another part of the chromosome. Often the enzyme coding section is lost, rendering the element immobile—until its carrier is crossed with another plant containing a complete element. This restores the element's reproductive machinery and it is able to relocate, causing genetic damage as it does so. There is an analogy with computer viruses, lengths of code that when activated are copied to another location in the computer's database, or even onto another computer, damaging the database in the process.

Transposable elements in the fruitfly *Drosophila* were discovered through a sudden and, at first, inexplicable increase in the mutation rate at many loci when flies from North America were crossed with those from other parts of the world. The effect was due to the mobilization of a "P-element," a piece of transposable DNA. Remarkably enough, crosses between modern American stocks and those collected in the same place 50 years earlier and held for many generations in the laboratory showed a similar effect. The older stocks lacked P-elements, which must hence have invaded *Drosophila melanogaster* quite recently. They may have done so from a South American species of fruit fly.

Unstable DNA of this kind is important in causing human mutations. It helps explain some previously baffling patterns of inheritance. Parts of the human genome may be far more fluid and open to change than classical geneticists ever imagined. The commonest single cause of inherited mental illness is the *fragile-X syndrome*. Children born with the disease may be severely retarded, although some show only minor symptoms. Fragile-X children usually have a distortion of the X chromosome—two small pieces that seem to be breaking away from one end of its long arm. The sex-linked nature of the defect means that boys are more likely to be affected than are girls. Families with the condition show several unusual features. Just as for any sex-linked locus, normal women may have affected sons. However, apparently normal fathers can have affected daughters. Even more oddly, in those families in which several generations are available for study, the severity of the disease can increase or decrease over two or three generations. What is going on?

The fragile-X mutation is unpredictable because mobile DNA is involved. Just next to the actual locus involved is a region in which a repeated triplet of the DNA bases CCG is common. In normal people there are between six and sixty repeats of the sequence. Those with fragile-X syndrome have several thousand—and the more copies of the CCG repeat, the more severe the symptoms. The rare males who pass on the disease without themselves showing much sign of it have around a hundred CCGs. Their phenotype is delicately poised on the boundaries of normality but—if the number of CCGs goes up by mutation as their sperm is formed—their children are in danger of being affected.

This surprising new form of mutation helps explain the genetical phenomenon of *anticipation*, described, but largely forgotten, many years ago. Within a lineage the symptoms of a disease may get worse with each generation. This effect is due to repeated mutation and to the accumulation of unstable DNA elements. Some families are luckier: the effects of the gene become less severe as one generation follows the last.

Such volatile mutation is involved in several other inborn errors. Huntington's disease (page 94) is due to the insertion of copies of a repeat of a CAG triplet into the working gene itself. Those with the disease have more copies of this triplet, and the age at which the symptoms first show themselves depends on just how many copies there are. The number of copies is more likely to increase by mutation if the gene is passed through the father than through the mother.

The idea that mutation can arise from the movement of sections of DNA with some autonomy has gained fresh impetus from the discovery that many cancers are due to mutations in certain genes (oncogenes and tumor suppressor genes; page 70) controlling cell division. In certain forms of the disease (such as colorectal cancers), there is genetic instability in the cancer cells with the expansion of tri-nucleotide repeats. What is cause and what is effect here is not clear. Quite often in cancer the damage is done because a virus imports foreign DNA into the cell, which disrupts the function of a working gene. Gene mutation is beginning to look like a much more dynamic and active process than seemed possible only a few years ago.

SUMMARY

1. Mutations are mistakes in copying the genetic message. The arrival of dominant mutations can be seen directly. More often mutation must be inferred using various screening tests. Conditional mutants, for example, show their presence only in certain environments. In other experiments two chromosomes that may each carry a recessive mutation are brought together by making crosses in the hope that the effect of the mutations will become manifest.

2. Mutagens increase the rate of mutation and include ionizing radiation, ultraviolet light and a variety of chemicals. The Ames test used bac-

teria to show that many chemicals known to be carcinogens were also bacterial mutagens, and it showed that certain chemicals were antimutagens.

3. We are not defenseless against mutation. A variety of enzymes including superoxide dismutase destroy mutagens. Damaged DNA can be repaired. Individuals in which of one or more DNA repair enzymes is congenitally absent have very high rates of both mutation and cancer.

4. There are many different kinds of mutation. They include changes in the number of chromosome sets (polyploidy) or the presence or absence of a particular chromosome (aneuploidy). The location of chromosome segments can change in the processes of translocation, deletion, duplication, and inversion.

5. DNA mutations include all these plus a variety of substitutions, sometimes leading to stop codons that terminate the synthesis of the protein chain. Many mutations are due to the insertion of repeated sequences of DNA, often mediated by the action of transposase enzymes that allow segments of genetic material to move around the genome. Some diseases, such as the fragile-X syndrome, are due to the interposition of short segments of mobile DNA. Sometimes more copies enter with each generation, causing the symptoms to become more extreme in a process called anticipation.

BOX 8.2

Motoneurone Disease and Superoxide Dismutase

Amyotrophic lateral sclerosis (ALS), also known as motoneurone disease, is a progressive debilitating nerve disease characterized by the death of more and more of the nerve cells called motoneurones (page 335). The baseball star Lou Gehrig was a famous sufferer. Families with a history of ALS are known, and in these the genetic lesion has been identified: the superoxide dismutase gene is abnormal.

Is familial ALS due to an absence of superoxide dismutase that cannot remove superoxide radicals, leaving the cell open to oxide radical damage? Surprisingly this is not the answer. The mutant enzyme performs as well as normal superoxide dismutase in destroying superoxide. Furthermore a cell that is engineered to express the mutant superoxide dismutase in addition to its own normal enzyme dies early. Somehow the mutant SOD is actively damaging to nerve cells.

This is as far as the story is known at present. Future discoveries may reveal more about the role of superoxide dismutase in normal cells and help in the development of a cure for this dreadful disease.

CHAPTER
9

DNA REPLICATION AND DNA REPAIR

The genetic material DNA must be faithfully replicated every time a cell divides to ensure that the information encoded in DNA is passed unaltered to a daughter cell. DNA molecules have to last a long time compared to RNA and protein. Some of your DNA molecules are as old as you are. The sugar-phosphate backbone of DNA is a very stable structure because deoxyribose is much more resistant to chemical attack than is the sugar ribose found in RNA. The bases themselves are protected from chemical attack because they are hidden within the DNA double helix. Cells have evolved safety mechanisms to safeguard the base sequence of their DNA. These ensure that mutation is kept to a minimum. Repair systems are essential for both cell survival and to ensure that the correct DNA sequence is passed on to daughter cells. This chapter describes how new DNA molecules are made during chromosome duplication and how the cell acts to correct base changes in DNA.

DNA REPLICATION

DNA Replication Is Semiconservative

During replication the two strands of the double helix unwind, and each acts as a *template* for the synthesis of a new strand. This generates two double-stranded daughter DNA molecules each identical to the parent molecule. The base sequence of the new strands are *complementary* in sequence to the template strands upon which they were built. This means that G, A, C, and T in an old strand cause C, T, G, and A, respectively, to be placed in the new strand. That replication does indeed happen in this semiconservative way was proved experimentally in 1958 by Matthew Meselson and Franklin Stahl.

They grew the bacterium *E. coli* in a medium containing the heavy isotope ^{15}N that could be incorporated into new DNA molecules. After several cell divisions they transferred the bacteria, now containing "heavy" DNA, to a medium containing only the lighter, normal isotope ^{14}N. Any newly synthesized DNA molecules would therefore be lighter than the original parent DNA molecules containing ^{15}N. The difference in density between the heavy and light DNAs allows their separation using very high speed centrifugation. The results of this experiment are illustrated in Figure 9.1. DNA isolated from cells grown in the ^{15}N medium had the highest density and migrated the furthest during centrifugation. The lightest DNA was found in cells grown in the ^{14}N medium for two generations, whereas DNA from bacteria grown for only one generation in the lighter ^{14}N medium had a density half way between these two. Because the heavy DNA was replaced by DNA of intermediate density after one cell division, the two heavy parental strands must have separated during replication, with each acting as a template for a newly synthesized light strand.

The DNA Replication Fork

Replication of a new DNA strand starts at specific sequences known as *origins of replication*. The small circular chromosome of *E. coli* has only one of these, whereas eukaryotic chromosomes, which are much larger, have many. At each origin of replication, the parental strands of DNA untwist to give rise to a structure known as the *replication fork* (Fig. 9.2). This unwinding permits each parental strand to act as a template for the synthesis of a new strand. The structure of the double-helix and the semiconservative nature of DNA replication pose a mechanical problem. How do the two strands unwind and how do they stay unwound so that each can act as a template for a new strand?

PROTEINS OPEN UP THE DNA DOUBLE-HELIX DURING REPLICATION

The DNA molecule must be opened up rapidly if replication is to succeed. The helix is a very stable structure and in a test tube the two strands separate only when the temperature reaches about 90°C. In the cell the combined actions of several proteins help to separate the two strands. Much of our knowledge of

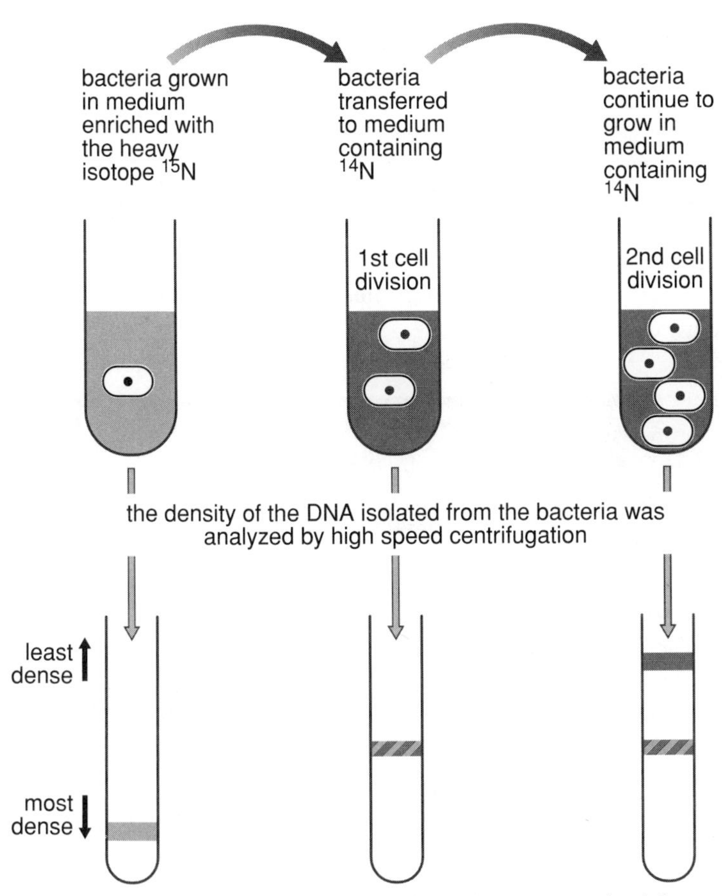

This experiment demonstrated that the two strands of the parental DNA unwind and each acts as template for a new daughter strand

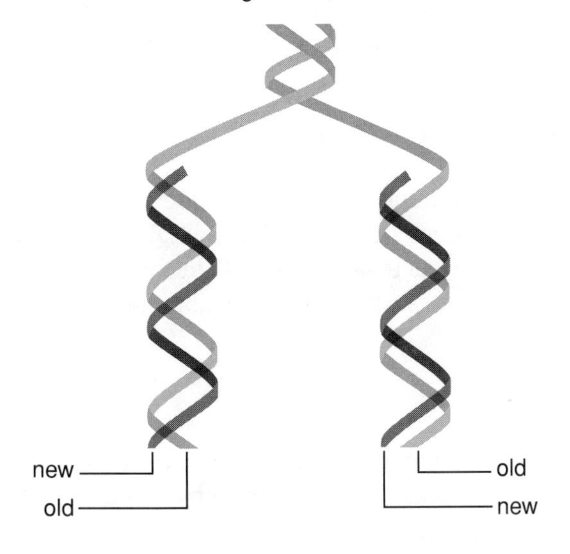

Figure 9.1
DNA replication is semi-conservative. The Meselson and Stahl experiment.

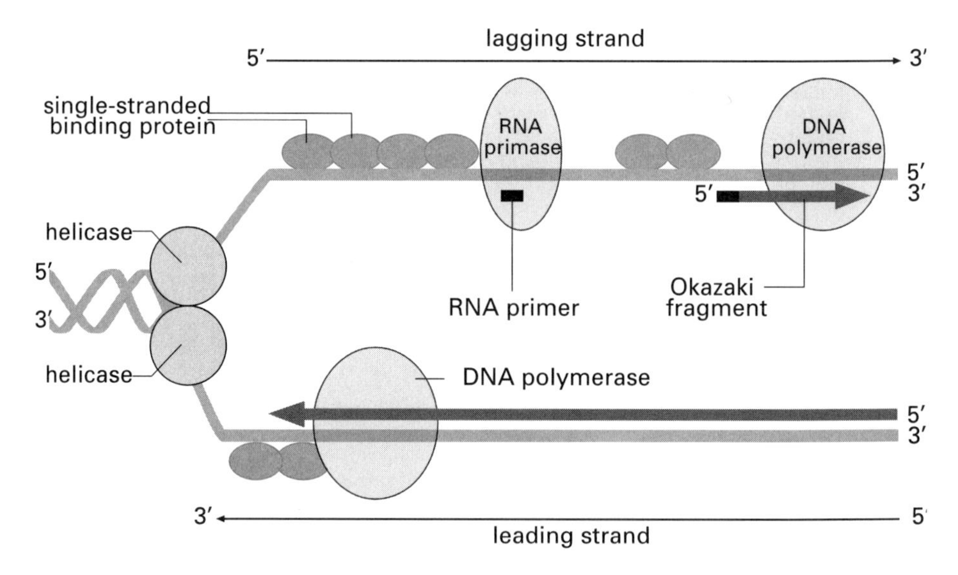

Figure 9.2
DNA replication. The helicases, and the replication fork, are moving to the left.

replication comes from studying *E. coli.* The proteins it uses to open up the double helix during replication include:

DNA Helicases

Two helicases are involved in unwinding the helix (Fig. 9.2). One attaches to one of the template strands and moves in the $5' \rightarrow 3'$ direction, the second attaches to the other strand and moves in the $3' \rightarrow 5'$ direction. The unwinding of the DNA double helix by the helicases is an ATP-dependent process.

Single-Stranded Binding Proteins (SSB)

As soon as the two parental strands unwind, they are engulfed by SSB. This protein binds to adjacent groups of 32 nucleotides. DNA covered by SSB is rigid, without bends or kinks. It is a good template for DNA synthesis (Fig. 9.2). SSB proteins are sometimes called helix-destabilizing proteins.

Topoisomerase I

The DNA helix must be able to rotate if it is to reduce the torsion caused by the formation of the replication fork. This is achieved by an enzyme called topoisomerase I which cleaves a phosphodiester bond in one strand of the helix. The unbroken strand passes through the nick in the broken strand and the phosphodiester bond is reformed (see Fig. 9.3). This results in progressive unwinding of the two template strands as replication proceeds. (Topoisomerase I, described here, is not the same as topoisomerase II, described in Box 5.1. Topoisomerase II cuts an entire chromosome double helix and passes another chromosome through the gap.)

THE BIOCHEMISTRY OF DNA REPLICATION

DNA Polymerase III

The synthesis of a new DNA molecule is catalyzed by DNA polymerase III. Its substrates are the four deoxyribonucleoside triphosphates, dATP, dCTP, dGTP, and dTTP (Figs. 5.1 and 5.2). DNA polymerase III catalyzes the formation of a

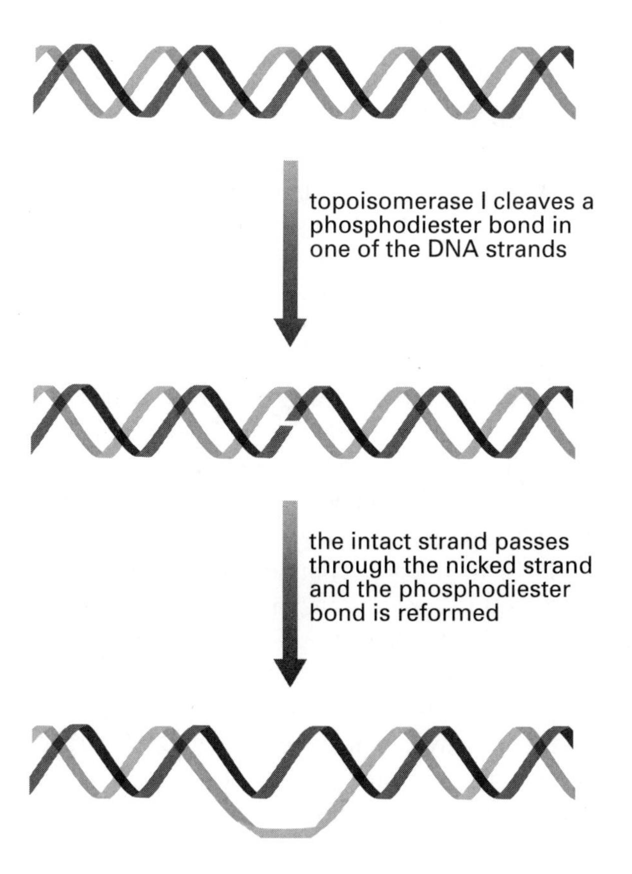

Figure 9.3
Topoisomerase I helps the DNA strand to unwind during DNA replication.

phosphodiester bond between the 3′-hydroxyl group of one sugar residue and the 5′ phosphate group of a second sugar residue (Fig. 5.3). The semi-conservative nature of DNA replication means that the base sequence of a newly synthesized DNA strand is dictated by the base sequence of its parental strand. If the sequence of the template strand is 3′ATCG5′, then that of the daughter strand is 5′TAGC3′.

DNA polymerase III can only add a nucleotide to a free 3′-hydroxyl group and therefore synthesizes DNA in the 5′ to 3′ direction. The template strand is read in the 3′ to 5′ direction. However, the two strands of the double helix are antiparallel (Fig. 9.2). They cannot be synthesized in the same direction because only one has a free 3′-hydroxyl group, the other has a free 5′-phosphate group. No DNA polymerase has been found that can synthesize DNA in the 3′ to 5′ direction, that is, by attaching a nucleotide to a 5′ phosphate, so the synthesis of the two daughter strands must differ. One strand, the leading strand, is synthesized continuously, while the other, the lagging strand, is synthesized discontinuously. DNA polymerase III can synthesize both daughter strands but must make the lagging strand as a series of short 5′ to 3′ sections. The fragments of DNA, called Okazaki fragments after Reiji Okazaki who discovered them in 1968, are then joined together (Fig. 9.2).

DNA Synthesis Requires an RNA Primer

DNA polymerase III cannot itself initiate the synthesis of DNA. The enzyme primase is needed to catalyze the formation of a short stretch of RNA complementary in sequence to the DNA template strand (Fig. 9.2). This RNA chain, the primer, is needed to prime (or start) the synthesis of the new DNA strand. DNA polymerase III catalyzes the formation of a phosphodiester bond between the 3′-hydroxyl group of the RNA primer and the 5′-phosphate group of the appropriate deoxyribonucleotide. Several RNA primers are made along the length of the lagging strand template. Each is extended in the 5′ to 3′ direction by DNA polymerase III until it reaches the 5′ end of the next RNA primer (Fig. 9.2). In prokaryotes the lagging strand is primed about every 1,000 nucleotides, whereas in eukaryotes this takes place each 200 nucleotides.

Once the synthesis of the DNA fragment is complete, the RNA primers are replaced by deoxyribonucleotides. In *E. coli* the enzyme DNA polymerase I removes ribonucleotides using its 5′ →3′ *exonuclease* activity and incorporates deoxyribonucleotides using its 5′ → 3′ polymerizing activity. Eukaryotic organisms probably use the enzyme ribonuclease H to remove the RNA primers. This hydrolyzes an RNA strand that is hydrogen bonded to a DNA strand. Synthesis of the lagging strand is completed by the enzyme *DNA ligase*, which catalyzes bond formation between the DNA fragments.

The Self-Correcting DNA Polymerase

The mammalian genome consists of about 3×10^9 base pairs of DNA. DNA polymerase III makes a mistake about every 1 in 10^4 bases and joins an incorrect deoxyribonucleotide to the growing chain. If unchecked, these mistakes would lead to a catastrophic mutation rate. Fortunately DNA polymerase III has an inbuilt proof-reading mechanism that corrects its own errors. If an incorrect base is inserted into the newly synthesized daughter strand, the enzyme recognizes the change in shape of the double-stranded molecule, which arises through incorrect base pairing, and DNA synthesis

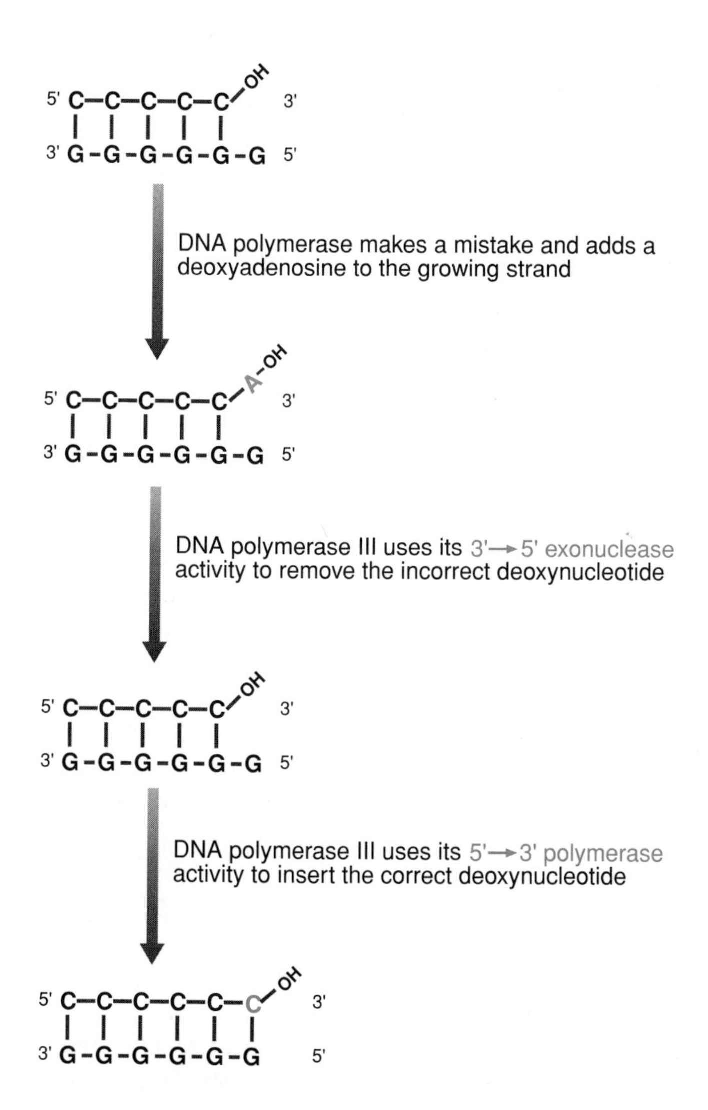

Figure 9.4
DNA polymerase III can correct its own mistakes.

stops (Fig. 9.4). DNA polymerase III then uses its *3′ to 5′ exonuclease* activity to remove the incorrect deoxyribonucleotide and replaces it with the correct one. DNA synthesis then proceeds. DNA polymerase III hence functions as a self-correcting enzyme.

DNA REPAIR

Exposure to chemicals and to radiation can change the sequence of bases in the DNA molecule. For a cell to survive and to pass on its genetic information unmutated, it must be able to correct these changes.

Spontaneous and Chemically Induced Base Changes

The most common damage suffered by a DNA molecule is depurination—the loss of adenine or guanine groups because the bond between the purine base and the sugar to which it is attached breaks (Fig. 9.5). Simple thermal disruption can break the bond. Within a human cell there are about 5 to 10 thousand depurinations every day. *Deamination* is a less frequent event; it happens about 100 times a day in every human cell. Collision of water molecules with the bond linking the amine group to cytosine sets off a spontaneous deamination that produces uracil (Fig. 9.5). Cytosine base pairs with guanine, whereas uracil pairs with adenine. If this change were not corrected then the C.G base pair would mutate to a U.A base pair the next time the DNA strand was replicated.

Ultraviolet light or chemical carcinogens such as benzopyrene, found in cigarette smoke, can also disrupt the structure of DNA. The absorption of ultraviolet light can cause two adjacent thymine residues to link and form a thymine dimer (Fig. 9.6). If uncorrected thymine dimers inhibit normal base pairing between the two strands of the double helix and block the replication process. Ultraviolet light has a powerful germicidal action and is widely used to sterilize equipment. One of the reasons why bacteria are killed by this treatment is because the formation of large numbers of thymine dimers prevents replication.

Repair Processes

If there were no way to correct altered DNA, the rate of mutation would be intolerable. *DNA repair enzymes* have evolved to detect and to repair altered DNA. DNA repair involves three stages: excision of the defective nucleotide,

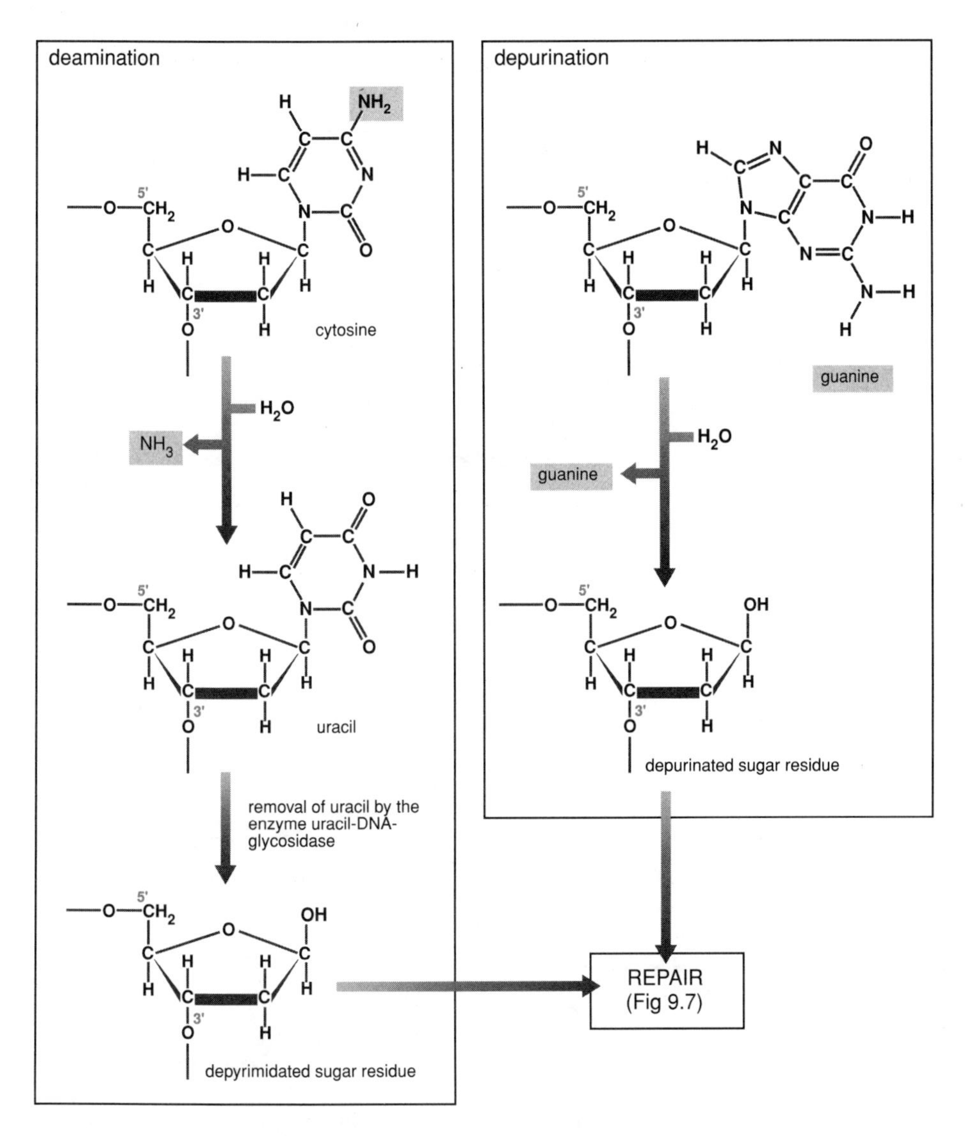

Figure 9.5

Spontaneous reactions corrupt the DNA database.

replacement of the missing nucleotide, and the formation of bonds between the newly inserted nucleotide and the DNA strand.

Deamination of cytosine generates uracil. The repair enzyme *uracil-DNA glycosidase* recognizes and removes uracil from DNA molecules. Perhaps cells use T instead of U in DNA because C is so easily converted to U. If U were always present, then the repair enzyme could not tell which U resulted from the deamination of C and which U's were part of the DNA molecule. Figure 9.7

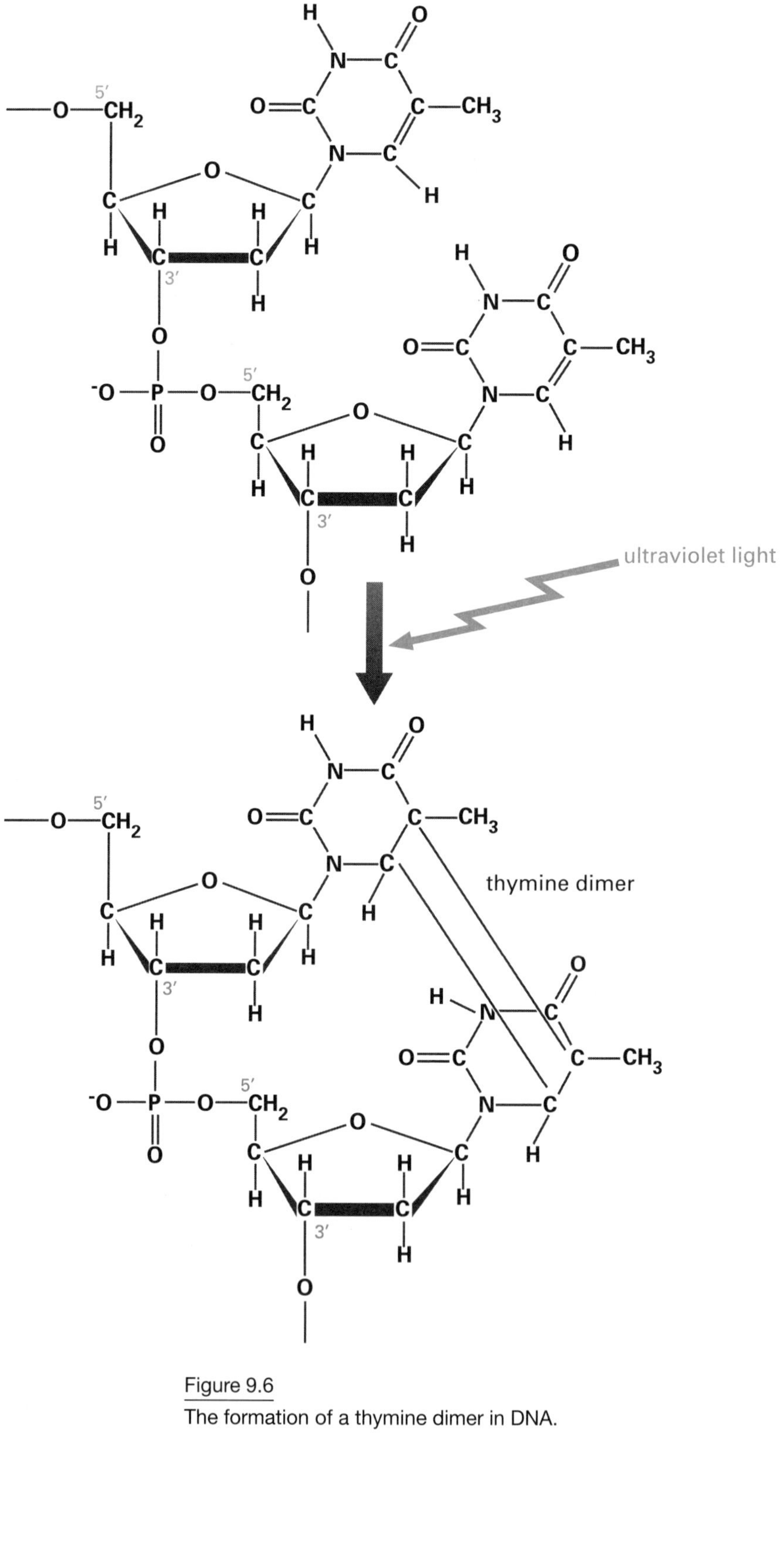

ultraviolet light

thymine dimer

Figure 9.6

The formation of a thymine dimer in DNA.

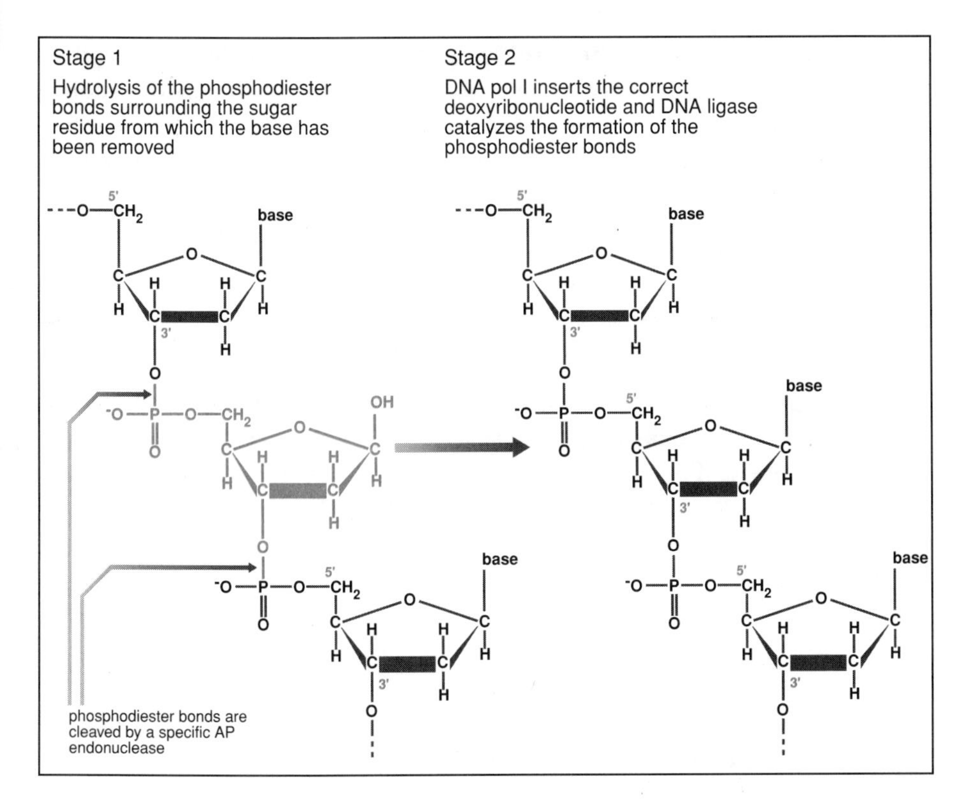

Figure 9.7
DNA repair.

shows that once uracil has been removed, the later steps in the repair process are the same as those used to replace a depurinated sugar. The phosphodiester bonds on either side of a depyrimidinated or depurinated sugar residues are cleaved by a repair *AP endonuclease*. This acts as the signal for DNA polymerase I to replace the damaged region with the correct deoxyribonucleotide. *DNA ligase* then seals the strand by catalyzing the reformation of the phosphodiester bonds.

Thymine dimers are removed by a process known as excision repair. A nuclease removes the damaged region, which is then replaced by the actions of DNA polymerase I and DNA ligase. The repair is quite complex and involves several enzymes—at least nine in humans.

SUMMARY

1. During replication each parent DNA strand acts as the template for the synthesis of a new daughter strand. The base sequence of the newly synthesized strand is complementary to that of the template strand.

2. Replication starts at specific sequences called origins of replication. The two strands untwist and form the replication fork. Topoisomerase and helicase enzymes unwind the double helix and single-stranded binding proteins keep it unwound during replication. DNA polymerase III synthesizes the leading strand continuously in the 5′ to 3′ direction. The lagging strand is made discontinuously in short pieces in the 5′ to 3′ direction. These are joined together by DNA ligase.

3. DNA polymerase is a self-correcting enzyme. It can remove an incorrect base using its 3′ to 5′ exonuclease activity and then replace it.

4. DNA repair enzymes can correct mutations. Uracil in DNA, resulting from the spontaneous deamination of cytosine, is removed by uracil-DNA glycosidase. The depyrimidinated sugar is cleaved from the sugar-phosphate backbone by AP endonuclease, and DNA polymerase then inserts the correct nucleotide. The phosphodiester bond is then reformed by DNA ligase.

BOX 9.1

DNA Repair Deficiencies—Bloom's Syndrome and *Xeroderma pigmentosum*

The final stage of DNA repair needs the enzyme DNA ligase to join the repaired deoxyribonucleotide to its neighbor in the DNA chain by catalyzing the formation of a phosphodiester bond (Fig. 9.5*b*). Some individuals lack a DNA ligase and are partially defective for DNA ligation. Hence their DNA is not repaired following mutation. Such persons have a greatly increased risk of developing skin and other cancers. This deficiency in DNA ligase is known as Bloom's syndrome.

People who suffer from the genetic disorder known as *Xeroderma pigmentosum* are deficient in one of the enzymes carrying out excision repair. As a result they are very sensitive to ultraviolet light. They contract skin cancer even when they have been exposed to sunlight for very short periods because thymine dimers produced by ultraviolet light are not excised from their genomes.

CHAPTER

10

TRANSCRIPTION AND THE CONTROL OF GENE EXPRESSION

Transcription (or RNA synthesis) is the process whereby the information held in the nucleotide sequence of DNA is transferred to RNA. The three major classes of RNA are *ribosomal RNA (rRNA)*, *transfer RNA (tRNA)*, and *messenger RNA (mRNA)*. All play key roles in protein synthesis (Chapter 12). Genes encoding mRNAs are known as protein-coding genes. A gene is said to be expressed when its genetic information is transferred to mRNA and then to protein. Two important questions are addressed in this chapter: How is RNA synthesized, and what factors control how much is made?

THE STRUCTURE OF RNA

RNA is a polymer made up of monomeric nucleotide units. RNA has a chemical structure similar to that of DNA (Chapter 5), but there are two major differences. First, the sugar in RNA is ribose instead of deoxyribose (Fig. 10.1). Second, although RNA contains the two purine bases adenine and guanine and the pyrimidine cytosine, the fourth base is different. The pyrimidine *uracil* (U) (Fig. 10.1) replaces thymine. The building blocks of RNA are the four ribonu-

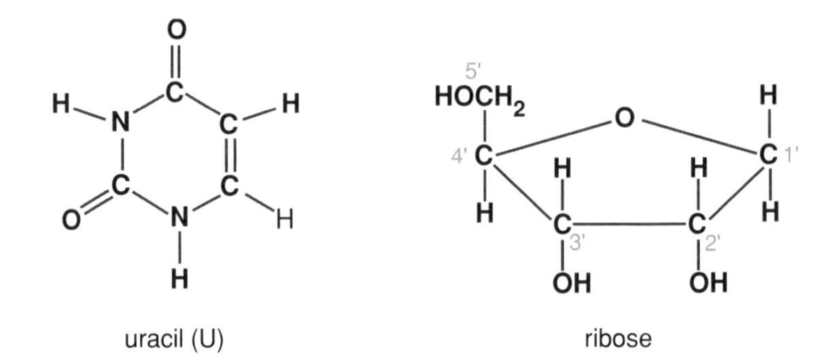

Figure 10.1
Structures of the base uracil and the sugar ribose.

cleoside triphosphates—adenosine 5′-triphosphate, guanosine 5′-triphosphate, cytidine 5′-triphosphate and uridine 5′-triphosphate. These four nucleotides are joined together by phosphodiester bonds (Fig. 10.2). Like DNA the RNA chain has direction. One end, the 5′ terminus, has a free phosphate group and the other end, the 3′ terminus, has a free hydroxyl group. RNA molecules are single stranded along much of their length, although they often contain regions that are double stranded due to intramolecular base pairing.

RNA POLYMERASE

In any gene only one DNA strand acts as the template for transcription. The sequence of nucleotides in RNA depends on their sequence in the DNA template. The bases T, A, G, and C in the DNA template will specify the bases A, U, C, and G, respectively, in RNA. DNA is transcribed into RNA by the enzyme RNA polymerase. Transcription requires that this enzyme recognize the beginning of the gene to be transcribed and catalyze the formation of phosphodiester bonds between nucleotides that have been selected according to the sequence within the DNA template (Fig. 10.2).

GENE NOTATION

Figure 10.3 shows the notation used in describing the position of nucleotides within and adjacent to a gene. The nucleotide in the template strand at which transcription begins is designated with the number +1. Transcription proceeds in the downstream direction, and nucleotides in the transcribed DNA are given successive positive numbers. Downstream sequences are drawn, by convention, to the right of the transcription start site. Nucleotides that lie to the left of this site are called the upstream sequences and are identified by negative numbers.

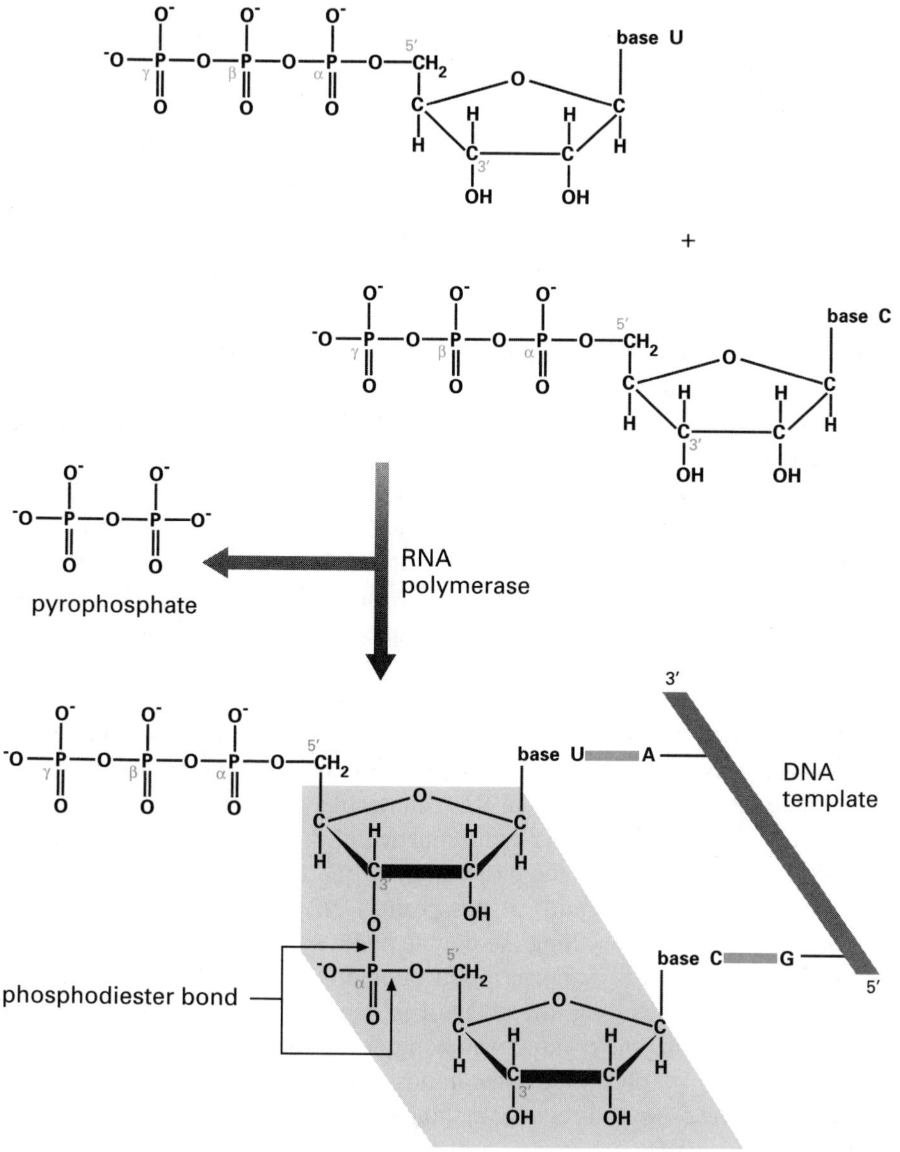

Figure 10.2
Synthesis of an RNA strand.

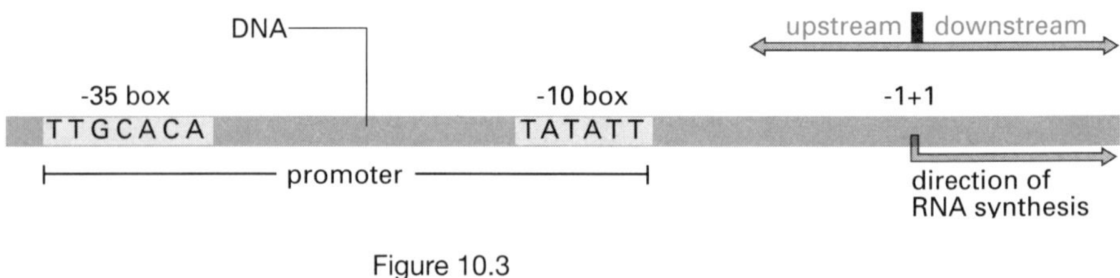

Figure 10.3
Numbering of a DNA sequence.

BACTERIAL RNA SYNTHESIS

All *E. coli* genes are transcribed by the same RNA polymerase. This enzyme is made up of several polypeptide chains. One of these, the sigma factor (σ), recognizes a specific DNA sequence called the promoter, which lies just upstream of the gene to be transcribed (Fig. 10.3). *E. coli* promoters contain two important sequences. One centered around nucleotide –10 usually has the sequence TATATT. This sequence is called the –10 box or the Pribnow box. The second, centered at nucleotide –35 often has the sequence TTGACA. This is the –35 box. The two sequences are usually separated by 17 bases.

To start transcription, RNA polymerase binds to the promoter sequence to form the closed promoter complex (Fig. 10.4). An open promoter complex develops when the two strands of DNA unwind enabling one strand to act as the template for the synthesis of an RNA molecule. The separation of the two DNA strands is helped by the AT rich sequence of the –10 box. There are only two hydrogen bonds between the bases adenine and thymine, so it is relatively easy to separate the two strands at this point. DNA unwinds and rewinds as RNA polymerase advances along the double helix, synthesizing an RNA chain as it goes. This produces a transcription bubble (Fig. 10.4). When the RNA chain is about ten bases long, the σ factor is released from RNA polymerase, and the rest of the enzyme continues moving down the DNA template catalyzing the formation of phosphodiester bonds between nucleotides. The RNA chain grows in the 5′ to 3′ direction, and the template strand is read in the 3′ to 5′ direction (Fig. 10.4).

E. coli has specific sequences, called *terminators*, at the ends of its genes that causes RNA polymerase to stop transcribing DNA. A terminator sequence consists of two regions rich in the bases G and C that are separated by about ten base pairs. This sequence is followed by a stretch of A bases. Figure 10.5 shows how the terminator halts transcription. When the GC rich regions are transcribed, a *hairpin loop* forms in the RNA with the first and second GC rich regions aligning and pairing up. Formation of this structure within the RNA molecule causes the transcription bubble to shrink because where the template DNA strand can no longer bind to the RNA molecule it reconnects to its sister DNA

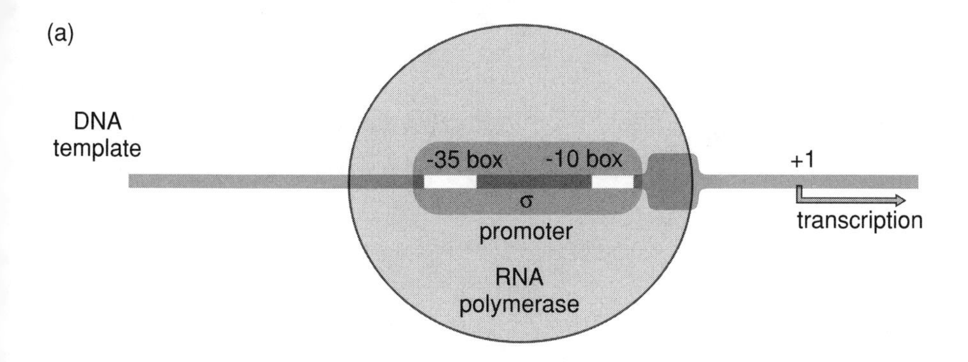

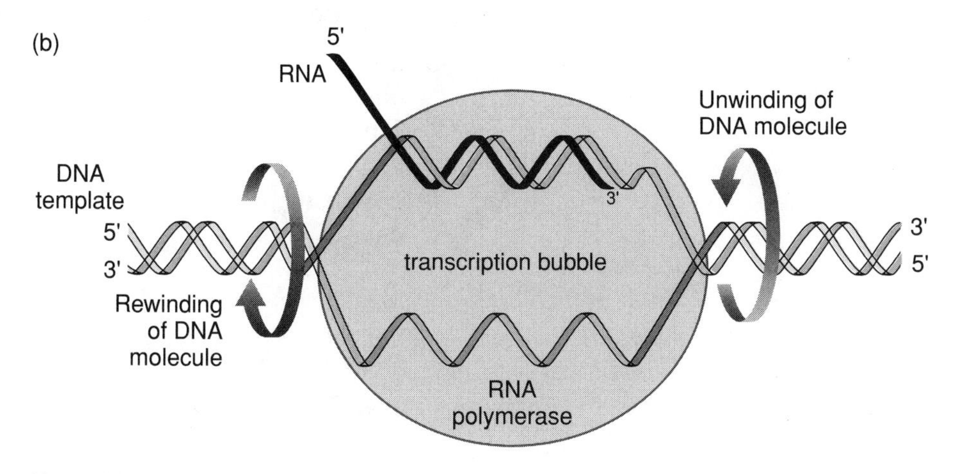

Figure 10.4

(*a*) RNA polymerase binds to the promoter. (*b*) DNA helix unwinds and RNA polymerase synthesizes an RNA molecule.

strand. The remaining interactions between the adenines in the DNA template and the uracils in the RNA chain have only two hydrogen bonds per base pair and are therefore too weak to maintain the transcription bubble. The RNA molecule is then released, transcription terminates, and the DNA double helix reforms.

Some *E. coli* genes contain different terminator sites. These are recognized by a protein known as rho that frees the RNA from the transcription complex.

CONTROL OF BACTERIAL GENE EXPRESSION

Many bacterial proteins are always present in the cell in a constant amount. However, the amount of others is regulated by the presence or absence of a par-

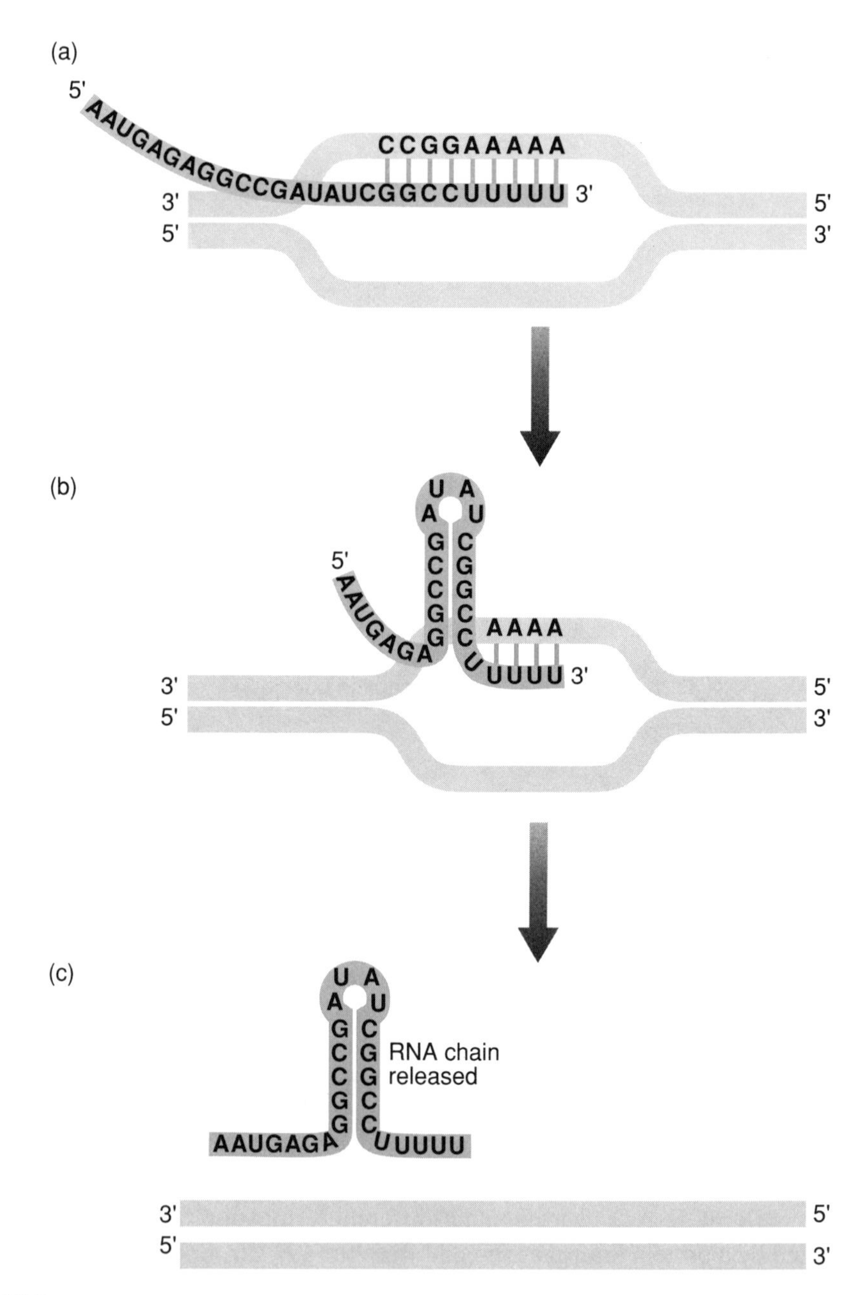

Figure 10.5

Transcription termination in *E. coli*: (*a*) RNA polymerase transcribes a GC-rich region of the template followed by a stretch of A's. (*b*) The GC rich region in the RNA forms a hairpin loop. (*c*) RNA is released and the DNA template reforms.

ticular nutrient. To grow and divide and not to waste energy, bacteria have to adjust quite quickly to changes in their environment. They do this by regulating the production of proteins required for breakdown or synthesis of particular compounds. Gene expression in bacteria is controlled mainly at the level of transcription. This is because bacterial cells have no nuclear envelope and RNA synthesis and protein synthesis are not separate but occur simultaneously. This is one reason why bacteria lack the more sophisticated control mechanisms that regulate gene expression in eukaryotes.

Each bacterial promoter usually controls the transcription of a cluster of genes coding for proteins that work together on a particular task. This collection of related genes is called an *operon* and is transcribed as a single mRNA molecule called a polycistronic mRNA. As shown in Fig. 10.6, translation of this mRNA produces the required proteins because there are several start (page 193) and stop codons along its length. Each start and stop codon specifies the region of RNA that will be translated into a particular protein. The organization of genes into operons helps bacteria to respond quickly to the environment and ensures that all the proteins necessary to metabolize a particular compound are made at the same time.

The three major factors involved in regulating transcription are (1) nu-

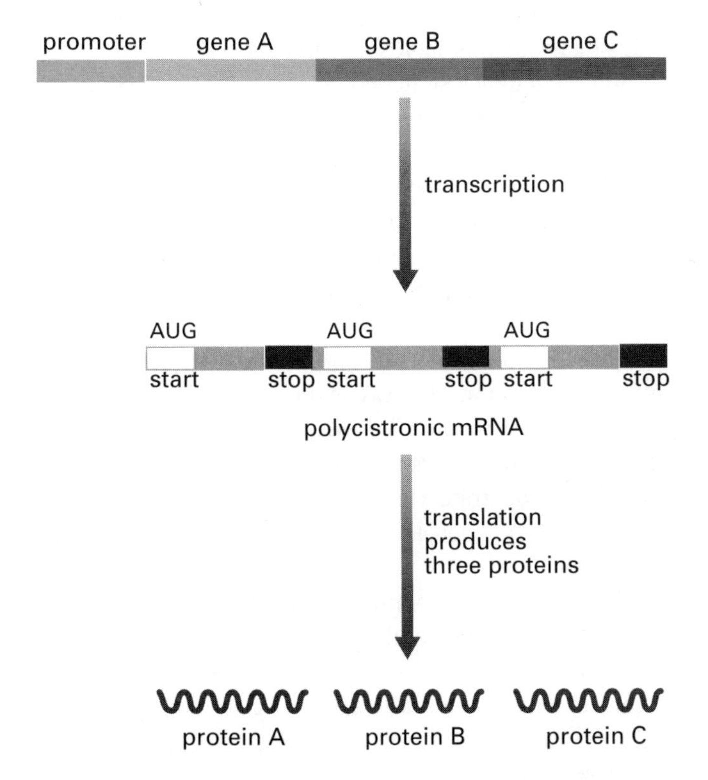

Figure 10.6

A bacterial operon is transcribed into a polycistronic mRNA.

cleotide sequences within or flanking a gene, (2) proteins that bind to these sequences, and (3) the environment. The human gut contains many million *E. coli* cells that must respond very quickly to the sudden appearance of a particular nutrient. For instance, most foods do not contain the disaccharide lactose (Fig. 10.7), but milk contains large amounts. Within minutes of our drinking a glass of milk *E. coli* in our intestines start to produce the enzyme β-galactosidase which cleaves lactose to glucose and galactose (Fig. 10.7). In general, the substrates of β-galactosidase are compounds like lactose that contain a β-galactoside linkage and are therefore called β-galactosides.

lac, an Inducible Operon

β-galactosidase is encoded by one of the genes that make up the lactose (*lac*) operon which is shown in Figure 10.8. It comprises three protein-coding genes called *lac z, lac y,* and *lac a.* β-galactosidase is encoded by the *lac z* gene. *Lac y* encodes β-galactoside permease, a carrier (page 292) that helps lactose get into the cell. The *lac a* gene codes for transacetylase. This protein is thought to remove compounds that have a structure similar to lactose but that are not useful to the cell.

The *lac* operon is an inducible operon because it is only transcribed into RNA when a β-galactoside sugar such as lactose is present. How is the transcription of the *lac z, lac y* and *lac a* genes switched on and off? When β-galactoside sugars are absent then a repressor protein (the product of the *lac i* gene) binds to a sequence in the *lac* operon known as the operator (Fig. 10.8). This lies next to the promoter so that, when the repressor is bound, RNA polymerase is unable to bind to the promoter. Thus in the absence of β-galactoside, the *lac z, lac y,* and *lac a* genes are not transcribed. However, if lactose appears, it is converted to allolactose by the tiny amount of β-galactosidase always present (Fig. 10.7). The repressor protein has a binding site for allolactose and undergoes a conformational change when bound to this compound. The repressor is now no longer able to bind to the operator. The way is then clear for RNA polymerase to bind to the promoter and to transcribe the operon. Thus in a short time the bacteria produce the proteins necessary for utilizing the new food source. The concentration of the substrate (lactose in this case) determines whether protein is synthesized. The *lac* operon is said to be under negative regulation by the repressor protein.

The transcription of the *lac* operon is controlled not only by the repressor protein but also by another protein, the catabolite activator protein (CAP). If both glucose and lactose are present, it is energetically more favorable to the cell to use glucose as the carbon source. The utilization of glucose requires no

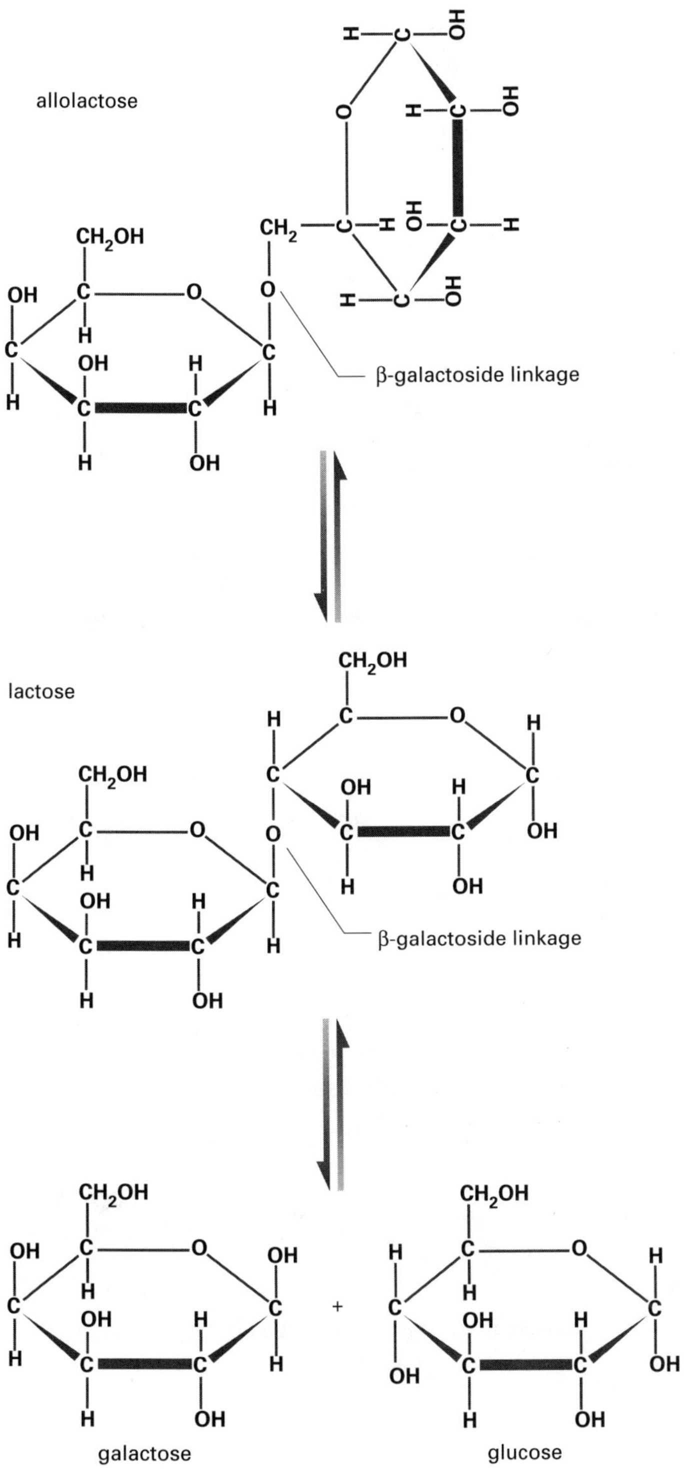

allolactose

β-galactoside linkage

lactose

β-galactoside linkage

galactose + glucose

Figure 10.7

Reactions catalyzed by β-galactosidase.

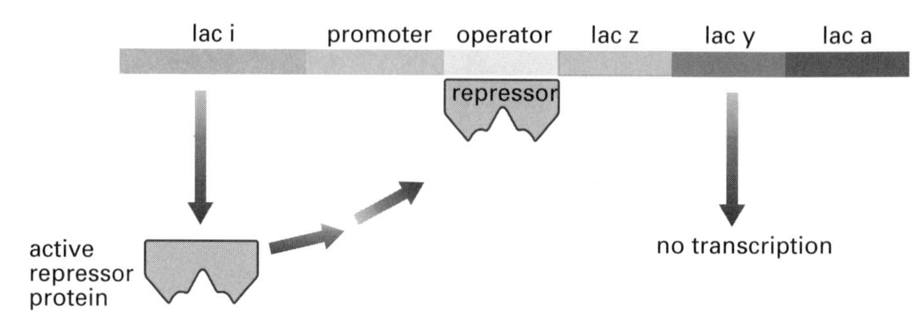

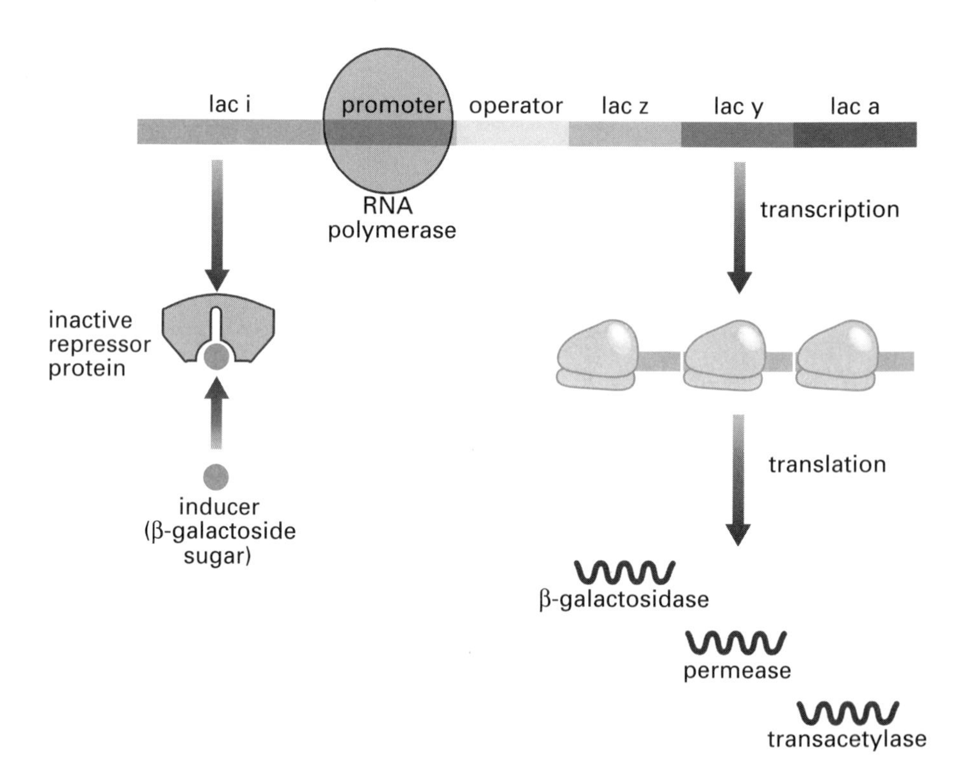

Figure 10.8

Transcription of the *lac* operon requires the presence of an inducer.

new RNA synthesis, since all the proteins necessary are already present in the cell. So even though lactose is present, the *lac* operon remains switched off. This happens because when glucose concentrations are high, the concentration of an intracellular messenger molecule called cAMP (Chapter 19) is low. For RNA polymerase to bind to the promoter sequence, it requires a complex formed between cAMP and the protein CAP. Low concentrations of cAMP mean that this complex cannot form. Only when the concentration of glucose falls does the concentration of cAMP rise and the CAP-cAMP complex form. The CAP protein is now able to bind to a sequence upstream of the *lac* operon promoter (Fig. 10.9). The presence of the CAP-cAMP complex increases the affinity of RNA polymerase for the *lac* operon promoter, so the *lac z, lac y,* and *lac a* genes can be transcribed. The *lac* operon is said to be under positive regulation by the CAP-cAMP complex.

To recap, the control of the *lac* operon is not simple. Several requirements need to be met before it can be transcribed. The repressor must not be bound to the operator, and the CAP-cAMP complex and RNA polymerase must be bound to their respective DNA binding sites. These requirements are only met when glucose is absent and a β-galactoside sugar such as lactose is present.

Other compounds such as isopropylthio-β-D-galactoside (IPTG) (Fig. 10.10) can bind to the repressor but are not metabolized. These gratuitous in-

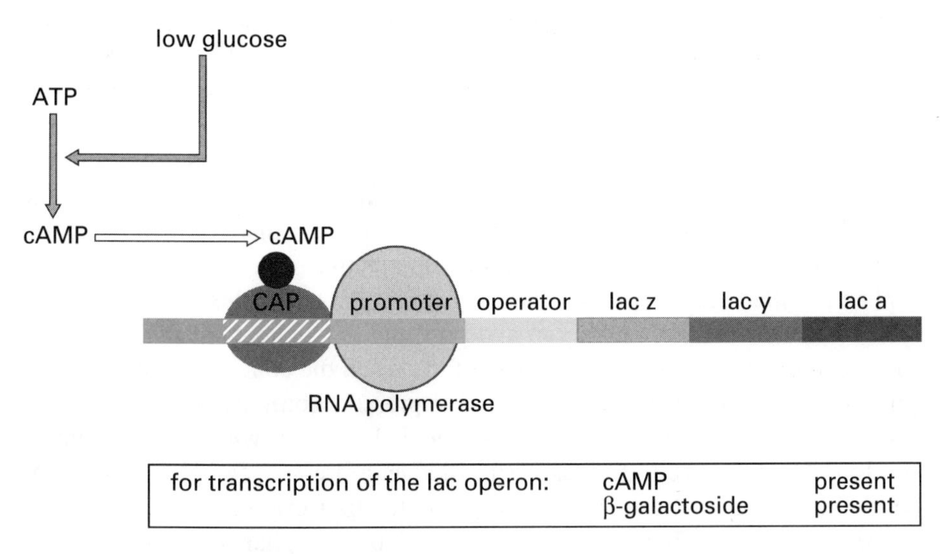

Figure 10.9

Positive regulation of the *lac* operon by cAMP.

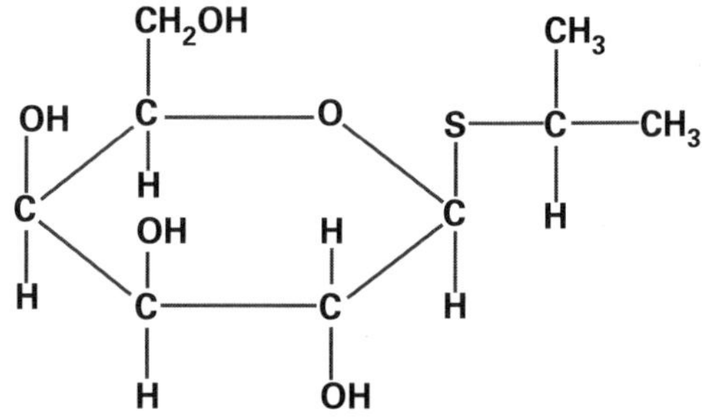

Figure 10.10

Isopropylthio-β-D-galactoside (IPTG) can bind to the *lac* repressor protein but is not metabolized.

IPTG (isopropyl thio-β-D-galactoside)

ducers are very useful in DNA research and in biotechnology. Chapter 11 deals with this and with some of the industrial applications of the *lac* operon.

trp, a Repressible Operon

Operons that code for proteins that synthesize amino acids are regulated in a different way to the *lac* operon. These operons are only transcribed if the amino acid is not present, and transcription is switched off if there is already enough amino acid around. In this way the cell carefully controls the concentration of free amino acids. The tryptophan (*trp*) operon is made up of five structural genes encoding enzymes that synthesize the amino acid tryptophan (Fig. 10.11). This is a repressible operon. The cell regulates the amount of tryptophan produced by preventing transcription of the *trp* operon mRNA when there is sufficient tryptophan about. As with the *lac* operon the transcription of the *trp* operon is controlled by a regulatory protein and a small molecule. The gene *trp r* (Fig. 10.11) encodes an inactive repressor protein that is called an aporepressor. Tryptophan binds to this to produce an active repressor complex. Tryptophan is called the corepressor. The active repressor complex binds to the operator sequence of the *trp* operon and prevents the attachment of RNA polymerase to the promoter sequence. Therefore, when the concentration of tryptophan in the cell is high, the active repressor complex will form, and transcription of the *trp* operon is prevented. However, when the amount of tryptophan in the cell decreases, the active repressor complex cannot be formed. RNA polymerase binds to the promoter, transcription of the *trp* operon proceeds, and the enzymes needed to synthesize tryptophan are produced. This is an example of negative feedback (page 279).

The regulation of many other operons also involves the interaction of specific regulatory proteins with specific small molecules.

(a) no tryptophan operon transcribed

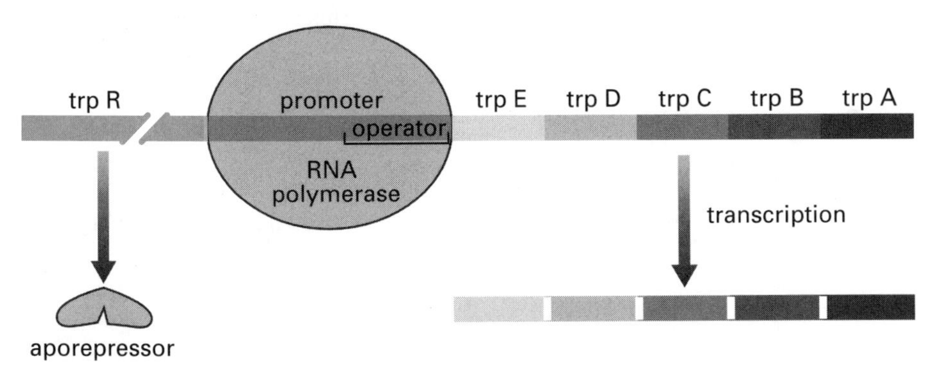

(b) tryptophan present operon repressed

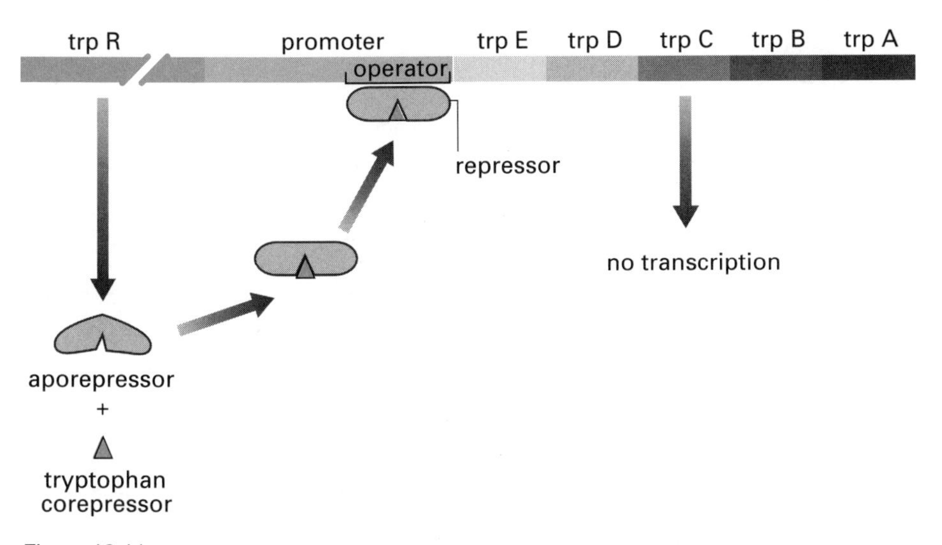

Figure 10.11

The transcription of the *trp* operon is controlled by the concentration of the amino acid tryptophan.

EUKARYOTIC RNA SYNTHESIS

Eukaryotes have three types of RNA polymerase. RNA polymerase I transcribes the genes that code for most of the ribosomal RNAs. All messenger RNAs are synthesized using RNA polymerase II. Transfer RNA genes are transcribed by RNA polymerase III. This last enzyme also catalyzes the synthesis of several small RNAs including the 5S ribosomal RNA. The chemical reaction catalyzed by these three RNA polymerases, the formation of phosphodiester bonds between nucleotides, is the same in eukaryotes and bacteria.

Messenger RNA Processing

A newly synthesized eukaryotic mRNA undergoes several modifications before it leaves the nucleus (Fig. 10.12). The first is known as capping. Very ear-

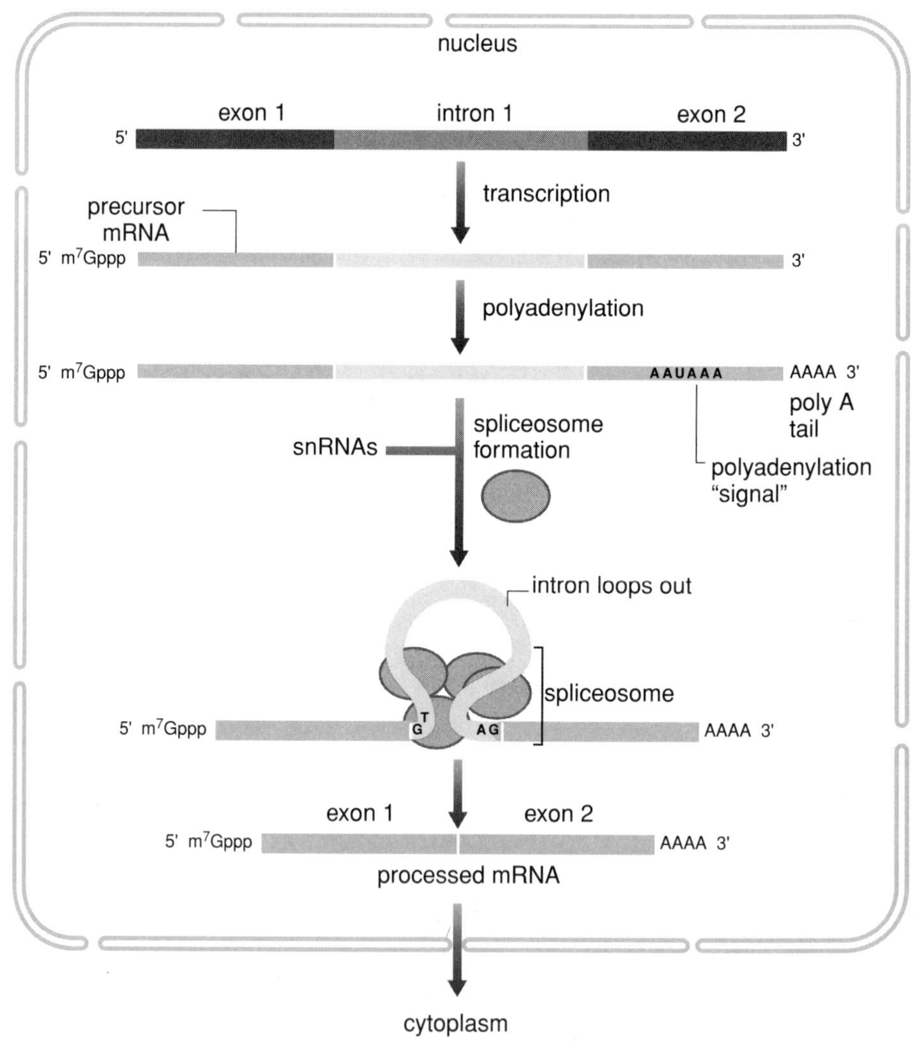

Figure 10.12
mRNA processing in eukaryotes.

ly in transcription the 5′-terminal triphosphate group is modified by the addition of a guanosine via a 5′–5′-phosphodiester bond. The guanosine is subsequently methylated to form a *7-methyl-guanosine cap*. The 3′ end of nearly all eukaryotic mRNAs are modified by the addition of a long stretch of adenosine residues, the *poly A tail* (Fig. 10.12). A sequence AAUAAA is found in most eukaryotic mRNAs about 20 bases from where the poly A tail is added, and is probably a signal for the polyadenylation process. The length of the poly A tail varies, it can be as long as 250 nucleotides. Unlike DNA, RNA is a very unstable molecule, and the capping of eukaryotic mRNAs at their 5′ ends and the addition of poly a A tail to their 3′ end increases the lifetime of mRNA molecules by protecting them from digestion by nucleases.

Many eukaryotic genes are split into exon and intron sequences (Chapter 7). The introns have to be removed and the exons joined together by a process known as RNA splicing before the mRNA can be used to make protein. Removal of introns takes place within the nucleus. Splicing is complex and not yet fully understood. It has, however, certain rules. Within a protein coding gene the first two bases following an exon are always GT, and the last two bases of the intron are always AG. Several small nuclear RNAs (snRNAs) are involved in splicing. These are complexed with a number of proteins to form a structure known as the spliceosome. One of the snRNAs is complementary in sequence to either end of the intron sequence. It is thought that binding of this snRNA to the intron by complementary base pairing brings the two exon sequences together, which causes the intron to loop out (Fig. 10.12). The proteins in the spliceosome remove the intron and join the exons together. Splicing is

BOX 10.1

Splice Mutations That Can Cause Disease—β^0 Thalassemia

Adult hemoglobin is made of two molecules of α globin and two molecules of β globin (page 231). The β-globin gene has three exons and two introns. Both introns begin with the sequence GT. This ensures that the introns are removed and the exons correctly spliced to form β-globin mRNA (Fig. 10.12). Subsequently a normal β-globin polypeptide is translated from this mRNA. However, some people are unable to produce a β-globin polypeptide and therefore suffer from β^0 thalassemia (the 0 after the β means there is no β globin present). This is because the GT at the beginning of either the first or second intron of their β-globin gene has been mutated to an AT. This sequence is no longer recognized by the spliceosome, and the β-globin mRNA is incorrectly spliced. The defective mRNA is not translated into β globin. Affected individuals consequently have far too little hemoglobin—they are severely anemic.

the final modification made to the mRNA in the nucleus. The mRNA is now transported to the cytoplasm for protein synthesis.

CONTROL OF EUKARYOTIC GENE EXPRESSION

Since most eukaryotes are multicellular organisms with many cell types, gene expression must be controlled so that different cell lineages develop differently and remain different. A brain cell is quite different from a liver cell because it contains different proteins even though the DNA in the two cells is identical. During development and differentiation, different sets of genes are switched on and off. Hemoglobin, for example, is only expressed in developing red blood cells even though the globin genes are present in all types of cell. Genetic engineering technology (Chapter 11) has made the isolation and manipulation of eukaryotic genes possible. This has given us some insight into the extraordinarily complex processes that regulate transcription of eukaryotic genes and allow a fertilized egg to develop into a multicellular, multi-tissue adult.

Unlike the situation in bacteria, the eukaryotic cell is divided by the nuclear envelope into nucleus and cytoplasm. Transcription and translation are therefore separated in space and in time. This means that the expression of eukaryotic genes can be regulated at more than one place in the cell. Although gene expression in eukaryotes is controlled primarily by regulating transcription in the nucleus, there are many instances in which expression is controlled at the level of translation in the cytoplasm or by altering the way in which the primary mRNA transcript is processed.

The interaction of RNA polymerase with its promoter is far more complex in eukaryotes than it is in bacteria. This section describes how the transcription of a gene encoding mRNA is transcribed by RNA polymerase II. In contrast to bacterial RNA polymerase, RNA polymerase II cannot recognize a promoter sequence. Instead, other proteins known as *transcription factors* bind to the promoter and guide RNA polymerase II to the beginning of the gene to be transcribed.

The promoter sequence of most eukaryotic genes encoding mRNAs contains an AT rich region about 25 base pairs upstream of the transcription start site. This sequence, called the TATA box, binds a protein called transcription factor IID (TFIID) (Fig. 10.13). Several other transcription factors then bind to TFIID and to the promoter region. These proteins provide a binding site for RNA polymerase II and ensure that the enzyme is correctly positioned for transcription. The complex formed between the TATA box, TFIID, the other transcription factors, and RNA polymerase is known as the *transcription preinitiation complex*. Although many protein coding genes contain a TATA box, some

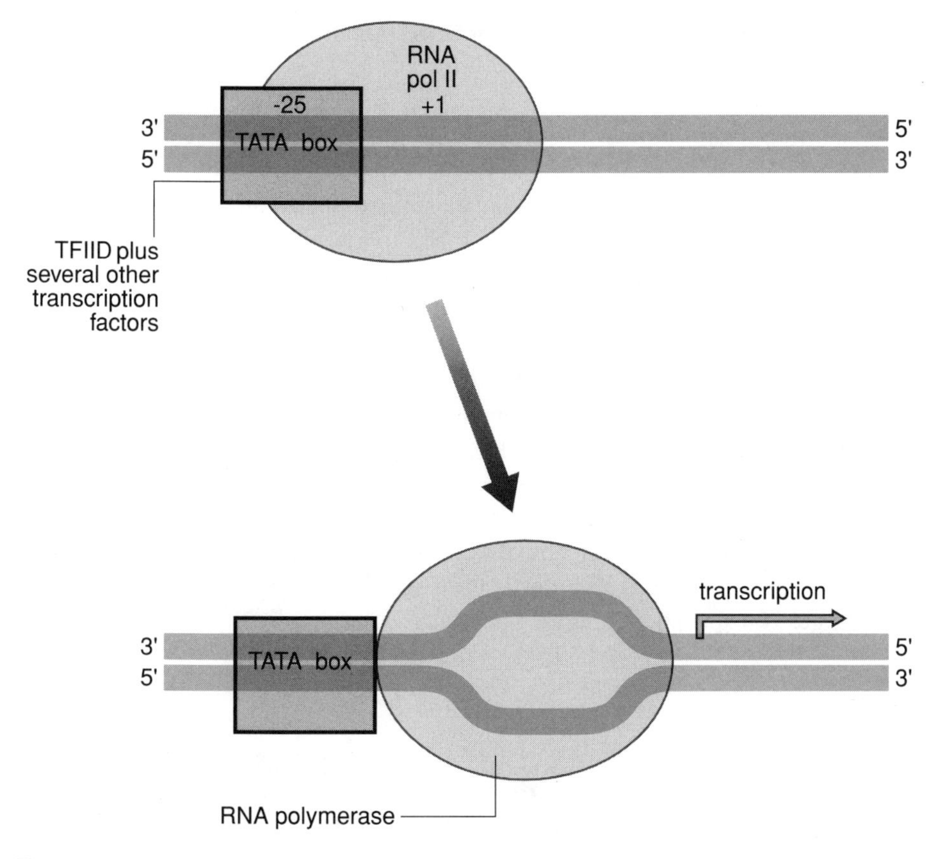

Figure 10.13

The preinitiation complex RNA polymerase is guided to the promoter by several accessory proteins.

do not. These TATA-less genes usually encode proteins that are needed in every cell and are hence called housekeeping genes.

Although the formation of the initiation protein complex is sometimes enough to produce a few molecules of RNA, the binding of other proteins to sequences next to the gene greatly increases the rate of transcription producing much more mRNA. These proteins are also called transcription factors, and the DNA sequences to which they bind are called *enhancers*, so named because their presence enhances transcription. Enhancer sequences often lie upstream of a promoter, but they have also been found downstream of a promoter. Enhancer sequences and the proteins that bind to them play an important role in determining whether a particular gene is to be transcribed. Some transcription factors bind to a gene to ensure that it is transcribed at the right stage of development. Other transcription factors are responsible for ensuring that a gene is only expressed in the correct tissue or in response to various signaling molecules such as steroid hormones.

Glucocorticoids are steroid hormones produced by the adrenal cortex. Glucocorticoids are known to increase the transcription of several genes important in carbohydrate and protein metabolism. Because they are uncharged, steroid hormones can pass through the plasmalemma by simple diffusion to enter the cytoplasm. There the hormone binds to a specific receptor molecule—the glucocorticoid receptor (Fig. 10.14). In the absence of hormone, this receptor remains in the cytoplasm and is inactive because it is complexed to an inhibitor protein. However, when the glucocorticoid hormone binds to its receptor, the inhibitor protein is displaced. The glucocorticoid receptor-hormone complex can now move into the nucleus. Here two molecules of the complex bind to a 15 base pair sequence known as the hormone response element (HRE) that lies upstream of the TATA box. The HRE is an enhancer sequence. The binding of the glucocorticoid receptor-hormone complex to the HRE stimulates transcription. Precisely how it does this is not yet fully understood.

The glucocorticoid receptor is an example of a zinc finger protein. The zinc finger is a specialized part, or domain, of the protein that fits into the ma-

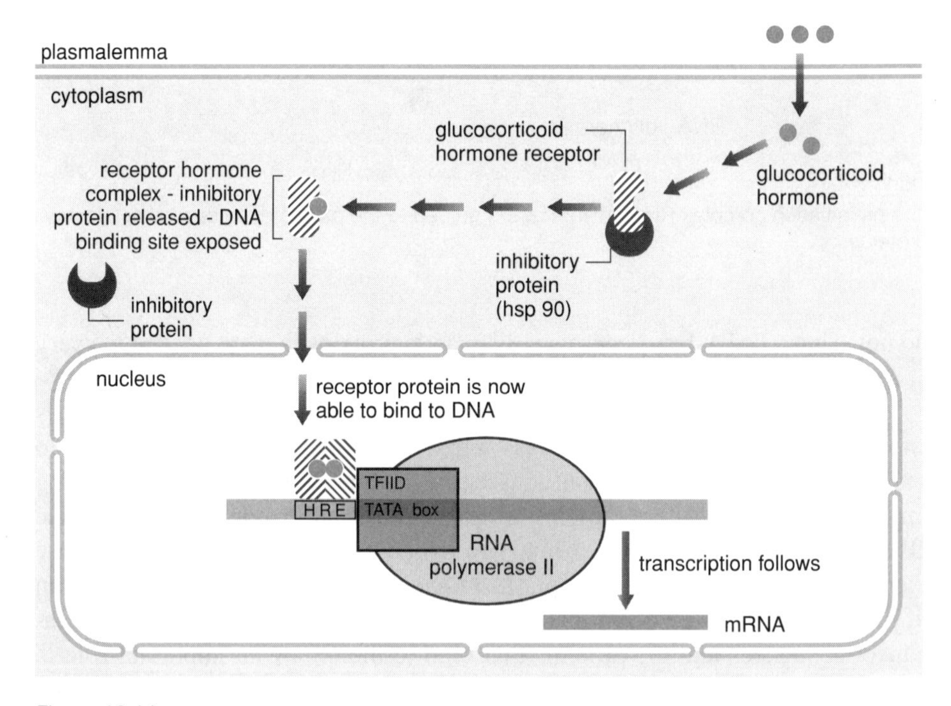

Figure 10.14

The steroid hormone receptor acts to increase gene transcription in the presence of hormone.

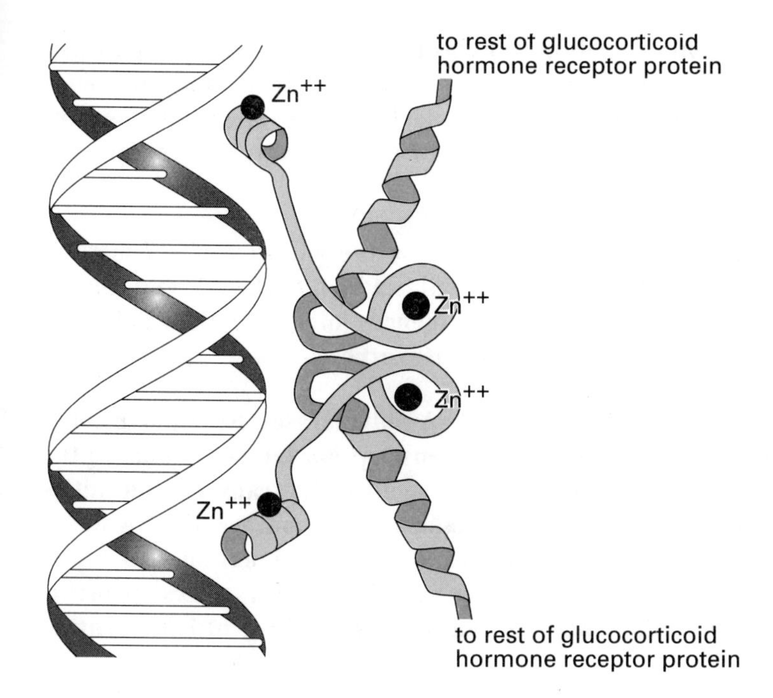

Figure 10.15
The zinc finger domains of a dimerized pair of steroid hormone receptors.

jor groove of DNA and binds to specific enhancer sequences. Figure 10.15 shows the zinc finger domains of two glucocorticoid receptors that are making up an active dimer. Zinc ions are included within the protein and help stabilize its structure. Like other transcription factors that interact with enhancer sequences the glucocorticoid receptor possesses three specific domains. It has a sorting signal that tells the cell to move the protein into the nucleus, the DNA-binding domain, in this case a zinc finger, and a third domain that interacts with the proteins bound to the TATA box. The glucocorticoid receptor has a fourth domain whose function is to bind the steroid hormone.

SUMMARY

1. DNA is transcribed into RNA by the enzyme RNA polymerase. The three types of RNA are ribosomal RNA (rRNA), transfer RNA (tRNA), and messenger RNA (mRNA). Uracil, adenine, cytosine, and guanine are the four bases in RNA.

2. In bacteria RNA polymerase binds to the promoter sequence just upstream of the start site of transcription. The enzyme moves down the DNA template and synthesizes a RNA molecule. RNA synthesis stops once the enzyme has transcribed a terminator sequence.

3. Bacterial genes encoding proteins for the same metabolic pathway are often clustered into operons. Some operons are induced in the presence of the substrate of the pathway, for example, the lactose operon. Others are repressed in the presence of the product of the pathway, for example, the tryptophan operon.

4. Eukaryotic mRNAs are modified by the addition of a 7-methylguanosine cap at their 5′ end. A poly A tail is added to their 3′ end. Intron sequences are removed, and the exon sequences joined together. This fully processed mRNA is then ready for protein synthesis.

5. In eukaryotes there are three RNA polymerases. RNA polymerase II needs the help of proteins known as transcription factors to bind to the promoter region. This group of proteins is called the transcription initiation complex, and this is sufficient to make a small number of RNA molecules. However, to make a lot of RNA in response to a signal, such as a hormone, other proteins bind to sequences called enhancers. These proteins interact with the initiation complex and increase the rate of RNA synthesis.

BOX 10.2

Blocking Calcineurin—How Immunosuppressants Work

The drug cyclosporin A is invaluable in modern medicine because it suppresses the immune response that would otherwise cause the rejection of transplanted organs. It does this by blocking a critical stage in the activation of T lymphocytes, one of the cell types in the immune system. T lymphocytes signal to other cells of the immune system by synthesizing and releasing the transmitter (page 318) interleukin 2. Transcription of the interleukin 2 gene is activated by a transcription factor called $NFAT_c$.

$NFAT_c$ has a sorting signal that would normally direct it to the nucleus, but in unstimulated cells this is masked by a phosphoryl group, so $NFAT_c$ remains in the cytoplasm and interleukin 2 is not made. However, when foreign proteins activate the T lymphocyte, they cause the concentration of calcium ions to rise in the cytosol (a rise in calcium is a common feature of cell stimulation to be described in Chapter 19). Calcium activates a phosphatase called calcineurin that removes the phosphoryl group from many substrates including $NFAT_c$. $NFAT_c$ then moves to the nucleus and activates interleukin 2 transcription. The released interleukin 2 activates other immune system cells that attack the foreign body. Cyclosporin blocks this process by inhibiting calcineurin. Consequently, even though calcium rises in the cytoplasm of the T lymphocyte, $NFAT_c$ remains phosphorylated and does not move to the nucleus.

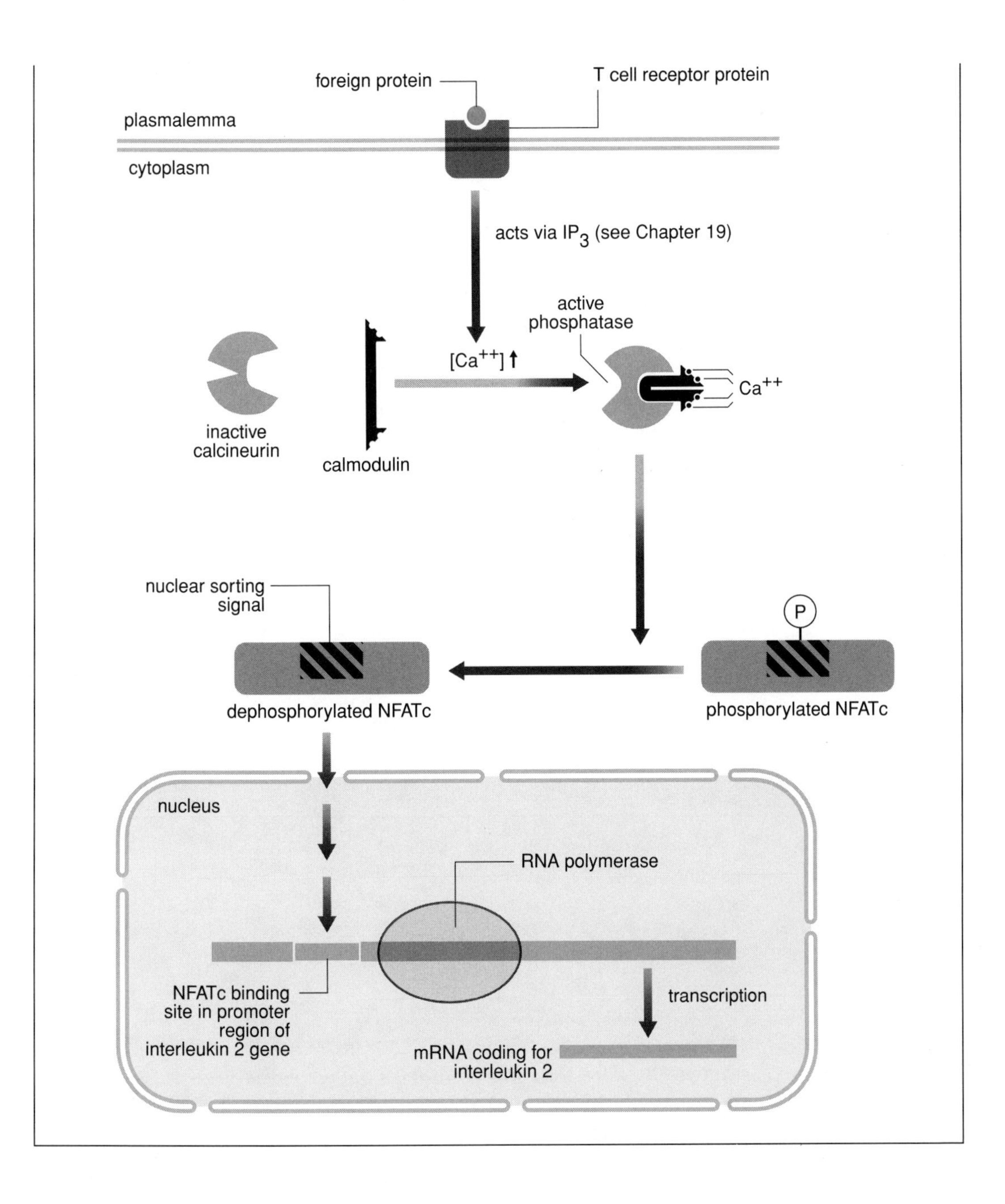
foreign protein
T cell receptor protein
plasmalemma
cytoplasm
acts via IP3 (see Chapter 19)
active
phosphatase
[Ca++] ↑
Ca++
inactive
calcineurin
calmodulin
nuclear sorting
signal
P
dephosphorylated NFATc
phosphorylated NFATc
nucleus
RNA polymerase
transcription
NFATc binding
site in promoter
region of
interleukin 2 gene
mRNA coding for
interleukin 2

CHAPTER 11

RECOMBINANT DNA AND GENETIC ENGINEERING

DNA is the cell's database. Within its base sequence is all the information necessary to encode RNA and protein. A number of biological and chemical methods now give us the ability to isolate DNA molecules and to determine their base sequence. Once we have the DNA and know the sequence, many possibilities open up. We can identify mutations that cause disease, make a human vaccine in a bacterial cell, or alter a sequence and hence the protein it encodes. Eventually we will know the entire base sequence of the human genome. The power of these techniques is already revolutionizing medicine and biology and, in future years, is likely to impact ever more strongly on industry and on the way we live. This chapter describes some of the important methods involved in recombinant DNA technology at the heart of which is DNA cloning.

DNA CLONING

Since DNA molecules are composed of only four nucleotides (A, T, C, G) their physical and chemical properties are very similar. Hence it is extremely difficult to purify individual species of DNA by classical biochemical techniques similar to those used successfully for the purification of proteins. However, we can use DNA cloning to help us to separate DNA molecules. A clone is a population of cells that arose from a single mother cell, and in the absence of mutation all members of a clone will be genetically identical. If a foreign gene or gene fragment is introduced into a cell and the cell then grows and divides repeatedly, many copies of the foreign gene can be produced and the gene is then said to have been cloned. A DNA fragment can be cloned from any organism. The basic approach to cloning a gene is to take the genetic material from the cell of interest, which in the examples we will describe is a human cell, and to introduce this DNA into bacterial cells. Clones of bacteria are then generated, each of which contains and replicates one fragment of the human genetic material. The clones that contain the gene we are interested in are then identified and grown separately. We therefore use a biological approach to isolate DNA molecules rather than physical or chemical techniques.

Creating the Clone

How do we clone a human DNA sequence? The haploid human genome has 1.5×10^9 base pairs of DNA, and the DNA content of each somatic cell is identical. However, each cell expresses only a fraction of its genes. Different types of cells express different sets of genes and thus their mRNA content is not the same. In addition, processed mRNA is shorter than its parent DNA sequence and contains no introns (see Chapter 10). Consequently, it is much easier to isolate a DNA sequence by starting with mRNA. We therefore start the cloning process by isolating mRNA from the cells of interest. *Reverse transcriptase*, an enzyme coded for by the genome of certain viruses, can copy mRNA into DNA. As the DNA is complementary in sequence to the mRNA template, it is known as *complementary DNA*, or *cDNA*. The way in which a cDNA molecule is synthesized is shown in Figure 11.1.

Most eukaryotic mRNA molecules have a string of As at their 3′ end, the poly A tail (page 159). A short synthetic run of T residues can therefore be used to prime the synthesis of DNA from an mRNA template using reverse transcriptase. The resulting double-stranded molecule is a hybrid containing one strand of DNA and one of RNA. The RNA strand is removed by digestion with the enzyme ribonuclease H. This enzyme cleaves—“nicks”—phosphodiester bonds in the RNA strand of the paired RNA-DNA complex. DNA polymerase (page 137) homes in on the nick and replaces a ribonucleotide with a deoxyribonucleotide. DNA ligase (page 138) then reforms any missing phosphodiester bonds. In this way a double-stranded DNA molecule is generated by the replacement of the RNA strand with a DNA strand. If the starting point had been mRNA isolated from liver cells, then a collection of cDNA molecules representative of all the mRNA molecules within the liver

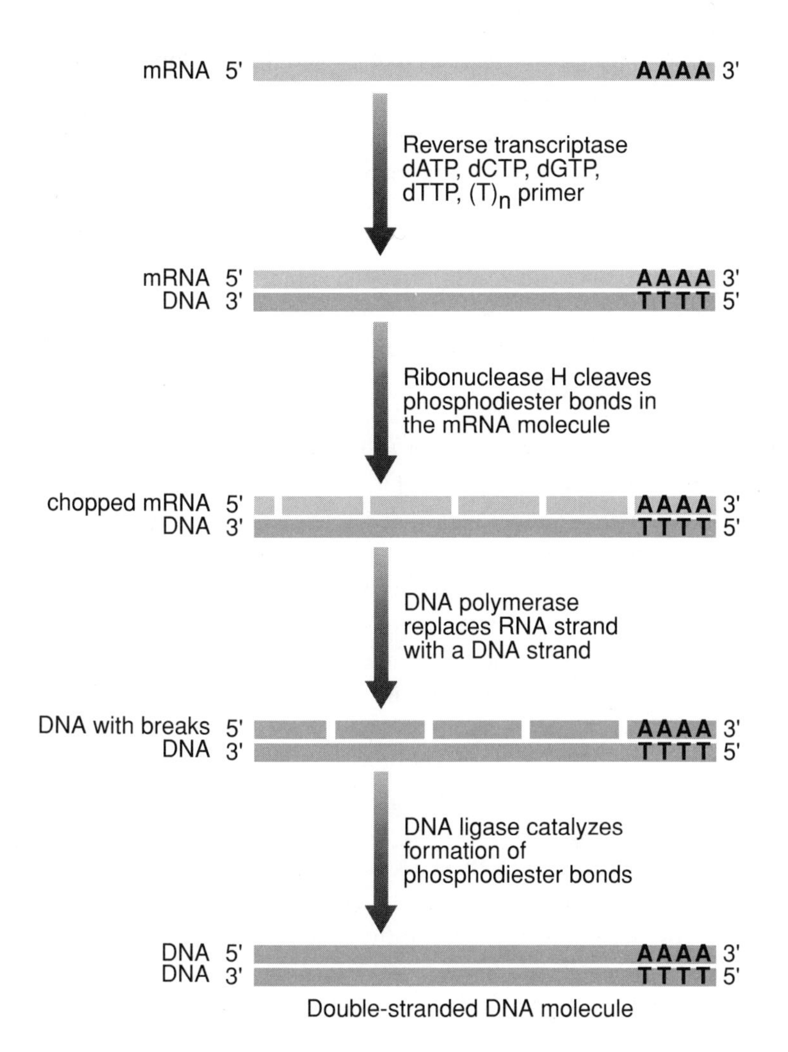

Figure 11.1
Synthesis of a double-stranded DNA molecule.

would have been produced. These DNA molecules have now to be introduced into bacteria.

Introduction of Foreign DNA Molecules into Bacteria

CLONING VECTORS. To ensure the survival and propagation of foreign DNAs they must be inserted into a vector that can replicate inside bacterial cells and be passed on to subsequent generations of the bacteria. The vectors used for cloning are derived from naturally occurring bacterial plasmids or bacteriophages.

Plasmids (page 80) are small circular DNA molecules found within bacteria. Each contains an origin of replication (page 134) and thus can replicate independently of the bacterial chromosome and produce many copies of itself. Plasmids often carry genes that confer antibiotic resistance on the host bacterium. The advantage of this to the scientist is that bacteria containing the plasmid can be selected for in a population of other bacteria simply by applying the antibiotic. Those bacteria with the antibiotic resistance gene will survive, whereas those without it will die. Figure 11.2 shows a typical plasmid cloning vector. Bacteriophages are viruses that infect bacteria and utilize the host cell's components for their own replication.

JOINING FOREIGN DNAs TO A CLONING VECTOR. Enzymes known as *restriction endonucleases* are used to insert foreign DNA into a cloning vector. Each restriction endonuclease recognizes a particular DNA sequence of (usually) 4 or 6 base pairs. The enzyme binds to this sequence and then cuts both strands of the double helix. Many restriction endonucleases have been isolated. The names and recognition sequences of a few of the common ones are shown in Table 11.1. Some enzymes such as *Bam*HI, *Eco*RI and *Pst*I make staggered cuts on each strand. The resultant DNA molecules are said to have sticky ends. Such fragments can associate by complementary base pairing to any other fragment of DNA generated by the same enzyme. Other enzymes such as *Sma*I cleave the DNA smoothly to produce blunt ends. DNA fragments produced in this way can only be joined to another fragment that has a blunt end.

Figure 11.3 illustrates how human DNA is inserted into a plasmid that contains a *Bam*HI restriction endonuclease site. A short length of DNA (an oligonucleotide) that includes a *Bam*HI recognition site is added to each end of the human DNA fragment. Both the human DNA and the cloning vector are cut with *Bam*HI. The cut ends are now complementary and will anneal together. DNA ligase then catalyzes the formation of a phosphodiester bond between the vector and the human DNA. The resultant molecule is known as a *recombinant* plasmid.

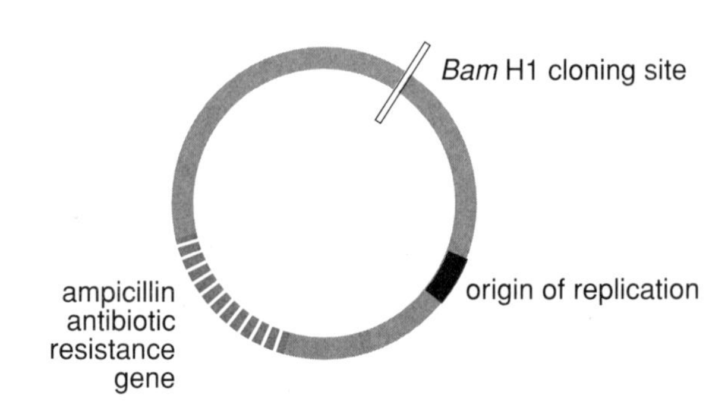

Figure 11.2
A plasmid-cloning vector.

Table 11.1 Recognition sites of some common restriction enzymes

Bacterial strain	Enzyme name	Recognition sequences and cleavage sites
Bacillus **am**yloliquefaciens **H**	*Bam* H1	G|GATCC CCTAG|G
Escherichia **co**li Ry13	*Eco* R1	G|AATTC CTTAA|G
Provedencia **st**uartii 164	*Pst* 1	CTGCA|G G|ACGTC
Serratia **ma**rcescens SB	*Sma* H1	CCC|GGG GGG|CCC
Rhodopseudomonas **s**ph**a**eroides	*Rsa* 1	GT|AC CA|TG

INTRODUCTION OF RECOMBINANT PLASMIDS INTO BACTERIA. Figure 11.4 summarizes how recombinant plasmids are introduced into bacteria such as *E. coli.* Bacteria are first treated with concentrated calcium chloride. This makes the cell wall more permeable and allows DNA to enter the cell. Cells that take up DNA in this way are said to be *transformed.* The transformation process is very inefficient and only a small percentage of cells actually take up the recombinant molecules. This means that it is extremely unlikely that any one bacterium has taken up two plasmids. The presence of an antibiotic resistance gene in the cloning vector (Fig. 11.2) makes it possible to select those bacteria that have taken up a molecule of foreign DNA (and its resistance gene), since only the transformed cells can survive in the presence of the antibiotic. The collection of bacterial colonies produced after this selection process is a clone *library.* Each cell in a single colony harbors an identical recombinant molecule, whereas other colonies contain plasmids carrying different DNA inserts. Isolating individual bacterial colonies will produce different clones of foreign DNA. If the starting DNA material used to produce these clones was a population of cDNA molecules, the collection of clones is called a cDNA library.

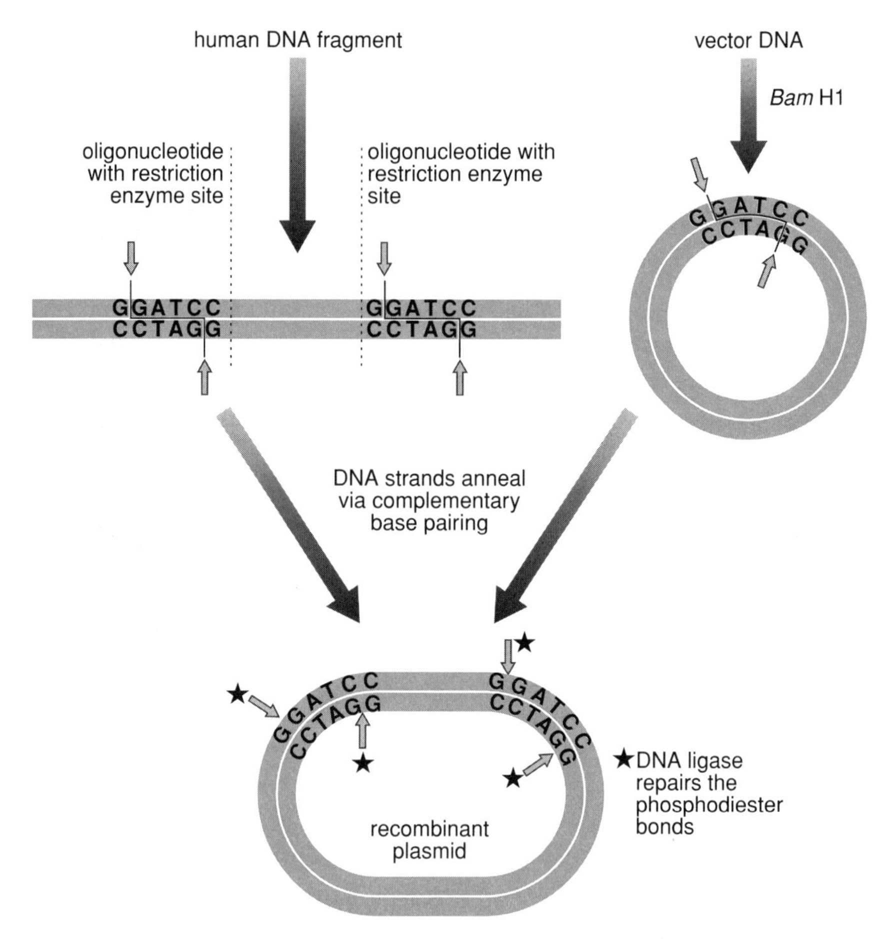

Figure 11.3

DNA molecules can be cut and joined at specific sites.

Selection of cDNA Clones

Having constructed a library—which may contain many thousands of different clones—the next step is to identify the bacteria that contain the cDNA of interest. There are many ingenious ways of doing this. One makes use of specific antibodies to detect the bacteria expressing the corresponding antigen. For this to work, the foreign DNA must be expressed in the bacterial cells, that is to say its information must be copied first into mRNA and then into protein. To ensure efficient expression, the vector contains a bacterial promoter sequence that is used to control transcription of foreign DNA. Such cloning vectors are

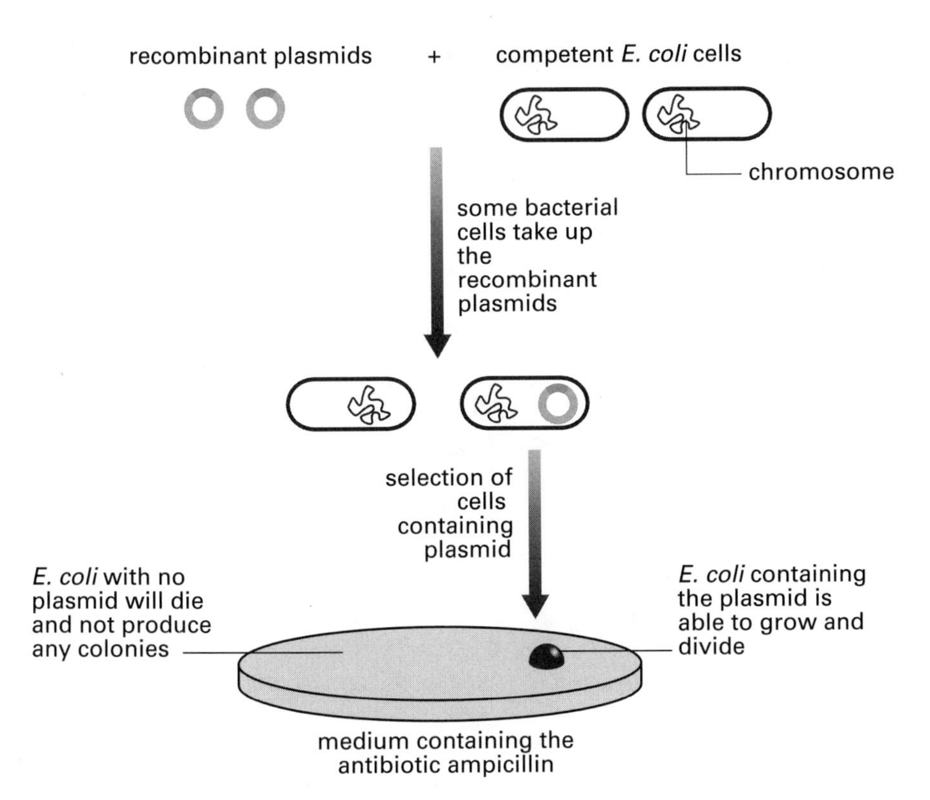

Figure 11.4

Introduction of recombinant plasmids into bacteria.

BOX 11.1

Cloning a Receptor Protein

Glutamate is one of the most important transmitters (page 318) in the brain. The gene coding for the glutamate receptor, the protein on the surface of nerve cells that binds glutamate, remained uncloned for a number of years. Success came with the use of a very clever cloning strategy, based on the function of the receptor. mRNA was isolated from brain cells and used as the template for the production of cDNA molecules. These were inserted into a plasmid expression vector. Following the introduction of these cDNAs into bacteria, a cDNA library representative of all the mRNAs in the brain was produced. The many thousands of cDNA clones in the library were then divided into pools. Each pool was then injected into a different *Xenopus* oocyte. The cDNAs were transcribed into RNA inside this cell. To see which of the oocytes had been injected with the cDNA for the glutamate receptor these cells were whole cell patch clamped (Box 17.2). Glutamate was applied to the oocytes, and the oocyte containing the mRNA for the glutamate receptor responded with an inward current of sodium ions (page 341).

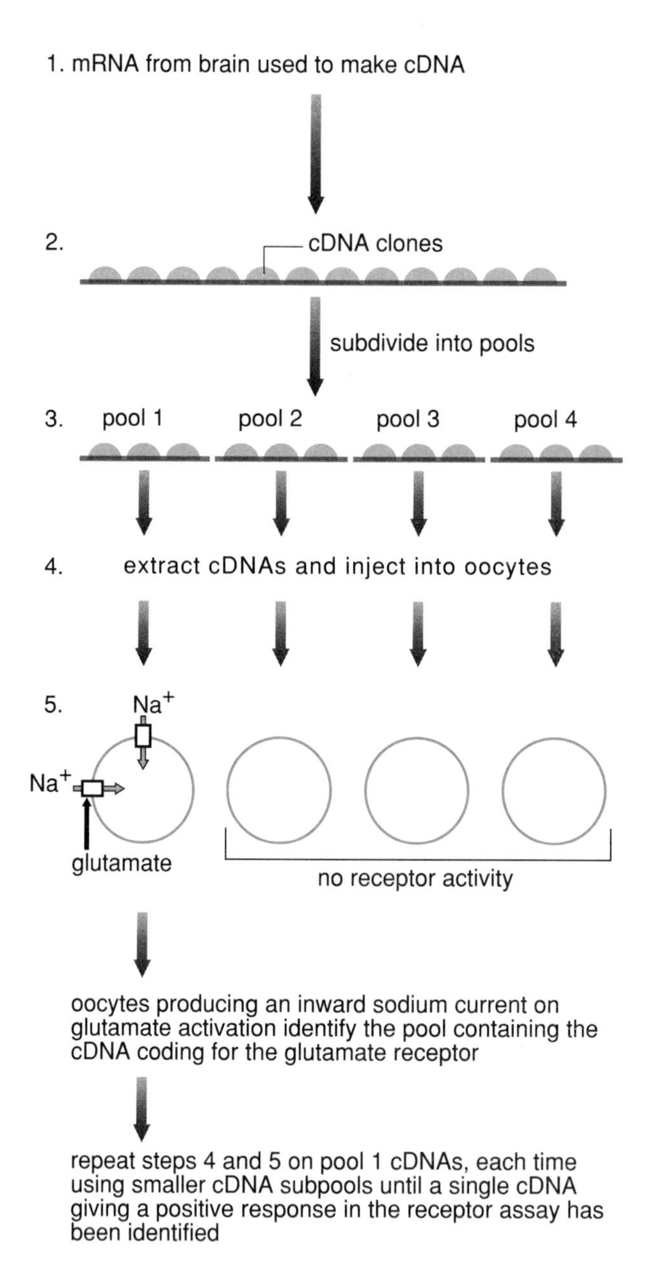

The pool of cDNAs giving this response was further divided into smaller pools. Each of these was rescreened for the presence of glutamate receptor activity. This was followed by several rounds of rescreening. For each round a further subdivision was made of the cDNAs into pools containing fewer and fewer cDNA molecules. Eventually each pool contains only a single cDNA so that the cDNA for the glutamate receptor could be identified. A number of other receptors have now been isolated using the strategy of a functional assay to identify the cDNA encoding that receptor.

known as expression vectors. The promoter of the *lac* operon is commonly used in this way (see Chapter 10).

The clone library is plated onto agar plates containing an inducer of the *lac* operon such as IPTG (page 156). The antibiotic is still present to ensure that bacteria without a plasmid die. A membrane is placed over the bacterial colonies and then lifted off. Bacterial proteins stick to the membrane, which can be incubated with the appropriate antibody. Bacteria that have synthesized the corresponding antigen are identified by carrying out a simple color test based on an enzyme reaction (Fig. 11.5).

Genomic DNA Clones

The approach described in the previous section permits the isolation of cDNA clones. cDNA clones have many uses, some of which are described below. However, if we want to investigate how transcription of the gene is regulated, we need to create genomic DNA clones that contain fragments of the chromosomal DNA including the promoter region. Genomic DNA clones are much larger than cDNA clones, since they contain both the exon and intron sequences. The cloning vectors used for the isolation of genomic sequences must therefore accommodate quite large pieces of DNA. Vectors based on the bacteriophage lambda are often used, as they will accept DNA fragments of about 20,000 base pairs. Large DNA fragments are usually generated by digesting genomic DNA with a restriction enzyme for a very short time. Not all the recognition sites for that enzyme are cleaved, and large fragments are produced by partial digestion. The genomic DNA fragments are joined to bacteriophage vectors in the usual way. Bacteria are then infected with the recombinant bacteriophage, ensuring that the efficiency of infection is low so that each bacterium will take up only a single bacteriophage particle. The collection of bacteria produced in this way is called a genomic DNA library (Fig. 11.6).

To select the genomic DNA sequence of interest, the library is plated onto a layer (or lawn) of cultured bacteria so that many copies of the recombinant bacteriophage can be produced. The recombinant bacteriophage multiply inside the host bacterial cell. The cells die and burst (or lyse), and the bacteriophage spread to the surrounding layer of bacteria and infect them. These cells lyse, in turn, and the process is repeated. The dead cells give rise to a clear area on the bacterial lawn called a plaque. Each plaque contains many copies of a recombinant bacteriophage that can be transferred to a membrane (Fig. 11.6). Specific DNA clones are selected by incubating the membrane with a radiolabeled cDNA probe complementary to the genomic sequence being searched for. This produces a radioactive area on the membrane that is identified by au-

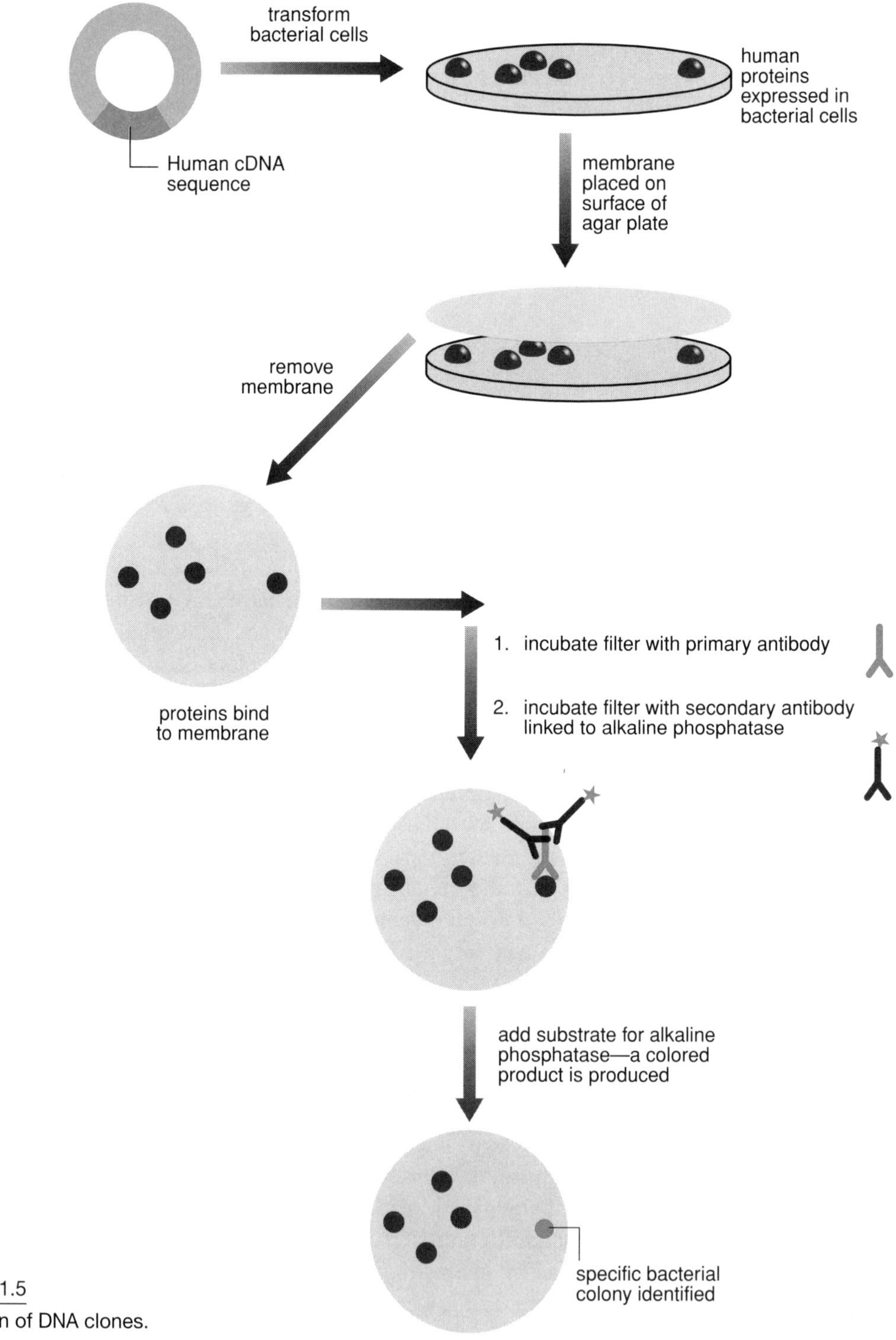

Figure 11.5

Selection of DNA clones.

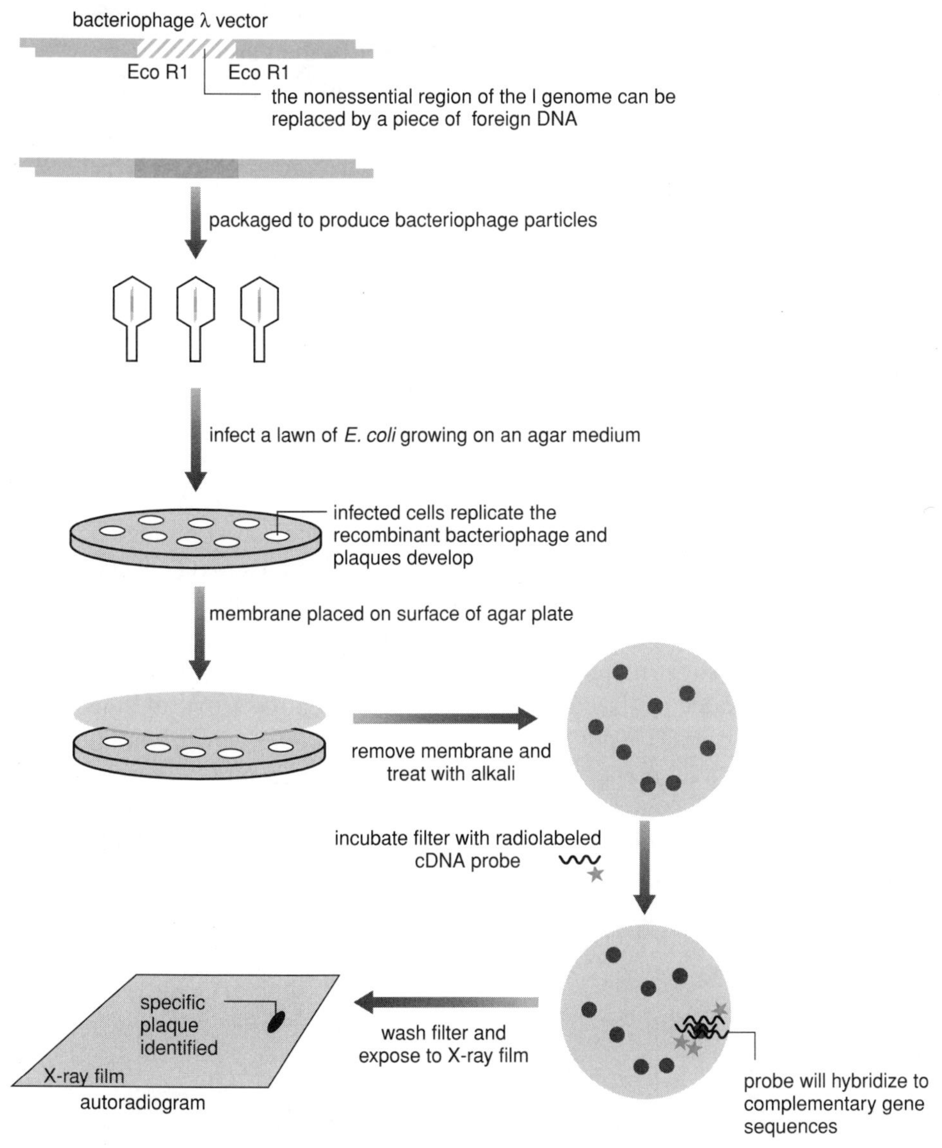

Figure 11.6
Selection of genomic DNA clones.

toradiography. The use of a cDNA sequence as a gene probe makes the task of isolating the corresponding genomic sequence much easier.

Unfortunately, for most genetic diseases we do not know what has gone wrong in the cellular machinery. If the protein product of a particular gene is unknown, it is impossible to isolate an appropriate cDNA clone using conventional techniques. To find the gene for a disease when the identity of the normal protein is not known is very difficult. Often the only clue comes from genetic linkage studies (page 107) that may identify the chromosome on which the gene is located. Gene technology has revolutionized genetics. The subject has been turned on its head taking it into a completely novel phase. Reverse genetics starts with a genome and works backward to identify the protein product. Although time-consuming and expensive, this approach has been used successfully to isolate and analyze many human genes, including the gene responsible for the disease cystic fibrosis as described in Chapter 21.

USES OF DNA CLONES

DNA Sequencing

The ability to determine the order of the bases within a DNA molecule has been one of the greatest technical contributions to molecular biology. DNA is made by the polymerization of the four deoxynucleotides dATP, dGTP, dCTP, and dTTP (Chapter 5). These are joined together when DNA polymerase catalyzes the formation of a phosphodiester bond between a free 3′ hydroxyl on the deoxyribose sugar moiety of one nucleotide and a free 5′ phosphate group on the sugar residue of a second nucleotide. However, the artificial dideoxynucleotides ddATP, ddGTP, ddCTP and ddTTP have no 3′ hydroxyl on their sugar residue (Fig. 11.7), and so if they are incorporated into a growing DNA chain, synthesis will stop. This is the basis of the dideoxy chain termination DNA sequencing technique devised by Frederick Sanger and for which he was awarded the Nobel prize in 1980.

This technique is illustrated in Figure 11.7. A cloned piece of DNA of

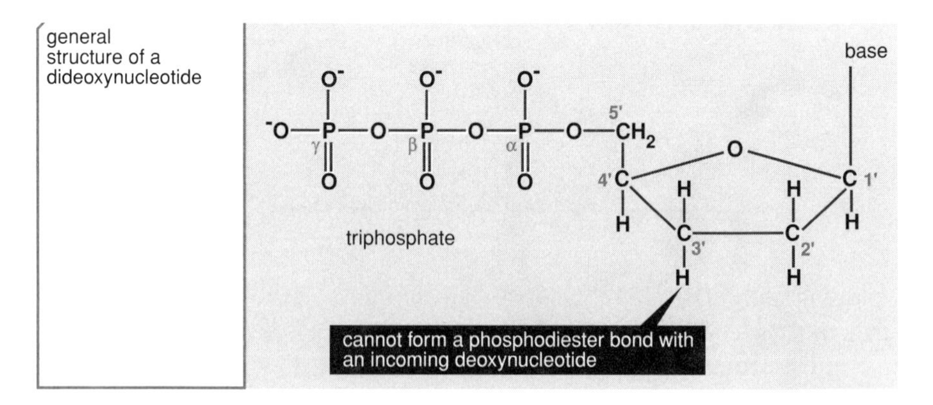

Figure 11.7

DNA sequencing and the dideoxy chain termination method.

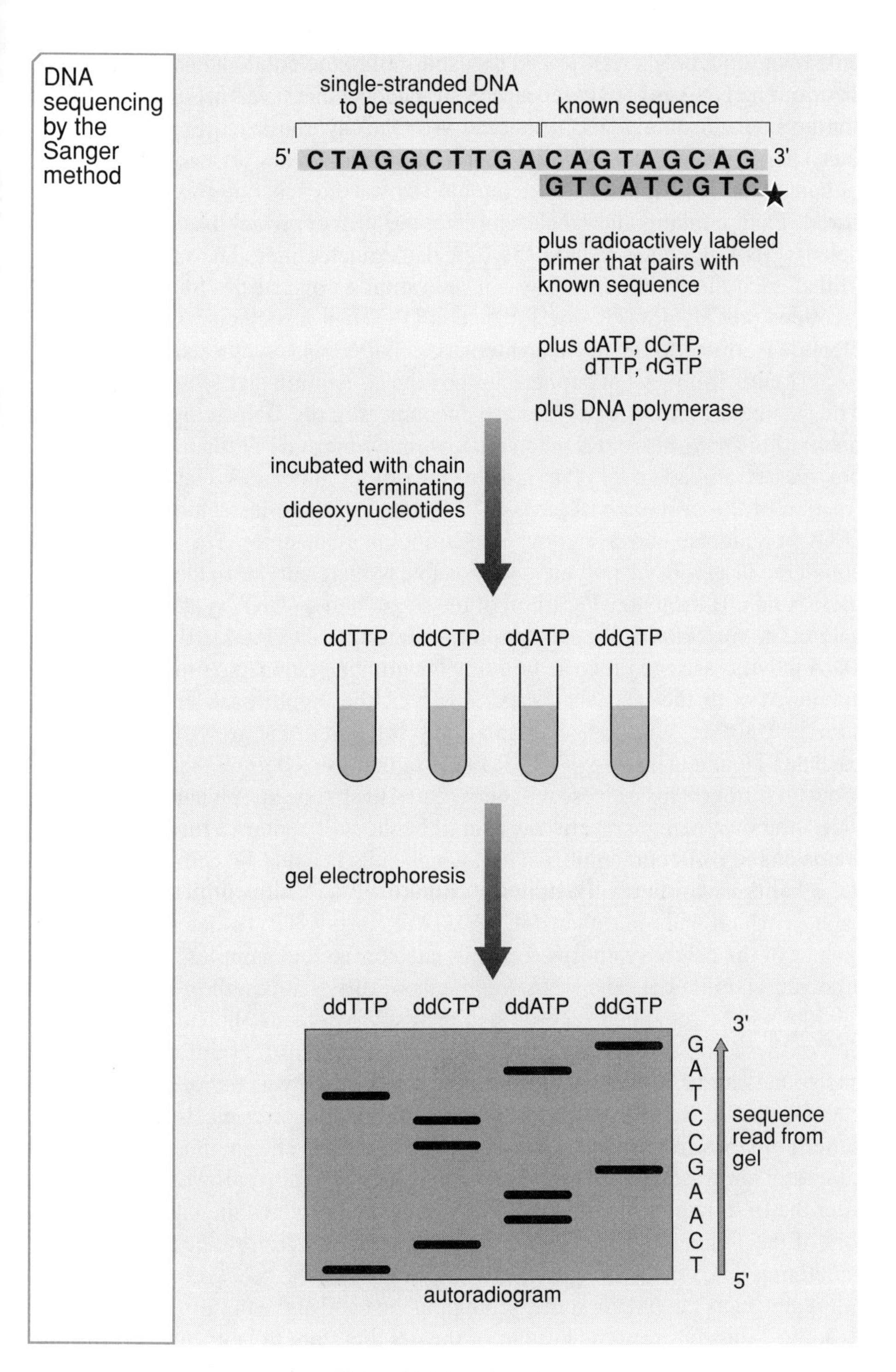

Figure 11.7 (*continued*)

unknown sequence is first joined to a short oligonucleotide whose sequence is known. The DNA is then made single-stranded so that it can act as the template for the synthesis of a new DNA strand. All DNA synthesis requires a primer; in this case a primer is provided that is complementary in sequence to the oligonucleotide attached to the template DNA. Four separate mixtures are prepared. Each contains the DNA template, the primer (which has been radiolabeled), DNA polymerase, and the four deoxynucleotides. The mixtures differ in that each also contains a low concentration of one of the four dideoxynucleotides ddATP, ddGTP, ddCTP, or ddTTP. When a molecule of dideoxynucleotide is joined to the newly synthesized chain, DNA synthesis will stop.

Let us follow what happens in the tube containing ddTTP in Figure 11.7. The first base that DNA polymerase encounters in the DNA template to be sequenced is an A. Since the tube contains much more dTTP than ddTTP, DNA polymerase will add a dTTP to most of the primer molecules. However, a small fraction of the primers will have ddTTP added to them instead of dTTP since DNA polymerase can use either nucleotide as a substrate. The next base encountered is a G. DNA polymerase is unable to attach dCTP to the ddTTP since there is no OH group on 3′ carbon of the sugar, and so DNA synthesis is terminated. The majority of the strands, however, had a dTTP added, and for these DNA polymerase can proceed, building the growing chain. No problems are encountered with the next six bases. However, the eighth base in the template strand is another A, and once again a small fraction of the growing strands will have ddTTP added instead of dTTP. In the same way as before, these strands can grow no further. This process will be repeated each time an A occurs on the template strand. When the reaction is over, the tube will contain a mixture of DNA fragments of different length, each of which ends in a ddTTP. Similarly, each of the other three tubes will contain a mixture of DNA chains of different length, each of which ends in either ddCTP, ddATP, or ddGTP. To determine the sequence of the newly synthesized chains, each of the four samples is loaded onto a polyacrylamide gel. The DNA fragments of different length in each reaction mixture will be separated according to their size by electrophoresis. The resolution of these polyacrylamide gels is so good that chains that differ by only one nucleotide can be separated. Because the primer used was radiolabeled, all the new DNA chains will carry a radioactive tag, so that after electrophoresis the pattern of DNA fragments on the gel can be detected. The smallest DNA molecules are that fraction in the tube containing ddTTP whose growth was blocked after the first base, T. These move the furthest and produce the band at the bottom of the T lane. DNA molecules one base larger were produced in the tube containing ddCTP—their growth was blocked after the second base, C. These move almost as far, but not quite, producing the band at the bottom of the C lane. Reading bands up from the bottom of the gel therefore tells us the sequence in which bases were added to the unknown strand: T, C, and so on. Because the new chain is complementary in sequence to its template strand the sequence of the template strand can be inferred.

Southern Blotting

In 1975 Ed Southern developed an ingenious technique, now known as Southern blotting, which can be used to detect specific genes (Fig. 11.8). Genomic DNA is isolated and digested with one or more restriction endonucleases. The resultant fragments are separated according to their size by agarose gel electrophoresis and transferred to a membrane filter (Fig. 11.8). This produces an exact replica of the pattern of DNA fragments in the agarose gel. The membrane filter is incubated with a cloned DNA fragment marked with a radioactive label. This gene probe will base pair to its complementary sequences on the membrane filter. This is called hybridization. As the gene probe is radiolabeled, the sequences to which it has hybridized can be detected by autoradiography. The pattern of DNA bands on the autoradiogram can be used to identify a particular gene. Mutations that change the pattern of DNA fragments—for instance, by altering a restriction enzyme recognition site or deleting a large section of the gene—can easily be detected by Southern blotting. This technique is hence useful in determining whether an individual carries a certain genetic defect or if a fetus is homozygous for a particular disorder. All that is needed is a small DNA sample, either from white blood cells or, in the case of a fetus, from the amniotic fluid in which it is bathed, or by removing a small amount of tissue from the chorion villus that surrounds the fetus in the early stages of pregnancy.

Forensic laboratories use Southern blotting to generate DNA fingerprints from samples of blood or semen left at the scene of a crime. A DNA fingerprint is a person-specific Southern blot. The gene probe used in the test is a sequence that is highly repeated very many times within the human genome—a minisatellite sequence (page 115). Everyone carries a different number of these repeated sequences, and because they lie side by side on the chromosome they are called VNTRs (Variable Number Tandem Repeats). When genomic DNA is digested with a restriction enzyme and then analyzed by Southern blotting, a DNA pattern of their VNTRs is produced. Except for identical twins (Fig. 11.9), it is extremely unlikely that two individuals will have the same DNA fingerprint profile.

In situ Hybridization

A cDNA clone can be used as a probe to detect if a gene is expressed in a particular tissue. This is very useful because in a tissue made of different cell

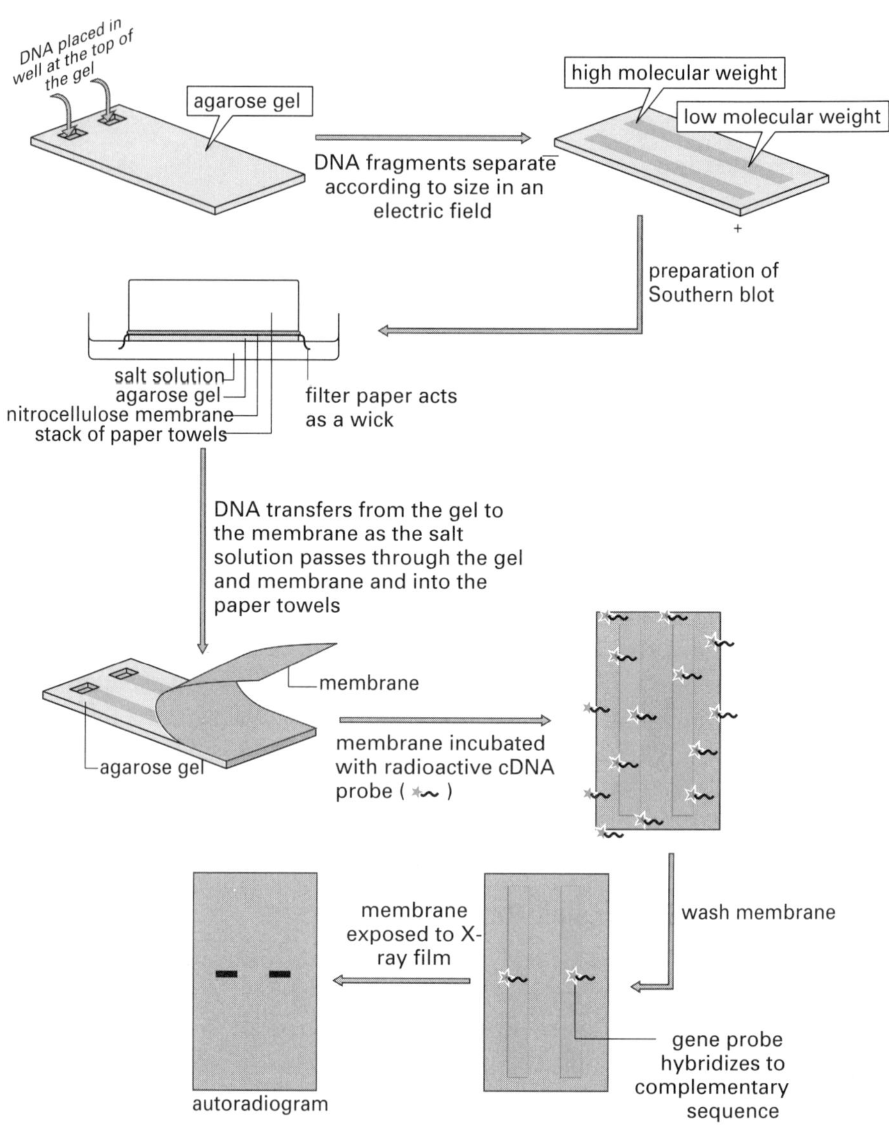

Figure 11.8

The technique of Southern blotting.

types we can identify the specific cell type in which a particular mRNA is made. A technique known as *in situ* hybridization is used to do this. We first have to make a radioactive DNA probe. In this case, because we want to detect mRNA we must radiolabel the template DNA strand. The DNA probe is applied to fixed tissue sections, and it will hybridize to its complementary mRNA

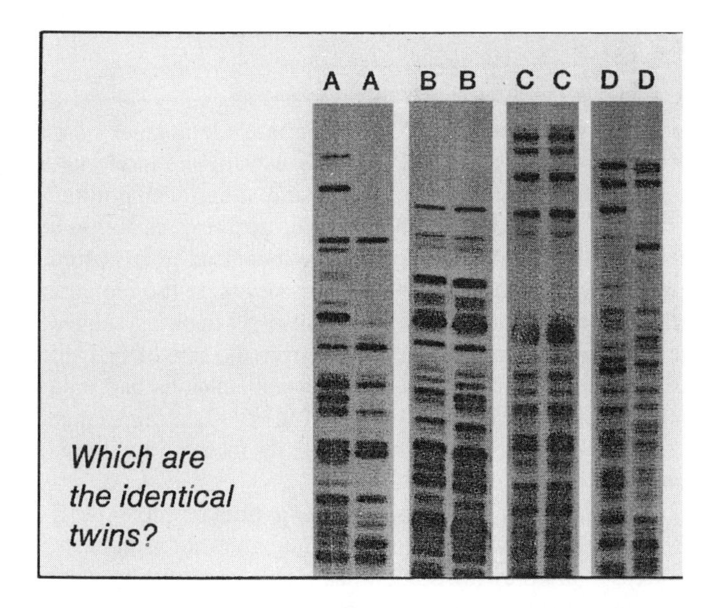

Figure 11.9
Each pair of lanes AA, BB, etc., is a genetic fingerprint from a pair of twins. Which twins are identical?

in those cells that express the gene. After the unhybridized probe is washed off, only cells that were expressing the gene of interest remain radioactive and can thus be identified by autoradiography.

Production of Mammalian Proteins in Bacteria

The large-scale production of proteins using cDNA-based expression systems has wide applications for medicine and industry. It is increasingly being used to produce polypeptide-based drugs, vaccines, and antibodies. Such protein products are called recombinant. For a mammalian protein to be synthesized in bacteria its cDNA must be cloned into an expression vector (as described on page 172). Insulin was the first human protein to be expressed from a plasmid introduced into bacterial cells, and it has now largely replaced insulin from pigs and cattle for the treatment of diabetes. Other products of recombinant DNA technology include growth hormone and Factor VIII, a protein used in the treatment of the blood clotting disorder hemophilia. Factor VIII was previously isolated from donor human blood. However, because of the danger of infection from viruses such as HIV, it is much safer to treat hemophiliacs with recombinant factor VIII. It should, in theory, be possible to express any human protein via its cDNA.

BOX 11.2

Transgenic Animals

A transgenic animal is produced by injecting a foreign gene into the nucleus of a fertilized egg. The egg is then implanted into a foster mother, and the offspring are tested to determine whether they carry the foreign gene. If they do, a transgenic animal has been produced. Transgenic mice are commonly used to investigate the regulation of gene expression in mammals. The production of the protein metallothionein is known to be increased by exposure to heavy metal ions. The first transgenic mice produced were used to investigate the regulation of this gene. The gene encoding rat growth hormone was fused to the 5′-flanking sequence of the metallothionein gene. Mice carrying the transgene grew to twice the size of their litter mates after they had drunk water containing zinc because the heavy metal ions had increased the production of growth hormone. This experiment proved that the 5′-flanking sequence of the metallothionein gene contains the sequences responsible for its increased expression when heavy metal ions are present.

Transgenic farm animals—such as sheep synthesizing human factor VIII in their milk—have been created. This is an alternative to producing human proteins in bacteria. Making transgenic animals is not easy, and they are sometimes sterile or suffer from other disorders. These problems arise because foreign DNA is integrated at random into the genome of the host animal. The next step must be to perfect techniques that permit genes to be inserted into the host genome in the appropriate chromosomal location.

Protein Engineering

The ability to change the amino acid sequence of a protein by altering the sequence of its cDNA is known as protein engineering. This is achieved through the use of a technique known as site-directed mutagenesis. A new cDNA is created that is identical to the natural one except for changes designed into it by the scientist. This DNA can then be used to generate protein in bacteria, to transfect cell lines, or to create transgenic organisms.

The first use of protein engineering is to study the protein itself. A comparison of the catalytic properties of the normal and mutated form of an enzyme helps to identify amino acid residues important for substrate and co-factor binding sites (Chapters 13, 14). This technique was also used to identify the particular charged amino acid residues responsible for the selectivity of ion channels (page 291). Now scientists are using protein engineering to generate new proteins as tools, not only for scientific research but for wider medical and industrial purposes (Box 12.2).

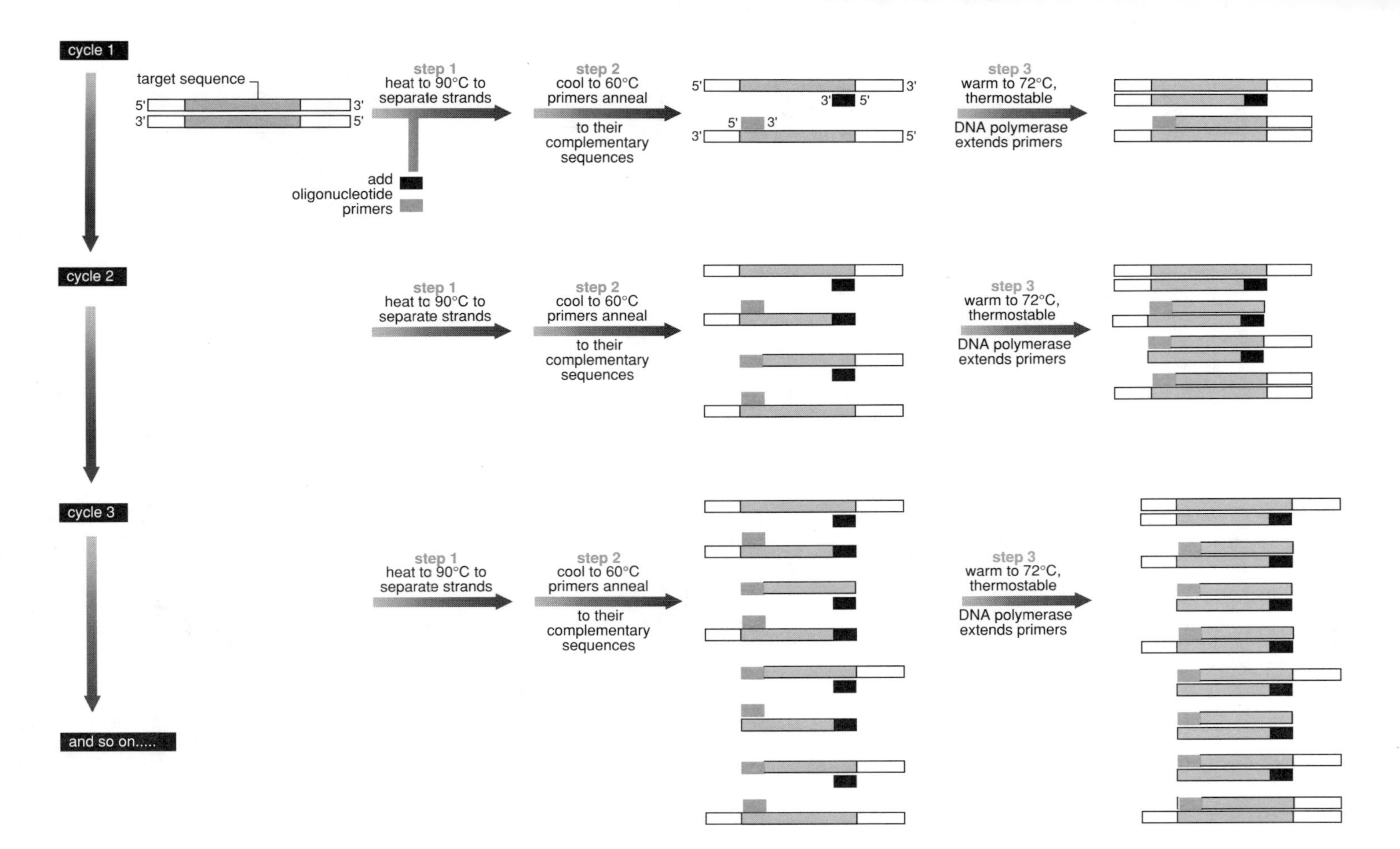

Figure 11.10
The amplification of a DNA sequence using the polymerase chain reaction.

Polymerase Chain Reaction

A technique, called the polymerase chain reaction (PCR), has revolutionized recombinant DNA technology. It can amplify DNA from as little material as a single cell and from very old tissue such as that isolated from Egyptian mummies, a frozen mammoth, and insects trapped in ancient amber. A simple saliva swab can yield enough DNA to determine carriers of a particular recessive genetic disorder. PCR is used to amplify DNA from fetal cells or from small amounts of tissue found at the scene of a crime.

Figure 11.10 shows how the PCR uses a thermostable DNA polymerase and two short oligonucleotide DNA sequences called primers. Each primer is complementary in sequence to one of the two strands of DNA to be amplified. The primers direct DNA polymerase to copy each of the template strands. The DNA duplex is heated to 90°C to separate the two strands (step 1). The mixture is cooled to 60°C to allow the primers to anneal to their complementary sequences (step 2). At 72°C the primers direct a thermostable DNA polymerase to copy each of the template strands (step 3). These three steps, which together constitute one cycle of the PCR, produce twice the number of original templates. The process of template denaturation, primer annealing, and DNA synthesis is repeated many times to yield many thousands of copies of the original target sequence. Even though PCR was introduced only in 1985, it is hard to imagine how scientists managed without it.

The applications of recombinant DNA technology are exciting and far-reaching. In the future it may be possible to correct genetic defects before or after birth by replacing mutated genes by a normal copy. The technology to carry out such experiments is very demanding, and important ethical questions are raised. In Chapter 21 we will discuss the potential use of gene therapy for the correction of the genetic disorder cystic fibrosis.

SUMMARY

1. DNA sequences can be cloned using reverse transcriptase, which copies mRNA into DNA to make a hybrid mRNA:DNA molecule. The mRNA is then changed into DNA by the enzymes ribonuclease H and DNA polymerase. The new double-stranded DNA molecule is called complementary DNA (cDNA).

2. Restriction endonucleases cut DNA at specific sequences. DNA molecules cut with the same enzyme can be joined together. To clone a cDNA or genomic DNA it is first joined to a cloning vector to form a recombinant molecule.

3. Recombinant DNA molecules are introduced into bacterial cells by the process of transformation. This produces a collection of bacteria (a library). Each bacterium contains a different DNA molecule. The DNA molecule of interest is then selected from the library using either an antibody or a nucleic acid probe.

4. There are many important medical, forensic, and industrial uses for DNA clones. These include:

- **Determination of the base sequence of the cloned DNA fragment.**
- **Southern blotting and genetic fingerprinting to analyze an individual's DNA pattern.**
- ***In situ* hybridization to detect specific cells making a particular mRNA.**
- **Synthesis of mammalian proteins in bacteria.**
- **Protein engineering to alter the DNA sequence encoding a particular protein.**
- **The polymerase chain reaction, which lets us produce many copies of a DNA molecule in a test tube.**

BOX 11.3

Transgenic Plants

Foreign genes can also be introduced into dicotyledonous plants—trees, potatoes, and turnips, for example—using the naturally occurring Ti plasmid from the bacterium *Agrobacterium tumefaciens.* The normal plasmid causes tumors (called galls on a plant), but this troublesome region of the plasmid can be cut out and replaced by foreign genes. This technique has been used to insert into maize a gene coding for a protein rich in the essential amino acid lysine. This transgenic maize will, it is hoped, greatly improve the nutritional status of people in countries in Africa, for example. In the same way, wheat has been made resistant to herbicides by the introduction of a bacterial gene whose protein product inactivates such chemicals. This means that a field can be treated with herbicide to kill weeds, leaving the original crop plants unharmed.

In supermarkets there is now a tomato with reduced amounts of an enzyme necessary for ripening. This tomato does not go soft on storage. The plant was created by inserting a gene which, when transcribed, produces a RNA complementary in sequence to the mRNA for the ripening protein. The two RNAs bond by complementary base-pairing, and so the translation of the normal mRNA is inhibited. This gene knockout technique is increasingly used to study the function of proteins in animals as well as in plants. A gene is indirectly "knocked out" by inhibiting the translation of its mRNA, and the effect on development and functioning of the animal is observed.

CHAPTER

12

TRANSLATION AND PROTEIN TARGETING

The genetic code dictates the sequence of amino acids in a protein molecule. The synthesis of proteins is quite complex. Three types of RNA are needed. Messenger RNA (mRNA) contains the code and is the template for protein synthesis. Transfer RNAs (tRNAs) are adapter molecules that carry amino acids to the mRNA. Ribosomal RNAs (rRNAs) form part of the ribosome that brings together all the components necessary for protein synthesis. Several enzymes also help in the construction of new protein molecules. This chapter describes how the nucleotide sequence of an mRNA molecule is translated into the amino acid sequence of a protein and how proteins are modified after they are made to enable them to reach their correct cellular location.

Translation (or protein synthesis) can be divided into four stages:

1. Attachment of an amino acid to its tRNA.
2. Initiation of protein synthesis.
3. Elongation of the polypeptide chain.
4. Termination of protein synthesis.

THE ATTACHMENT OF AN AMINO ACID TO ITS tRNA

Amino acids are not directly incorporated into protein on a messenger RNA template. A second type of RNA molecule, transfer RNA (tRNA), is needed. It brings the correct amino acid, as specified by the codons of the mRNA, to the mRNA chain.

tRNA molecules have both an amino acid attachment site and an *anticodon* (3 bases that are complementary in sequence to a codon) (Fig. 12.1). The anticodon recognizes its codon in the mRNA molecule and they attach by hydrogen bonds to form antiparallel base pairs.

Transfer RNA, the Anticodon, and the Wobble

Although 61 codons specify the 20 different amino acids, there are not 61 tRNAs. Instead the cell economizes; the codons for some amino acids differ only in the third position of the codon. Table 5.1 shows that when an amino acid is encoded by only two different triplets, the third bases will be either C and U or A and G, for example aspartate (GAC, GAU) and glutamine (CAA, CAG). The wobble hypothesis suggests that the pairing of the first two bases in the codon and anticodon follows the rules—G bonds with C, and A bonds with U—but the base pairing in the third position is not as restricted, it can wobble. Both U and C in the third position of the codon can hydrogen bond with a G in the anticodon, and vice versa. Thus only one tRNA molecule is

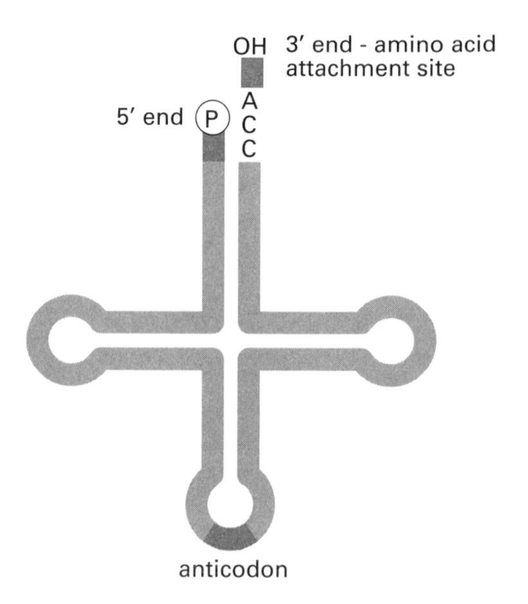

Figure 12.1
Transfer RNA (tRNA).

required for two codon sequences. The anticodon of some tRNAs contains the modified nucleoside inosine (I). I can base pair with any of U, C, or A in the third position of the codon. Some tRNA molecules can therefore base pair with as many as three different codons provided the first two bases of the codon are the same. Isoleucine (AUU, AUC, AUA) is an example.

The attachment of an amino acid to its correct tRNA molecule is illustrated in Figure 12.2. This process occurs in two stages, both catalyzed by the enzyme aminoacyl tRNA synthetase. During the first reaction the amino acid is joined, via its carboxyl group, to an adenosine monophosphate (AMP) molecule. This complex remains bound to the aminoacyl tRNA synthetase until the amino acid interacts with the correct tRNA molecule to form an aminoacyl tRNA. There are at least 20 aminoacyl tRNA synthetases, one for each amino acid and its specific tRNA.

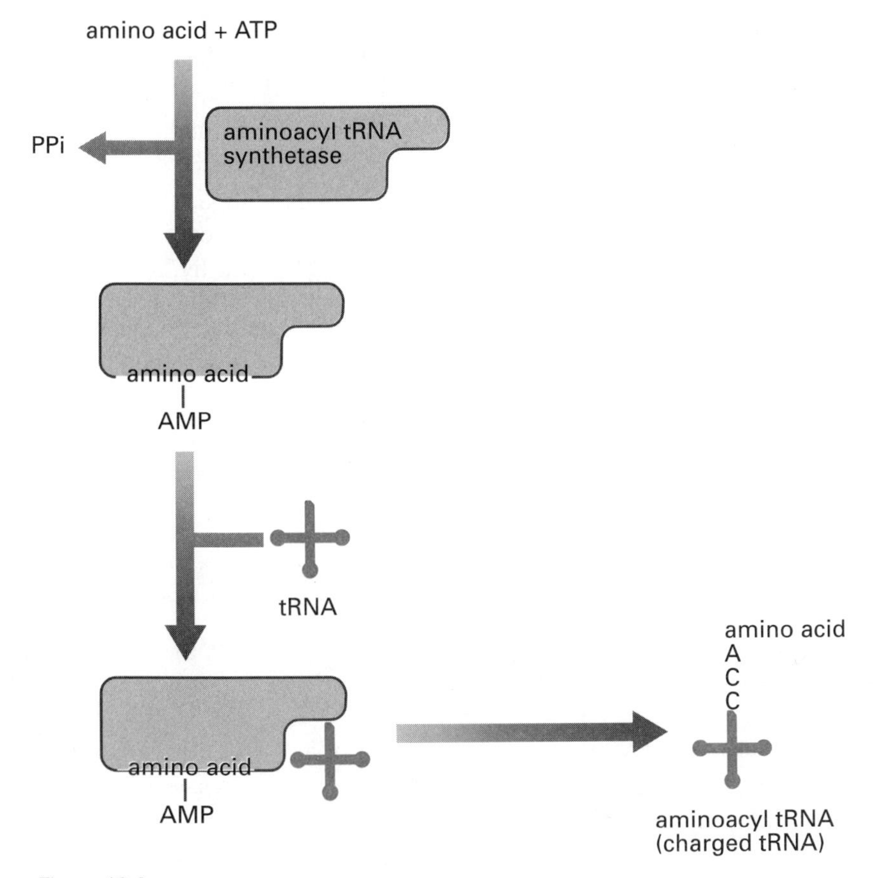

Figure 12.2

The attachment of an amino acid to its tRNA. PPi represents pyrophosphate (see Fig. 10.2, page 147).

All tRNA molecules have at their 3′ end the nucleotide sequence CCA. An amino acid is joined to its tRNA through an ester bond between its carboxyl group and either the 2′- or 3′-hydroxyl group of the ribose of the terminal adenosine (A) on the tRNA to form an *aminoacyl tRNA*. This step is often referred to as amino acid activation because the energy of the ester bond can be used in the formation of a lower energy peptide bond between two amino acids. A tRNA that is attached to an amino acid is known as a charged tRNA.

THE RIBOSOME

The ribosome is the cell's factory for protein synthesis. Each ribosome consists of two subunits, one large and one small, that join together on the mRNA molecule. In bacteria the ribosome consists of 3 ribosomal RNA (rRNA) molecules and about 55 proteins. In eukaryotic organisms the ribosome is larger and comprises 4 rRNAs and about 80 proteins. The ribosomal subunits and their RNAs are named using numbers that describe how fast they sediment in a centrifuge. These are called S values and are shown in Table 12.1. Some of the proteins of the ribosome are enzymes that are needed for protein synthesis.

The formation of a peptide bond (page 210) between two amino acids takes place on the ribosome. The ribosome has binding sites for the mRNA template and for two charged tRNAs. An incoming tRNA with its linked amino acid occupies the *aminoacyl site (A-site)* and the tRNA attached to the growing polypeptide chain occupies the *peptidyl site (P-site)*.

Table 12.1 Ribosomal subunits and their rRNAs

	Subunit	S Value	rRNAs
Bacterial	Large	50S	23S, 5S
	Small	30S	16S
Eukaryotic	Large	60S	28S, 5.8S, 5S
	Small	40S	18S

Note: The S value (Svedberg) refers to the sedimentation coefficient of the intact ribosome or its subunits. It is a measure of how rapidly a particle sediments in an ultracentrifuge.

BACTERIAL PROTEIN SYNTHESIS

Chain Initiation

The first amino acid incorporated into a new bacterial polypeptide is always a modified methionine, formyl methionine (fmet) (Fig. 12.3). Methionine first attaches to a specific tRNA molecule, $tRNA^{fmet}$, and is then modified by the addition of a formyl group that attaches to its amino group. $tRNA^{fmet}$ has the anticodon sequence 5′ CAU 3′ that binds to its complementary codon, the universal start codon AUG.

Ribosome-Binding Site

For protein synthesis to take place, a ribosome must first attach to the mRNA template. AUG is not only the start codon for protein synthesis, it is used to code for all the other methionines in the protein. How does the ribosome recognize the correct AUG codon at which to begin synthesis? All bacterial mRNAs have at their 5′ end a stretch of nucleotides—the untranslated (or leader) sequence—which do not code for the protein. These nucleotides are nevertheless essential for the correct placing of the ribosome on the mRNA. Bacterial mRNA molecules usually have a nucleotide sequence similar to 5′ GGAGG 3′ whose center is about 8 to 13 nucleotides upstream (5′ to) the AUG start codon. This sequence is complementary to a short stretch of sequence, 3′ CCUCC 5′, found at the 3′ end of the 16S rRNA molecule in the small ribosomal subunit. The mRNA and the rRNA interact by complementary base pairing to place the small ribosomal subunit in the correct position to start protein synthesis. The sequences on the

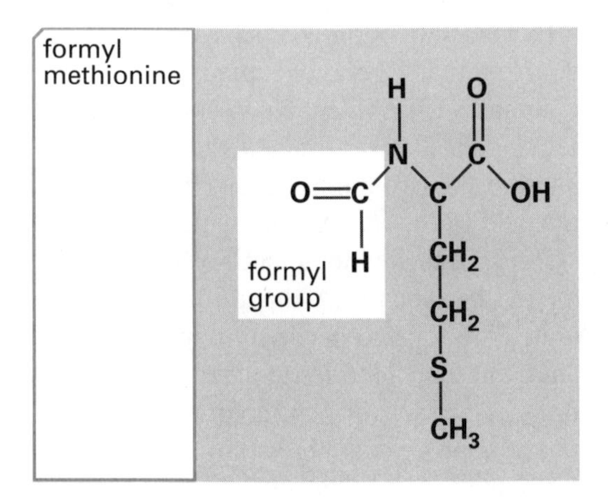

Figure 12.3
Formyl methionine.

mRNA molecule are called the ribosome-binding site. This is sometimes referred to as the Shine–Dalgarno sequence after the two scientists who found it.

The 70S Initiation Complex

The initiation phase of protein synthesis involves the formation of a complex between the ribosomal subunits, a mRNA template, and $tRNA^{fmet}$ (Fig. 12.4). A 30S subunit attaches to the ribosome-binding site as described above. $tRNA^{fmet}$ then interacts with the AUG initiation codon, and finally the 50S ribosomal subunit attaches. The ribosome is now complete, and the first tRNA and its amino acid are in place in the P-site of the ribosome. A 70S initiation complex has been formed, and protein synthesis can begin. The ribosome is orientated so that it will move along the mRNA in the 5′ to 3′ direction, the direction in which the information encoded in the mRNA molecule is read.

Three proteins called initiation factors 1, 2, and 3, together with guanosine triphosphate (GTP) are needed to help the 70S initiation complex to form (Fig. 12.4). Initiation factors 1 and 3 are attached to the 30S subunit. Initiation factor 3 helps in the recognition of the ribosome-binding site on the mRNA. Initiation factor 2 specifically recognizes $tRNA^{fmet}$ and binds it to the ribosome. When the 50S subunit attaches, the three initiation factors are released and GTP is hydrolyzed.

Elongation of the Protein Chain

The synthesis of a protein begins when an aminoacyl-tRNA enters the A-site of the ribosome (Fig. 12.5). The identity of the incoming aminoacyl tRNA is determined by the codon on the mRNA. If, for example, the second codon is 5′ UUC 3′, then phenylalanyl $tRNA^{phe}$ (whose anticodon is 5′ GAA 3′) will occupy the A-site. The P-site has of course already been occupied by $tRNA^{fmet}$ during the formation of the initiation complex (Fig. 12.5). Now that both the A- and P-sites are occupied, the enzyme *peptidyl transferase* catalyzes the formation of a peptide bond (page 210) between the two amino acids (fmet and phe in this example). The dipeptide is attached to the tRNA occupying the A-site. Because the tRNA in the P-site is no longer attached to an amino acid, it is released from the ribosome and it can be reused. The tRNA-dipeptide now moves to occupy the P-site (leaving the A-site free for another incoming aminoacyl tRNA). The movement of the ribosome, three nucleotides at a time, relative to the mRNA, is known as translocation. The process of peptide bond formation is followed by translocation and the whole process is repeated until the ribosome reaches a stop signal and protein synthesis terminates.

Proteins are synthesized beginning at their amino or N terminus (page

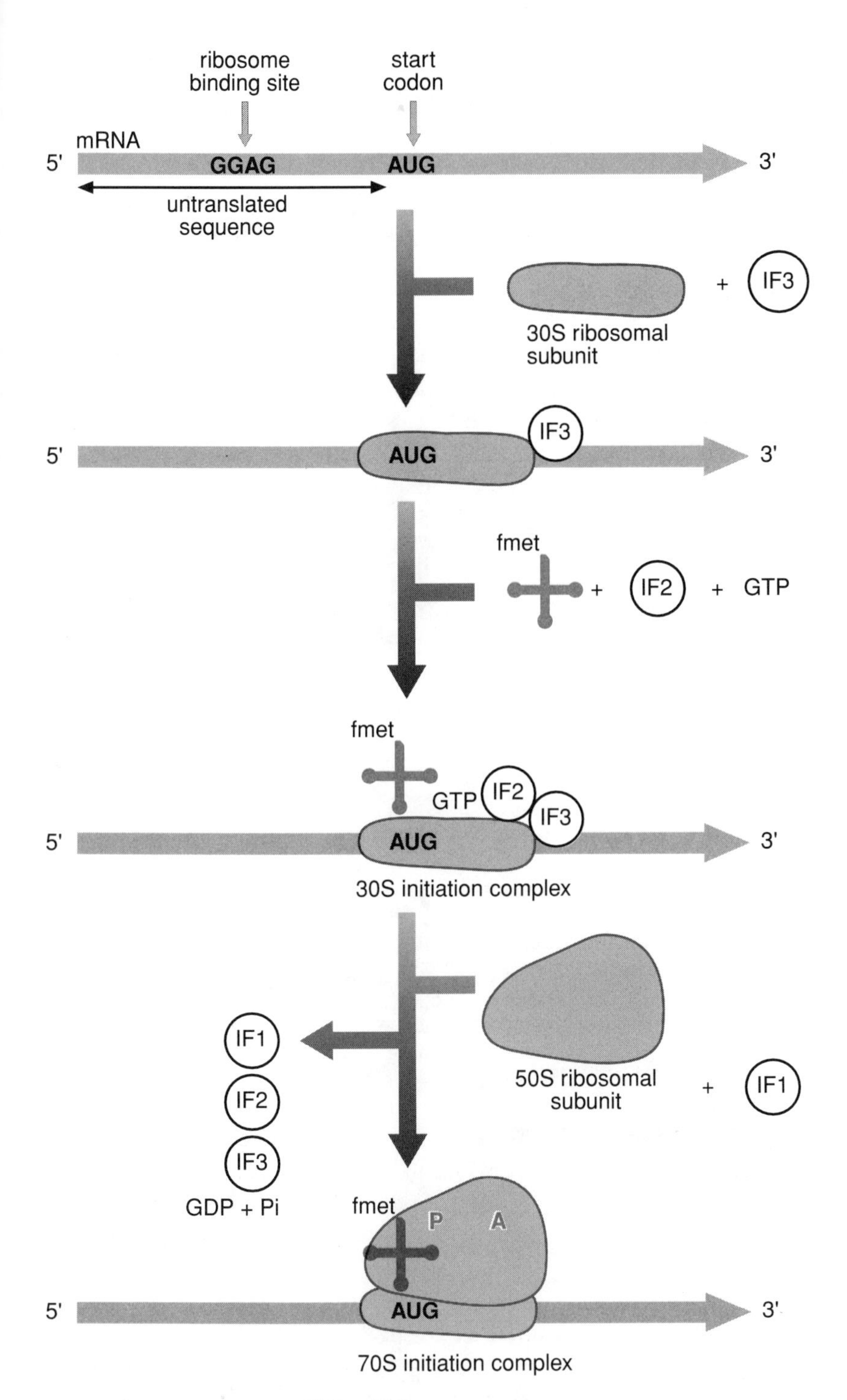

Figure 12.4
The 70S initiation complex.

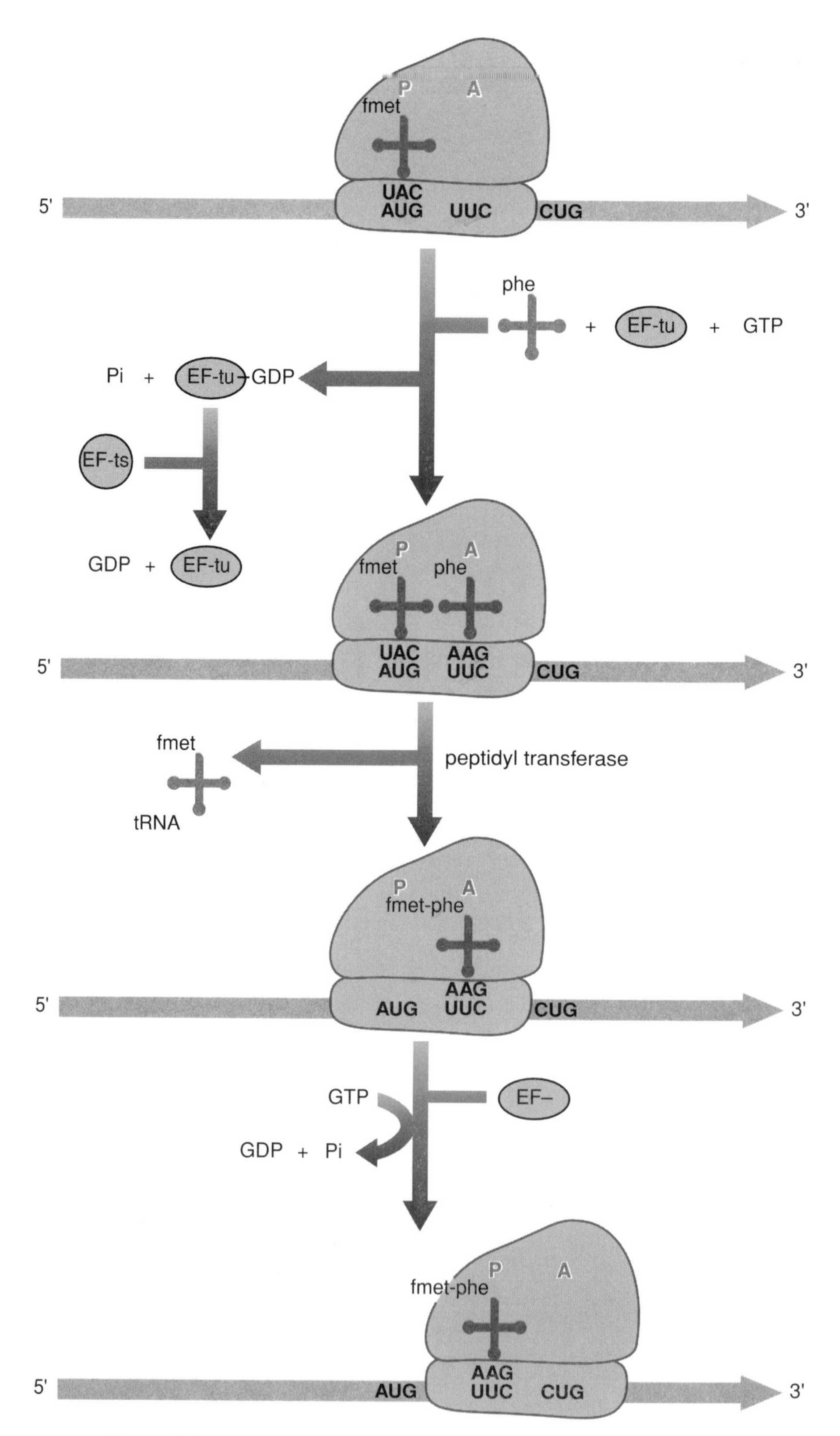

Figure 12.5

Elongation of the protein chain. Pi represents an inorganic phosphate ion.

211). The first amino acid hence has a free amino group. The last amino acid in the chain has a free carboxyl group and is known as the carboxyl or C terminus (page 211).

Elongation of a polypeptide chain needs the help of three proteins called *elongation factors*. These proteins are needed to speed up the process of protein synthesis. To bind to the A-site, the aminoacyl-tRNA must form a complex with elongation factor tu and a molecule of GTP. On hydrolysis of GTP to GDP and an inorganic phosphate ion (Pi) (page 252), the aminoacyl tRNA is able to enter the A-site. Elongation factor tu is released from the aminoacyl-tRNA by the action of a second protein, elongation factor ts. This removes the GDP bound to elongation factor tu, and the protein is recycled. The movement of the ribosome needs the help of a third protein elongation factor G and the energy supplied by the hydrolysis of a second GTP molecule.

Some bacteria cause disease because they inhibit eukaryotic protein synthesis. Diphtheria was once a wide spread and often fatal disease caused by infection with the bacterium *Corynebacterium diphtheriae.* This organism produces an enzyme (diphtheria toxin) that inactivates eukaryotic elongation factor 2 (the equivalent of the bacterial elongation factor G). Diphtheria toxin transfers ADP-ribose from NAD^+ (nicotinamide adenine dinucleotide) (page 253) to elongation factor 2. This inhibits protein synthesis because the protein is now inactive and is unable to assist in the movement of the ribosome along the mRNA template.

The Polyribosome

More than one polypeptide chain is synthesized from an mRNA molecule at any given time. Once a ribosome has begun translocating along the mRNA, the start AUG codon is free, and another ribosome can bind. A second 70S initiation complex forms. Once this ribosome has moved away, a third ribosome can attach to the start codon. This process is repeated until the mRNA is covered with ribosomes. Each of these spans about 80 nucleotides. The resultant structure is called the polyribosome or polysome and is visible under the electron microscope (Fig. 12.6). This is an efficient structure as many protein molecules are made at the same time.

Termination of Protein Synthesis

There are three codons, UAG, UAA, and UGA, that have no corresponding tRNA molecule. These codons are called *stop codons*. Instead of interacting with tRNAs, the A-site occupied by one of these codons is filled by proteins known as chain *release factors*. In the presence of these factors the newly syn-

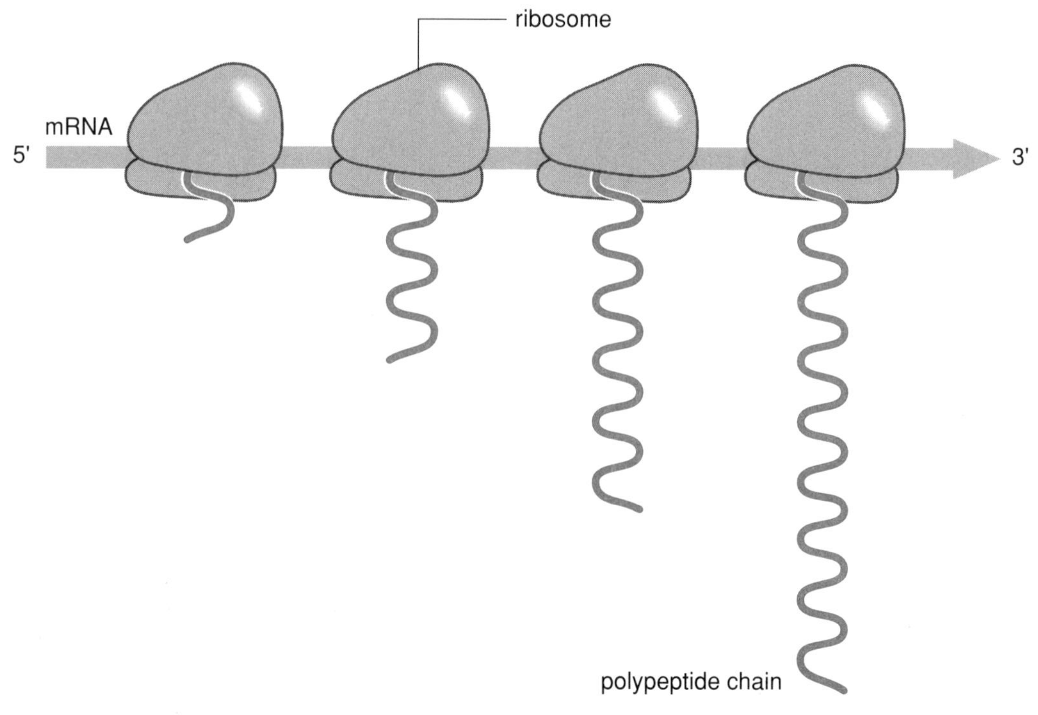

Figure 12.6
A polyribosome.

thesized polypeptide chain is freed from the ribosome, and the mRNA, tRNA, and the 30S and 50S ribosomal subunits dissociate (Figure 12.7). Release factor 1 causes polypeptide chain release from UAA and UAG, and release factor 2 terminates chains with UAA and UGA. A third protein, release factor 3, cooperates with the other two to stop protein synthesis. When the A-site is occupied by a release factor, the enzyme peptidyl transferase is unable to add an amino acid to the growing polypeptide chain and catalyzes the hydrolysis of the bond joining the polypeptide chain to the tRNA molecule. The carboxyl (COOH) end of the protein is therefore freed from the tRNA, and the protein is released.

The Ribosome Is Recycled

It would be wasteful if a complex structure such as the ribosome were used only once. When protein synthesis is over, the two ribosomal subunits dissociate and enter the cell's ribosomal pool. They can be reused when needed.

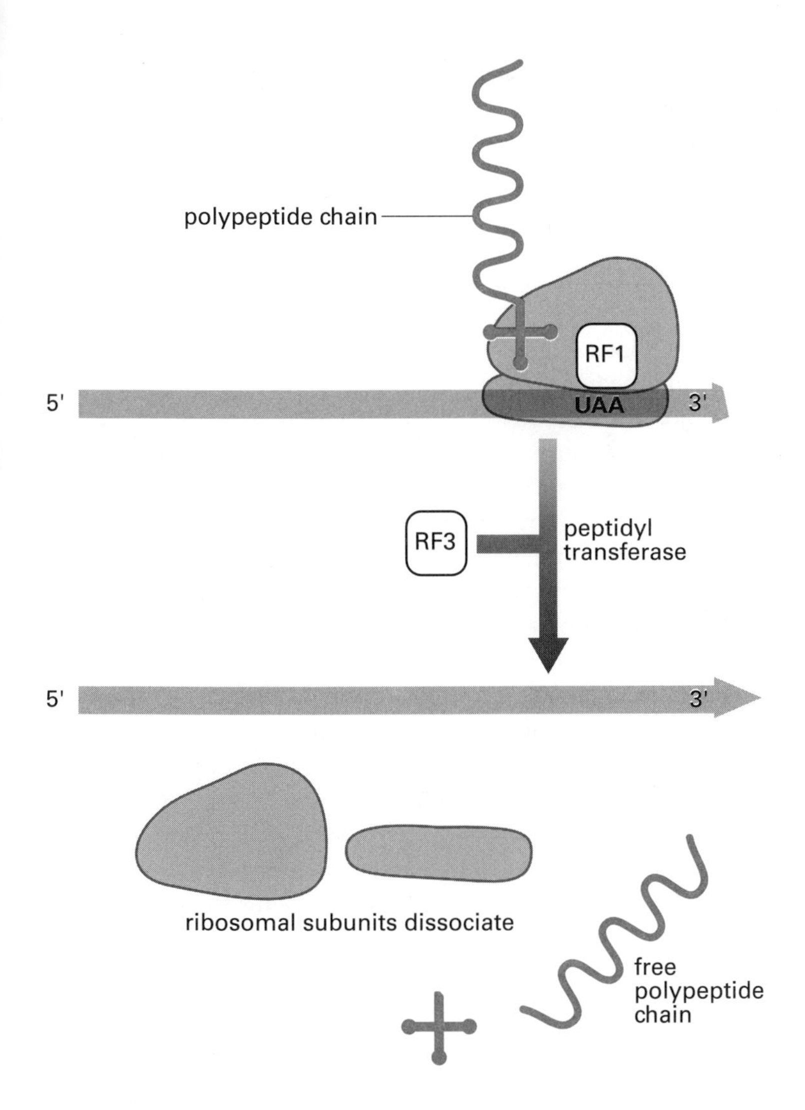

Figure 12.7
Termination of protein synthesis.

EUKARYOTIC PROTEIN SYNTHESIS IS A LITTLE MORE COMPLEX

Elongation of the polypeptide chain and the termination of protein synthesis in eukaryotes does not differ very much from that described for bacteria. However, the initiation of protein synthesis is more complex in eukaryotes. Their proteins always start with methionine instead of the formyl methionine used in bacterial protein synthesis. A special transfer RNA, $tRNA^{met}{}_i$ is used to initiate protein synthesis from the AUG start codon. The methionine is often removed from the protein after synthesis. Eukaryotic mRNAs do not contain the bacterial Shine–Dalgarno sequence for ribosome binding. The recognition of the AUG codon that specifies the start site for translation requires at least nine proteins. The details of this complex process are beyond the scope of this book.

Antibiotics and Protein Synthesis

Protein synthesis can be blocked by antibiotics, a property that is extensively exploited in research and medicine. Many antibiotics only inhibit protein synthesis in bacteria and not in eukaryotes. They are therefore extremely useful in the treatment of infections because the invading bacteria will die but protein synthesis in the host organism remains unaffected. Examples are chloramphenicol, which blocks the peptidyl transferase reaction, and tetracycline, which inhibits the binding of an aminoacyl-tRNA to the A-site of the ribosome. Hence both of these antibiotics block chain elongation. Streptomycin, on the other hand, inhibits the formation of the 70S initiation complex because it prevents $tRNA^{fmet}$ from binding to the P-site of the ribosome.

Puromycin causes the premature release of polypeptide chains from the ribosome and acts on both bacterial and eukaryotic cells. This antibiotic has been widely used in the study of protein synthesis. Puromycin can occupy the A-site of the ribosome because its structure resembles the 3′ end of an aminoacyl-tRNA (Fig. 12.8). Puromycin blocks protein synthesis because peptidyl transferase uses it as a substrate and forms a peptide bond between the growing polypeptide and the antibiotic. The protein no longer remains tightly bound to the ribosome and is released.

PROTEIN TARGETING

Proteins must be delivered to the appropriate site within the cell to carry out their correct functions. This process is known as protein targeting. Figure 12.9 shows a route map of protein movement within the cell. Address labels called *sorting signals* are added to proteins to direct them to the right place. Specific amino acid sequences within a protein act as sorting signals. Extra address labels are provided by the addition of sugar residues to the protein.

In eukaryotic cells proteins that will remain within the cytosol or that will end up in mitochondria, chloroplasts and peroxisomes are synthesized in free-floating polyribosomes. Other proteins are synthesized on polyribosomes attached to the rough endoplasmic reticulum (page 36). Proteins contain a short stretch of amino acids (about 20–30 residues) called the signal sequence, which targets them to the endoplasmic reticulum (Figure 12.10). Their synthesis starts on free polyribosomes. When the signal sequence emerges from the

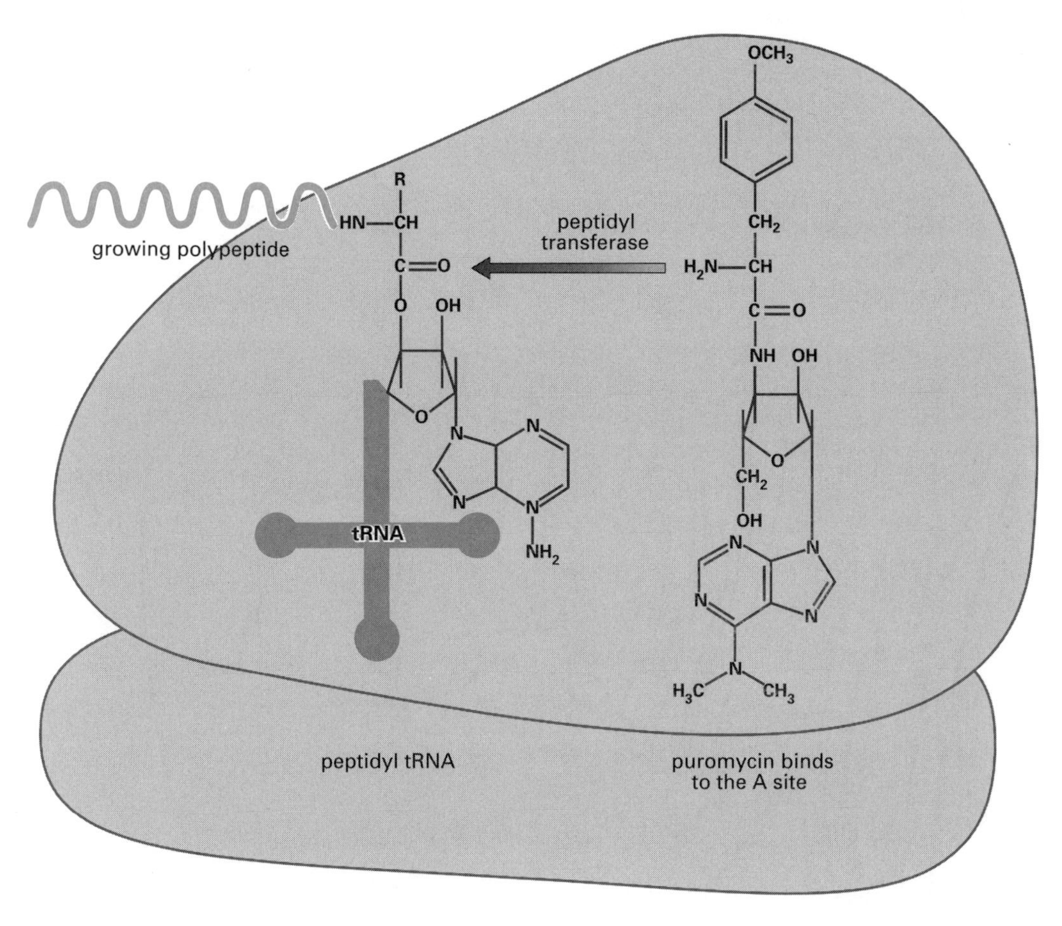

Figure 12.8
The antibiotic puromycin inhibits protein synthesis.

ribosome, it is recognized and bound by a signal recognition particle that is made up of a small RNA molecule and several proteins. The signal recognition particle then binds to a specific receptor on the endoplasmic reticulum membrane called the signal recognition particle receptor (or the docking protein). The ribosome then binds to a ribosome receptor protein on the membrane. The signal recognition particle is released and protein synthesis continues.

Signal sequences are rich in nonpolar, hydrophobic amino acids (page 216), and this helps them move into the endoplasmic reticulum membrane. As the polypeptide grows, it moves through the membrane. Secretory proteins (those synthesized in the cell and then released into the extracellular medium) have their signal sequences cleaved by an enzyme called signal peptidase as they pass into the lumen of the endoplasmic reticulum. Other proteins remain

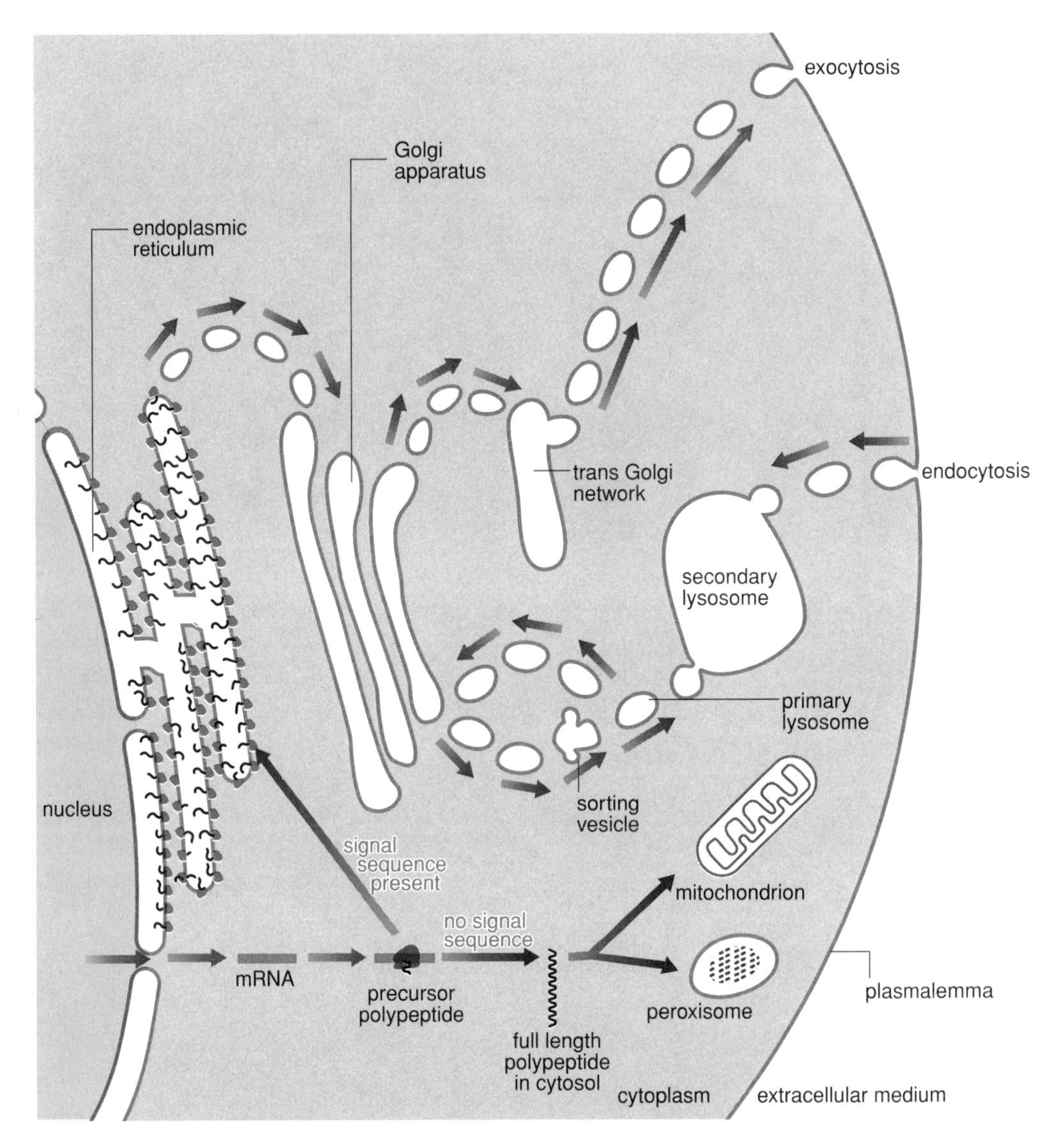

Figure 12.9
Proteins targeted to specific places within a cell or to another cell.

attached to their signal sequences and therefore become integral membrane proteins.

Most polypeptides synthesized at the rough endoplasmic reticulum are glycosylated (i.e., they have sugar residues added to them) as soon as the growing polypeptide chain enters the lumen of the endoplasmic reticulum (Fig. 12.10). A pre-made oligosaccharide comprising fourteen six-carbon sugars (2

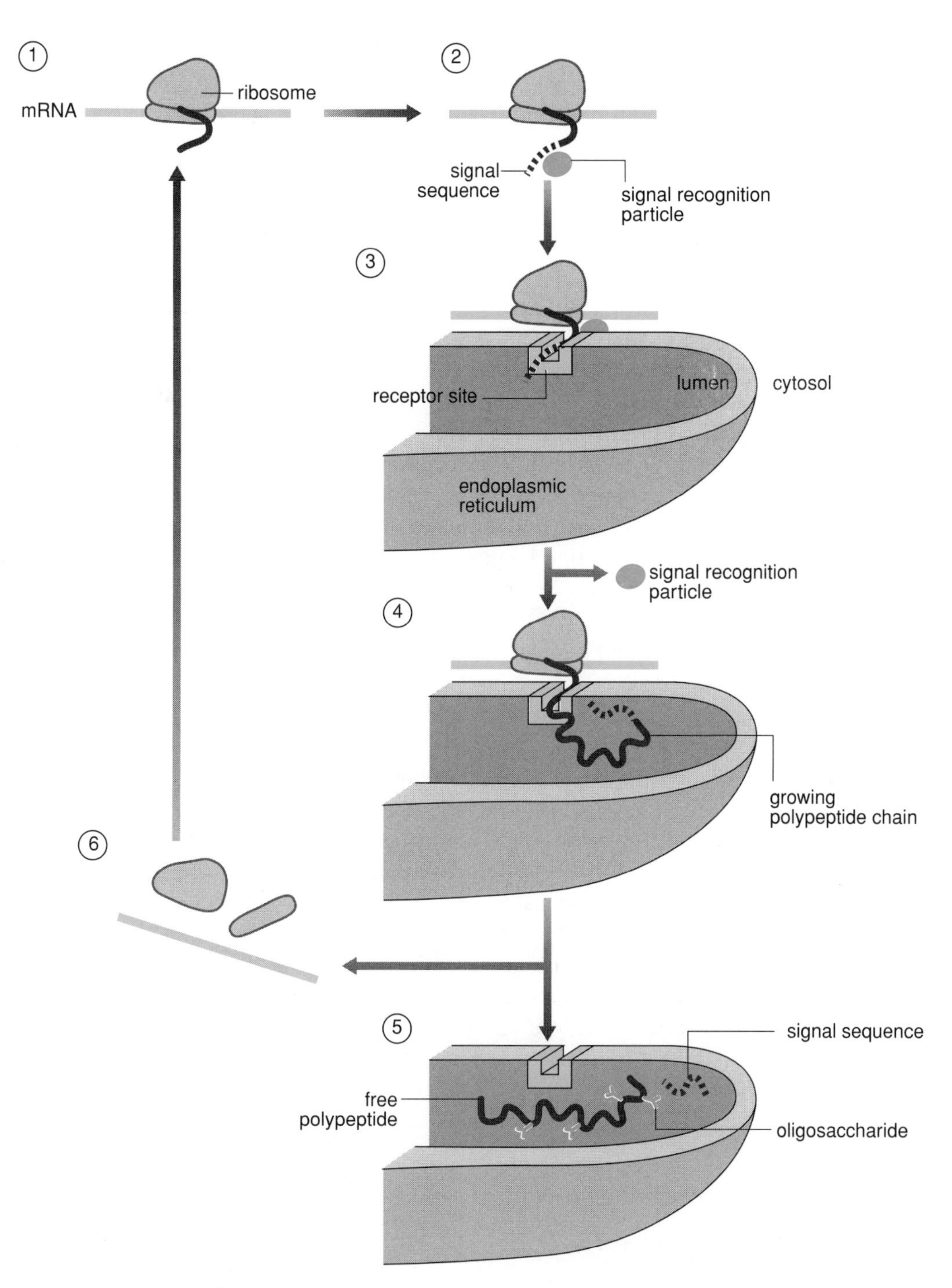

Figure 12.10

Synthesis of a protein on membrane-bound polyribosomes.

N-acetyl glucosamines, 9 mannoses, and 3 glucoses) is added to an asparagine residue by the enzyme oligosaccharide transferase. The three glucose residues are then removed. Proteins not destined to be retained in the endoplasmic reticulum are now ready to be transported to the Golgi apparatus.

Golgi Apparatus

The Golgi apparatus is the distribution point where proteins made at the rough endoplasmic reticulum are directed to their final destination. Structurally the Golgi apparatus is made up of one or more stacks of flattened discs or cisternae. The asymmetry of the stack is reflected in the morphology of the membranes from which it is formed. The cis cisternae are made of membranes 5.5 nm thick, like those of the endoplasmic reticulum, while the trans cisternae have 10 nm thick membranes, like the plasmalemma. Each cisterna has a central flattened region where the luminal space, as well as the gap between adjacent cisternae, is uniform. The margins of each cisterna are often dilated, particularly at the trans face of the stack, and are often fenestrated (have holes through them).

Small, spherical vesicles are always found in association with the Golgi apparatus, especially with the cis Golgi where they are seen attached to the dilated margins of the cisternae. These vesicles are referred to as transfer vesicles; some of them carry proteins from the endoplasmic reticulum to the Golgi stacks, others transfer proteins between the stacks, i.e., from cis to middle and from middle to trans. In cells that secrete proteins, condensing vesicles on the margins of the trans cisternae concentrate the proteins, then bud off. Once this process is complete, these vesicles are referred to as secretion granules, and they are transported via the trans Golgi network to the cell membrane for release. In addition to being concentrated, exported proteins undergo another important modification during their passage through the Golgi stack. Specific blocks of sugar molecules are added to the protein, a process known as glycosylation. Most secreted proteins are glycosylated. To achieve this, the oligosaccharide already attached to proteins as they enter the Golgi apparatus is processed as the protein moves through the sacs of the Golgi. Six of the mannose residues are removed and other sugar residues added.

Trans Golgi Network

Beyond the trans face of the Golgi complex, lies the trans Golgi network (TGN) of sacks and tubules. Although there is some final processing of proteins in the TGN, most of the proteins reaching this compartment have received all the modifications necessary to make them fully functional or to specify their final destination. Three types of vesicles bud off from the different parts of the TGN membrane: constitutive exocytotic vesicles, regulated exocytotic vesicles, and transport vesicles destined to fuse with lysosomes.

The first two categories of vesicles, though functionally different, look very much alike and are directed to the cell surface. When a constitutive exocytotic vesicle comes into contact with the plasmalemma, the two membranes fuse, and, in a process like endocytosis in reverse—and called, appropriately enough, exocytosis—the contents of the vesicle are expelled to the extracellular medium, and the vesicle membrane becomes a part of the plasmalemma. Proteins that the cell exports as soon as it has made them are packaged into constitutive exocytotic vesicles. An example is collagen (page 11).

Regulated exocytotic vesicles package proteins that the cell releases only under certain conditions. They accumulate in the cytoplasm until they receive a specific signal, usually a rise in the concentration of calcium ions in the surrounding cytosol, whereupon exocytosis proceeds rapidly.

Notice that in exocytosis the membrane of the vesicle becomes incorporated into the plasmalemma; consequently the integral proteins and lipids of the vesicle membrane become the integral proteins and lipids of the cell membrane. This is the way that proteins destined to be integral to the plasmalemma reach their final destination.

Targeting Proteins to the Lysosome

Proteins are transported in vesicles from one organelle to another. Just how this is done in such a precise manner is not yet understood. One of the best examples of targeting comes from the proteins destined for the lysosome (Fig. 12.11). These proteins are synthesized on the endoplasmic reticulum and are glycosylated as soon as they enter the lumen of this organelle. Some of the mannose residues are then phosphorylated in the cis Golgi to form mannose-6-phosphate. This phosphorylation signal indicates that the protein is destined for the lysosome. The trans sacs of the Golgi contain mannose-6-phosphate receptors. The receptors are localized to special areas of the trans sac that are coated with the molecule clathrin (page 37). These regions of membrane bud off from the Golgi apparatus to form a transport vesicle which then fuses with a sorting vesicle. In the low pH (5.0) environment of the sorting compartment, the lysosomal protein can no longer bind to its receptor. The phosphate group is usually removed by a phosphatase. Vesicles containing the receptor bud off from the sorting vesicle and deliver the mannose-6-

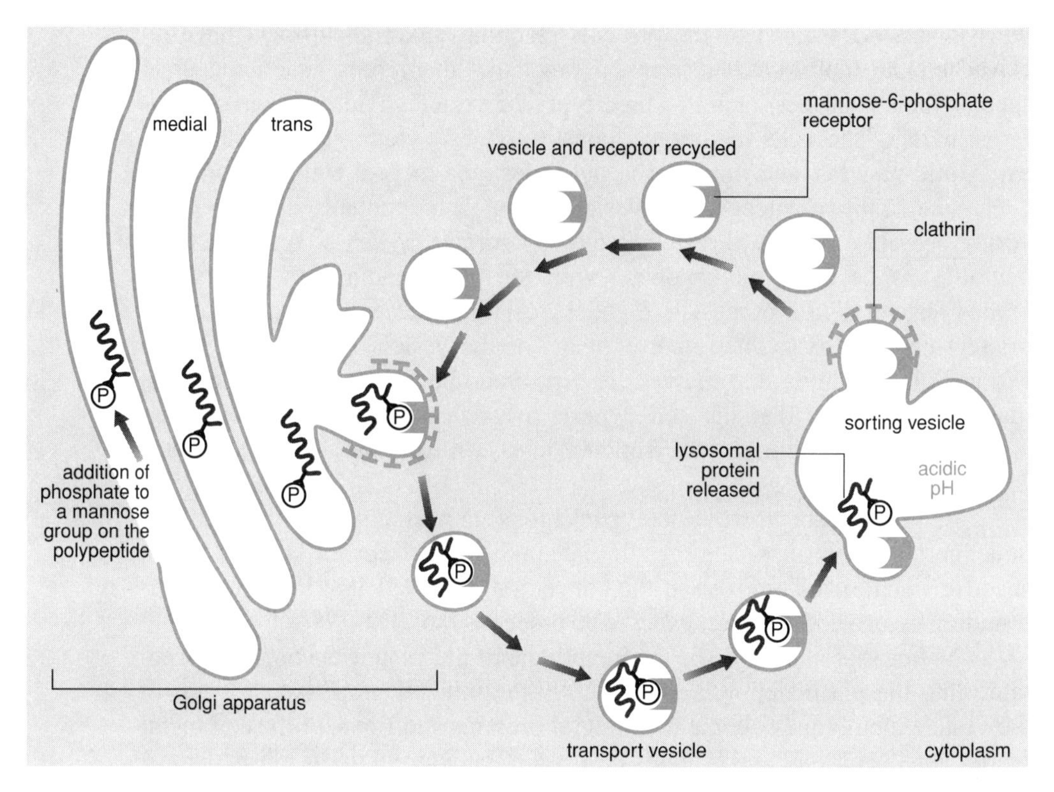

Figure 12.11

The mannose-6-phosphate receptor that helps target proteins destined for lysosomes.

BOX 12.1

Inclusion Cell Disease

Inclusion cell disease is a severe inherited disorder affecting a number of organs especially the liver. Some of the cells of affected individuals contain large vesicles filled with toxic wastes such as glycolipids. The immediate cause is that the lysosomes of these patients do not contain the normal degradative enzymes. However, people with the disease synthesize their lysosomal proteins normally. What is missing is the enzyme responsible for the phosphorylation of mannose. Thus, although the degradative enzymes are made, they are not targeted to the lysosome. Instead, they are secreted from the cell. This example illustrates that protein synthesis alone is not enough for correct protein function—the protein must also be delivered to the correct place in the cell for it to do its job.

phosphate receptors to the trans sacs of the Golgi. The remainder of the vesicle fuses with a lysosome.

SUMMARY

1. During protein synthesis the genetic code in a mRNA is translated into a sequence of amino acids. An amino acid attaches to the $3'$ end of a tRNA to form an aminoacyl-tRNA. Protein synthesis takes place on the ribosome which binds both to mRNA and tRNA. It has two tRNA binding sites, the P-site and the A-site.

2. Initiation of protein synthesis in bacteria involves the binding of the 30S ribosomal subunit to the mRNA. $tRNA^{fmet}$ binds to the initiation codon, and then the 50S ribosomal subunit attaches and the 70S initiation complex is formed.

3. Protein synthesis begins when a second aminoacyl-tRNA occupies the A-site. Each incoming amino acid is specified by the codon on the mRNA. The anticodon on the tRNA hydrogen bonds to the codon, thus positioning the amino acid on the ribosome. A peptide bond is formed, by peptidyl transferase, between the amino acids in the P- and A-sites. The newly synthesized peptide occupies the P-site, and another amino acid is brought into the A-site. This process of elongation requires a number of proteins (elongation factors); as it continues, the peptide chain grows. When a stop codon is reached, the polypeptide chain is released with the help of proteins known as release factors.

4. More than one ribosome can attach to an mRNA. This forms a polyribosome, and many protein molecules can be made simultaneously from the same mRNA.

5. Proteins that are to be secreted from the cell are synthesized on ribosomes associated with the rough endoplasmic reticulum and then pass via the Golgi apparatus and the trans Golgi network to the cell surface. Secreted proteins are extensively modified in each of these compartments.

6. Proteins have to end up in the right place in the cell to carry out their particular function. Sorting signals, often in the form of sugar molecules, are added to those proteins made on polyribosomes attached to the endoplasmic reticulum, and these proteins subsequently move to the Golgi apparatus where sugar residues may be added or removed. Such proteins are transported in vesicles that fuse with the appropriate organelle.

BOX 12.2

Designer Probes

The protein aequorin, found in the cytoplasm of particular cells in a jellyfish, gives out light when it binds calcium ions. It has been used for many years to detect changes in the calcium concentration in the cytosol, but recently protein engineering (page 184) has allowed scientists to modify the aequorin gene to produce probes that detect calcium changes in the nucleus (illustrated), the mitochondria, and the endoplasmic reticulum. In each case the strategy has been the same: DNA coding for the appropriate sorting signal has been attached to one end of the aequorin gene and the new gene injected into the cells of interest. The cells produce the protein, detect the sorting signal, and move the protein to the specified region. Light emission from the cells then gives information about calcium changes in the organelle of interest.

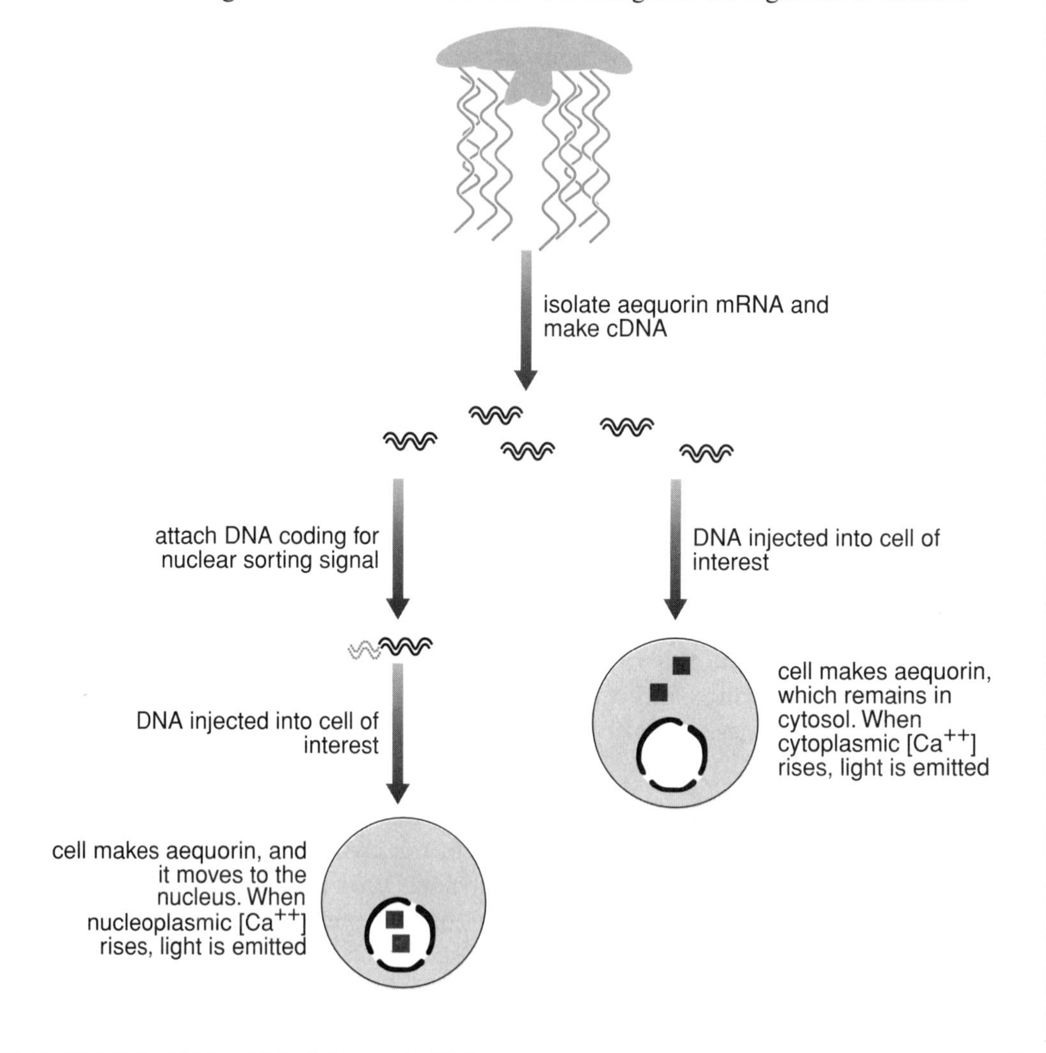

CHAPTER

13

PROTEIN STRUCTURE

We are all, regardless of build, made of water plus more or less equal amounts of fat and protein. Although the DNA in our cells contains the information necessary to make our bodies, DNA itself is not a significant part of our body mass. Nor is it a chemically interesting molecule, in the sense that one length of DNA is much the same as another in terms of shape and chemical reactivity. The simplicity of DNA arises because it is a polymer made up of only four fairly similar monomers, and this is appropriate because the function of DNA is simply to remain as a record and to be read during transcription (Chapter 10). In contrast, proteins made using the instructions on DNA vary enormously in physical characteristics and function. Silk, hair, the lens of an eye, an immunoassay (such as found in a pregnancy test kit), cottage cheese: these examples are all just protein plus more or less water, but they are different because the proteins they contain are different. Proteins carry out almost all of the functions of the living cell including of course the synthesis of new DNA. Neither growth nor development would be possible without proteins.

Most proteins have functions that depend upon their ability to recognize other molecules by binding (Box 2.2). This recognition depends on specific three-dimensional binding sites that make multiple interactions with the ligand, the molecule being bound. To do this a protein must itself have a specific three-dimensional structure. Each of the huge number of protein functions demands its own protein structure. Evolution has produced this diversity by using a palette of 20 monomers, each with its own unique chemical properties, as the building blocks of proteins. Huge numbers of very different structures are therefore possible.

POLYMERS OF AMINO ACIDS

Translation produces linear polymers of α-amino acids. If there are fewer than 50 amino acids in the polymer, we call the molecule a peptide. More, and it is a *polypeptide*. Proteins are polypeptides, and most have dimensions of a few nanometers (nm), although structural proteins like the keratin in hair are much bigger. The molecular weight of proteins can range from 5,000 to hundreds of thousands.

The Amino Acid Building Blocks

Leucine and γ-amino butyrate (Fig. 13.1) are both amino acids.

We name organic acids by labeling the carbon that bears the carboxyl group α, the next one β, and so on. When we add an amino group, making an amino acid, we state the letter of the carbon to which the amino group is attached. Hence, the name γ-amino butyrate, and the description of leucine as an α-amino acid. The 20 amino acids in proteins and peptides are of the α type. They have the general structure

$$^{-}OOC—\overset{\displaystyle R \atop |}{C}H—NH_3^+$$

where R is called the side chain. It is the side chain that gives each amino acid its unique properties.

Peptidyl transferase (page 194) joins the amino group of one amino acid

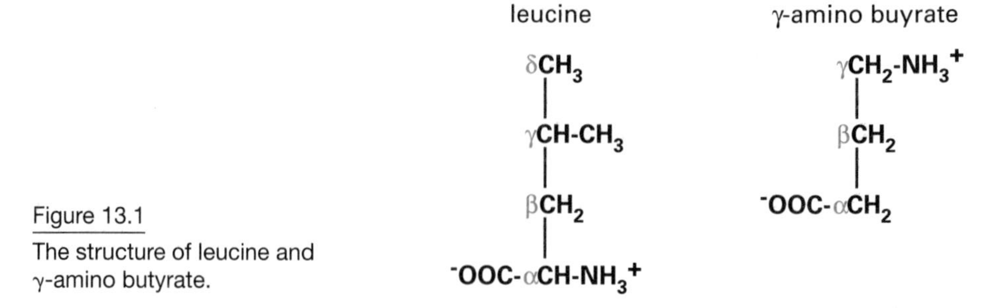

Figure 13.1

The structure of leucine and γ-amino butyrate.

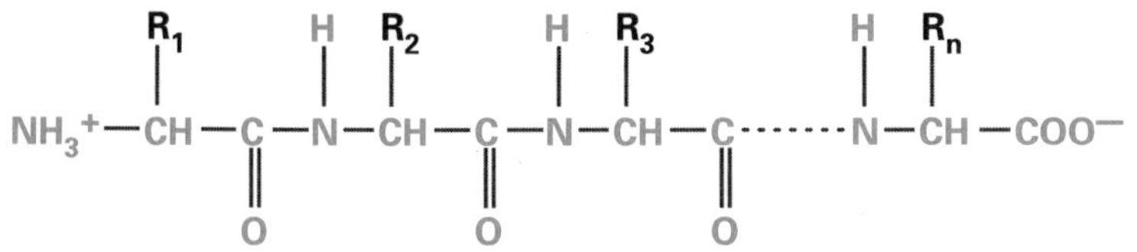

$$NH_3^+-\underset{\displaystyle R_1}{CH}-\underset{\|}{C}(=O)-\underset{H}{N}-\underset{R_2}{CH}-C(=O)-\underset{H}{N}-\underset{R_3}{CH}-C(=O)\cdots\cdots\underset{H}{N}-\underset{R_n}{CH}-COO^-$$

Figure 13.2
The general structure of a polypeptide.

to the carboxyl of the next. This is a peptide bond. The complete polypeptide has the form shown in Figure 13.2.

The backbone, a series of peptide bonds, is shown in blue. At the left-hand end is a free amino group; this is known as the N terminal or amino terminal. At the right-hand end is a free carboxyl group; this is known as the C terminal or carboxy terminal. Peptides are always written this way, with the C terminal on the right.

The properties of individual polypeptides are conferred by the *side chains* of their constituent amino acids. Many different properties are important—size, electrical charge, the ability to participate in particular reactions—but the most important is the affinity of the side chain for water. Side chains that interact strongly with water are hydrophilic. Those that do not are called hydrophobic. We have already encountered the 20 amino acids (Table 5.1, page 84), but we will now describe each in turn, beginning with the most hydrophilic and ending with the most hydrophobic. Each amino acid has a three-letter abbreviation, which is usually obvious, and a one-letter code, which is often quite arbitrary.

asp *D* *Aspartate*

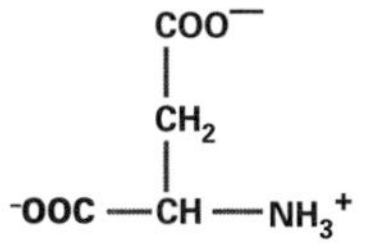

$$\begin{array}{c} COO^- \\ | \\ CH_2 \\ | \\ {}^-OOC-CH-NH_3^+ \end{array}$$

glu *E* *Glutamate*

$$\begin{array}{c} COO^- \\ | \\ CH_2 \\ | \\ CH_2 \\ | \\ {}^-OOC-CH-NH_3^+ \end{array}$$

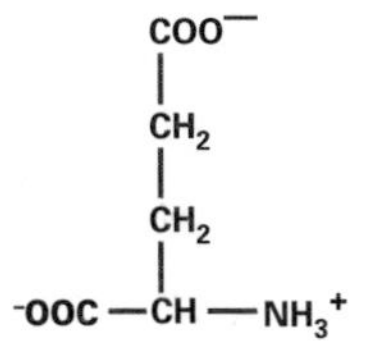

At the pH of cytoplasm, these side chains bear negative charges, which interact strongly with water molecules (page 16). A polypeptide made entirely of these amino acids is very soluble in water.

lys *K* *Lysine*

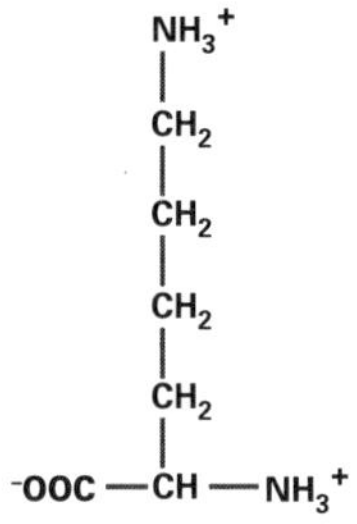

arg *R* *Arginine*

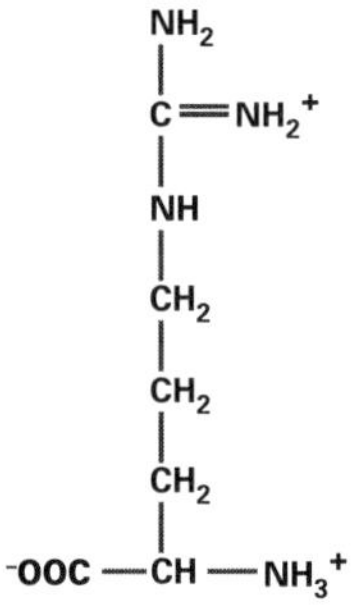

At the pH of cytoplasm, these side chains bear positive charges, which interact strongly with water molecules.

his *H* *Histidine*

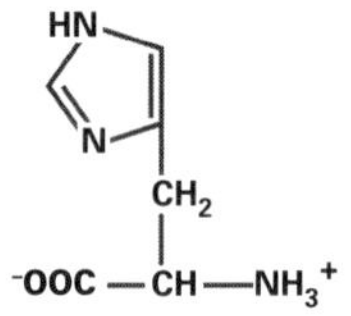

cys *C* *Cysteine*

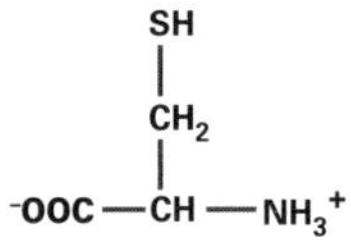

The side chain of histidine is basic and will accept an H^+:

histidine protonation

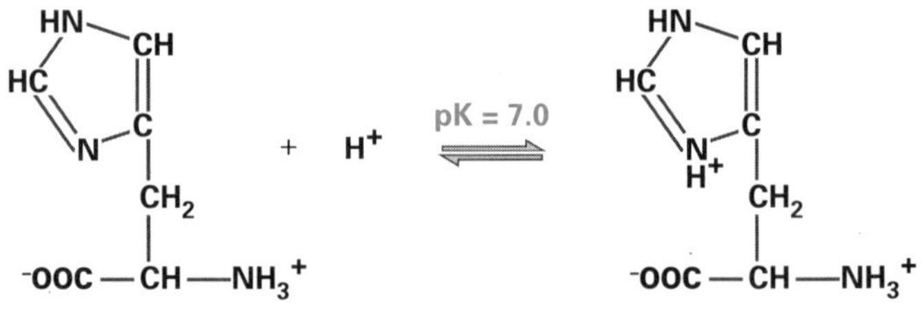

pK = 7.0

The pK_a (Box 1.4, page 18) of histidine protonation is 7.0, so at neutral pH about half the histidine side chains bear a positive charge. The fact that histidine is equally balanced between protonated and unprotonated forms gives it important roles in enzyme catalysis. At neutral pH a polypeptide made entirely of histidine will be very soluble in water.

The side chain of cysteine is weakly acidic, giving up an H^+ and becoming negatively charged

cysteine deprotonation

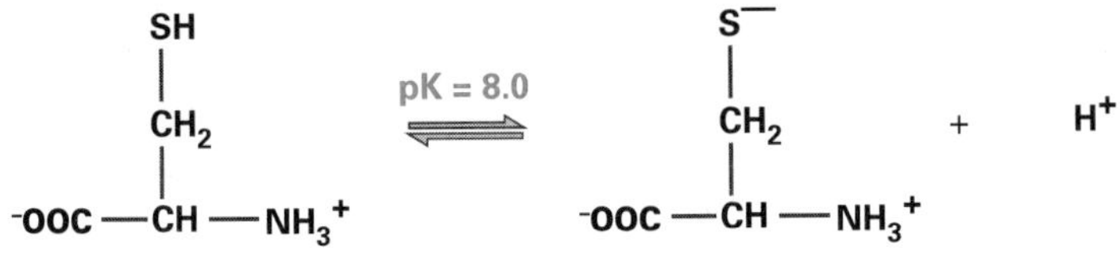

$$^{-}OOC{-}CH(CH_2SH){-}NH_3^+ \overset{pK = 8.0}{\rightleftharpoons} {}^{-}OOC{-}CH(CH_2S^-){-}NH_3^+ + H^+$$

The pK_a is 8, so at neutral pH most (about 90%) of cysteine side chains still have their hydrogen attached. Even so, the charge on the remaining 10% means that a polypeptide made entirely of cysteine will be very soluble in water.

ser *S* *Serine*

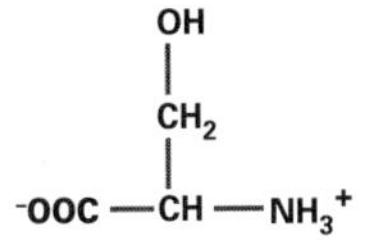

$$^{-}OOC{-}CH(CH_2{-}OH){-}NH_3^+$$

thr *T* *Threonine*

$$^{-}OOC{-}CH(CH(OH){-}CH_3){-}NH_3^+$$

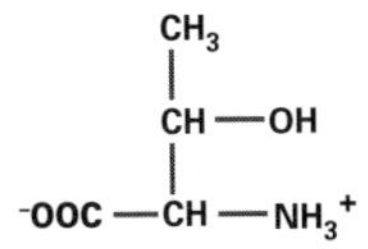

tyr *Y* *Tyrosine*

$$^{-}OOC{-}CH(CH_2{-}C_6H_4{-}OH){-}NH_3^+$$

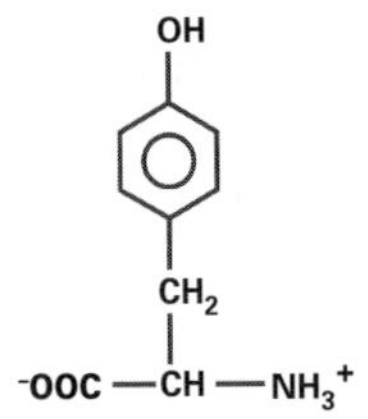

These amino acids have hydroxyl (–OH) groups that can hydrogen bond with water molecules (Box 1.4). A polypeptide composed entirely of these amino acids is soluble in water.

asn *N* *Asparagine*

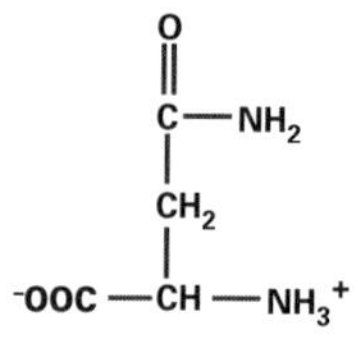

$$\begin{array}{c} O \\ \| \\ C-NH_2 \\ | \\ CH_2 \\ | \\ {}^{-}OOC-CH-NH_3^+ \end{array}$$

gln *Q* *Glutamine*

$$\begin{array}{c} O \\ \| \\ C-NH_2 \\ | \\ CH_2 \\ | \\ CH_2 \\ | \\ {}^{-}OOC-CH-NH_3^+ \end{array}$$

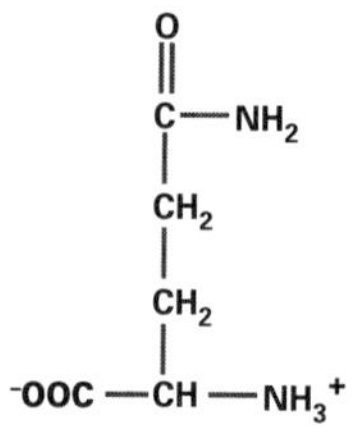

Chemists recognize these as the amides of aspartate and glutamate. They are hydrophilic.

gly *G* *Glycine*

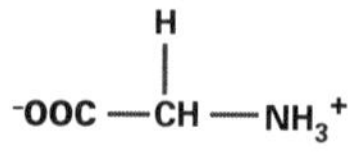

$$\begin{array}{c} H \\ | \\ {}^{-}OOC-CH-NH_3^+ \end{array}$$

Glycine has no side chain and is therefore relatively indifferent to its surroundings.

leu *L* *Leucine*

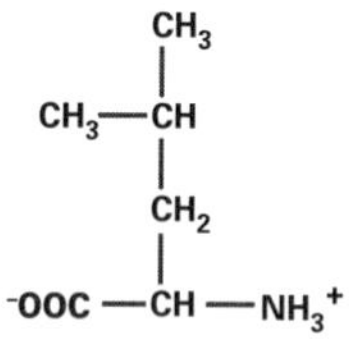

$$\begin{array}{c} CH_3 \\ | \\ CH_3-CH \\ | \\ CH_2 \\ | \\ {}^{-}OOC-CH-NH_3^+ \end{array}$$

ile *I* *Isoleucine*

$$\begin{array}{c} CH_3 \\ | \\ CH_2 \\ | \\ CH_3-CH \\ | \\ CH_2 \\ | \\ {}^{-}OOC-CH-NH_3^+ \end{array}$$

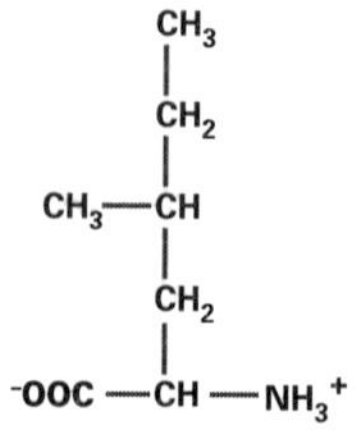

val *V* *Valine*

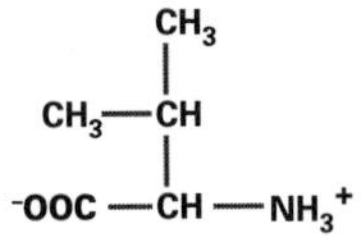

CH_3
$CH_3—CH$
$^-OOC—CH—NH_3^+$

ala *A* *Alanine*

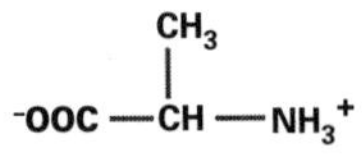

CH_3
$^-OOC—CH—NH_3^+$

phe *F* *Phenylalanine*

CH_2
$^-OOC—CH—NH_3^+$

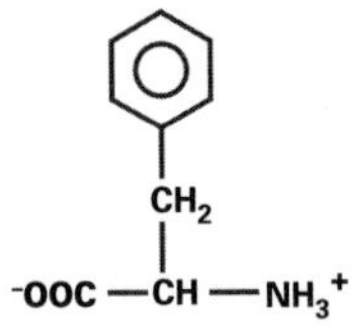

These five amino acids have side chains of carbon and hydrogen only. The side chains cannot interact with water; they are hydrophobic. A polypeptide composed entirely of these amino acids does not dissolve in water but will dissolve in olive oil.

trp *W* *Tryptophan*

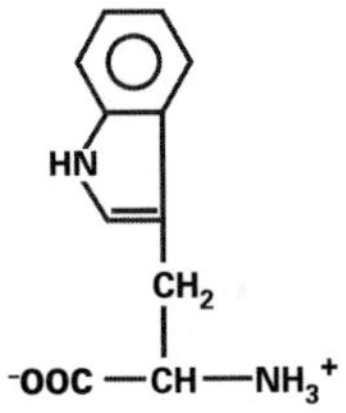

HN
CH_2
$^-OOC—CH—NH_3^+$

Another weakly hydrophobic amino acid.

met *M* *Methionine*

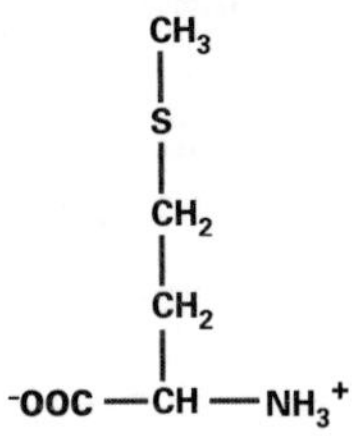

CH_3
S
CH_2
CH_2
$^-OOC—CH—NH_3^+$

The sulfur atom in methionine is tucked away in the middle of the molecule and cannot interact with the environment, so methionine is hydrophobic.

pro *P* *Proline*

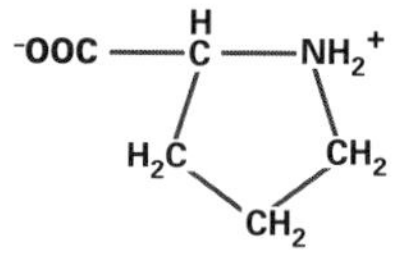

$$^{-}OOC{-}\underset{}{\overset{H}{C}}{-}NH_2^+ \quad \text{(ring: } H_2C{-}CH_2{-}CH_2\text{)}$$

Proline is not really an amino acid at all—it does not have the structure

$$^{-}OOC{-}\overset{\overset{\displaystyle R}{|}}{C}H{-}NH_3^+$$

In fact proline is an imino acid, but biologists give it honorary amino acid status. The side chain is hydrophobic.

There is an apparent paradox here. Leucine, isoleucine, valine, alanine, phenylalanine, methionine, and proline have hydrophobic side chains, and a protein made entirely of these amino acids will not dissolve in water. How is it then that these amino acids are present in cytoplasm and can be picked up by their respective tRNAs? The answer is simple: As free amino acids they all bear two charges, a positive on the amino group and a negative on the carboxyl group. Like all small ions, they are soluble in water. Formation of the peptide bond removes these charges.

The Unique Properties of Each Amino Acid

Although we have classified side chains on the basis of their affinity for water, their other properties are important.

CHARGE. If a side chain is charged, it will be hydrophilic. But charge has other effects too. A positively charged residue such as lysine will attract a negatively charged residue such as glutamate. If the two residues are buried deep within a folded polypeptide, where neither can interact with water, then it will be very difficult to tear them apart. We call such an electrostatic bond inside a protein a salt bridge. Negatively charged residues will attract positively charged ions out of solution, so a pocket on the surface of a protein lined with negatively charged residues will, if it is the right size, form a binding site for a particular positively charged ion like sodium or calcium.

UV ABSORBANCE. Tyrosine, phenylalanine, and tryptophan absorb ultraviolet light. When we are exposed to UV, cells in our skin make the compound melanin from tyrosine. Melanin helps protect the DNA from UV damage (page 140).

DISULFIDE BRIDGING. Under oxidizing conditions two cysteine residues link together in a disulfide bridge (Fig. 13.5). Proteins made for export to the extracellular medium often have disulfide bridges which confer additional rigidity on the protein molecule. If all the peptide bonds in a polypeptide are hydrolyzed, any cysteines that were linked by disulfide bridges remain connected in dimers called—very confusingly—cystine molecules.

PHOSPHORYLATION. Phosphorylation is the attachment of a phosphoryl group:

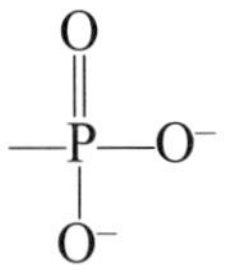

```
     O
     ‖
  —P—O⁻
     |
     O⁻
```

Usually the phosphoryl group comes from ATP. All the amino acids that have side chains with hydroxyl or carboxyl groups (S, T, Y, D, and E) can be phos-

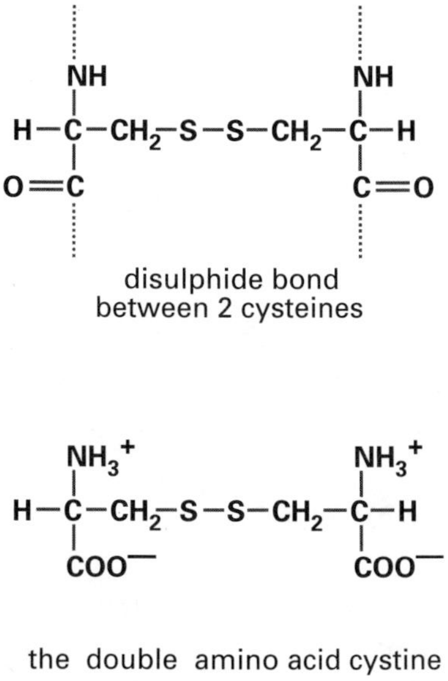

```
       :                           :
      NH                          NH
       |                           |
  H—C—CH2—S—S—CH2—C—H
       |                           |
  O═C                          C═O
       :                           :
```

disulphide bond
between 2 cysteines

```
      NH3+                        NH3+
       |                           |
  H—C—CH2—S—S—CH2—C—H
       |                           |
      COO−                        COO−
```

the double amino acid cystine

Figure 13.5

Oxidation of adjoining cysteines which produces a disulfide bridge; proteolysis produces cystine.

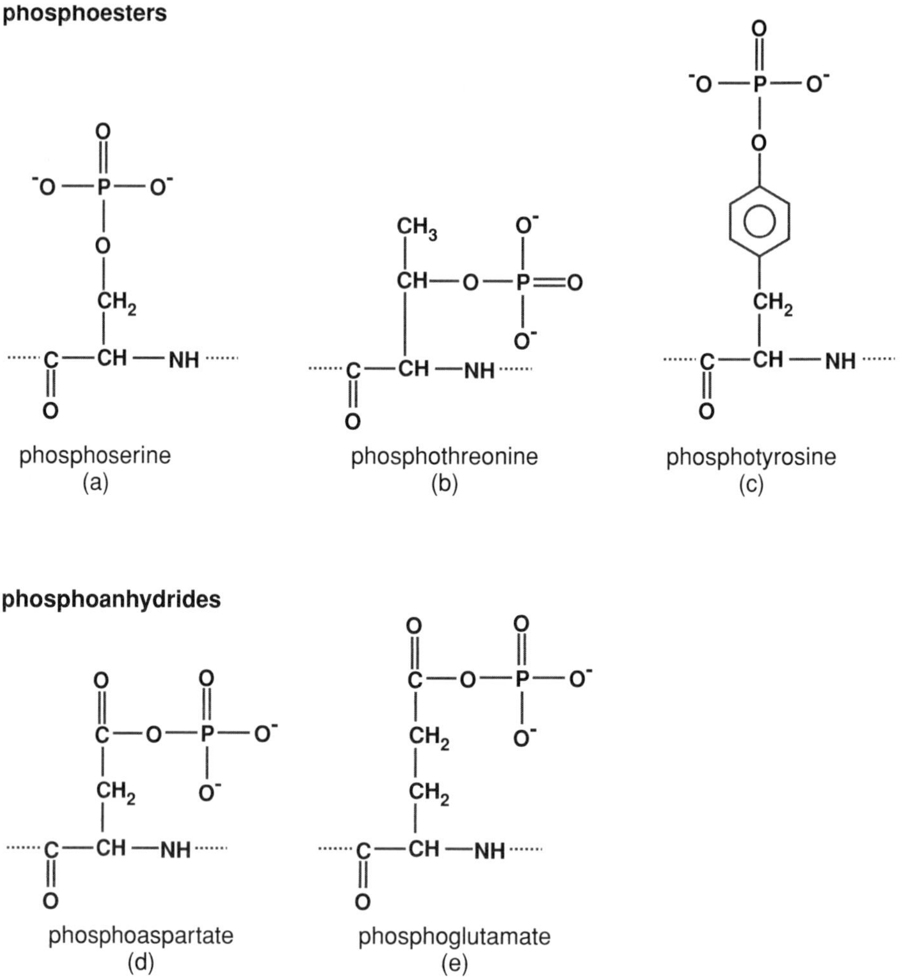

phosphoesters

phosphoserine
(a)

phosphothreonine
(b)

phosphotyrosine
(c)

phosphoanhydrides

phosphoaspartate
(d)

phosphoglutamate
(e)

Figure 13.6
Phosphorylation of serine, threonine, tyrosine, aspartate, and glutamate.

phorylated (Fig. 13.6). Phosphorylation can modify the ability of a protein to bind a second protein (Box 2.2) or can cause a protein to change its tertiary structure (Fig. 14.4).

OTHER AMINO ACIDS ARE FOUND IN NATURE

γ-amino butyrate (Fig. 13.1) is an intracellular transmitter in the brain (page 343). It is an amino acid that plays a vital role in the body, but it is not used as a building block in the synthesis of polypeptides. There are many other nonprotein amino acids. Some are components of ubiquitous biochemical pathways; others are unique to particular species.

Very early in the evolution of life, the palette of amino acids available to make polypeptides became fixed at the 20 that we have described. Almost the entire substance of all living things on earth is either polypeptide, composed of these 20 monomers, or other molecules synthesized by enzymes that are themselves composed of these 20 monomers. Natural selection has directed evolution within the constraints imposed by the palette. If a different palette had been used by the first cells, today's organisms might have very different powers and limitations.

THE THREE-DIMENSIONAL STRUCTURES OF PROTEINS

The polypeptides within our bodies are folded into complex three-dimensional shapes and are called proteins. The shape can be fully defined by stating the position and orientation of each amino acid, and such knowledge lets us produce three-dimensional images of the protein (Fig. 13.7). However, it is helpful to think of protein structure at different levels of complexity.

The *primary structure* of a protein is the sequence of amino acids. Lysozyme is an enzyme that attacks bacterial cell walls. It is found in secretions such as tears and in the white of eggs. Lysozyme has the following primary structure:

(NH_2)KVFGRCELAAAMKRHGLDNYRGYSLGNWVCAAKFESNFNTQA
TNRNTDGSTDYGILQINSRWWCDNGRTPGSRNLCNIPCSALLSSDITAS
VNCAKKIVSDGDGMN AWVAWRNRCKGTDVQAWIRGCRL(COOH)

Numbering is always from the amino terminal end (protein synthesis occurs in this direction too; page 194). The 129 amino acids of hen egg white lysozyme are shown in linear order in Figure 13.7*a* with the disulphide bridges shown between cysteines.

Lysozyme was the first enzyme to have its three-dimensional structure fully determined (in 1965). If we look at all the atoms that form the molecule (Fig. 13.7*b*), we see little except an irregular surface. If the amino acid side chains are stripped away and the path of the peptide-bonded backbone drawn (Fig. 13.7*c*), we see that some regions of the protein are ordered in a repeating pattern. Two patterns are common to many proteins: the α helix and the β sheet. Figure 13.7 *d* redraws the peptide backbone of lysosome to emphasize these patterns, with the lengths of peptide participating in β sheets represented

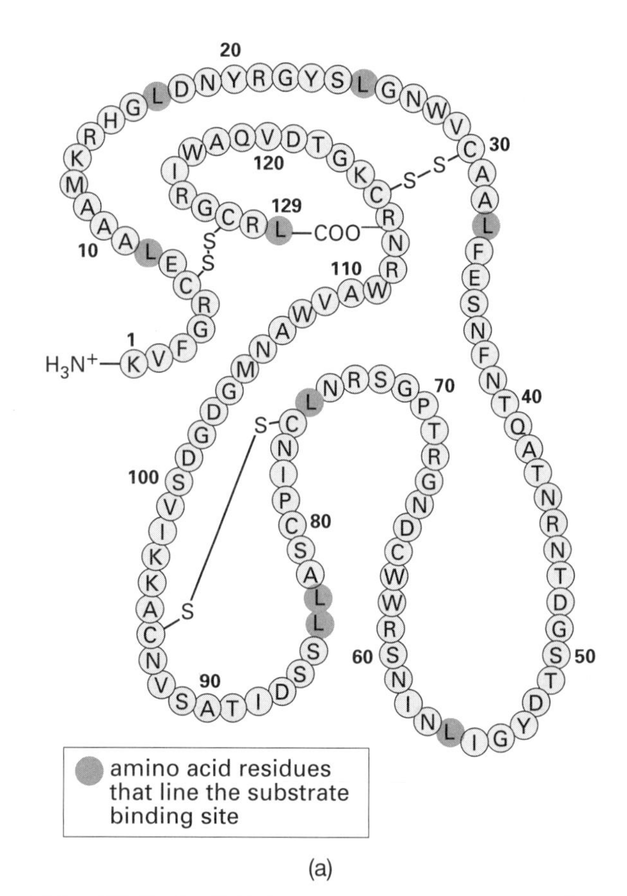

(a)

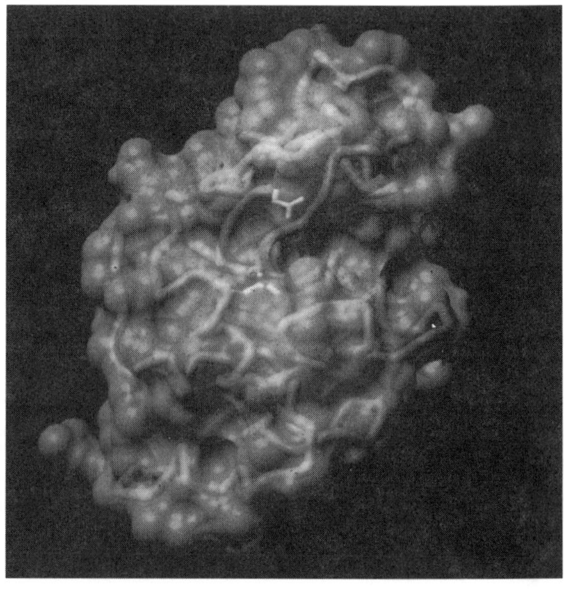

(b)

Figure 13.7

Lysozyme: (*a*) Linear map, (*b*) space-filling model.

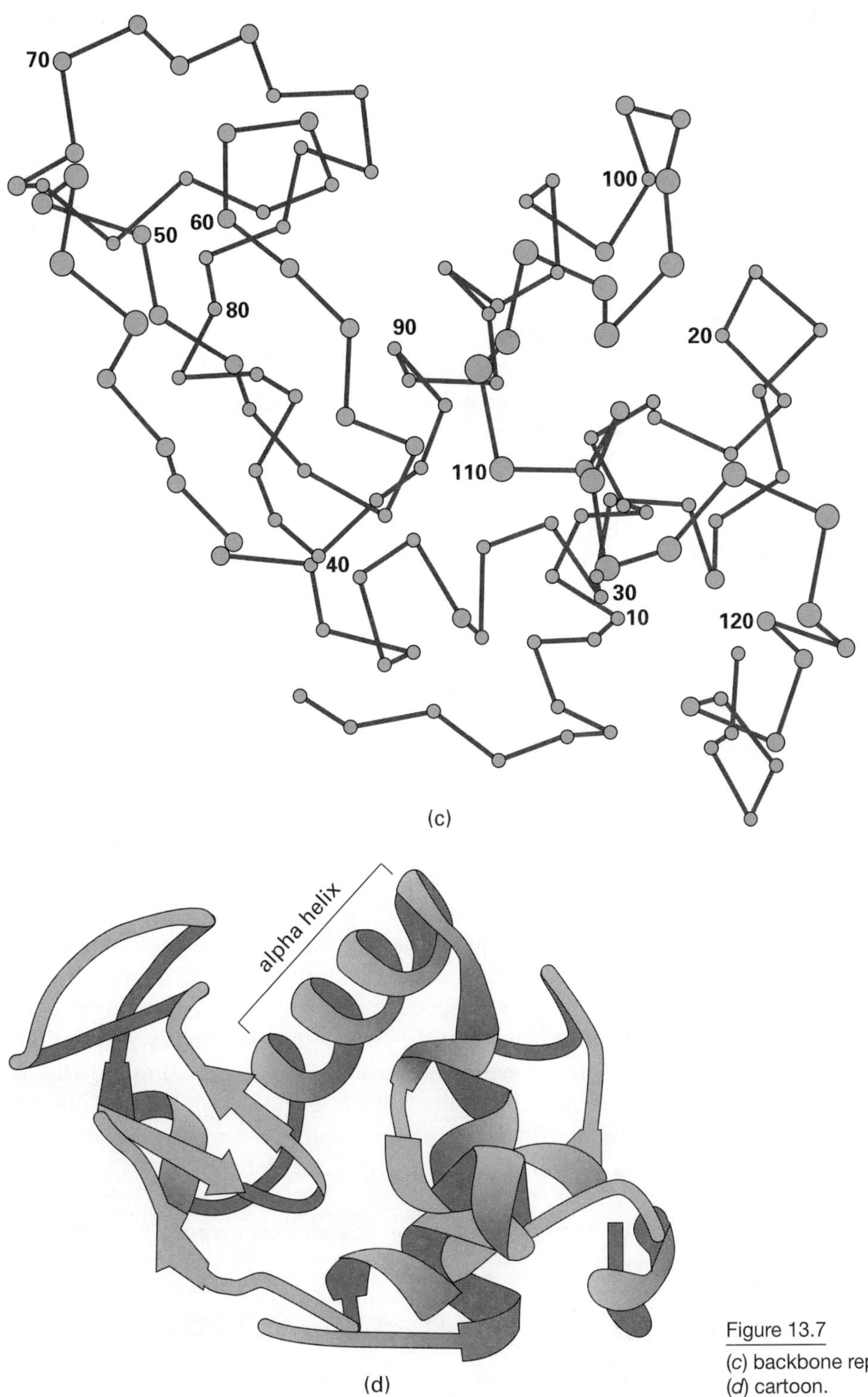

Figure 13.7
(*c*) backbone representation, (*d*) cartoon.

as arrows. Collectively these repeating patterns are known as *secondary structures*. There are other regions of the protein that do not have any such ordered pattern.

In an *α helix* the polypeptide chain twists around in a spiral, each turn of the helix taking 3.6 amino acid residues. This allows the nitrogen atom in each peptide bond to form a hydrogen bond (Box 2.1) with the oxygen of the peptide bond four ahead of it in the polypeptide chain (Fig. 13.8). All of the peptide bonds in the helix can form hydrogen bonds, producing a rod with the amino acid side chains pointing outward.

In a *β-sheet* lengths of polypeptide run alongside each other, and hydrogen bonds form between the strands. This generates a sheet that has the side chains protruding above and below it. Along a single strand the side chains alternate up then down, up then down. Because the actual geometry prevents them from being completely flat, they are sometimes called β-pleated sheets. A polypeptide chain can fold back on itself in two ways: Either all of the strands in the β sheet are running in the same direction (Fig. 13.9*a*) forming a parallel β sheet or they can alternate in direction (Fig. 13.9*b*) making an antiparallel β sheet.

In structural proteins like the keratin in hair or the fibroin in silk, the whole polypeptide chain is ordered into one of these secondary structures. Such fibrous proteins have relatively simple repeating shapes and so cannot be used for precise binding. Most proteins have large regions without secondary structure, the precise packing of the amino acids being unique to the protein. The amino acid residues are still forming hydrogen and other bonds (Box 13.2), but they are not doing so in an ordered, simple manner. There can be grooves, clefts, or protrusions on the surface of the protein where particular amino acid side chains are positioned to form sites that bind ligands and catalyze reactions. The whole three-dimensional arrangement of the amino acids in the protein is the *tertiary structure*.

Tertiary structures are unique to each protein. However, common motifs occur in tertiary structures. Many proteins with different functions show a β barrel structure, for instance, where a core of β sheet is surrounded by α helices. Often the tertiary structure can be seen to divide into discrete regions. The calcium-binding protein calmodulin shows this clearly (Fig. 13.10): Its single chain is organized into two domains, one shown in blue, the other in black, joined by only one strand of the polypeptide chain. In calmodulin the two domains are very similar, and the modern gene probably arose through duplication of an ancestral gene that was half as big. Domains are easier to see than to define. 'A separately folded region of a single polypeptide chain' is as good a definition as any.

Domains may be similar or different in both structure and function. The catabolite activating protein (CAP) of *E. coli* (page 152) binds to a specific se-

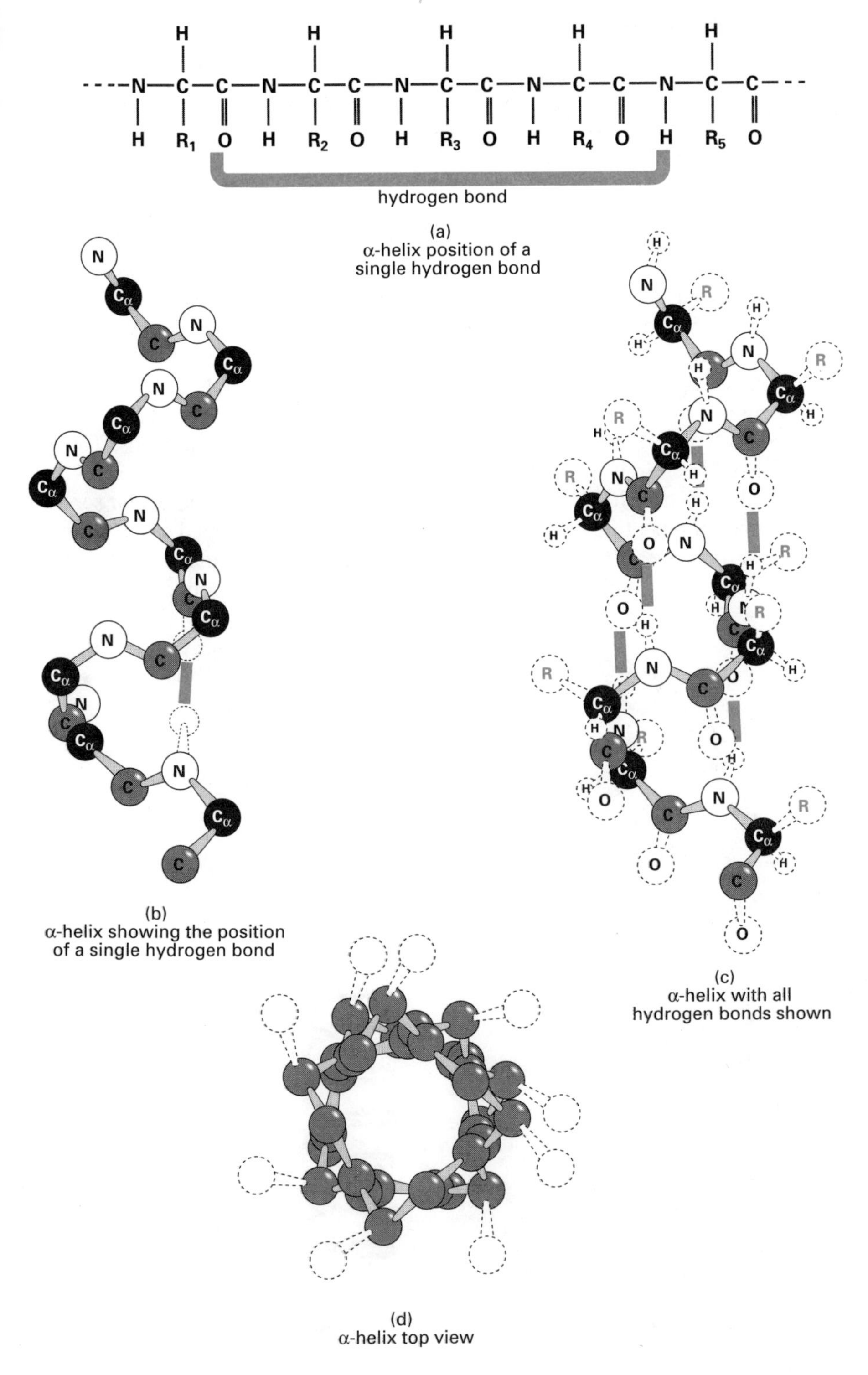

(a)
α-helix position of a single hydrogen bond

(b)
α-helix showing the position of a single hydrogen bond

(c)
α-helix with all hydrogen bonds shown

(d)
α-helix top view

Figure 13.8

The α-helix: (*a*) The nitrogen atom in each peptide bond forming a hydrogen bond with the oxygen of the peptide bond four ahead, (*b*) an α helix, with one hydrogen bond drawn, (*c*) an α helix, with all the hydrogen bonds drawn, (*d*) an end-on view of an α helix.

BOX 13.1

Chirality and D-Amino Acids

A chiral structure is one in which the mirror image of the object cannot be superimposed on the object. The shape of your whole body is not chiral. Your mirror image could be rotated through 180°, so that it faces into the mirror, and step back and be superimposed on you. Your right hand, however, is chiral: its mirror image is not a right hand any more but a left hand. L-amino acids are exclusively used in proteins and predominate in the metabolism of amino acids. However, D-amino acids are found in nature. D–alanine occurs in bacterial cell walls and some antibiotics such as valinomycin and Gramicidin A contain D-amino acids. These molecules are synthesized in entirely different ways from proteins.

Helices are chiral too. The α helix found in proteins is right handed, like a regular screw thread. Reflect a length of polypeptide in a mirror, and you will get a left handed α helix composed of D-amino acids. Because of the bond angles, it turns out that L-amino acids fit nicely in a right-handed α helix, while D-amino acids fit nicely in a left–handed α helix.

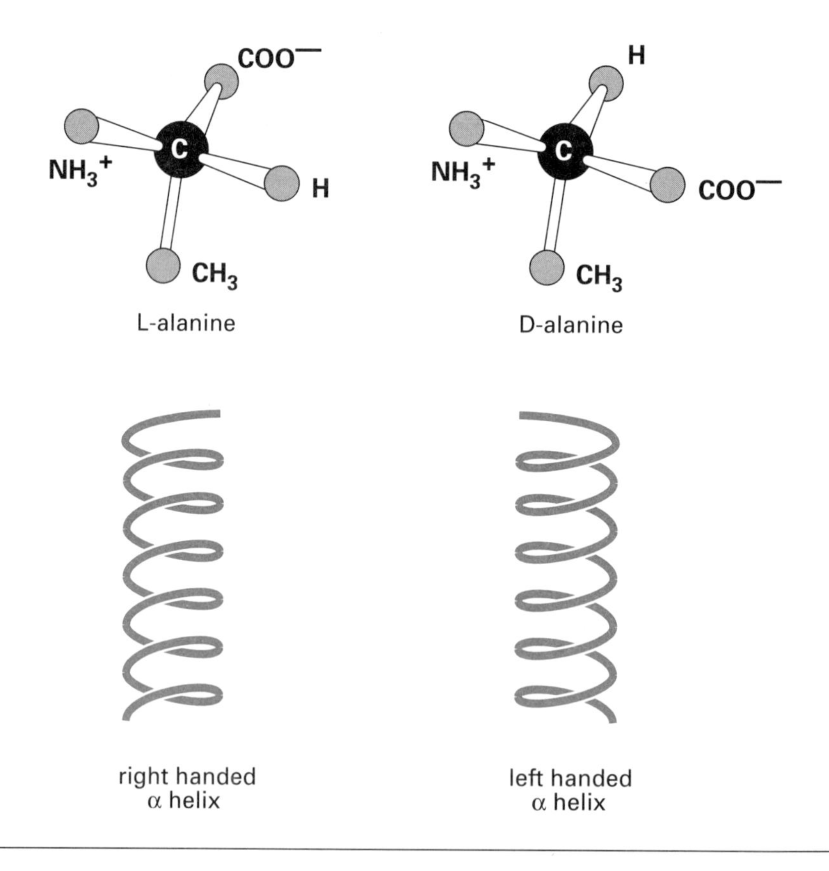

quence of bases on DNA assisting RNA polymerase to bind to its promoter and initiate transcription of the *lac* operon. It does this only when it has bound cAMP (in *E. coli* a signal that glucose is low). One of the domains of CAP has the job of binding cAMP. Another recognizes DNA sequences using a *helix-turn-helix* motif (Fig. 13.11). One of these helices fits into the major groove of the DNA where it can make specific interactions with the exposed edges of the bases.

Other proteins interact with DNA via zinc fingers. In Chapter 10 we described how the glucocorticoid receptor (Fig. 10.14) is made up of four domains. The first binds steroid hormone, the second enables the protein to move into the nucleus, the third is the zinc finger that binds to DNA, and the fourth interacts with the proteins bound to the TATA box. The zinc finger itself consist of an α helix and two strands of β sheet held together by bonds between amino acids and a zinc ion. The finger fits into the major groove of DNA.

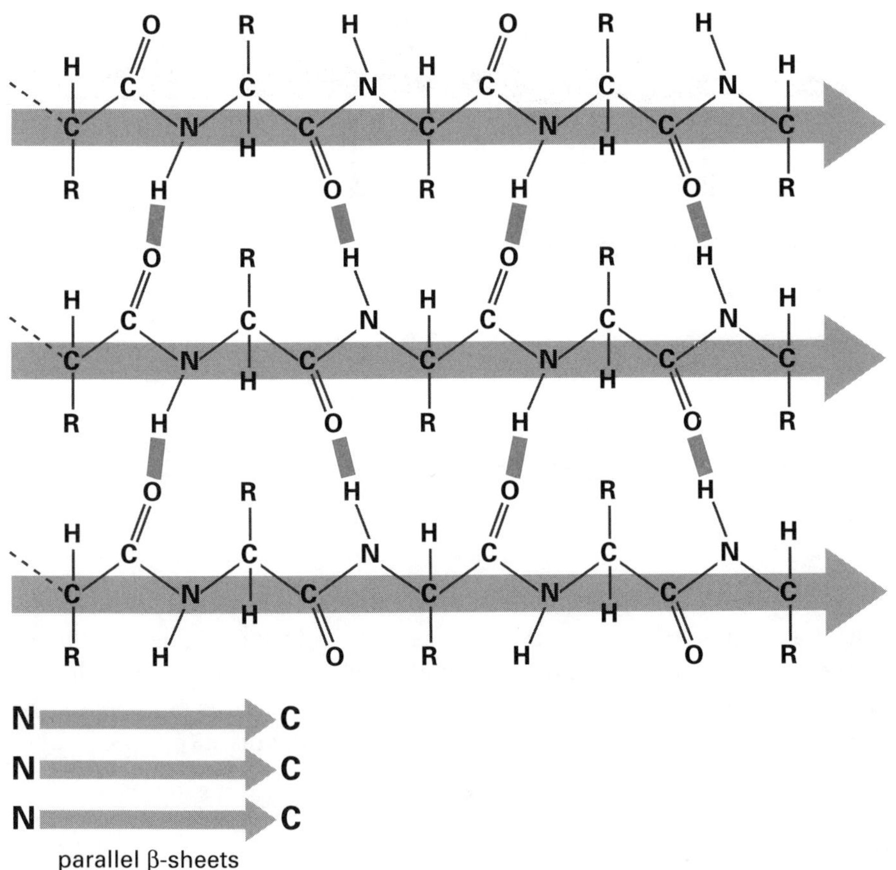

Figure 13.9*a*
Parallel β sheet.

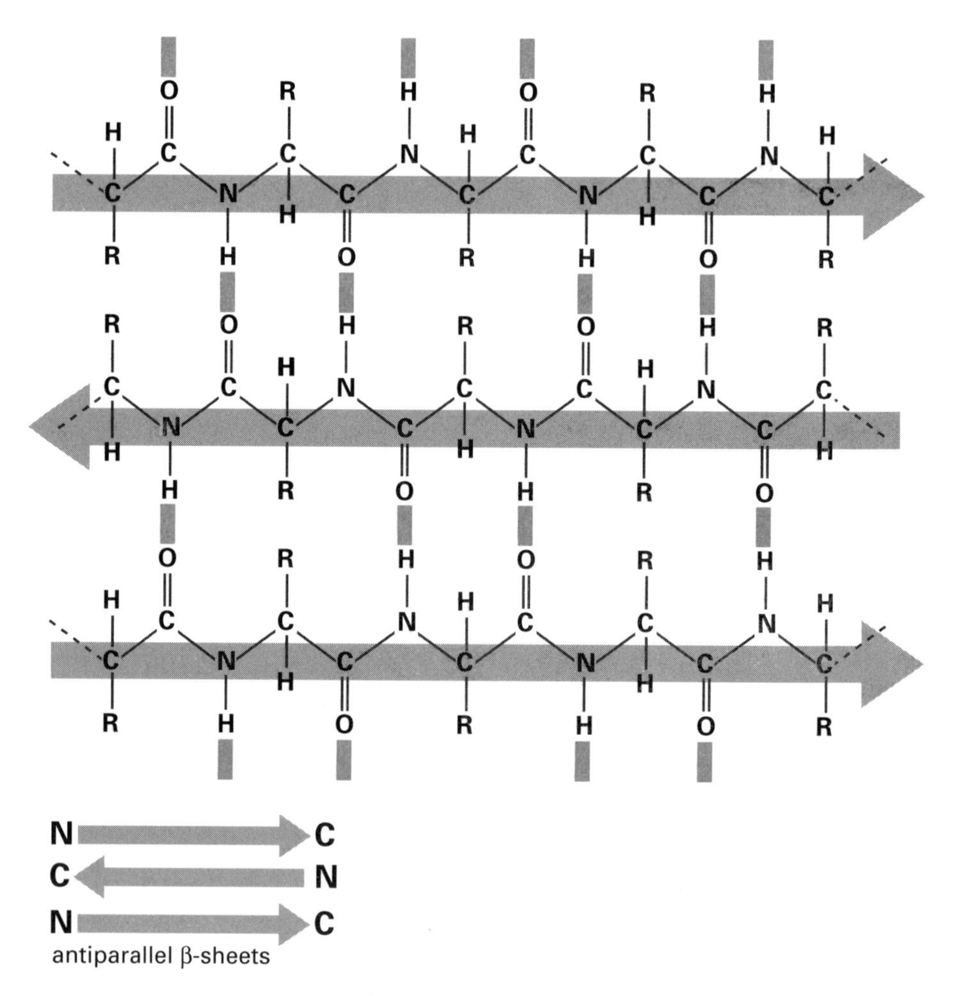

Figure 13.9*b*
Antiparallel β sheet.

Domains usually correspond to exons in the gene. It is therefore relatively easy for evolution to create new proteins by mixing and matching domains from existing proteins. The cAMP-gated channel (page 325) arose in the ancestor of the vertebrates when a duplication/translocation event spliced the cAMP-binding domain from cAMP-dependent protein kinase (page 328) onto the end of the voltage-gated sodium channel (page 309).

Proteins that pass through membranes have special tertiary structures. The polypeptide chain may cross the membrane one or many times, but each

BOX 13.2

The Forces That Mold the Shape of Proteins

HYDROGEN BONDS

Hydrogen has a valency of one and forms single covalent bonds with other atoms. However, if that atom is oxygen, nitrogen, or sulfur, then it can be a donor and share its hydrogen with a second oxygen, nitrogen, or sulfur (the acceptor) somewhere else in the same molecule or in a neighboring one. The donor and acceptor must be within a fixed distance of one another (typically 0.3 nm), with the hydrogen on a straight line between them. In an α helix the nitrogen atom within the peptide bond shares its hydrogen with the oxygen of the peptide bond four ahead of it in the polypeptide chain.

ELECTROSTATIC INTERACTIONS

If positive and negative amino acid residues are buried deep within a protein, where neither can interact with water, then they will attract each other, and it will be very difficult to tear them apart. Such an electrostatic bond inside a protein is a *salt bridge.*

Polar groups such as hydroxyl and amide groups are dipoles: They have an excess of electrons at one atom and a compensating deficiency at the other. The partial charges of dipoles will be attracted to other dipoles and to fully charged ions.

VAN DER WAALS FORCES

These are relatively weak close-range interactions between atoms. Imagine two atoms sitting close together. At a given instant more of the electrons of one atom may be on one side and this exposes the positive charge on the nucleus. This positive charge attracts the electrons of the adjacent atom thus exposing its nuclear charge, which would attract the electrons of another atom and so on. The next instant the electrons will have moved so we have a situation of flucturating attractions between atoms. These forces are important in the interiors of proteins and membranes and in the specific binding of a ligand to its binding site.

HYDROPHOBIC INTERACTIONS

A polypeptide with hydrophilic and hydrophobic residues will spontaneously adopt a configuration in which the hydrophobic residues are not exposed to water, either by sitting in a lipid bilayer (Fig. 2.2) or by adopting a globular shape in which the hydrophobic residues hide in the center of the protein.

DISULPHIDE BONDS GIVE SOME PROTEINS EXTRA STABILITY

Extracellular proteins often have disulfide bonds between specific cysteine residues. These tend to lock the molecule into its conformation. Although relatively few proteins contain disulfide bonds, these proteins are more stable and are therefore easy to purify and study. For this reason many of the first proteins studied in detail (e.g., the digestive enzymes chymotrypsin and ribonuclease and the bacterial cell wall degrading enzyme lysozyme) had disulfides.

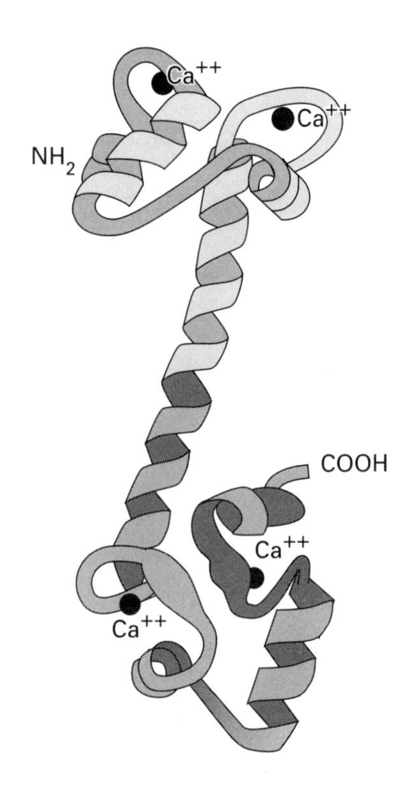

Figure 13.10
Calmodulin is composed of two very similar domains.

time it does so it exposes hydrophobic side chains that interact with the hydrophobic interior of the lipid bilayer. The commonest membrane-spanning structure is a 22 amino acid α helix. Figure 13.12 shows an ion channel (page 291) coded for by the cystic fibrosis gene (Chapter 21). Hydrophobic side chains on the transmembrane α helices interact with the hydrophobic interior of the membrane, while hydrophilic residues point inward and form a water-filled hole running through the center of the protein. Chloride ions, which cannot pass through the hydrophobic interior of the membrane, can pass from extracellular fluid to cytoplasm, and vice versa by passing through the water-filled center of the channel.

Many proteins associate to form multiple molecular structures that are held together tightly but not covalently by the same interactions that form the tertiary structure (Box 13.2). Microfilaments (page 53) are formed when thousands of identical actin monomers join together, end to end. CAP is only active when it dimerizes (Fig. 13.11). Hemoglobin (Fig. 13.13) is formed from four individual polypeptide chains, two α and two β (page 115). We describe the three-dimensional arrangement of the protein units as they fit together to form the multimolecular structure as the *quaternary structure*.

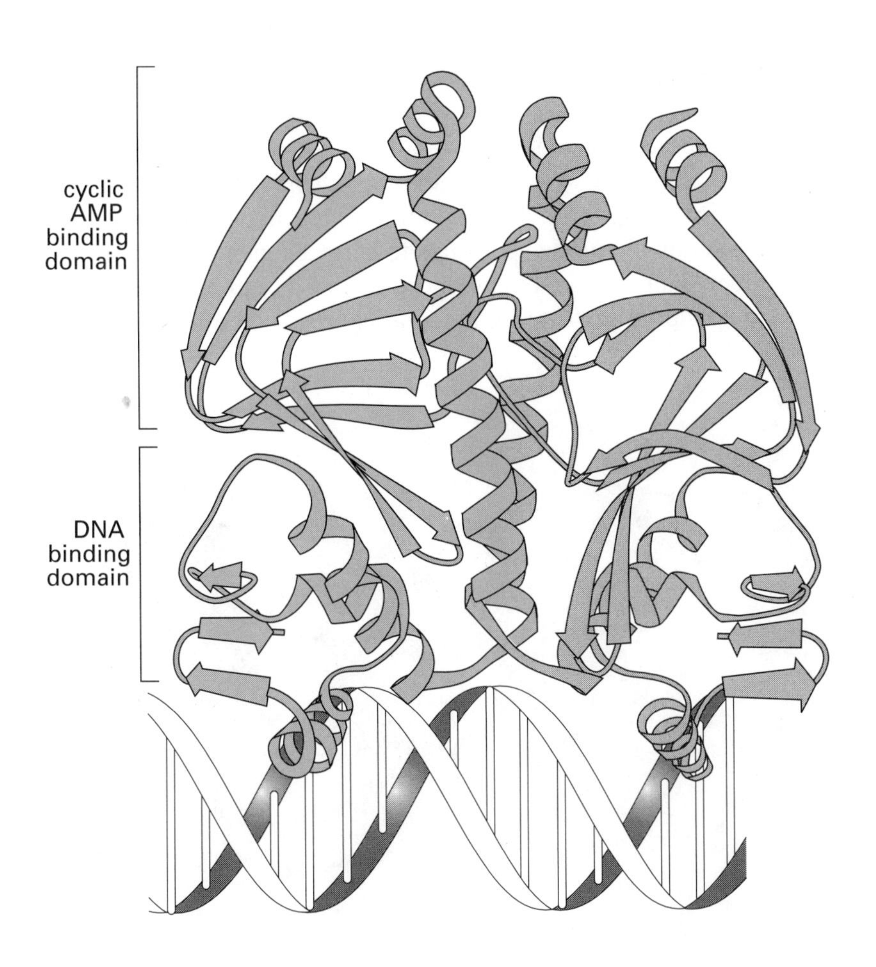

Figure 13.11

Active CAP is a dimer of two identical α helix-turn-helix DNA-binding proteins, one shown in blue, one in black.

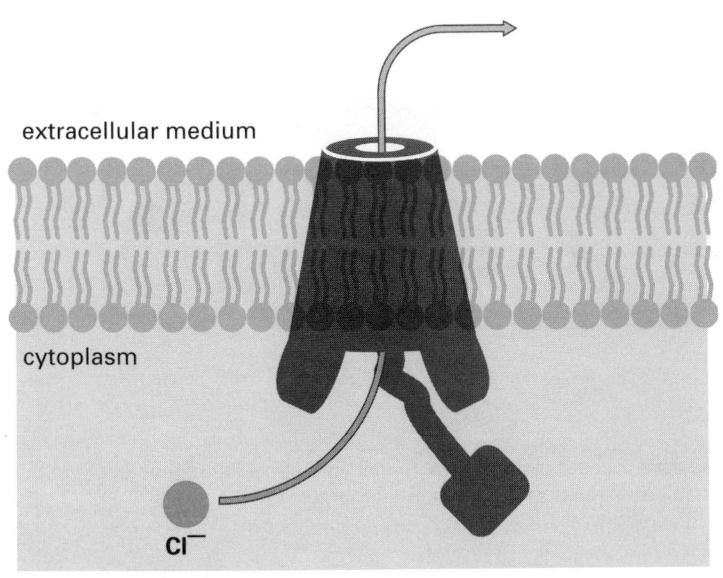

Figure 13.12

The cystic fibrosis gene product, a channel protein.

PROSTHETIC GROUPS

The 20 protein amino acids cannot perform all the functions required of proteins. Many proteins therefore associate with other chemical species that have the required chemical properties. Hemoglobin, which transports oxygen in our blood, uses iron-containing heme groups to carry oxygen molecules (Fig. 13.13). The general name for a nonpolypeptide that is strongly associated with a protein and helps it perform its function is prosthetic group.

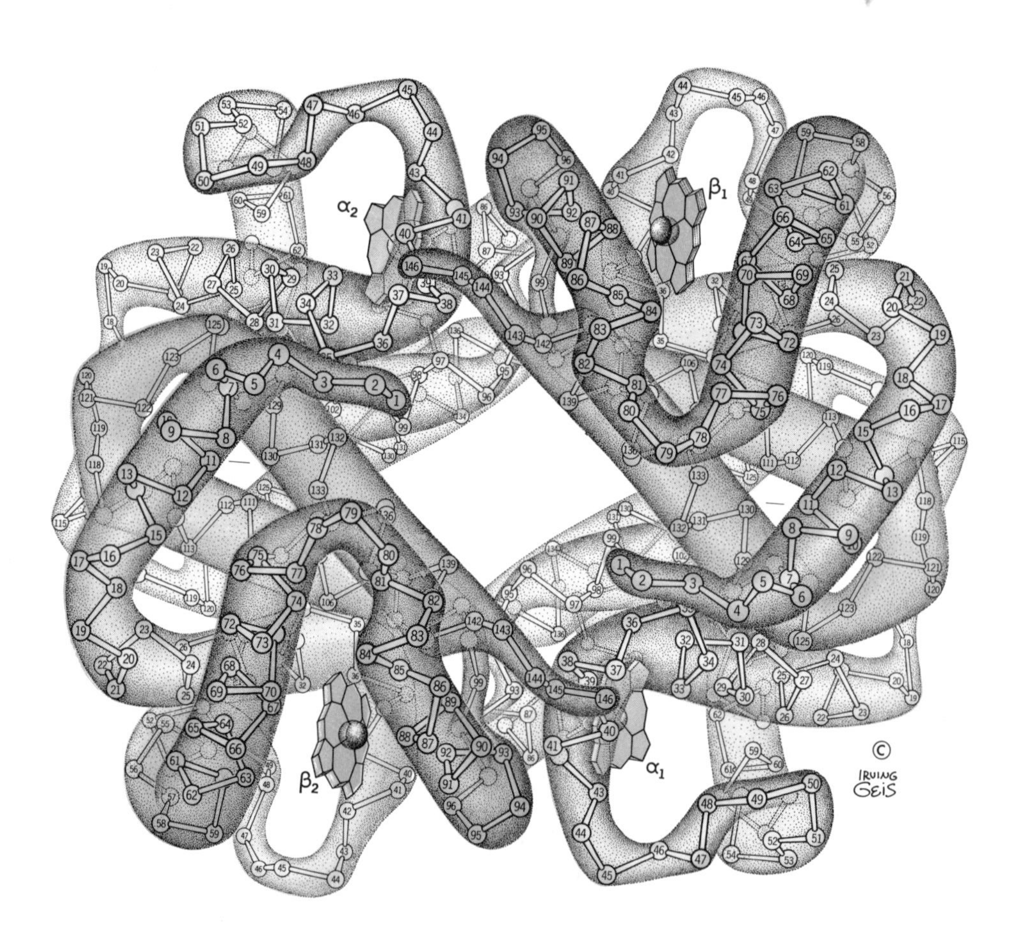

Figure 13.13

Hemoglobin formed from four subunits that each contain a heme as a prosthetic group. The subunits at top left and bottom right are α; the other two are β.

THE PRIMARY STRUCTURE CONTAINS ALL THE INFORMATION NECESSARY TO SPECIFY HIGHER-LEVEL STRUCTURES

Protein structures are stable and functional over a small range of environmental conditions. Outside of this range the pattern of bonds that stabilize the tertiary structure is disrupted, and the molecule denatures—activity disappears as the molecule loses its structure. Denaturation may be caused by many factors, which include excessive temperature (Box 13.3), change of pH, and detergents. Concentrated solutions of urea (8 mol $liter^{-1}$) have long been known to denature proteins. Physical chemical techniques have shown that all of the higher levels of structure are lost and that the polypeptide chains adopt random, changing conformations. Reagents that do this are known as chaotropic. If the solution is rapidly diluted, the protein may precipitate, but if the urea is removed more slowly, the protein refolds regaining its structure and biological activity. This shows that the sequence of amino acids contains all of the information necessary to specify the final structure. The refolding of an urea denatured protein cannot be random. Even a small protein with 100 amino acids would take some 10^{50} years to try all of the possible structural conformations available. The universe is not old enough. The fact that refolding does happen and happens often on a time scale of seconds tells us that there must be a folding pathway and the process is not random. Secondary structures seem to form first and then fold as units. In the cell, folding is helped by other proteins called chaperones.

Temperature-sensitive mutants (Box 4.1) make a protein that does not fold into the proper high-level structure as readily as does the normal protein. At the permissive temperature, the protein folds correctly and retains its function. Warming to the restrictive temperature deranges the mutant protein's proper high level structure, and stops it working. Siamese cats (page 101) have a temperature-sensitive mutation affecting the enzyme that makes fur pigment.

BOX 13.3

Cooking Is Chemistry!

Many cooking processes involve protein denaturation. Heat is the most obvious agent, but others are important too. Proteins may be denatured by changing the pH (often with vinegar or lemon juice). The familiar stiff foam resulting from beating egg whites (chiefly protein) forms because the protein molecules are unfolded at surfaces and by the shearing of the beater as air is beaten in: the denatured polypeptide chains entangle with one another and form a "web," which stabilizes the foam.

SUMMARY

1. Polypeptides are linear polymers of α-amino acids linked by peptide bonds. There are 20 amino acids coded for by the genetic code. They differ in the properties of their side chains which range from hydrophobic groups to charged and uncharged hydrophilic groups.

2. Proteins are polypeptides that have a complex three-dimensional shape. It is convenient to consider these structures as having three levels. The primary structure is the linear sequence of the amino acid monomers. In some parts of the structure regular, repeated foldings of the peptide chain can be seen: these are secondary structures. The most common secondary structures are held together by hydrogen bonds between the carbonyl oxygens and the hydrogens on the nitrogens of the peptide bonds. There are two common types of secondary structure:

- **In the α helix the chain coils upon itself making a springlike spiral with hydrogen bonds running parallel to the length of the spiral.**
- **In β sheets the hydrogen bonds are between extended strands of polypeptide that run alongside of one another.**

The final, complex folding of a protein is its tertiary structure. Interactions between side chains stabilize the tertiary structure.

3. Some proteins have a quaternary structure that is an association of subunits, each of which has a tertiary structure.

BOX 13.4

Protein Folding Gone Awry: Mad Cow Disease

When liquid water is cooled, the molecules pack themselves in such a way that they solidify into the crystal we know as ice. In the novel *Cat's Cradle,* Kurt Vonnegut imagines an alternative crystalline form of water called ice-9. In the story, ice-9 is such a stable way of arranging water molecules that the melting point of the crystal is 46°C. However, since cooled water crystallizes not as ice-9 but as regular ice, nobody gave ice-9 a second thought. Then, in the climax of the book, someone throws a small crystal of ice-9 into the sea, the water molecules of the sea start adding themselves to the crystal, the crystal of ice-9 grows, and after a few days all the oceans and rivers of the world are solid ice-9. End of the world as we know it.

This is just a story, but mad cow disease is real, and the cause is similar. Our brains contain a protein called PrP. If this is denatured and then allowed to refold, it adopts the shape found in normal brains, denoted PrP^c. PrP^c is soluble in cytosol. However, if PrP is allowed to contact an alternatively folded form, denoted PrP^{Sc}, then slowly but surely all the PrP protein around agglomerates as larger and larger lumps of PrP^{Sc}. In the brain, these growing lumps of PrP^{Sc} destroy the brain cells, leaving empty holes—hence the general name for diseases of this type, spongiform encephalopathies. The disease has different names in different species: scrapie in sheep, mad cow disease in cattle, Creutzfeldt-Jakob disease in humans—but the cause is the same in all cases. Contact with some PrP^{Sc} causes all the PrP in the victim's brain slowly to change to this alternatively folded form.

The bovine form of PrP is similar enough to the sheep form that bovine PrP will be altered by small lumps of sheep PrP^{Sc}. Cows that eat material from the brains of sheep with scrapie may therefore get mad cow disease. Are the human and bovine forms of PrP dissimilar enough that eating material containing bovine PrP^{Sc} poses no threat to humans? The British government were advised that they were. Evidence collected since then indicates that they were probably wrong.

CHAPTER

14

HOW PROTEINS WORK

The three-dimensional structures of proteins generate binding sites for other molecules. This reversible binding is central to most of the biological roles of proteins, whether the protein is a cell adhesion molecule that binds a similar molecule on another cell (Box 2.2) or a transcription factor that binds to DNA (Fig. 13.11). One special class of proteins, enzymes, have sites that not only bind another molecule but then catalyze a chemical reaction involving that molecule.

DYNAMIC PROTEIN STRUCTURES

It is easy to get the impression that protein structures are fixed and immobile. In fact protein molecules are always flexing and changing their structure slightly about their lowest energy state. A good term for this is "breathing."

Many proteins have two low energy states in which they spend most of their time, like a sleeper who, though twisting and turning throughout the night, nevertheless spends most time lying on their back or side. An example is the glucose carrier. This is a transmembrane protein that is stable in one of two configurations (Fig. 14.1). In one, it has a pocket that is open to the extracellular fluid. In the other, the pocket is open to the cytoplasm. By switching between the two states, the glucose carrier carries glucose into and out of the cell.

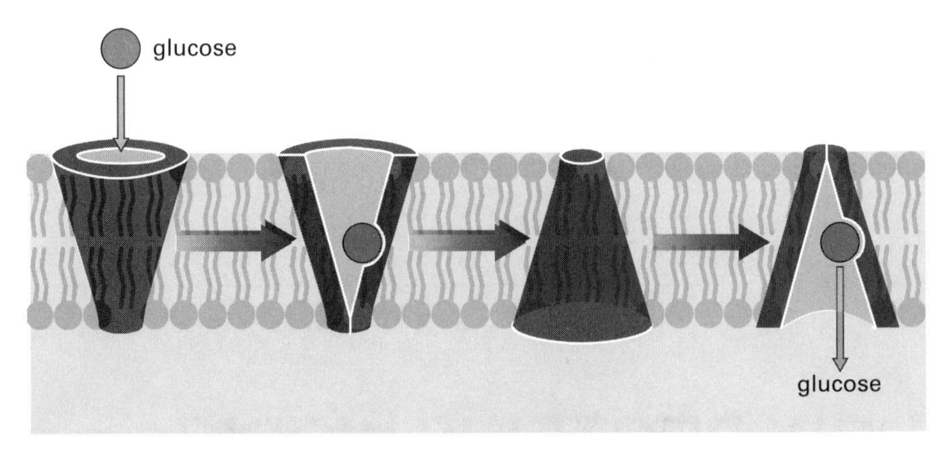

Figure 14.1
The glucose carrier switches easily between two shapes.

Allosteric Effects

The glucose transporter is able to bind a ligand—the glucose molecule—in either of its low energy configurations. In contrast, the *lac* repressor (page 152) can only bind its ligand, the operator region of the *lac* operon, in one configuration. On its own, the protein predominantly adopts this configuration, so transcription is prevented. When the *lac* repressor binds allolactose, it is locked into a second, inactive form that cannot bind to DNA (Fig. 10.8). Transcription is no longer repressed, although it requires the cAMP-CAP complex to proceed (page 155). This type of interaction, in which binding of one ligand at one place affects the ability of a protein to bind another ligand at another location, is called allosteric, and it is usually a property of multiple subunit proteins with a quaternary structure.

Hemoglobin (Fig. 13.13) is an example of a protein where allosteric effects play an important role. Each heme prosthetic group, one on each subunit of the four, can bind an oxygen molecule. We can get an idea of what one subunit on its own can do by looking at myoglobin (Fig. 14.2*a*), a related molecule that moves oxygen within the cytoplasm. Myoglobin has just one polypeptide chain and one heme. The blue line in Fig. 14.2*b* shows the oxygen-binding curve for myoglobin. Starting from zero oxygen, the first small increase in oxygen concentration produces a large amount of binding to myoglobin; the next increase in oxygen produces a slightly smaller amount of binding, and so on, until myoglobin is fully loaded with oxygen. A curve of this shape is known as hyperbolic. The black line in Fig. 14.2*b* shows the oxygen-binding curve for hemoglobin. Starting from zero oxygen, the first small increase in oxygen concentration pro-

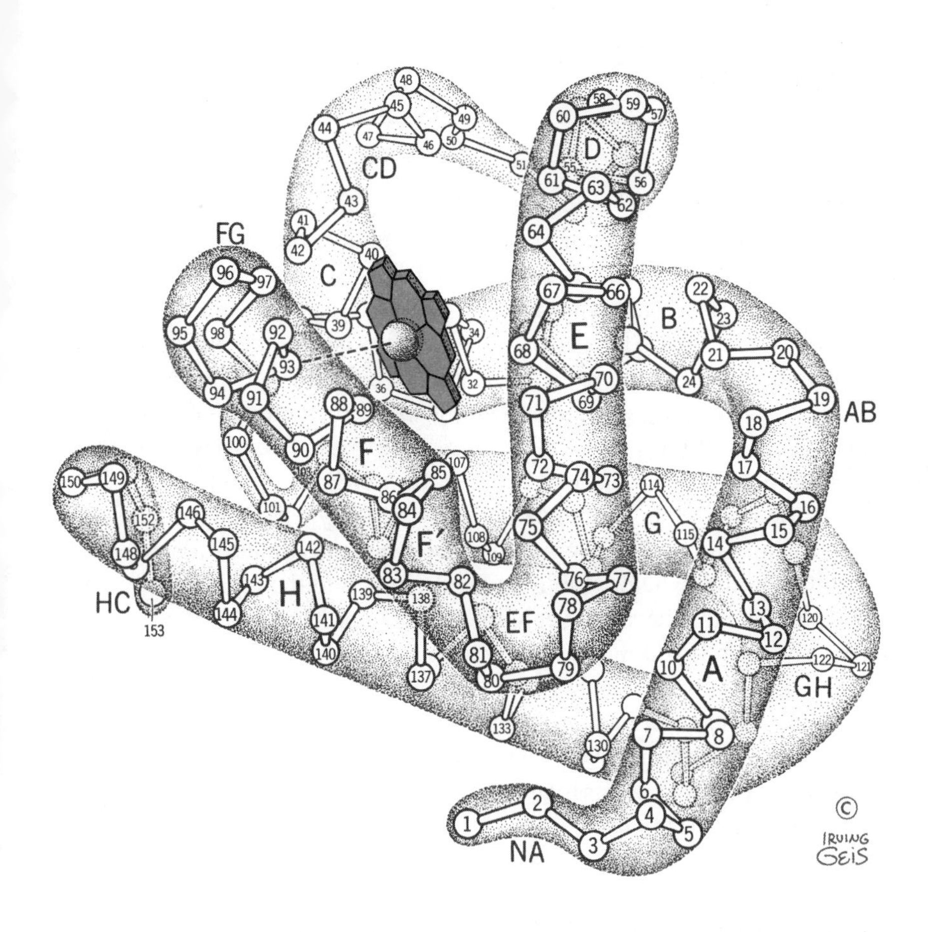

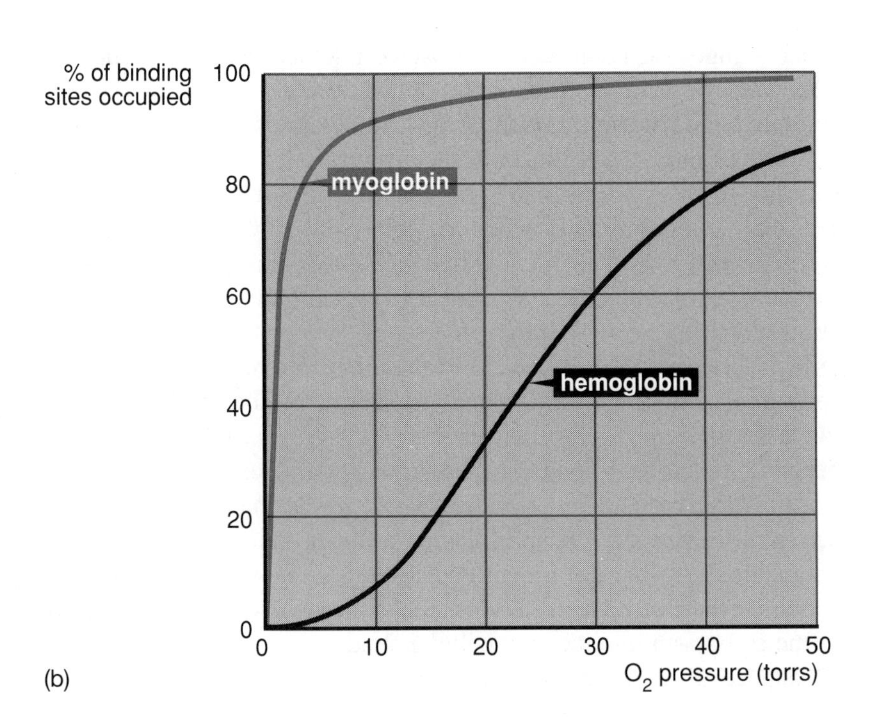

(b)

Figure 14.2

Myoglobin is a single polypeptide chain with a single heme group. Its oxygen loading curve is different from that of hemoglobin. One torr is equal to one mm of mercury or 133 Nm^{-2}.

duces hardly any binding to hemoglobin. The next increase in oxygen produces much more binding and so the curve gets steeper and steeper before leveling off again as the hemoglobin becomes fully loaded. The explanation for this behavior is that the hemoglobin subunits can exist in two shapes, only one of which has a high affinity for oxygen. The way the four subunits fit together means that they must all be in one form or the other—this type of behavior is called cooperativity. When the oxygen concentration is low, most of the hemoglobin molecules have their subunits in the low-affinity form. As oxygen increases, it begins to bind to hemoglobin—slowly at first, because the hemoglobin is in the low-affinity form. As the hemoglobin subunits begin to bind oxygen, they tend to switch to the high-affinity form. The cooperativity between the subunits means that all four subunits within the hemoglobin molecule switch to the high-affinity form, even those subunits that have not yet bound an oxygen. Thus, once some oxygen loading has occurred, adding further amounts of oxygen gets easier and the graph (Fig. 14.2b) gets steeper. It is not terribly important when hemoglobin is loading up with oxygen in our lungs, but it is very useful is when hemoglobin moves into a region of the body with a lower oxygen concentration and begins to give up oxygen. As it does so, its affinity for oxygen falls, so it rapidly dumps all its oxygen for the tissues to use.

Chemical Changes that Shift the Preferred Shape

Anything that changes the pattern of electrostatic interactions within a protein will alter the relative energy of its states. If the protein contains histidine residues, merely changing the pH will do this. In solutions with a pH greater than 7, most of the histidine residues in the protein will be uncharged (page 213). In solutions with a pH less than 7, the histidine residues will bear a positive charge. A protein conformation that brought two histidine residues close together would therefore be stable in alkaline conditions, but in acid conditions the two residues would each bear a positive charge and repel, destabilizing the configuration (Fig. 14.3).

Another mechanism that is used to change the relative energy of protein states is phosphorylation, the addition of a negatively charged phosphoryl group by a class of enzymes called protein kinases. Proteins can be phosphorylated on serine, threonione, tyrosine, aspartate or glutamate. The calcium ATPase (Fig. 14.4) is a transmembrane protein whose lowest energy state forms a pocket open to the cytoplasm. Phosphorylation of the protein changes the electrostatic interactions so that the lowest energy state is now one that forms a pocket open to the extracellular fluid. This mechanism is used to force calcium ions out of the cytoplasm into the extracellular fluid.

The distances moved within a protein as it is phosphorylated are only a

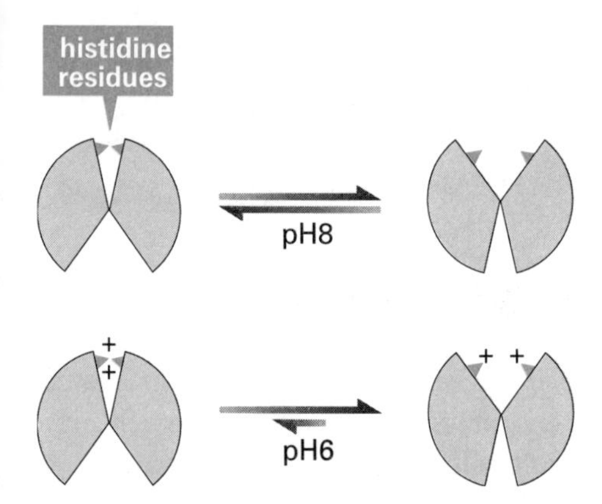

Figure 14.3

A pH change that alters the charge on histidine will alter the balance of forces within a protein.

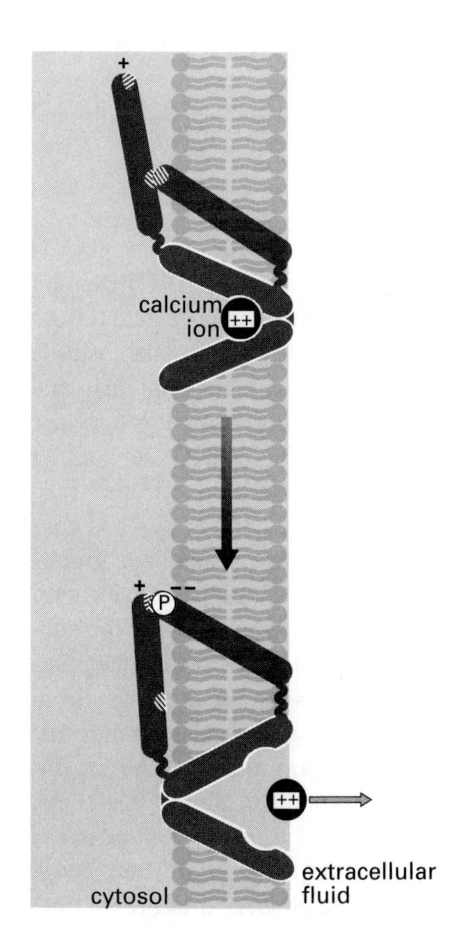

Figure 14.4

Phosphorylation changes the charge pattern, and hence the pattern of forces within the calcium ATPase, forcing a change of shape.

few nanometers. However, when repeated many times and amplified by lever systems, they can produce movements of micrometers or even meters. The force that beats a flagellum (page 46) or kicks a leg is in both cases produced by phosphorylation-induced protein shape changes.

ENZYMES ARE PROTEIN CATALYSTS

Life depends on complex networks of chemical reactions. These are mediated by enzymes. Enzymes are catalysts of enormous power and high specificity. Consider a lump of sugar, it is combustible but quite difficult to set alight. A chemical catalyst would speed up its combustion, and we would end up with heat, a little light, carbon dioxide, and water. Swallowed and digested, the sucrose is broken down eventually to carbon dioxide and water by the action of at least 22 different enzymes. The energy released is used to drive other reactions in the body.

At a basic level the reaction carried out by an enzyme can be expressed as

$$E + S \rightleftarrows ES \rightleftarrows EP \rightleftarrows E + P$$

E is the enzyme, which binds the substrate S to form the complex ES. The region of the protein where the substrate binds and the reaction occurs is called the acitve site. This binding is specific, often highly so. The enzyme β-galactosidase (page 152) is moderately specific, and will split not only lactose but also any other disaccharide joined with a β-galactoside bond. By contrast phosphorylase kinase (page 281) acts with absolute specificity on a single substrate, another enzyme called glycogen phosphorylase—none of the thousands of other proteins in the cell can substitute. In general the specificity of an enzyme is conferred by the shape of the active site and by particular amino acid side groups that interact with the substrate. Binding of substrate produces the enzyme:substrate complex ES; the catalytic function of the protein then converts the substrate to product P, still bound to the protein in the complex EP. Finally, the product dissociates from the enzyme. Chemical engineers use catalysts made of many materials, and within cells there are even catalysts called ribozymes that are made of RNA. However, only proteins, with their enormous repertoire of different shapes and different side chains, can produce catalysts of high selectivity.

The catalytic rate constant k_{cat} (also known as the *turnover number*) of an enzyme gives us an idea of the enormous catalytic power of most enzymes. It is defined as the number of molecules of substrate converted to product per molecule of enzyme per unit time (equally it is moles of substrate converted per mole of enzyme per unit time). Many enzymes have k_{cat} values around

1000 s^{-1}. The reciprocal of k_{cat} is the time taken for a single event. Thus, if k_{cat} is 1,000 s^{-1}, one substrate molecule will be converted to product every millisecond. Some enzymes achieve very much higher rates. Catalase has a k_{cat} of 4×10^7 s^{-1} and so takes only 25 nsec to split each molecule of hydrogen peroxide into oxygen and water.

The Initial Velocity of an Enzyme Reaction

The fundamental experiment in the study of enzyme behavior is measurement of the appearance of product as a function of time. This is often called an enzyme assay. As the reaction progresses the rate at which the concentration of product is increasing slows, eventually to zero (Figure 14.5). All reactions are in principle reversible, so at any one time the overall reaction rate (the rate of product formation) is the sum of the rate at which the enzyme makes the product, i.e. its forward reaction, and the rate at which the enzyme breaks the product down—the back reaction. Depending on the particular reaction the rate may become zero because all the substrate has been used up, or because the back reaction, becoming steadily larger, has become equal to the forward reaction. (i.e. the reaction has reached equilibrium).

We can simplify the analysis if we consider the start of the reaction, when there is no product around and therefore no back reaction. The rate of reaction at time zero (the *initial velocity* v_o sometimes called the initial rate) is found by plotting a graph of product concentration as a function of time and measuring the slope at time zero (Fig. 14.5). The initial velocity v_o expresses how fast the enzyme can work under these ideal, conditions. Initial velocity is usually expressed as the rate at which the concentration of product increases (moles per litre per second).

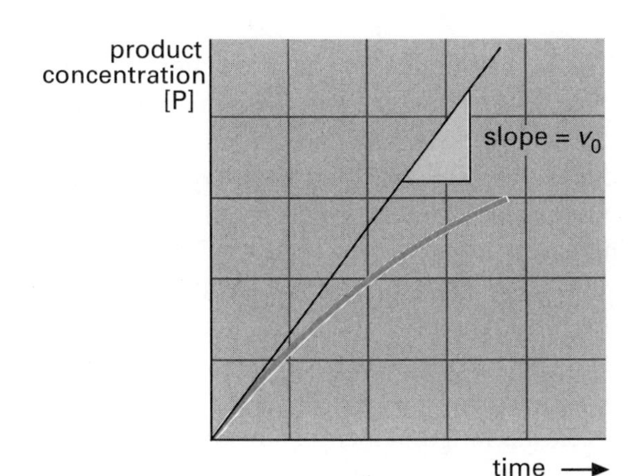

Figure 14.5
The initial velocity is the slope of the straight line at the start of the reaction.

Most enzymes are such very effective catalysts that they must be assayed under conditions where the concentration of enzyme is always very much less than the concentration of substrate. Otherwise, the reaction would be over in seconds or milliseconds.

Enzymes are studied for many reasons: Understanding of how they achieve their catalytic excellence is of fundamental interest and practical application as we increasingly use them in industrial processes and seek to design enzymes for particular tasks. Measurements of enzyme concentrations and properties allow us to study processes within cells and organisms. The starting point of any study of an enzyme is to determine its activity, measured as k_{cat}, and its affinity for its substrate. Affinity is measured by the Michaelis constant K_M. We must now see how this is measured.

The Effect of Substrate Concentration

Let us consider a series of experiments designed to see how the initial velocity of an enzyme reaction varies with the concentration of the substrate (always much greater than that of the enzyme). Each of the smaller graphs in Fig. 14.6 shows the result of one of these experiments. As we increase the substrate concentration, we find that at first the velocity increases with each increase in substrate concentration but that, as the substrate concentration becomes larger, the increases produced get smaller and smaller (Fig. 14.6). The initial velocity approaches a maximum value which is never exceeded. Reactions that show this sort of dependence on substrate are said to show saturation kinetics.

How can we explain this relationship? The reaction sequence can be simplified to

$$E + S \rightleftarrows ES \rightarrow E + P$$

(We are using initial velocities, so we can ignore any back reaction.) The enzyme and substrate must collide in solution, and the substrate must bind to the enzyme's active site to form the ES complex. The chemical reaction takes place in the ES complex, and the product is released. In experiments with higher and higher substrate concentration, there is ever more ES present, and this increasing ES concentration gives an increasing rate of product release. There is a limit set by the amount of enzyme present. At high substrate concentration virtually all of the enzyme is present as ES, and the observed rate is limited by the ES $\rightarrow$ E + P step. Thus, as the substrate concentration increases, the reaction rate levels off as it approaches the maximal velocity or V_m.

The full relationship between the initial velocity and substrate concentration is described by two parameters: V_m, the maximal velocity, and the

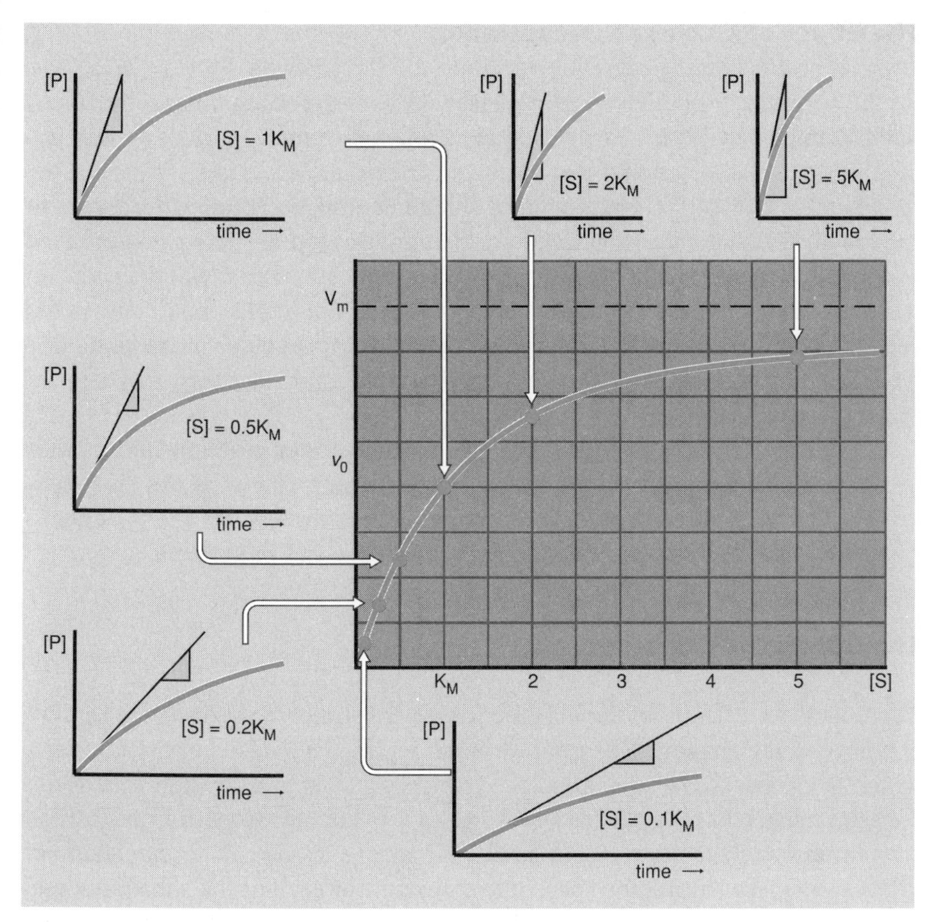

Figure 14.6
v_0 measured in a number of reaction tubes ([E] is constant and always less than [S]) forms a hyperbolic curve when plotted as a function of substrate concentration.

Michaelis constant, K_M, which is defined as the substrate concentration that gives an initial velocity half as great as V_m. Enzymes with a low K_M have a high affinity for the substrate. Even at low concentrations of substrate, substrate will bind to the enzyme and the reaction will proceed.

Mathematically saturation kinetics can be described by the equation

$$v_o = \frac{V_m[S]}{K_M + [S]}$$

where [S] is the substrate concentration. This is the Michaelis–Menten equation, named after Maud Menten and Leonor Michaelis who propounded a general theory of enzyme action in 1913.

The Effect of Enzyme Concentration

As the concentration of enzyme increases, so does the reaction rate. This obvious fact means that one can estimate how much enzyme is present by measuring the rate of the reaction that it catalyzes. Clinicians make use of this to assay for enzymes in blood. If enzymes that should be trapped within cells appear in the blood, this indicates that cells are being damaged and their contents are leaking out. The presence of enzymes from a particular organ tells the clinician which tissue is affected. For instance trypsinogen is normally found only in the cells of the pancreas and in the intestine lumen. Its appearance in the blood of a fetus is a sign that cells in the pancreas are dying, and is an early indicator of cystic fibrosis (page 349).

This principle, that more enzyme means a faster reaction rate, is also used when purifying an enzyme from a crude tissue preparation. As the preparation becomes a purer sample, less is required to give the same V_m. We quantify purity as the specific activity, the V_m per unit mass of preparation.

The Specificity Constant

K_M reflects the affinity of the enzyme for the substrate: k_{cat} reflects the catalytic ability of the enzyme. The ratio of these, k_{cat}/K_M, is the specificity constant, which is a measure of how good the enzyme is at its job. A high specificity constant means that the reaction goes fast (k_{cat} is big) and does not need a high concentration of substrate (K_m is small). When an enzyme can accept a number of substrates, measuring the specificity constant for each of the substrates, and looking to see which is biggest, will tell us which is the preferred substrate for the enzyme.

A reaction cannot go faster than the rate at which enzyme and substrate actually collide. Diffusion limited values of k_{cat}/K_M fall in the range 10^8 to 10^{10} L mol^{-1} s^{-1}. Many enzymes have values in this range (Table 14.1).

Table 14.1 Values of k_{cat}/K_M for a number of enzymes

Enzyme	$k_{cat}(s^{-1})$	$K_M$mol $liter^{-1}$	k_{cat}/K_M liter mol^{-1} s^{-1}
Lysozyme	0.5	6×10^{-6}	8.3×10^{3}
Tyrosyl-tRNA synthetase	7.6	9×10^{-4}	8.4×10^{3}
Ribonuclease	7.9×10^{2}	7.9×10^{-3}	1×10^{5}
Fumarase	8×10^{2}	5×10^{-6}	1.6×10^{8}
Urease	1×10^{4}	2.5×10^{-2}	4×10^{5}
Acetylcholinesterase	1.4×10^{4}	9×10^{-5}	1.6×10^{8}
Catalase	4×10^{7}	1.1	4×10^{7}

Note: Fumarase and acetylcholinesterase have achieved "catalytic perfection," since they both show k_{cat}/K_M in the 10^8 to 10^{10} liter mol^{-1} s^{-1} range, which shows that the reaction rate is limited only by diffusion.

ENZYME CATALYSIS

A catalyst speeds the rate of a reaction but does not alter the position of the equilibrium E + S ⇄ E + P. Reaction speed is increased by lowering the activation energy barrier between the reactants and products (Fig. 14.7). Each enzyme has evolved to achieve this end for its particular reaction. Clearly the binding of the substrate is as central to catalysis as it is to specificity. The active site contains precisely positioned amino acid side chains that promote the reaction. Koshland proposed that the enzyme's binding site was shaped not so much to fit the substrate but to fit a molecule that was halfway in structure between substrate and product, and therefore to reduce the activation energy of

(a)
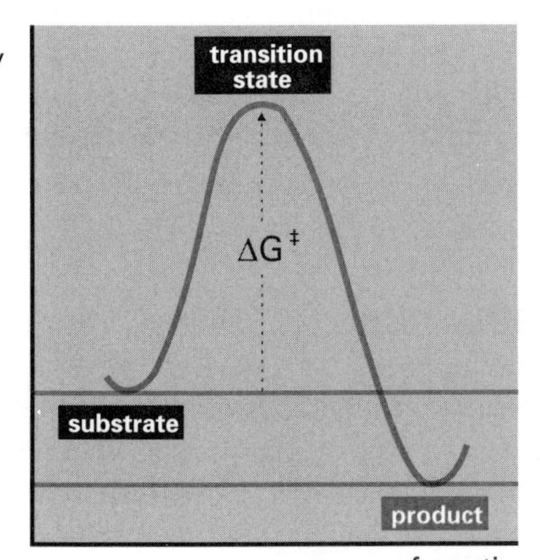

(b)
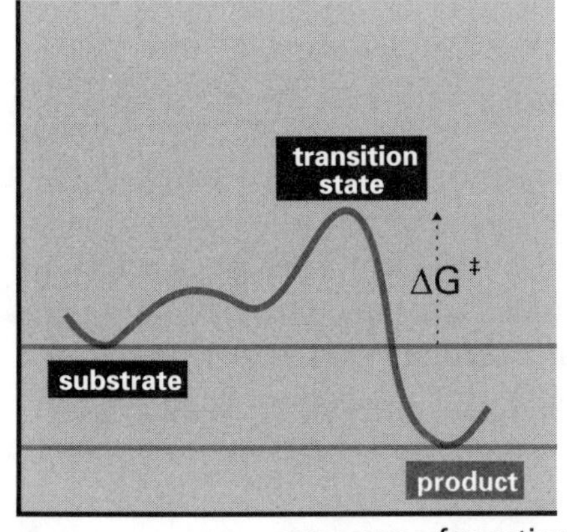

Figure 14.7
Catalysts act by reducing the activation energy ($\Delta G^{\ddagger}$) of a reaction.

the transition. This induced fit seems to be a common feature in many enzyme mechanisms.

COFACTORS AND PROSTHETIC GROUPS

Enzymes that are called upon to perform reactions that are outside the repertoire of the 20 possible protein amino acid sides chain recruit other chemical species to help them do the job. The aminotransferases provide a good example. These enzymes are central to amino acid metabolism (page 269). They catalyze the interconversion of amino acids and oxo-acids by moving an amino group (Fig. 14.8). Pyridoxal phosphate (derived from vitamin B_6) is bound by the protein and accepts the amino group from the amino acid donor, which then leaves the enzyme converted to an oxo-acid. The oxo-acid substrate binds, accepts the amino group from the pyridoxal phosphate, and leaves having been converted to an amino acid. Pyridoxal phosphate ends up exactly as it started. These helper chemicals are called cofactors when, like pyridoxal phosphate, they are not tightly bound to the enzyme. We already know the term for a molecule that is tightly bound to a protein and helps it perform its job—it is a prosthetic group (page 230).

ENZYMES CAN BE REGULATED

In this chapter we have discussed the catalytic role of enzymes in isolation. In fact enzymes are like other proteins in showing the complex behavior described in Chapter 13—multiple states, multiple binding sites, quaternary structure, phosphorylation, and so on. For instance, some important enzymes have quaternary structures and show cooperativity between the active sites. Such enzymes will not follow the Michaelis–Menten equation, but instead the curve of initial velocity against substrate concentration will be sigmoid, in the same way that the binding of oxygen to hemoglobin is (see Fig. 14.2*b*). Other enzymes have only one catalytic site, but binding of a different ligand at a different regulatory site affects the catalytic action. $p34^{cdc2}$, the enzyme that triggers mitosis (page 63), is such an allosterically regulated enzyme. On its own in solution, it spends the vast majority of the time in an inactive configuration. Even when it is in the active configuration the catalytic site is blocked by a phosphoryl group on one specific tyrosine residue, Y15. In order to assume the active configuration the protein must be phosphorylated at a second site, threonine T167, and the protein ligand, cyclin B must bind (Fig. 4.4). This complex control means that $p34^{cdc2}$ can act as a check point. If the phosphoryl group has been removed from Y15 AND the phosphoryl group has been added to T167 AND cyclin B is present at a high enough concentration, THEN it is safe to proceed into mitosis.

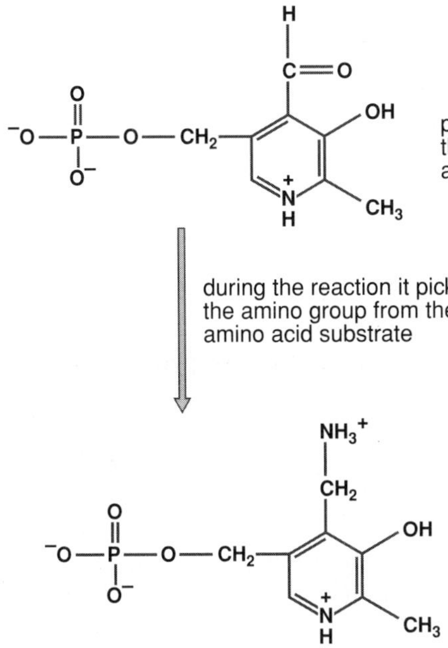

pyridoxal phosphate is the cofactor for aminotransferases

during the reaction it picks up the amino group from the amino acid substrate

pyridoxamine

alanine aminotransferase catalyzes the reaction:

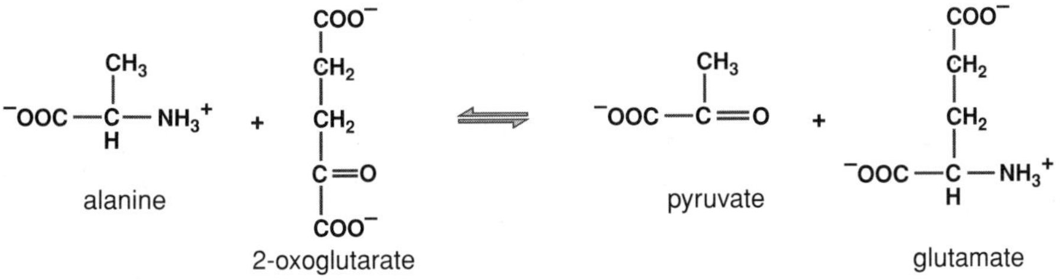

alanine + 2-oxoglutarate ⇌ pyruvate + glutamate

the reaction is in two stages:

E-pyridoxal phosphate + alanine ⇌ E-pyridoxamine + pyruvate

E-pyridoxamine + 2-oxoglutarate ⇌ E-pyridoxal phosphate + glutamate

Figure 14.8

Aminotransferase enzymes use a cofactor that participates in the reaction but ends up unchanged. E represents the enzyme molecule.

SUMMARY

1. Most protein functions arise from their ability to specifically bind other molecules in a reversible fashion. Shape changes, triggered by the binding of other molecules, mediate protein movements and function. Allostery is a special case, where a shape change induced by the binding of ligand at one site changes the affinity of another site.

2. Enzymes are highly specific biological cataysts. The turnover number, or catalytic rate constant, k_{cat}, is the maximum number of substrate molecules that can be converted to product per molecule of enzyme per unit time.

3. The initial velocity (i.e., when product is absent) of many enzyme reactions shows a hyperbolic dependence on substrate concentration. At high substrate concentration the initial velocity approaches a limiting value V_m as the enzyme is saturated with substrate. The substrate concentration that gives an initial velocity equal to half V_m is the Michaelis constant, K_M. This indicates the enzyme's affinity for the substrate.

4. The initial velocity is related to substrate concentration by the Michaelis–Menten equation

$$v_0 = \frac{V_m[S]}{K_M + [S]}$$

and is directly proportional to the enzyme concentration. The specificity constant, k_{cat}/K_M, provides a "figure of merit" for the enzyme: The higher it is, the better the enzyme is at doing its job.

5. Some enzymes use cofactors to carry out reactions that require different properties from those of the 20 amino acid side chains.

6. Allosteric interactions within enzymes allows control of reactions within the cell.

BOX 14.1

Rapid Reaction Techniques

Everyday enzyme kinetics is carried out with low concentrations of the enzyme, since higher concentrations would give velocities far too fast to measure. Our initial velocity measurements are made long after the enzyme-substrate complex has formed. It would be very interesting to be able to observe the actual formation of enzyme substrate complexes (more generally to observe the formation of protein-ligand complexes). These reactions are rapid, occurring on a millisecond time scale or less.

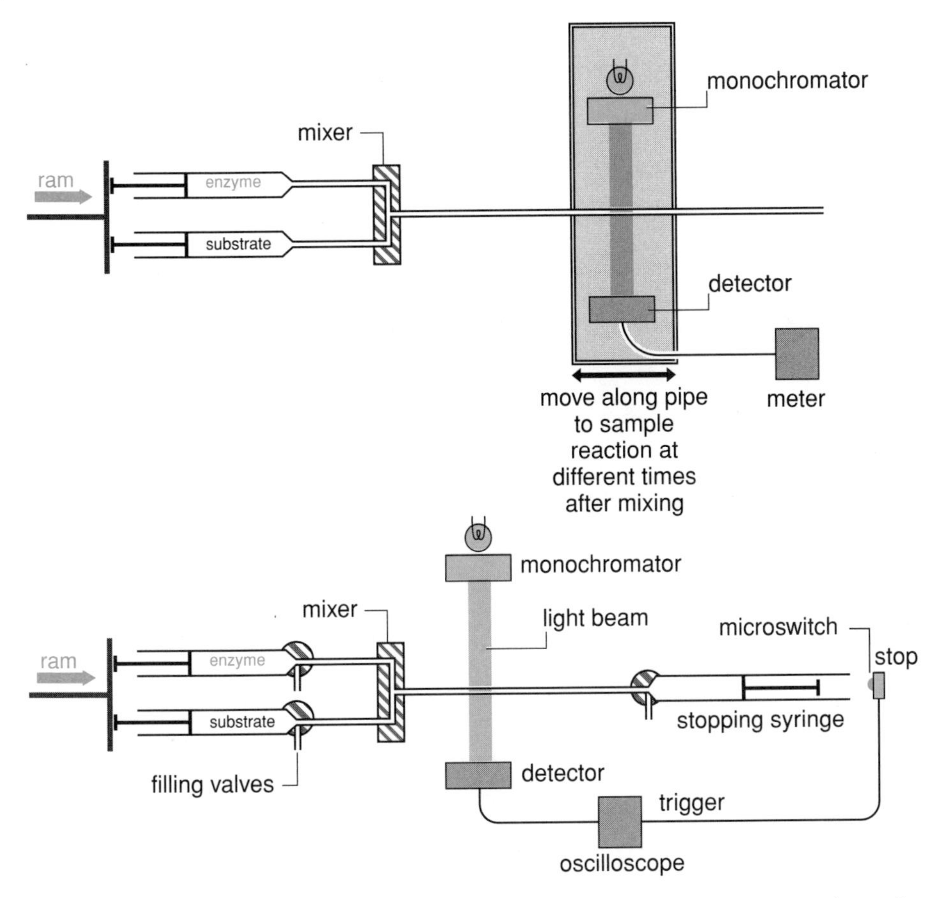

If the enzyme and substrate are mixed rapidly and flow along an observation tube, the reaction mixture gets older as it moves along the tube. Observations made at different distances along the tube give different reaction times.

It is not difficult to force solutions through a tube of 1 mm diameter at 10 m s^{-1} (this uses 7.85 mliter s^{-1}). In such a case each cm along the tube corresponds to 1 ms after mixing. Such a continuous flow method is very wasteful of reagents and only practical when large quantities of the protein are available. It was devised for studies of the oxygenation of hemoglobin by Hartridge and Roughton in 1923.

A less wasteful method is stopped flow. Here the enzyme and substrate are contained in two small syringes driven by a ram. The two solutions are mixed and passed through an observation chamber into a syringe. The plunger of this stopping syringe moves back until it hits a stopping plate. Observations are made as the mixture in the observation chamber ages (usually a high-speed recording device is triggered by the stopping syringe). Stopped flow allows observation of reactions down to about 0.1 ms after mixing. This approach has been used to examine how calmodulin binds calcium and how the calcium–calmodulin–complex then binds to its target proteins such as calcineurin.

CHAPTER

15

ENERGY TRADING WITHIN THE CELL

In the nonliving world, complex things degrade naturally to simpler things: gradients of temperature or concentration disappear, and uniformity triumphs. Living things do not follow these trends. Cells are complex and divide to make other complex cells: A fertilized egg differentiates to make a whole complex organism. This escape from the behavior of nonliving systems is allowed because living systems take energy from their environment and use it to grow, to reproduce, and to repair themselves. All the processes that occur within a living cell are ultimately driven by energy taken from the outside world.

We can use an analogy with the world of economics. It is unlikely that other people would spontaneously repair our houses, or feed us, or give us this book, but we can drive these otherwise unlikely processes by spending money we have been given. In a similar way cells can drive otherwise unlikely processes by using up one of four *energy currencies* that are then replaced using energy taken from the outside world.

CELLULAR ENERGY CURRENCIES

The scientific way of saying that a process will occur spontaneously is to say that the change in *Gibbs free energy*, expressed as ΔG, is negative. One reaction that we will meet again is the hydrolysis of glucose-6-phosphate to yield glucose and a phosphate ion. Phosphate ions are usually called "inorganic phosphate," Pi for short, to emphasize that they are not part of carbon-containing molecules.

$$\text{Glucose-6-phosphate} + H_2O \rightarrow \text{Glucose} + H_2PO_4^-$$

$$\Delta G = -14 \text{ kJmol}^{-1}$$

Since this reaction has a negative ΔG, it proceeds spontaneously, releasing 14 kJ of energy for every mole of glucose-6-phosphate hydrolyzed. Another reaction we will meet again is the hydrolysis of nucleotides, the building blocks of DNA and RNA (Chapter 5). Simply losing the terminal phosphoryl group from the nucleotide ATP releases 50 kJmol^{-1}:

$$\text{Adenosine triphosphate} + H_2O \rightarrow \text{Adenosine diphosphate} + H_2PO_4^-$$

$$\Delta G = -50 \text{ kJmol}^{-1}$$

The reverse of these reactions will of course not proceed spontaneously. For instance, cells need to phosphorylate glucose to make glucose-6-phosphate but cannot use the reaction

$$\text{Glucose} + H_2PO_4^- \rightarrow \text{Glucose-6-phosphate} + H_2O$$

$$\Delta G = +14 \text{ kJmol}^{-1}$$

The reaction will not proceed spontaneously because it has a positive ΔG. Crucially, though, an unfavorable (positive ΔG) reaction can occur if it is tightly coupled to a second reaction that has a negative free energy change (negative ΔG), so the overall change for the reactions put together is negative. Thus cells phosphorylate glucose by carrying out the following reaction:

$$\text{Glucose} + \text{Adenosine triphosphate} \rightarrow$$

$$\text{Glucose-6-phosphate} + \text{Adenosine diphosphate}$$

$$\Delta G = -36 \text{ kJmol}^{-1}$$

Adenosine triphosphate, or ATP, has given up the energy of its hydrolysis to drive an otherwise energetically unfavorable reaction forward. We call ATP a cellular currency to draw an analogy with money in human society. Just as we can spend money to cause someone to do something they would not otherwise do, such as give us food or build us a house, the cell can spend its energy currency to cause processes that would otherwise not occur. However, the analogy is not exact because energy currencies are not hoarded. There is a continuous turnover of ATP to ADP and back again. ATP is therefore not an energy store but simply a way of linking reactions. It can be thought of as a truck that carries metabolic energy to where it is needed and returns empty to be refilled. The number of trucks is small, but the amount moved can be large. An average person hydrolyzes about 50 kg of ATP per day but makes exactly the same amount from ADP and inorganic phosphate. We will see how this happens in this chapter. The cell has four energy currencies: NADH, ATP, the hydrogen ion gradient across the mitochondrial membrane, and the sodium gradient across the plasmalemma.

Nicotine Adenine Dinucleotide (NADH)

This, the most energy rich of the four currencies is shown in Fig. 15.1. NADH is a combination of two nucleotides. On top is the familiar nucleotide 5′–adenosine monophosphate that we first met as a component of nucleic acids (page 74). On the bottom is the second nucleotide made of a ribose sugar and a base called nicotinamide (derived from vitamin B1, niacin). The complete molecule is a strong reducing agent. It will readily react to allow two hydrogen atoms to be added to molecules, in the general reaction $NADH + H^+ + X \rightarrow NAD^+ + H_2X$. Molecular oxygen is plentiful in most cells, and if the two hydrogen atoms are used to reduce oxygen to water, then energy is released. Every mole of NADH that is used in this way releases 220 kJ of energy.

Nucleoside Triphosphates (ATP plus GTP, CTP, TTP, and UTP)

ATP, the second most energy rich of the four currencies, is shown in Fig. 15.1. In earlier chapters we have met many chemical processes in the cell that are driven by ATP hydrolysis. When one mole of ATP is hydrolyzed 50 kJ of energy are released. NADH and ATP take part in so many reactions within the cell that they are often called *coenzymes*, meaning molecules that act as second substrates for many enzymes as they do particular jobs.

The γ phosphoryl group is easily transferred between nucleotides in reactions such as this one:

$$ATP + GDP \rightleftharpoons ADP + GTP$$

So as far as energy is concerned, we can regard GTP, CTP, TTP and UTP as equivalent to the most commonly used nucleotide energy currency, ATP.

NADH

adenine

ribose

$CONH_2$

nicotinamide

ribose

ATP

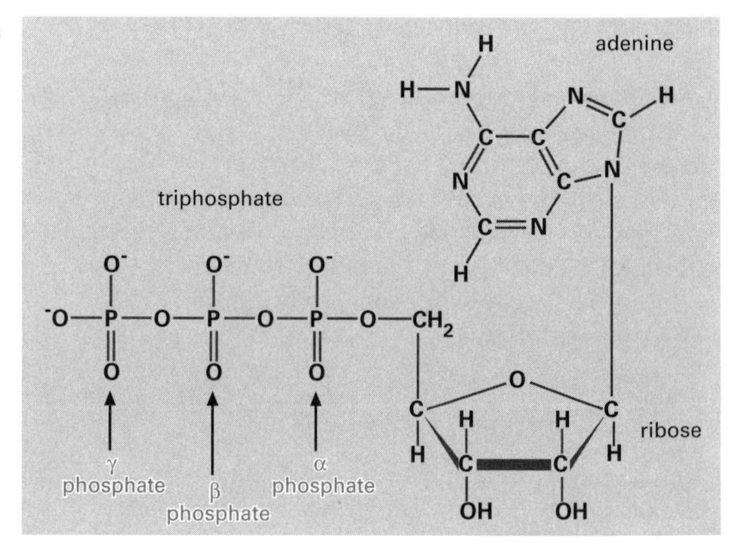

Figure 15.1

The nicotinamide adenine dinucleotide (NADH) and adenosine triphosphate (ATP) energy currencies.

The Hydrogen Ion Gradient across the Mitochondrial Membrane

The endosymbiotic theory states that mitochondria are derived from bacteria that evolved to live in eukaryotic cells (page 10). The bacterial cytosol is usually about one pH unit more alkaline than the world outside; that is, H^+ ions are ten times more concentrated outside than inside. If they could move freely across the bacterial cell membrane, H^+ ions would rush in down this gradient. Furthermore there is a voltage difference across the membrane: The inside is about 160 mV more negative than the extracellular medium. Transmembrane voltages are always referred to in terms of the internal voltage relative to that outside: in this case, –160 mV. The transmembrane voltage attracts the positively charged H^+ ions into the bacterium. Any combination of a concentration gradient and a voltage gradient is called an electrochemical gradient. For hydrogen ions at the bacterial membrane the electrochemical gradient is large and inward. Should H^+ ions be allowed to rush into the bacterium, they would release energy—about 20 kJ for every mole that enters. One important process driven by this energy is bacterial motility (Box 3.2).

Figure 15.2 is a representation of a mitochondrion inside a eukaryotic cell.

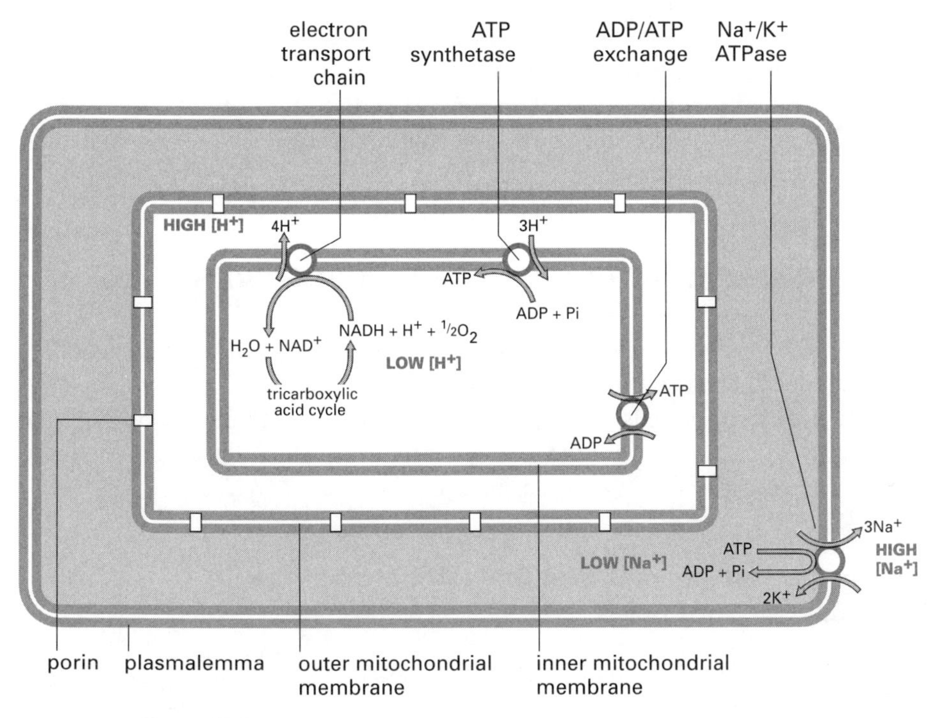

Figure 15.2

The sites in the cell where the energy currencies are interconverted.

This is not an accurate picture of what a mitochondrion looks like (see Fig. 2.4) but rather emphasizes the topology and the function. In the center is shown the mitochondrion with its two membranes. In the very middle is a volume equivalent to bacterial cytosol that is called the mitochondrial matrix. Next comes the inner mitochondrial membrane, then the outer mitochondrial membrane. The intermembrane space is the small space between the two mitochondrial membranes. The blue region is the cytosol, bound by the plasmalemma. A large channel (page 291) in the outer mitochondrial membrane called *porin* lets through all small ions and molecules, so the ionic composition of the intermembrane space is the same as that of cytosol. However, there is a large electrochemical gradient across the inner mitochondrial membrane. When H^+ ions move in from the intermembrane space to the mitochondrial matrix, they release 20 $kJmol^{-1}$ of energy, exactly as in the mitochondrion's proposed bacterial ancestors.

The Sodium Gradient across the Plasmalemma

Unlike bacteria, most eukaryotic cells do not have an H^+ electrochemical gradient across their plasmalemmae. Rather, it is sodium ions that are more concentrated outside the cell than inside (Fig. 15.2). This chemical gradient is supplemented by a voltage gradient. The cytosol is between 70 and 90 mV more negative than the extracellular medium, that is, the transmembrane voltage of the plasmalemma is between –70 and –90 mV. There is therefore a large inward electrochemical gradient for sodium ions. If sodium ions are allowed to rush down this gradient, they release energy—approximately 14 kJ for every mole of Na^+ entering the cytosol.

THE ENERGY CURRENCIES ARE INTERCONVERTIBLE

A company that buys raw materials in America and Japan, spending dollars and yen, and then sells products in Europe, receiving pounds and marks, simply converts from pounds and marks to dollars and yen to pay its bills. In the same way cells convert from the energy currency in which they are in credit to the energy currency that they are using up.

Exchange Mechanisms that Convert between the Four Energy Currencies

The cell has mechanisms that transfer energy between the four currencies. The conversions are summarized in Fig. 15.3. In a typical animal cell, oxidation of fuel molecules in the mitochondria (by the Krebs cycle, page 264) tops up the

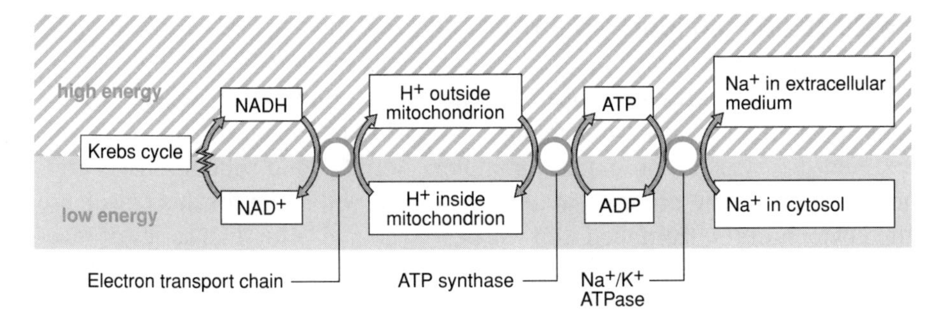

Figure 15.3
Energy flow between the currencies in a normal animal cell.

supply of NADH. The cell then converts this energy currency into the other three. All the interconversions are reversible. Figure 15.2 shows where each conversion mechanism is located.

Consider what happens if sodium ions move out of the extracellular medium and into the cytosol of a eukaryotic cell—for example, when a nerve cell transmits the electrical signal called an action potential (Chapter 18). The sodium gradient is now depleted: the cell holds less of this energy currency than it did before. However, it still has plenty of energy in the form of ATP. The sodium/potassium ATPase allows the conversion of energy as ATP to energy as sodium gradient. ATP is hydrolyzed to ADP as part of the process, giving up energy that is used to push Na^+ out of the cytosol to its higher energy state in the extracellular medium.

The cell has now used some ATP, but it still holds plenty of energy in another gradient—the H^+ gradient across the mitochondrial membrane. ATP synthase allows interconversion of these two energy currencies. H^+ ions move into the mitochondrion, giving up energy that is then used to make ATP from ADP and inorganic phosphate.

The energy of the H^+ gradient is now depleted. However, the cell still has plenty of energy in its NADH account. The electron transport chain allows interconversion of energy as NADH to energy as H^+ gradient. NADH is used to reduce molecular oxygen to water, releasing energy that is used to push H^+ ions out of the mitochondrion to their higher energy state in the cytosol.

Each of the four energy conversion systems is a protein structure called a carrier because it carries solute across the membrane. The sodium/potassium ATPase carries sodium and potassium ions, while both ATP synthase and the electron transport chain carry H^+ ions. There are many carriers in the cell with a wide variety of functions, some of which we will meet later in this chapter and in the next. The three that convert between the energy currencies are vital and are evolutionarily ancient.

The Electron Transport Chain

The electron transport chain consists of three large multiunit proteins intrinsic to the inner mitochondrial membrane—NADH dehydrogenase, the b-c_1 complex, and the cytochrome oxidase complex. Some of the subunits have iron-containing prosthetic groups that give them a red color, and for this reason are called cytochromes, from the Latin words for cell and color. In a series of reactions in which electrons are passed between the three protein complexes, the chain catalyzes the chemical reaction:

$$NADH + H^+ + \tfrac{1}{2}\,O_2 = NAD^+ + H_2O$$

This reaction releases energy, 220 kJ for every mole of NADH used. The energy is used to carry at least four H^+ ions from their low energy state in the mitochondrial matrix to their high energy state in the cytosol. Figure 15.4 summarizes the reaction in terms of energy currencies. The circle symbolizes the linkage between the energy released in the conversion of NADH to NAD^+ and the energy used to drive H^+ out of the mitochondrial matrix. If the electron transport chain simply allowed NADH to reduce oxygen to water, then the reaction's energy would be released as heat. Instead, the enzymic function of the electron transport chain is tightly coupled to its function as a carrier that moves H^+ ions. The energy of NADH is thus converted to the energy of the H^+ gradient.

ATP Synthase

This is a single protein in the inner mitochondrial membrane. Once again, its carrier action is tightly coupled to its action as an enzyme. In an animal cell under normal conditions, ATP synthase allows H^+ to move into the mitochondrial matrix. The energy released is used to synthesize ATP from ADP and phosphate. Three H^+ must enter to allow a single ATP to be made. Figure 15.5 summarizes the reaction in terms of energy currencies. Once again, the circle symbolizes the linkage between the enzyme and carrier functions.

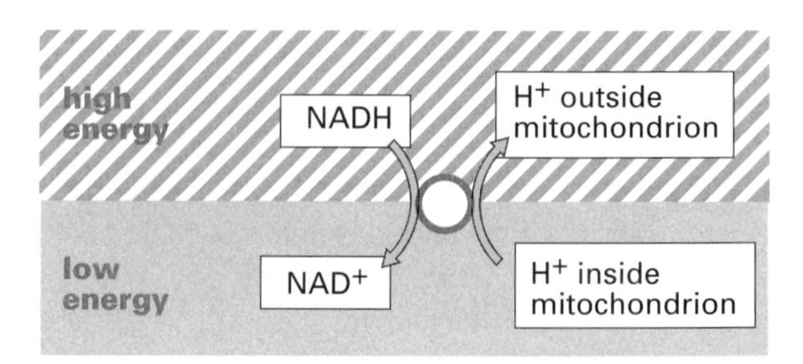

Figure 15.4
Currency conversion: The electron transport chain converts between NADH and the H^+ gradient.

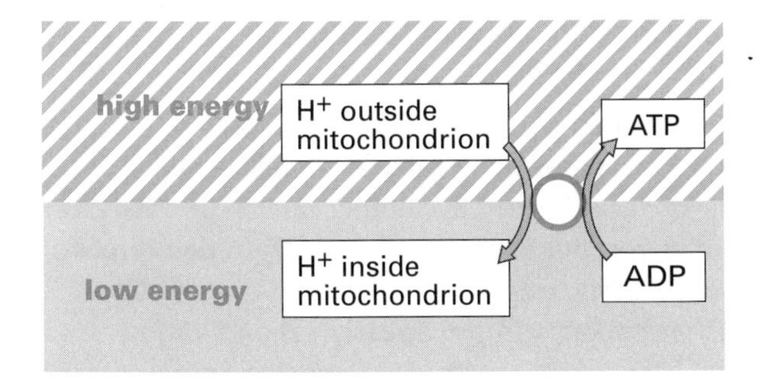

Figure 15.5
Currency conversion: ATP synthase interconverts the H^+ gradient and ATP.

Sodium/Potassium ATPase

This is a single protein in the plasmalemma, with a carrier action that is tightly linked to an enzymic one. Under normal conditions it hydrolyzes ATP. The energy released drives sodium ions up their electrochemical gradient out of the cell. The Na^+/K^+ ATPase also moves potassium ions the other way, into the cytosol. For every ATP hydrolyzed three Na^+ ions are moved out and two K^+ ions are moved in.

BOX 15.1

Highly Toxic Chemicals That Interfere with Energy Conversion

The four carriers NADH, ATP, the hydrogen ion gradient across the mitochondrial membrane, and the sodium gradient across the plasmalemma run the energy currency market and are vital to the cell. Chemicals that interfere with them are very toxic. One is cyanide, which stops the action of the electron transport chain. Another is digitalis, from foxgloves, which blocks the Na^+/K^+ ATPase.

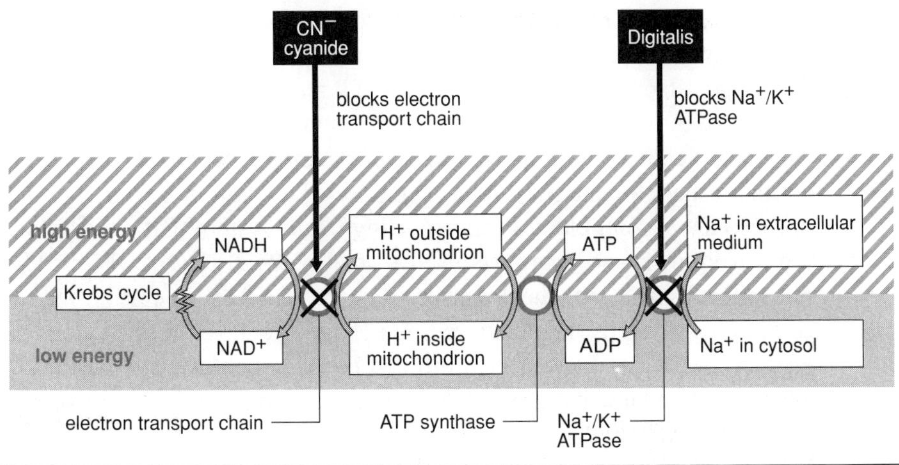

ADP/ATP Exchanger

ATP synthase makes ATP inside the mitochondrion. The earliest eukaryotic cells to incorporate bacteria needed a mechanism to enable ATP to leave the bacterium for use in the cytosol. This job is performed by another carrier, the ADP/ATP exchanger. It has no enzymic action but simply moves ADP in one direction across the mitochondrial inner membrane and ATP in the opposite direction. In most eukaryotic cells the carrier operates in the direction shown in Fig. 15.2. Carriers such as the Na^+/K^+ ATPase together with many synthetic processes use up ATP in the cytosol, producing ADP. ADP enters the mitochondria by the ADP/ATP exchanger and is reconverted to ATP by ATP synthase. ATP then leaves the mitochondrion with the help of the ADP/ATP exchanger.

All Carriers Can Change Direction

In a normal animal cell, the primary source of energy is the Krebs cycle (page 264). This regenerates NADH from NAD^+, making at the same time a small amount of ATP. Because the NADH currency is always being topped up, while the others are being used, the direction of operation of the energy conversion systems is usually that shown in Fig. 15.2. However, all the carriers are reversible.

Yeast cells in a wine barrel, or muscle cells in the leg of a sprinter, are anaerobic; there is no oxygen available. In this situation cells can make NADH, but the electron transport chain cannot drive H^+ out of the mitochondria because there is no molecular oxygen waiting to be reduced by NADH. Instead, the cells energy needs are met by anaerobic glycolysis (page 268), which makes ATP. Figure 15.6 shows how the cell maintains the amounts of energy currencies. Any drain on the mitochondrial H^+ gradient is counteracted by ATP synthase running in the opposite direction from that shown in Figure 15.2. ATP is hydrolyzed, and H^+ ions are pushed out of the mitochondrion. The ADP/ATP exchanger also reverses its direction. ATP is regenerated by anaerobic respiration in the cytosol and is used up by ATP synthase in the mitochondrial matrix.

Plant cells contain both mitochondria and chloroplasts. The chloroplast is another organelle that may have been derived in evolution from a prokaryote. It possesses a second electron transporting chain system that can capture the energy in light. Some energy is used to make NADPH ($NADP^+$, NAD^+ with a phosphoryl group attached, is used in place of NAD^+ in certain reactions, e.g. page 275), while the rest is stored in a H^+ gradient. As in mitochondria the energy in the H^+ gradient is then utilized by ATP synthase to make ATP (photosynthetic phosphorylation), which is carried out of the chloroplast by the ADP/ATP exchanger.

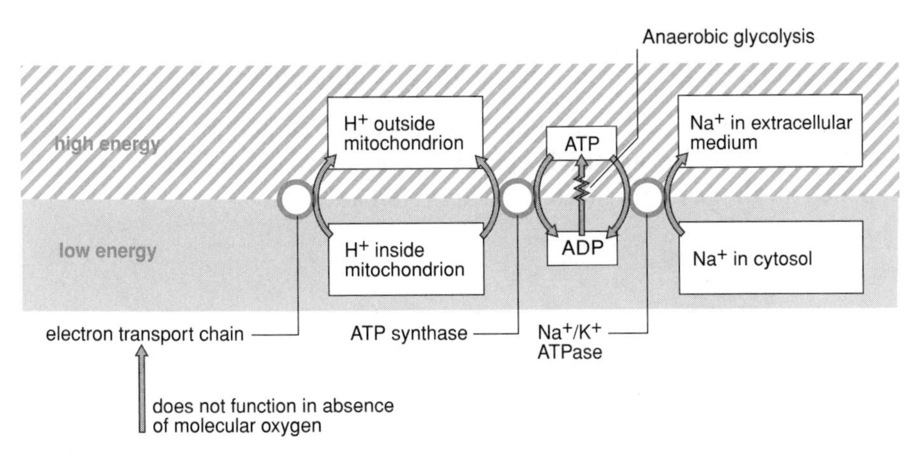

Figure 15.6

Energy flow between the currencies in an anaerobic cell.

Chloroplasts do not contain the enzymes of the Krebs cycle, so at night they have no internal source of energy and are dependent on their host cell. ATP from the cytosol (most of it created by mitochondria) enters the chloroplast by the ADP/ATP exchanger, which is carrying ADP and ATP in the opposite direction from that shown in Fig. 15.2. ATP synthase hydrolyzes ATP and maintains the H^+ gradient.

In a healthy cell none of the energy currencies are allowed to run down. The direction in which energy moves between the four currencies depends on the primary source of energy for that cell.

SUMMARY

1. Reactions with a positive Gibbs free energy change (ΔG) can be caused to happen in a cell by linking them with a second reaction with a larger, negative ΔG. The second reaction drives the first.

2. The majority of such reactions in the cell are driven by one of four energy currencies—NADH, ATP, the hydrogen ion gradient across the mitochondrial membrane, and the sodium gradient across the plasmalemma.

3. The electron transport chain in the mitochondrial inner membrane converts the energy in NADH into energy in the hydrogen ion gradient.

4. ATP synthase in the mitochondrial inner membrane converts between energy in the hydrogen ion gradient and energy in ATP.

5. The sodium, potassium ATPase in the plasmalemma converts between the energy in ATP and energy in the sodium gradient.

6. In a healthy cell, none of the energy currencies are allowed to run down. The direction in which energy moves between the four currencies depends on the primary source of energy for the cell.

BOX 15.2

Brown Fat

Triglycerides are stored within specialized fat cells in the body. Most fat cells are composed of a droplet of lipid surrounded by a thin layer of cytoplasm with a nucleus and a few mitochondria. The resulting tissue is white in color and simply releases or stores fatty acids in response to the needs of the organism. This is the kind of fat that is typically found around our kidneys and under the skin.

A second kind of fat is found in babies. Brown fat cells not only have stored triacylglycerols but are also rich in mitochondria, the cytochromes of the mitochondria giving the brown color. Brown fat is a heat-generating tissue. A channel selective for H^+ called thermogenin is found in the inner mitochondrial membrane. As fast as the electron transport chain pushes H^+ out of the mitochondrial matrix, they flow through thermogenin back down their electrochemical gradient into the matrix. In other cells this flux would only occur through ATP synthase and would be tightly coupled to the production of ATP from ADP. The presence of thermogenin uncouples the phosphorylation of ADP from the flow of electrons to oxygen so that the electron transport chain can work flat out even though there is no ADP available, as long as the cell contains triglycerides that can be oxidized to regenerate NADH (see β oxidation, page 269). This generates a lot of heat and helps the infant maintain body temperature. Large blocks of brown fat are found in animals that hibernate.

A similar uncoupling mechanism is used by some plants to generate heat. Some arum lilies rely on carrion-eating flies for pollination. Uncoupled mitochondria at the base of the flower generate sufficient heat to volatilize the evil-smelling odorants used to attract the flies.

CHAPTER 16

METABOLISM

All the processes that occur within a living cell are ultimately driven by energy taken from the outside world. Green plants take energy directly from sunlight. Other organisms take compounds made using sunlight and break them down to release energy, a process called *catabolism*. The most common way of breaking down these food compounds is to oxidize them, that is, to burn them. The energy released can then be used for the building and repair processes termed *anabolism*. The collective term for all the reactions going on inside a cell is metabolism. All metabolic reactions share some general features:

1. They are catalyzed by enzymes.
2. They are universal in that all organisms show remarkable similarity in the main pathways.
3. They involve relatively few types of chemical reaction.
4. They are controlled, often by modulation of the activity of key regulatory enzymes.
5. They are compartmentalized within cells. One organelle carries out one set of reactions, while another organelle carries out another set. In higher organisms this compartmentalization is carried further, so different reactions take place in different body organs.

THE KREBS CYCLE: THE CENTRAL SWITCHING YARD OF METABOLISM

Figure 16.1 is an overview of the metabolic pathways within a cell. At the center is a cycle of reactions within the mitochondrial matrix named after its discoverer, Hans Krebs. (It is also known as the tricarboxylic acid (TCA) cycle or citric acid cycle). Foods we eat—sugars, fats, proteins—are converted into acetate, CH_3COO^-. The acetate is not free but carried by a coenzyme called coenzyme A. Acetate bound to coenzyme A—acetyl-CoA for short—is then fed into the cycle and completely oxidized to carbon dioxide and water. In the process the energy currency NADH is produced. The reactions can be summarized as

$$CH_3CO\text{-}CoA + (3NAD^+ + FAD) + GDP + Pi + 3H_2O \rightarrow$$
$$CoA\text{–}H + (3NADH + FADH_2) + GTP + 2CO_2 + 3H^+$$

Pi represents an inorganic phosphate ion. FAD stands for Flavin adenine dinucleotide. $FADH_2$, the reduced form, does not carry as much energy as NADH does but like NADH is used to drive H^+ up its electrochemical gradient out of the mitochondrial matrix. We will first describe the cycle itself and then look at each of the metabolic pathways leading into and out of the cycle. Figure 16.2 shows the reaction sequence in detail.

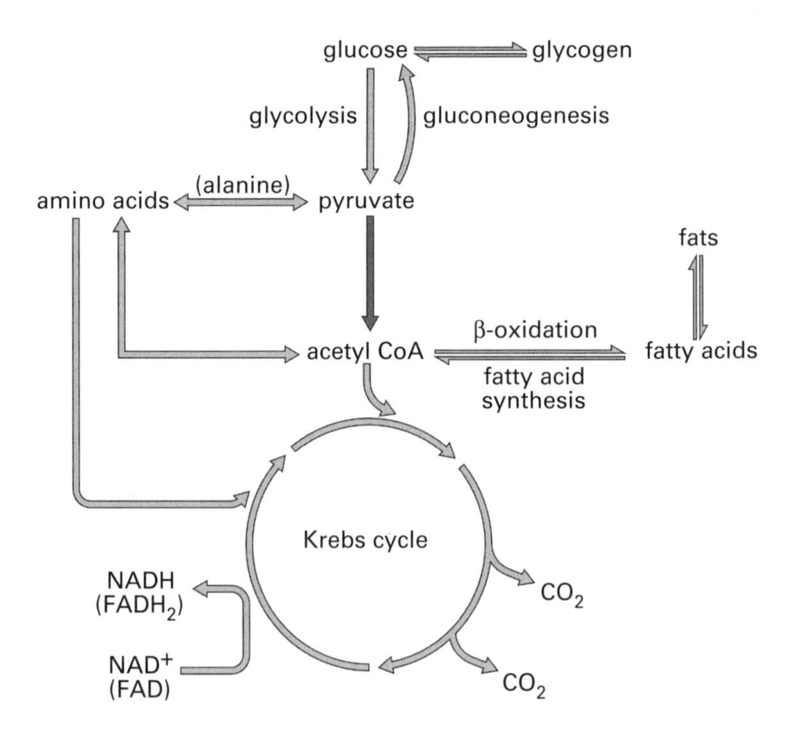

Figure 16.1
An overview of metabolism.

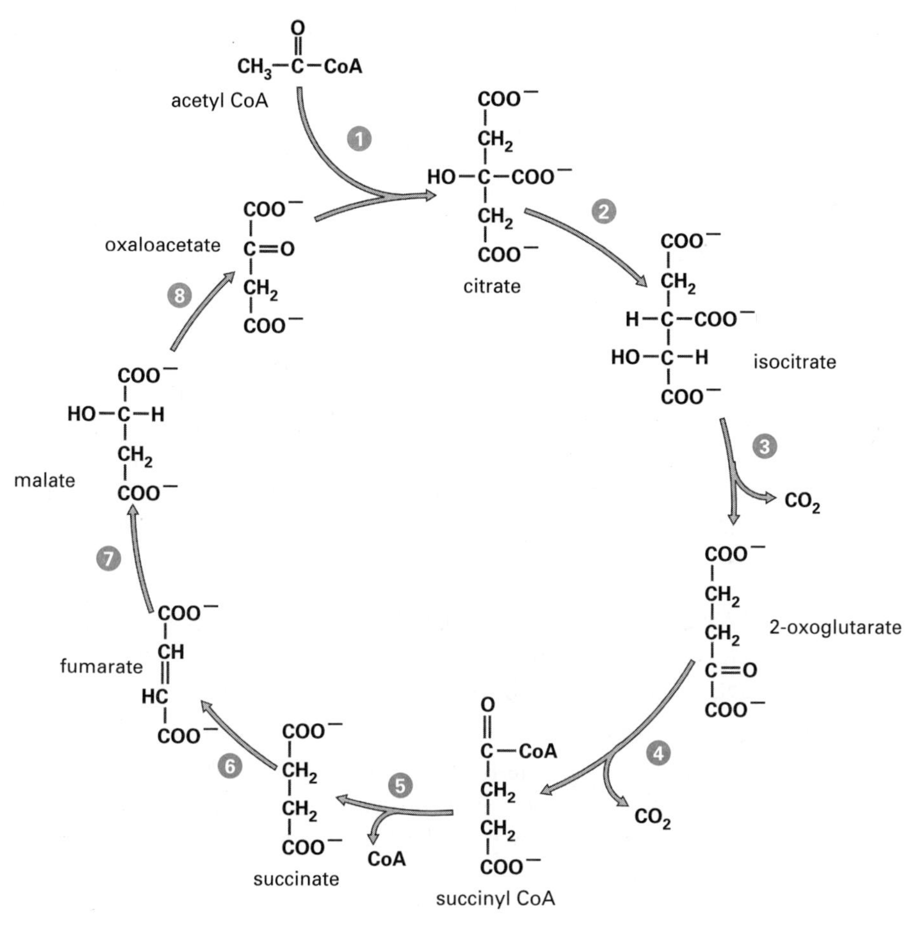

Figure 16.2
The Krebs cycle.

Step 1. A molecule of acetyl-CoA enters the cycle, and the acetate it carries is transferred to the four carbon molecule oxaloacetate, making citrate.

Step 2. Citrate is rearranged to isocitrate.

Step 3. This is the first oxidation step. One carbon leaves as CO_2. One NAD^+ is reduced to NADH, and the five-carbon acid 2-oxoglutarate is left.

Step 4. The second oxidation step. One carbon leaves as CO_2. One NAD^+ is reduced to NADH, and the four-carbon acid that is left, succinate, ends up attached to coenzyme A.

Step 5. Succinate leaves coenzyme A. This reaction is linked to phosphorylation of the nucleotide GDP to GTP. γ phosphoryl groups can be swapped between nucleosides, so this GTP can be used to regenerate ATP from ADP.

Step 6. Succinate is oxidized to fumarate. The oxidant is not NAD^+ but FAD.

Step 7. H_2O is added to fumarate giving malate.

Step 8. Malate is oxidized to oxaloacetate. One NAD^+ is reduced to NADH. The oxaloacetate is now available to accept another acetyl group from acetyl-CoA.

FROM GLUCOSE TO PYRUVATE: GLYCOLYSIS

Most of the cells in our body are bathed in glucose supplied by the blood. The breakdown of glucose, glycolysis, takes place in the cytoplasm. The pathway is shown in Fig. 16.3.

Step 1. Free glucose is phosphorylated on carbon number six to produce glucose 6-phosphate. As we discussed earlier (page 252), this reaction, catalyzed by hexokinase, is driven by the energy in ATP.

Step 2. Glucose 6-phosphate is isomerized to fructose 6-phosphate.

Step 3. Fructose 6-phosphate is phosphorylated to produce fructose 1,6-bisphosphate. Another ATP is used to drive this reaction. This reaction is catalyzed by phosphofructokinase.

Step 4. The fructose 1,6-bisphosphate is now cut in half. Each of the resulting halves has one phosphate attached. The two products are dihydroxyacetone phosphate and glyceraldehyde 3-phosphate.

Step 5. The next step interconverts dihydroxyacetone phosphate and glyceraldehyde 3-phosphate. This allows both halves to be used in the next step.

The glucose molecule has now been converted into two molecules of glyceraldehyde 3-phosphate. Each of the following reactions occurs twice for every molecule of glucose fed into the pathway:

Step 6. In this step inorganic phosphate becomes incorporated into an organic molecule; no ATP or other high energy phosphate compound is used. Glyceraldehyde phosphate is oxidized to diphosphoglycerate, while NAD^+ is reduced to NADH.

Step 7. One of the phosphoryl groups is transferred to ADP, regenerating the energy currency ATP. Together, steps 6 and 7 have carried out a process called substrate level phosphorylation. ATP has been

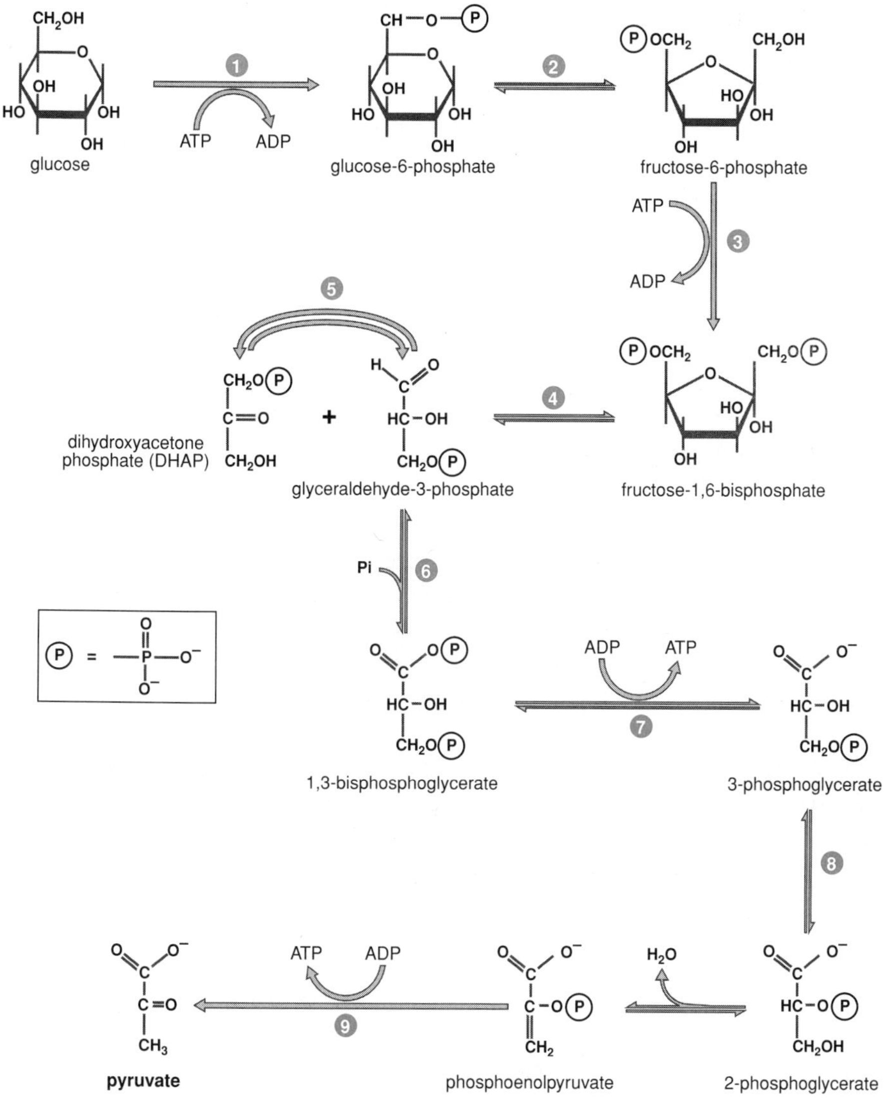

Figure 16.3
Glycolysis breaks glucose down into pyruvate.

regenerated from ADP and inorganic phosphate by biochemical reactions alone, without any involvement of ATP synthase. We are left with phosphoglycerate.

Step 8. Phosphoglycerate is converted to phosphoenolpyruvate.

Step 9. Another substrate level phosphorylation, produces pyruvate and regenerates an ATP from ADP and inorganic phosphate.

Pyruvate can be used in a number of ways. If it is to be used to feed the Krebs cycle, it moves into the mitochondrial matrix where it is oxidized and decarboxylated to acetyl-CoA by the complex enzyme pyruvate dehydrogenase, giving an NADH (Fig. 16.4). The acetyl-CoA can enter the Krebs cycle or can be used to synthesize fatty acids and some other molecules. Alternatively, if the components of the Krebs cycle are running low, as may happen when biosynthetic pathways use one or other component of the Krebs cycle as their starting material, new oxaloacetate is made from pyruvate. Overall, glycolysis has used two ATP molecules but made four: a net gain of 2 ATP per glucose.

Glycolysis without Oxygen

The leg muscles of a sprinter cannot be supplied with oxygen rapidly enough to power the mitochondria. Muscle cells need to make ATP by a method that does not require oxygen. As we have seen, glycolysis itself produces two ATPs. Can the cell let the pyruvate pile up and use it when oxygen is available? No, because glycolysis as far as pyruvate also reduces an NAD^+ to NADH. The cell does not have any oxygen to use to oxidize the NADH back to NAD^+, so the cell would rapidly run out of NAD^+ if glycolysis simply stopped at pyruvate. Instead, NADH is used to reduce pyruvate to lactate. The buildup of acid lactate in poorly oxygenated muscles causes the pain of cramp. When we stop using the muscle, the blood carries the lactate away to the liver, where it is reoxidized to pyruvate.

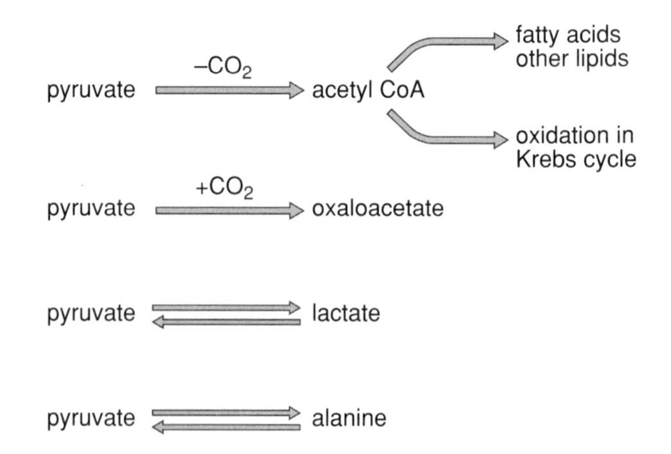

Figure 16.4
Pyruvate used in a number of ways.

Glycogen Can Feed the Glycolytic Pathway

In many cells, especially liver and muscle, the glucose polymer glycogen (Box 2.1) is an important carbohydrate store. The cell could utilize this store to feed the glycolysis pathway by cleaving off a glucose monomer to give free glucose. Hexokinase would then transfer a phosphoryl group from ATP to the glucose to yield glucose 6-phosphate. However, the enzyme glycogen phosphorylase does both jobs: cleavage and phosphorylation, in one pass (Fig. 16.5).

Notice that breaking up glycogen this way is more energy efficient because inorganic phosphate, not ATP, is used as the source of the phosphoryl group. Another enzyme moves the phosphoryl group from carbon one to carbon six to give glucose 6-phosphate.

FROM FATS TO ACETYL-COA: LIPOLYSIS AND β OXIDATION

Many cells use the fatty acids released when fats (triacylglycerols) (Fig. 2.1) are hydrolyzed, a process called lipolysis. The fatty acids are coupled to coenzyme A to give acyl-CoA. A spiral of reactions in the mitochondrial matrix called β oxidation oxidizes acyl-CoA. Each turn of the spiral shortens the fatty acyl chain by two carbons releasing acetyl-CoA and giving an NADH and a $FADH_2$ for each acetyl-CoA released. The pathway is shown in Fig. 16.6.

AMINO ACIDS AS ANOTHER SOURCE OF METABOLIC ENERGY

Protein forms a considerable part of the animal diet. It is broken down to free amino acids during digestion. The amino acids can be used for the biosynthesis of new proteins in the cell, but those in excess of this need can serve as metabolic fuels. There is a problem: Mammals do not store amino acids and cannot directly metabolize the amino acids. The way around this difficulty is to remove the amino groups. The resulting amino acid "carbon skeletons" may be readily converted to intermediates for the Krebs cycle or to acetyl-CoA. Some may be used to synthesize glucose (page 272).

Since there are 20 amino acids, there are many different steps, but the overall effect is that amino groups from all types of amino acids are transferred to make either glutamate or aspartate. The movement of amino groups is car-

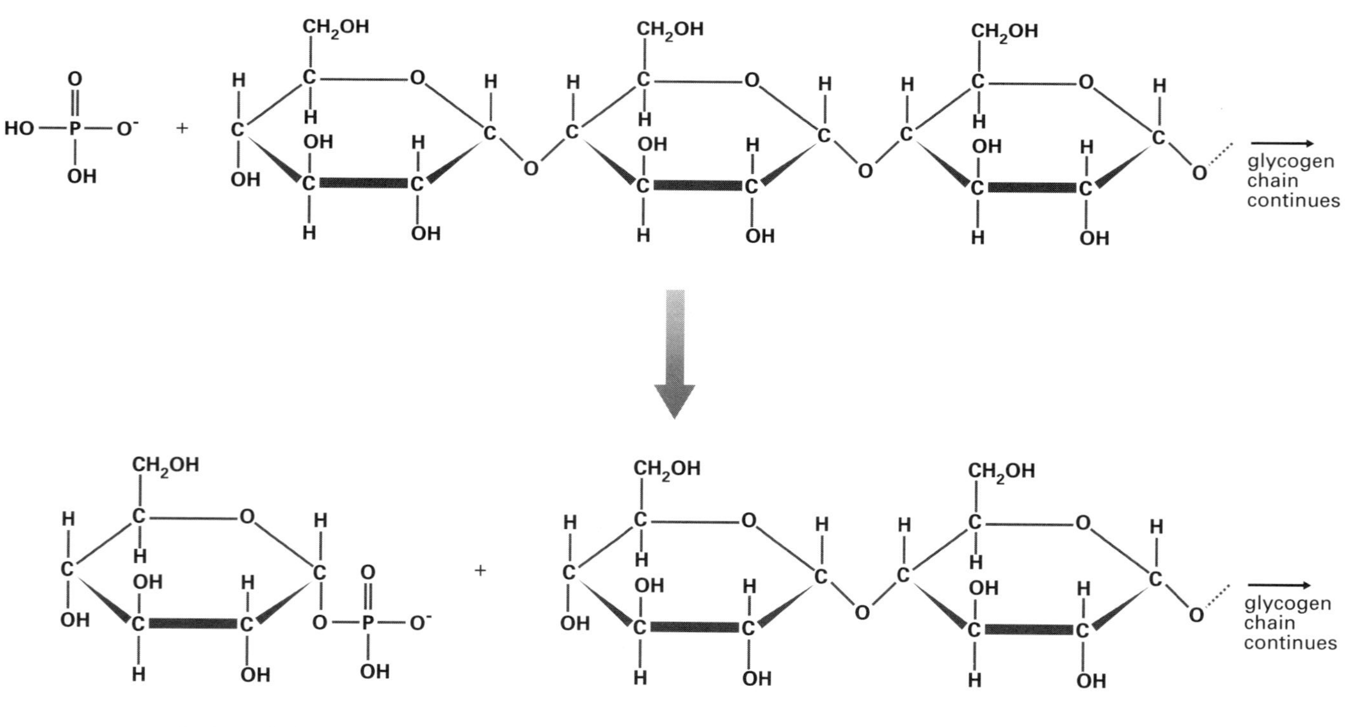

Figure 16.5
Glycogen phosphorylase cleaves a glucose monomer off glycogen and phosphorylates it.

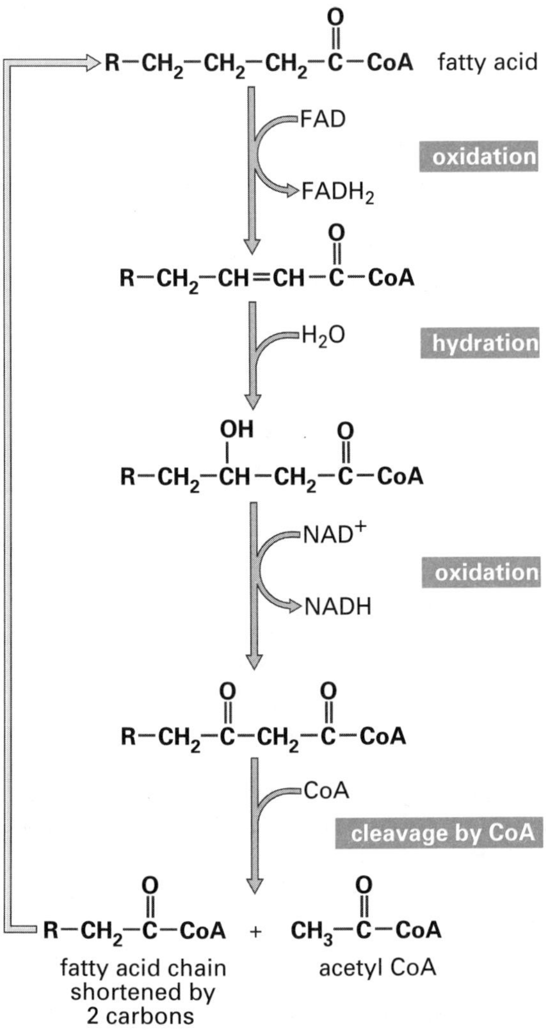

$R-CH_2-CH_2-CH_2-C(=O)-CoA$ fatty acid

FAD → $FADH_2$ oxidation

$R-CH_2-CH=CH-C(=O)-CoA$

H_2O hydration

$R-CH_2-CH(OH)-CH_2-C(=O)-CoA$

NAD^+ → NADH oxidation

$R-CH_2-C(=O)-CH_2-C(=O)-CoA$

CoA cleavage by CoA

$R-CH_2-C(=O)-CoA$ + $CH_3-C(=O)-CoA$

fatty acid chain shortened by 2 carbons

acetyl CoA

Figure 16.6
β oxidation of fatty acids produces acetyl-CoA.

ried out by a group of enzymes called aminotransferases (also known as transaminases) which interconvert 2-oxoacids and amino acids (page 246).

One in every 10,000 humans does not possess phenylalanine hydroxylase, the enzyme necessary to break down phenylalanine. If these individuals take in phenylalanine above the amounts needed for protein production or excretable in the urine, the resulting buildup of phenylalanine and its transamination product phenylpyruvate is toxic, with especially serious effects on brain development in children. The condition, phenylketonuria, is now detected by routine screening of new born infants and the subjects given an artificial diet low in phenylalanine until adulthood.

BOX 16.1

Diabetes, Starvation, Ketone Bodies, and the Odor of Sanctity

Although excellent energy stores, fats pose a solubility problem. Evolution has provided animals with proteins used for the transport of fats and fatty acids but has also provided soluble circulating fuels derived from fatty acids. These are called *ketone bodies*.

The fundamental ketone body is acetoacetate which the liver synthesizes from acetyl-CoA. Acetoacetate can be reduced to the second ketone body: 3-hydroxybutyrate. These two molecules are important circulating fuels in mammals. Heart muscle, for instance, prefers ketone bodies to glucose as a fuel source.

$$CH_3\text{–}CH(OH)\text{–}CH_2\text{–}COOH \underset{NADH}{\overset{NAD^+}{\rightleftharpoons}} CH_3\text{–}C(=O)\text{–}CH_2\text{–}COOH \longrightarrow CH_3\text{–}C(=O)\text{–}CH_3 + CO_2$$

3-hydroxybutyrate — acetoacetate — acetone

During starvation the body adapts by shifting to greater reliance on fat metabolism, since the only source of glucose is gluconeogenesis which must use body protein amino acids as a source of precursors. It is clearly far better to use fat reserves than to degrade body protein, so those tissues that can shift to using ketone bodies as their main fuel source do so. Even the brain changes over to ketone body utilization.

Diabetes mellitus is characterized by a similar shift to using ketone bodies. Diabetes arises when either no insulin is produced (Type 1) or cells are unable to respond to this hormone (Type 2). Both types result in the body switching to a starvation metabolism. Acetoacetate is chemically unstable. It slowly loses carbon dioxide to form acetone. When levels of ketone bodies are high, sufficient acetone is present to give the breath the fruity smell of acetone. This pathological condition is described as ketosis and is characteristic of untreated diabetes.

Medieval saints were given to mortification of the body by voluntary starvation. Was the so-called "odor of sanctity" simply acetone on the breath of starving and thus ketotic saints?

MAKING GLUCOSE: GLUCONEOGENESIS

Such is the importance of glucose that there is a pathway for its synthesis from other molecules. Glucose may be synthesized from the carbon skeletons of some amino acids, from lactate or from glycerol. Gluconeogenesis makes use of some of the enzymes of glycolysis but uses different enzymes to bypass the steps that are not freely reversible. Figure 16.7 shows the glycolytic pathway

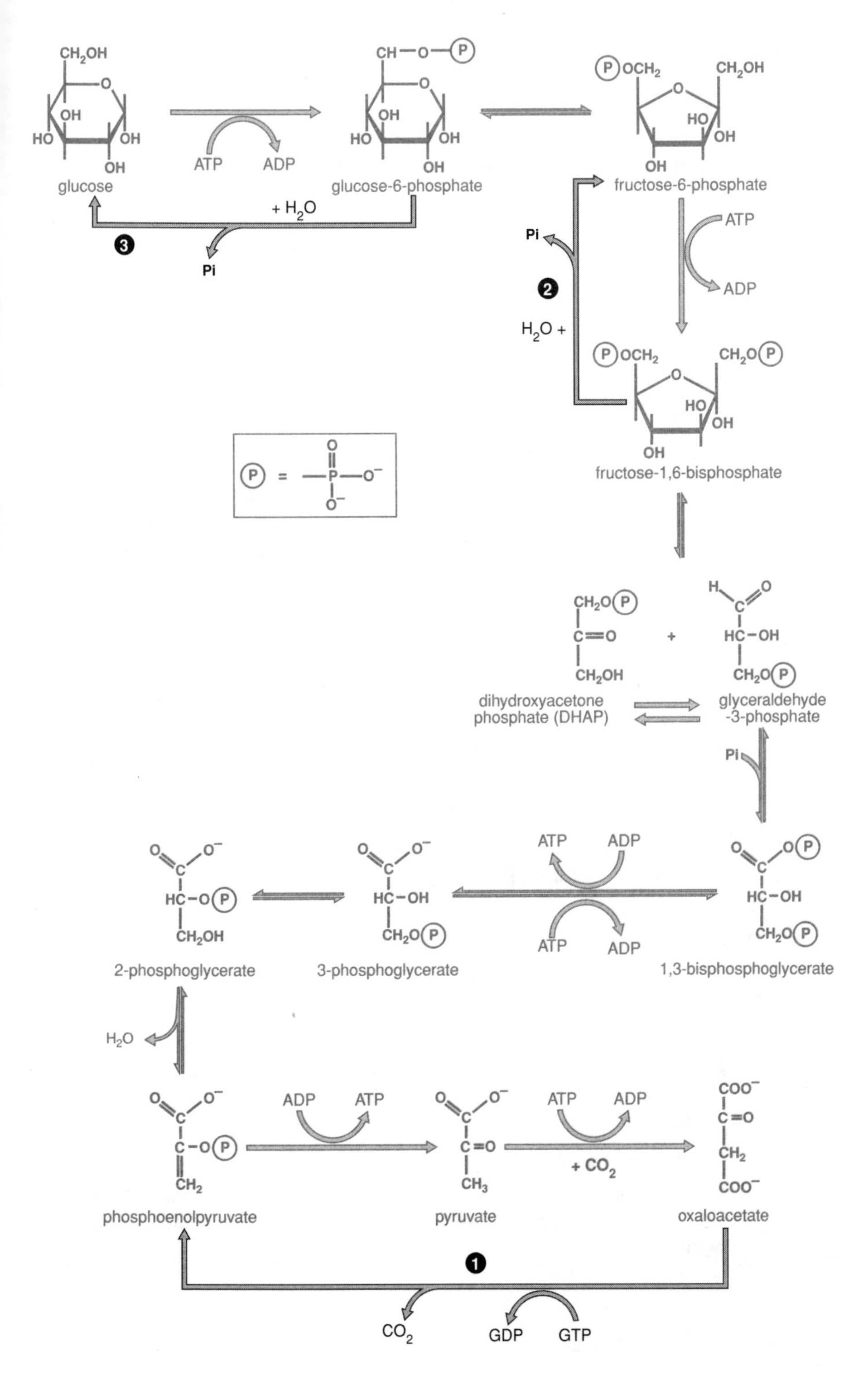

Figure 16.7

Gluconeogenesis allows glucose to be made from pyruvate.

(page 266) in blue. The new reactions that allow the entire pathway to run in reverse are shown in black.

New step 1. Oxaloacetate is moved from mitochondria into the cytosol and converted to phosphoenolpyruvate and CO_2. The reaction is driven by the hydrolysis of a GTP.

From phosphoenolpyruvate all of the glycolytic reactions are reversible until fructose bisphosphate is reached. The reaction catalyzed by phosphofructokinase does not reverse, and this step is avoided by

New step 2. One of the phosphoester bonds on fructose bisphosphate is hydrolyzed, and the phosphoryl group leaves as inorganic phosphate.

The interconversion of fructose phosphate and glucose phosphate is easily reversible. The final production of free glucose is accomplished by

New step 3. The other phosphoester bond is hydrolyzed, and the phosphoryl group leaves as inorganic phosphate.

Gluconeogenesis is an expensive process: The conversion of two pyruvates to a glucose molecule has used 4 ATP, 2 GTP, and 2 NADH.

Adults don't usually make a great deal of new protein, and some of the amino acids that we eat are used to make glucose. They are fed into the Krebs cycle, and oxaloacetate is then used as the precursor for gluconeogenesis. Fat cannot be used to make glucose because acetyl-CoA cannot be converted to oxaloacetate.

Excess glucose can be stored as the polymer glycogen (Box 2.1). Glucose will not spontaneously polymerize, so the reaction is coupled to the hydrolysis of nucleoside triphosphates—not only ATP but also UTP. If phosphorylated glucose is not available, then it is made by hexokinase using ATP (page 252). Glucose 1-phosphate then reacts with UTP (Fig. 16.8). Glycogen synthase then transfers the glucose from UDP-glucose to the growing glycogen chain (Fig. 16.9).

MAKING FATTY ACIDS

All cells need to be able to make fatty acids for membrane lipids. Fat cells synthesize large amounts of fat from acetyl-CoA in times of plenty. The process takes place in the cytosol using a multienzyme complex (in bacteria) or a mul-

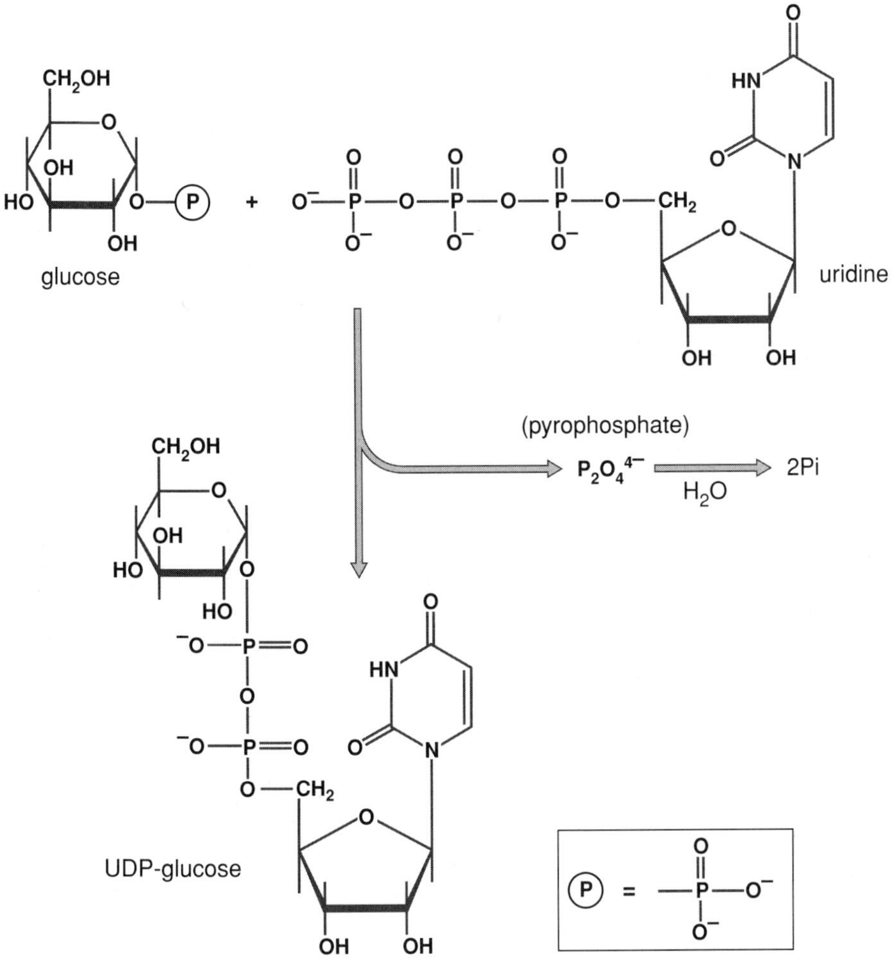

Figure 16.8
Uridine diphosphate glucose synthesized from UTP and glucose 1-phosphate.

tidomain protein (in eukaryotes). In both cases the growing fatty acid chain is not released: It swivels from enzyme to enzyme or domain to domain in the array adding two carbons for each cycle. The fatty acid product is released when it reaches a limiting size (16 carbons). This process is not a reversal of β oxidation (page 269). It uses entirely different enzymes and is separately regulated. Like all biosyntheses it is reductive, and the reducing power comes not from NADH but from the closely related dinucleotide NADPH.

Initially acetyl-CoA is carboxylated to malonyl-CoA (Fig. 16.10). From then on, however, fatty acid synthesis does not use free coenzyme A to carry the growing chain but instead uses a protein called acyl carrier protein (ACP).

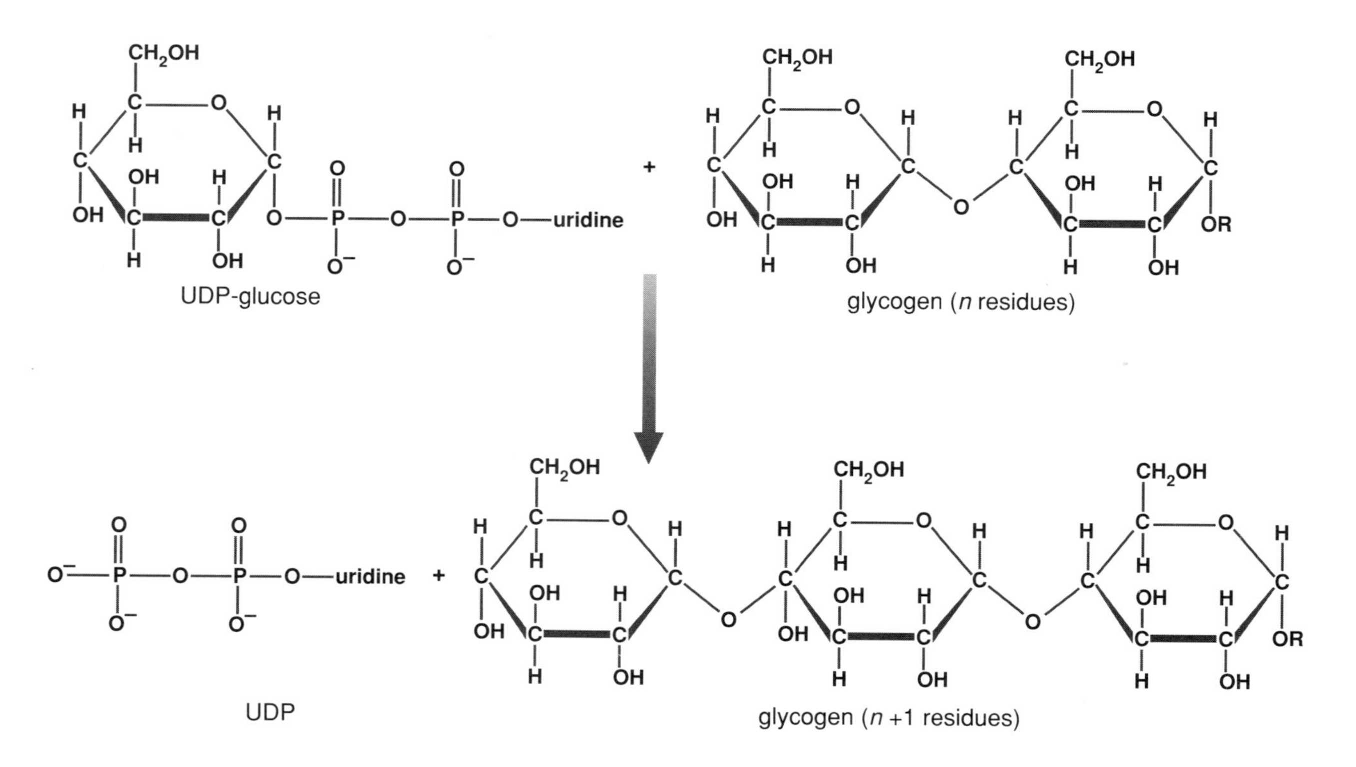

Figure 16.9
Glycogen synthesized from UDP-glucose monomers. R represents the remainder of the long glycogen molecule.

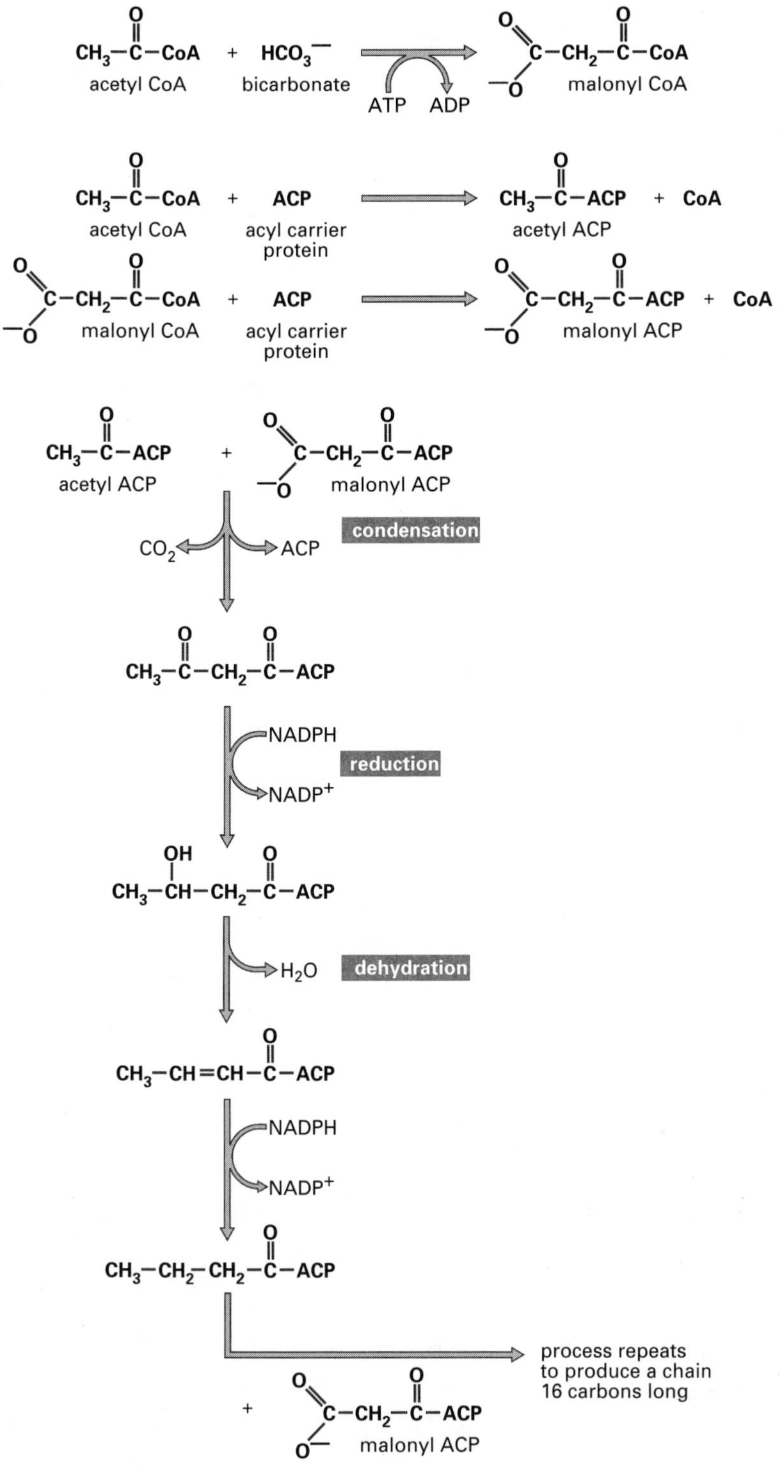

$CH_3-C(=O)-CoA$ + HCO_3^- → (ATP → ADP) $^-O-C(=O)-CH_2-C(=O)-CoA$

acetyl CoA + bicarbonate → malonyl CoA

$CH_3-C(=O)-CoA$ + ACP → $CH_3-C(=O)-ACP$ + CoA

acetyl CoA + acyl carrier protein → acetyl ACP + CoA

$^-O-C(=O)-CH_2-C(=O)-CoA$ + ACP → $^-O-C(=O)-CH_2-C(=O)-ACP$ + CoA

malonyl CoA + acyl carrier protein → malonyl ACP + CoA

$CH_3-C(=O)-ACP$ + $^-O-C(=O)-CH_2-C(=O)-ACP$

acetyl ACP + malonyl ACP

↓ condensation (CO_2, ACP released)

$CH_3-C(=O)-CH_2-C(=O)-ACP$

↓ reduction (NADPH → $NADP^+$)

$CH_3-CH(OH)-CH_2-C(=O)-ACP$

↓ dehydration (H_2O released)

$CH_3-CH=CH-C(=O)-ACP$

↓ (NADPH → $NADP^+$)

$CH_3-CH_2-CH_2-C(=O)-ACP$

\+ $^-O-C(=O)-CH_2-C(=O)-ACP$ (malonyl ACP) → process repeats to produce a chain 16 carbons long

Figure 16.10

Synthesis of fatty acids.

Malonyl-CoA transfers its malonyl residue to ACP. This is condensed with a molecule of acetyl-ACP (made from acetyl-CoA) to give a four carbon molecule with the release of ACP and CO_2. The four-carbon acetoacetyl-ACP is reduced to hydroxybutyryl-ACP. Water is removed leaving a double bond, which is again reduced to give butyryl-CoA. Another malonyl-ACP is condensed with this, and the cycle continues. Finally a chain 16 carbons long has been made (palmitic acid). At this point it is hydrolyzed from the ACP. Overall 14 NADPH molecules, 1 acetyl-CoA and 7 malonyl-CoA molecules have been used to make palmitic acid. Palmitic acid is used by enzymes on the endoplasmic reticulum that make longer chains and which can introduce double bonds. However, mammals cannot synthesize all the different kinds of fatty acids that they need for their membranes and must obtain some essential fatty acids in food (Box 2.3).

SYNTHESIS OF AMINO ACIDS

Bacteria and plants can synthesize all 20 amino acids. Some types of bacteria are able to fix atmospheric nitrogen N_2 by converting it to ammonia NH_3, which can then be incorporated into amino acids. The process uses 16 ATP for each N_2 molecule. Plants possess nitrate reductase systems that allow them to convert nitrate to NH_3. Ultimately all of our nitrogen is derived by the action of plants and bacteria, which are of course responsible for providing our organic carbon as well.

Even given NH_3, animals cannot synthesize all the protein amino acids. They can use aminotransferases (page 246) to move an amino group from one molecule to another. Thus, if fed some amino acids, they can make some others. However, some amino acids are completely essential because their carbon skeletons cannot be synthesized. Humans require histidine, isoleucine, leucine, lysine, methionine, phenylalanine, threonine, tryptophan, and valine in their diet.

CARBON FIXATION IN PLANTS

We have already described how chloroplasts use the energy of light to oxidize H_2O to give O_2 and generate ATP and NADPH. A series of so-called dark reactions in the chloroplast stroma use ATP and NADPH to grab carbon dioxide from the air and build it into molecules, a process called carbon fixation. The initial step is carried out by the enzyme ribulose bisphosphate carboxyase, said to be the most abundant protein on earth. CO_2 is combined with the phosphorylated five-carbon sugar ribulose bisphosphate to give a six-carbon intermediate that immediately splits to two molecules of 3-phosphoglycerate. Further reactions use NADPH and ATP to convert these three-carbon units to fructose-6-phosphate. Figure 16.11 summarizes this complex pathway.

CONTROL OF ENERGY PRODUCTION

Feedforward and Feedback

We have seen how an energy currency that runs low is topped up by conversion from another currency. However, this is not enough to ensure a constant energy supply. Therefore the cell has more mechanisms that ensure that the supply of cellular energy is accelerated or slowed as appropriate. These mechanisms are of two types: Feedforward and feedback. We will introduce the terms by analogy with real money. Consider a bank teller. During the day, people deposit mostly checks but draw out cash to spend. As time passes, the stock of bank-

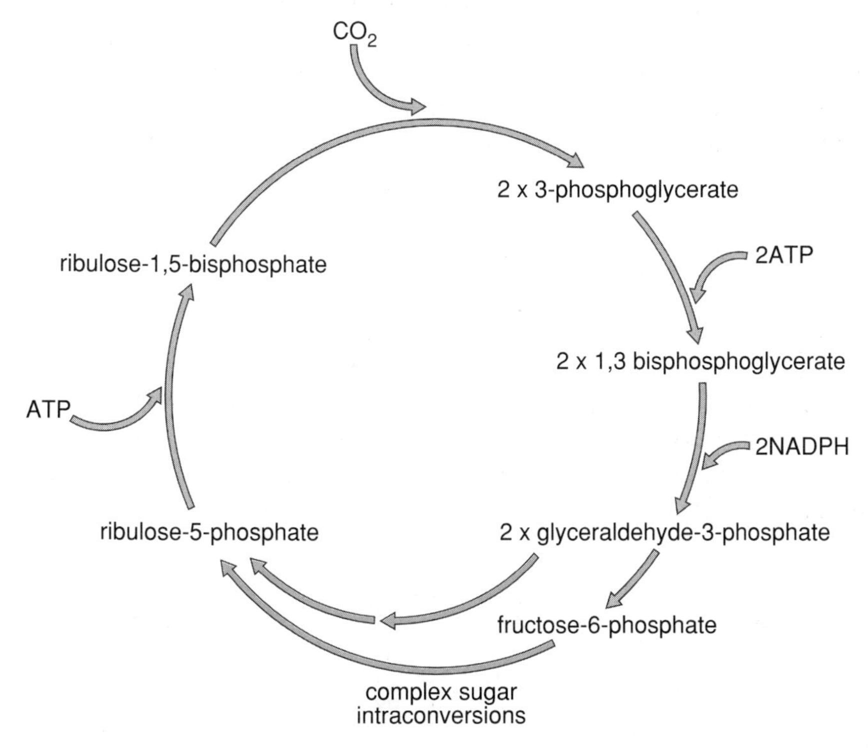

Figure 16.11
The dark reactions of photosynthesis capture carbon dioxide from the air.

notes and change in the till gets low. The teller signals to the supervisor, who opens the bank vault, takes out more cash, and refills the teller's till. This is an example of negative feedback. In general, negative feedback is said to occur when a change in some parameter activates a mechanism that reverses the change in that parameter. In this case a downward change in the level of the stack of banknotes in the till activates the mechanism that caused the supervisor to get cash from the vault and reverse the change by increasing the size of the cash pile. We have already met a very analogous negative feedback system in the control of tryptophan synthesis (page 156). A downward change in the concentration of tryptophan in the cell activates the mechanism that causes the cell to make more tryptophan. Positive feedback is less common in both biology and banking. It is said to occur when a change in some parameter activates a mechanism that accelerates the change. In banking this occurs when a rumor starts that a bank is about to fail. The lower the bank's reserves of money get, the more its depositors rush to take their money out, before it is too late. We will see a biological example of positive feedback when we discuss the action potential (Chapter 18).

What about *feedforward*? The bank is especially busy at lunchtime, with lots of cash withdrawn between 12:30 and 2:00. During this time everyone is rushed off their feet. The supervisors do not wait until tellers signal that they are short of cash but instead open the vaults at 12 noon and bring out enough cash to see the tellers through the lunchtime rush. They are preparing for a future drain on cash by stocking up the tills before the drain occurs—this is feedforward.

Negative Feedback Control of Glycolysis

Phosphofructokinase catalyzes the first irreversible step in glycolysis after the paths from glucose and glycogen converge (3 in Fig. 16.3). It is allosterically regulated by ATP (Fig. 16.12). When ATP concentrations are high, ATP binds to phosphofructokinase and locks it into an inactive configuration. When ATP levels fall, an increasing number of phosphofructokinase molecules lose their bound ATP and switch to an active configuration. Fructose 1,6-bisphosphate is

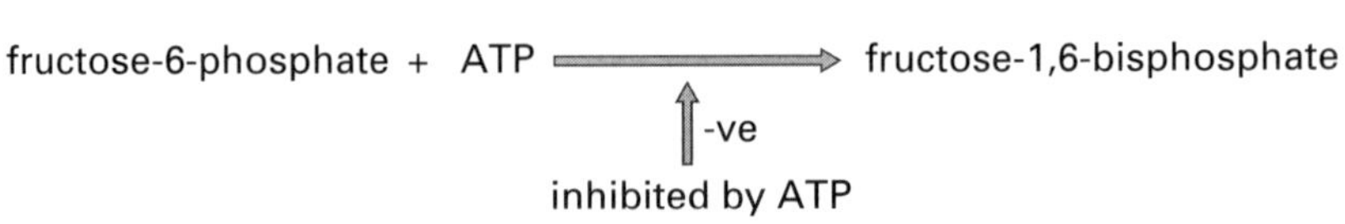

fructose-6-phosphate + ATP ⟹ fructose-1,6-bisphosphate

-ve

inhibited by ATP

Figure 16.12

Phosphofructokinase is regulated by the binding of ATP at a regulatory site (separate from the active site).

produced, feeding the glycolytic pathway that in turn feeds the mitochondria with pyruvate for the production of ATP.

When ATP concentration falls, phosphofructokinase is activated, and a mechanism initiated that produces more ATP. This is negative feedback. Changes in the amount of ATP act, through its allosteric action on phosphofructokinase, to reverse the change in the amount of ATP.

Feedforward Control in Muscle Cells

When a signal goes out from our brains to the muscles in our legs to tell them to start working, it causes the endoplasmic reticulum to release calcium ions into the cytosol. Calcium is acting as an intracellular messenger, a topic we will cover in more detail in Chapter 19. The rise of calcium activates several processes. It causes the cytoskeleton to contract, using the energy released by ATP hydrolysis to do mechanical work. At the same time other calcium ions pass down their electrochemical gradient from the cytosol into the mitochondrial matrix through a calcium channel. Once there, calcium activates three key enzymes: pyruvate dehydrogenase, oxoglutarate dehydrogenase, and malate dehydrogenase. The cell has not waited until ATP concentration starts to fall before activating the mitochondria, so this is feedforward control. Meanwhile, in the cytosol, calcium ions bind to the protein calmodulin (Fig. 13.10), which in turn binds to an enzyme called phosphorylase kinase (Fig. 16.13). The complex of calcium ions, calmodulin, and phosphorylase kinase activates glycogen phosphorylase, which proceeds to break down glycogen to release glucose-1-phosphate, which is then fed into the glycolytic pathway. Once again, the cell has not waited until glucose concentration falls before activating glycogen breakdown, so this is **feedforward control**.

In fact, muscles can begin to break down glycogen even before the message goes out from the brain to tell them to contract. When the brain realizes that we are in a dangerous situation and might be going to have to run, it causes the release of the hormone adrenaline from the adrenal gland above the kidneys (page 338). Adrenaline binds to an integral membrane protein of the skeletal muscle cells called the β-adrenergic receptor. This causes the production of another intracellular messenger called cAMP within the cytosol of the muscle cell. This topic is dealt with in more detail in Chapter 19. cAMP then activates cAMP-dependent protein kinase (Fig. 16.13). This enzyme, which is given the short name of protein kinase A, phosphorylates phosphorylase kinase and turns on the latter enzyme even when cytosolic calcium is low. Thus, even before we know for sure that we need to run, the muscles are breaking down glycogen and making the glucose that they will need if running becomes necessary.

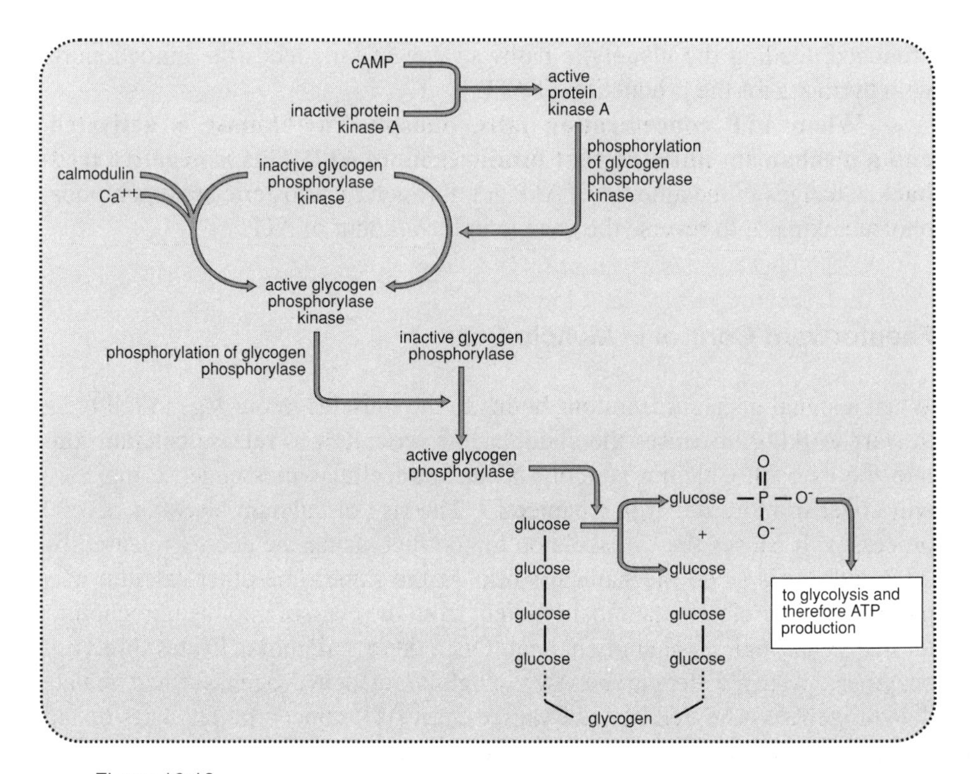

Figure 16.13

Calcium and cyclic AMP both activate glycogen breakdown in muscle and liver.

SUMMARY

1. Metabolism is the collective term for all of the reactions going on inside a cell. These reactions are divided into catabolic—those that break down chemical compounds to provide energy—and anabolic—those that build up complex molecules from simpler ones.

2. The Krebs cycle is at the center of the cell's metabolism. It can act to oxidize two carbon units derived from carbohydrates, fats, or amino acids and is also a central switching yard for the molecules in metabolism.

3. Glycolysis converts glucose to pyruvate. If the pyruvate is reduced to lactate, glycolysis can continue in the absence of oxygen.

4. Glycogen acts as a reserve of glucose for glycolysis.

5. Fats are concentrated fuel stores. Their fatty acid components are oxidized to two carbon units by β oxidation.

6. Amino groups must be removed before excess dietary amino acids can be used as fuels. This is done by transferring them to make aspartate or glutamate, and thence to urea for excretion.

7. Gluconeogenesis allows the synthesis of glucose from noncarbohydrate precursors.

8. Glycogen synthesis and fatty acid synthesis follow routes different from those used during breakdown.

9. Plants use light energy to fix carbon dioxide into sugars and to oxidize water to oxygen.

10. Metabolic reactions are controlled by feedforward and feedback mechanisms which make use of allosteric control and covalent modification of key enzymes.

CHAPTER

17

IONS AND VOLTAGES

We described in Chapter 2 how membranes are composed of phospholipids arranged so that their hydrophobic tails are directed toward the center of the membrane, while the polar hydrophilic head groups face out. Membranes are a barrier to the movement of many solutes. In particular, small hydrophilic solutes such as ions and sugars cannot pass through membranes easily because, to do so, they would have to lose the cloud of water molecules that forms their hydration shells (Box 1.4). Two consequences follow from the fact that membranes are barriers. First, the composition of the liquid on one side of a membrane can be different from the composition of the liquid on the other side. Indeed, by allowing cells to retain proteins, sugars, ATP and many other solutes, the barrier property of the cell membrane makes life possible. Second, the cell must make proteins called channels and carriers whose job it is to help hydrophilic solutes across the membrane. This chapter describes the properties of membranes, with particular emphasis on their role in energy storage.

THE POTASSIUM GRADIENT AND THE RESTING VOLTAGE

Ions are electrically charged. This fact has two consequences for membranes. First, the movement of ions across a membrane will tend to change the voltage across that membrane. If positive ions leave the cytosol, they will leave the cytosol with a negative voltage, and vice versa. Second, a voltage across a membrane will exert a force on all the ions present. If the cytosol has a negative voltage, then positive ions such as sodium and potassium will be attracted in from the extracellular medium. In this chapter we will begin to address the question of how ions and voltages interact by considering the resting voltage.

Potassium Channels Make the Plasmalemma Permeable to Potassium Ions

The Na^+/K^+ ATPase uses the energy of ATP hydrolysis to drive sodium ions out of the cell and, at the same time, brings potassium ions in to the cell. It ensures that potassium is much more concentrated in the cytosol than outside—typically 140 mmol $liter^{-1}$ in the cytosol but only 5 mmol $liter^{-1}$ in the extracellular medium.

The potassium channel (Fig. 17.1) is a protein found in the plasmalemmae of almost all cells. It is a tube that runs all the way through the membrane. Potassium ions, which cannot pass through the lipid bilayer of the plasmalemma, pass through the potassium channel easily. Other ions cannot go through. The precise shape of the internal tube, and the position of charged amino acid side chains within the tube, blocks their movement. The channel is selective for potassium.

There is an apparent paradox here. Potassium can pass through the potassium channel, yet this ion is much more concentrated inside the cell than outside. Why doesn't all the potassium rush out? To explain why, we must

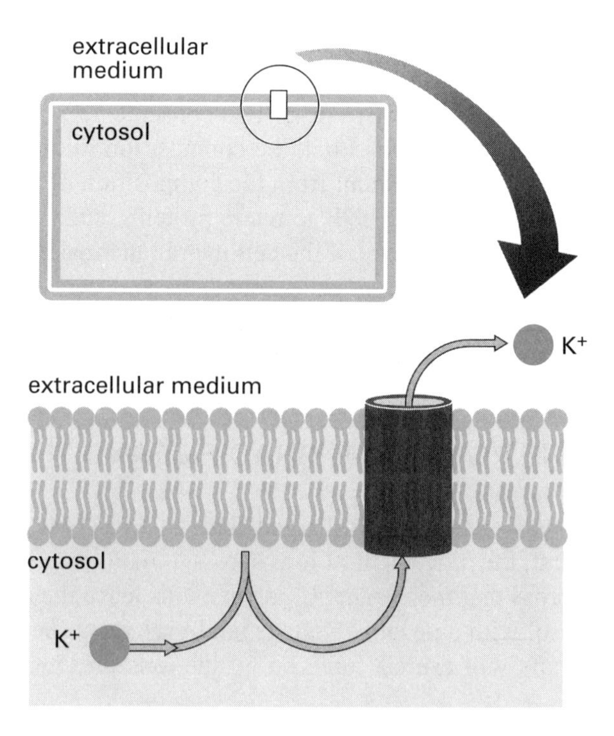

Figure 17.1

The positively charged potassium ion cannot cross the lipid bilayer but passes easily through a water-filled tube in the potassium channel.

think about the effects of ion movement on transmembrane voltage. First a word on nomenclature. All cells have a voltage across their membrane when they are not being stimulated. This *resting voltage* is about -80 mV in a relaxed skeletal muscle cell. As soon as the muscle is stimulated to contract, there is a sudden sharp change in the transmembrane voltage—the action potential (a process described in more detail in Chapter 18). Many cells never change their transmembrane resting voltage.

Concentration Gradients and Electrical Voltage Can Balance

We can now return to the problem of potassium ions. A few do escape from the cell through the channel. As they do so, they carry out their positive charge and leave the cytosol with a negative voltage that attracts positively charged ions like potassium. There is still a tendency for potassium ions to leave the cell down the concentration gradient, but there is now an electrical force pulling the positively charged potassium ions back inside. We have met a similar situation before, where we described how a concentration gradient and an electrical force combine to form an electrochemical gradient down which H^+ ions will rush into bacteria or mitochondria (page 255). However, in the case of potassium ions at the plasmalemma, the concentration gradient and the electrical force act in opposite directions. As potassium ions continue to leave the cell, carrying out their positive charge and leaving the cytosol at a more and more negative voltage, the electrical force pulling them back in gets increasingly strong. Soon the opposing electrical and concentration gradients are equal, and the overall electrochemical gradient for potassium is zero. Potassium ions then stop leaving the cell, even though they are much more concentrated inside than outside.

For every ion present on both sides of a membrane, it is possible to calculate the transmembrane voltage that will exactly balance the concentration gradient. This voltage is called the *equilibrium voltage* for that ion at that membrane. Box 17.1 gives details of the calculation. For potassium at the plasmalemma, the equilibrium voltage in a normal animal cell is about -90 mV.

The departure of potassium ions through the potassium channels, leaving negative charge behind, produces the resting voltage across the plasmalemma. Because the potassium channels are the major pathway by which ions can cross the plasmalemma of an unstimulated cell, the resting voltage has a value close to the potassium equilibrium voltage. In some cells, such as white blood cells, potassium channels are the only channels in the plasmalemma, and potassium ions move out until the transmembrane voltage tending to pull them back in exactly balances their tendency to move out down their concentration gradient. The resting voltage of white blood cells therefore has a value equal to the potassium equilibrium voltage, about -90 mV. In other cells the situation is more complicated. In muscle cells, for instance, the

BOX 17.1

The Nernst Equation

ion I at concentration $[I]_{out}$

ion I at concentration $[I]_{in}$

cytosol at voltage V

An ion that can pass across a membrane is acted on by two forces. The first derives from the concentration gradient. The ion tends to diffuse from a region where it is at high concentration to one where it is at low concentration. The second force derives from the transmembrane voltage. In the case of positively charged ions such as Na^+ and K^+, the ions tend to move toward a negative voltage. Negatively charged ions such as Cl^- tend to move toward a positive voltage. For each ion there is a value of the transmembrane voltage for which these forces balance and the ion will not move. The ion is said to be at equilibrium, and this value of the transmembrane voltage is called the equilibrium voltage for that ion at that membrane.

When the forces balance, then ions that move into the cell will neither gain nor lose energy. This way of describing equilibrium is useful because it allows us to set equivalent the effects of the two very different gradients, concentration and voltage. For concentration the free energy possessed by a mole of ions I by virtue of its concentration is

$$G = G_0 + RT\log_e[I] \quad \text{joules}$$

Where G_0 is the standard free energy, R is the gas constant (8.3 Jmol^{-1} per degree), and T is the absolute temperature. A mole of I passing into the cytosol therefore moves from a region where it had a free energy of

$$G_{\text{outside}} = G_0 + RT\log_e[I_{\text{outside}}] \quad \text{joules}$$

to one where its free energy is

$$G_{\text{inside}} = G_0 + RT\log_e[I_{\text{inside}}] \quad \text{joules}$$

One mole of ions I moving inward therefore gains by virtue of the concentration gradient free energy equal to

$$RT\log_e[I_{\text{inside}}] - RT\log_e[I_{\text{outside}}] \quad \text{joules}$$

Now consider the electrical force. The definition of a volt is that one coulomb of charge moving across a membrane with a transmembrane voltage of V volts gains V joules of free energy. However, we are working in moles, not coulombs. One mole of ions has a charge of zF coulombs, where z is the charge on the ion in elementary units. For Na^+ and K^+ z is 1; for Ca^{++}

z is 2, and for Cl^- z is -1. F is a number which relates the coulomb to the mole. It has the value 96,500. One mole of ions I moving inward gains by virtue of the transmembrane voltage free energy equal to

$$zFV \text{ joules}$$

This does not mean that an ion always gains energy from the transmembrane voltage when it moves inward: The term zFV can just as easily be negative as positive.

When the effects of concentration and voltage just balance a mole of ions moving inward, neither gains nor loses free energy. Hence at equilibrium

$$RT \log_e[I_{\text{inside}}] - RT \log_e[I_{\text{outside}}] + zFV_{\text{eq}} = 0$$

This expression can be simplified to

$$V_{\text{eq}} = \frac{RT}{zF} \log_e\left(\frac{[I_{\text{outside}}]}{[I_{\text{inside}}]}\right) \quad \text{volts}$$

This is the Nernst equation. At a typical mammalian body temperature of 37°C, the Nernst equation can be written in the more convenient form

$$V_{\text{eq}} = 62 \log_{10}\left(\frac{[I_{\text{outside}}]}{[I_{\text{inside}}]}\right) \quad \text{mvolts}$$

Note the change in the type of logarithm and the units.

resting voltage is -80 mV; in nerve cells it is -70 mV. Even in these cells, though, the major influence on the resting voltage is potassium movement through its channels, so the resting voltage does not deviate very far from the potassium equilibrium voltage.

The resting voltage set up by potassium movement turns the action of the Na^+/K^+ ATPase into an energetically asymmetrical one. Consider one conversion cycle: One molecule of ATP is hydrolyzed, three sodium ions are pushed out of the cell, and two potassium ions move in. Very little of the energy of ATP hydrolysis is used up in moving the two potassium ions into the cell because the electrochemical gradient for this ion is close to zero. Although potassium is being moved up a concentration gradient, it is also being pulled in by the negative voltage of the cytosol, and the two forces cancel. In contrast, pushing the three sodium ions out of the cell requires more energy than would be required if the cytosol were at the same voltage as the extracellular fluid. The sodium ions are positively charged, so they are attracted by the negative voltage inside the cell, which combines with the concentration gradient to form a large inward electrochemical gradient. All the energy re-

BOX 17.2

Measuring the Transmembrane Voltage

In 1949 Ling and Gerard discovered that when a fine glass micropipette filled with an electrically conducting solution impaled a cell, the plasmalemma sealed to the glass, so the transmembrane voltage was not discharged. The voltage difference between a wire inserted into the micropipette and an electrode in the extracellular fluid could then be measured. By passing current through the micropipette, the transmembrane voltage could be altered.

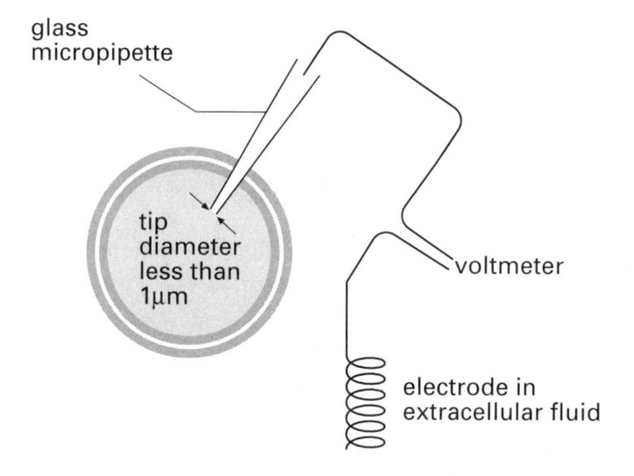

Twenty-five years later Neher and Sakmann showed that the micropipette did not have to impale the cell. If it just touched the cell, a slight suction caused the plasmalemma to seal to the glass. The technique, called cell attached patch clamping, can measure currents through the few channels present in the tiny patch of membrane within the pipette.

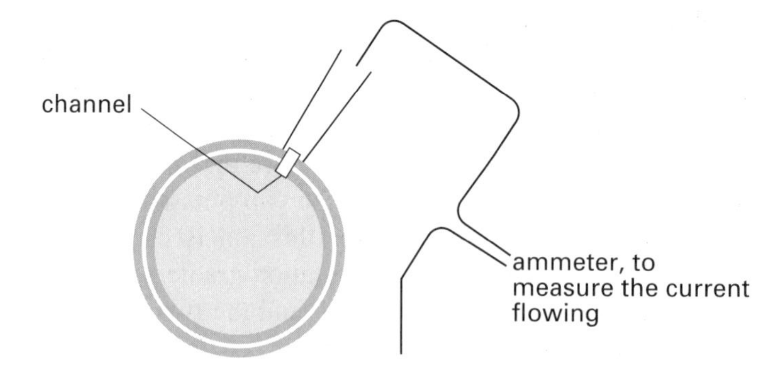

Stronger suction bursts the membrane within the pipette. The transmembrane voltage can now be measured.

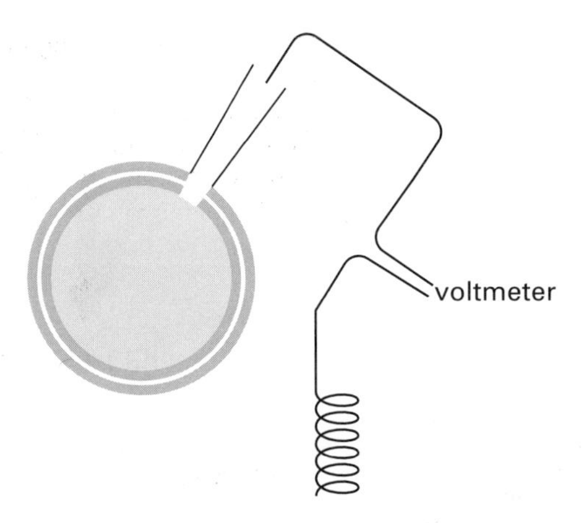

Alternatively, current can be passed through the micropipette to change the transmembrane voltage—this is the whole cell patch clamp technique. In 1991 Neher and Sakmann received the Nobel prize for medicine.

leased by ATP hydrolysis is needed to push the three sodium ions up this large electrochemical gradient. The presence of the potassium channels, and the resting voltage that they set up, means that almost all the energy of ATP hydrolysis by the Na^+/K^+ ATPase is stored in the sodium gradient, while potassium ions are close to equilibrium.

THE CHLORIDE GRADIENT

Chloride ions are at a lower concentration in the cytosol than in the extracellular fluid. Typically their concentration in the cytosol is 5 mmol $liter^{-1}$ compared with 100 mmol $liter^{-1}$ in the extracellular fluid. This is because of the resting voltage set up by the potassium channels. Chloride ions are repelled by the negative voltage of the cytosol. They leave until their tendency to enter the cell down their concentration gradient exactly matches their tendency to be repelled by the negative voltage of the cytosol (Fig. 17.2).

GENERAL PROPERTIES OF CHANNELS

Channels are integral membrane proteins that form water-filled tubes through the membrane. We have already discussed two. Channels that, like the potassium channel, are selective for particular ions can set up a transmembrane voltages. The gap junction channel (page 29) is much less selective than the potassi-

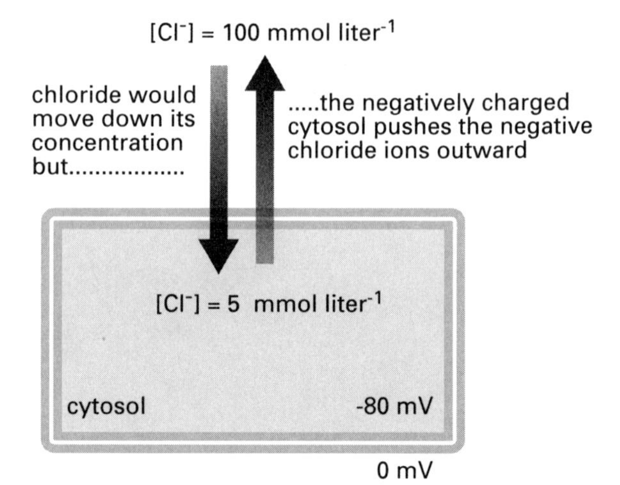

Figure 17.2
Chloride is close to equilibrium across most plasmalemmae.

um channel. It forms a tube, 1.5 nm in diameter, through which any solute of a molecular weight up to 1,000 can pass. The gap junction channel is not always open. It opens only when it connects with a second gap junction channel on another cell, forming a tube through which solutes can pass from the cytosol of one cell to the cytosol of the other. Channels that are sometimes open and sometimes shut are said to be *gated*. When one gap junction channel contacts another on another cell, its gate opens and solute can pass through; at other times the gate is shut. The usefulness of gating is obvious: if the gap junction channels not contacting others were open, many solutes, including ATP and sodium, would leak out into the extracellular fluid exhausting the cell's energy currencies.

A third channel important in energy conversion is porin in the outer mitochondrial membrane. It forms a very large diameter tube that allows all solutes of molecular weight up to 10,000 to pass. Porin is not gated but is always open. This is why the outer mitochondrial membrane is permeable to most solutes and ions. Box 17.3 lists all the different types of channels described in this book. It represents only a small fraction of the total number known.

GENERAL PROPERTIES OF CARRIERS

We have already met the three carriers that interconvert the four energy currencies of the cell (page 256). Carriers are like channels in that they are integral membrane proteins that allow solute to cross the membrane, and like channels, they form a tube across the membrane. However, there is a critical difference. In carriers the tube is never open all the way through; it is always closed at one or other end. Solutes can move into the tube through the open end. When the carrier changes shape, so that the end that was closed is open, the solute can move into the solution on the other side of the membrane.

The Glucose Carrier

One of the simplest carriers is the glucose carrier (Fig. 17.3). It switches freely between a form that is open to the cytosol and a form that is open to the extracellular medium. Inside the tube is a site to which a glucose molecule can bind. On the left, a glucose molecule is entering the tube and binding to the site.

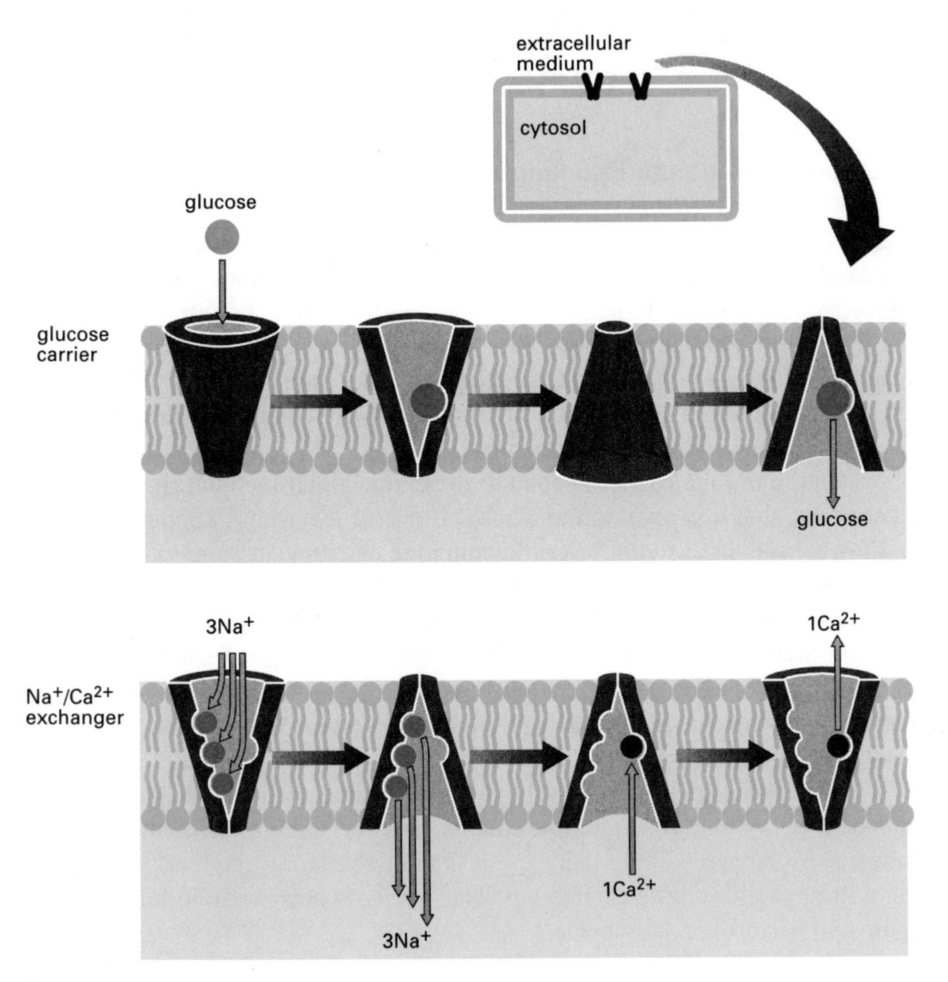

Figure 17.3

The glucose carrier and the sodium-calcium exchanger are sometimes open to the cytosol, and sometimes open to the extracellular medium.

Sometimes glucose leaves the binding site before the carrier switches shape. In the figure we see the other possibility: the carrier switches shape before the glucose has left. The binding site is now open to the cytosol, and the glucose can escape into the cytosol. It has been carried across the plasmalemma. **Unlike channels, carriers never form open tubes all the way across the membrane. Instead, they bind one or more molecules or ions, then change shape to carry the molecule or ions across the membrane.**

The glucose carrier is very simple, whereas other carriers are more complex. The sodium/calcium exchanger and the calcium ATPase do much the same job—they push calcium ions up their concentration gradient out of the cell—but they take their energy from different currencies. The sodium/calcium exchanger uses the sodium concentration gradient while the calcium ATPase uses ATP. All cells have one of these carriers to keep the concentration of calcium in the cytosol low, many cells have both.

The Sodium/Calcium Exchanger

Figure 17.3 shows that like the glucose carrier, the sodium/calcium exchanger can exist in two shapes, one open to the extracellular medium and one open to the cytosol. Inside the tube are three sites that can bind sodium ions and one site that can bind a calcium ion. The sodium/calcium exchanger is not free to switch between its two shapes at any time, instead, it only switches if either all the sodium sites are filled and the calcium site is empty, or if the calcium site is filled and all the sodium sites are empty.

On the left the carrier is open to the extracellular medium. It can switch its shape so that it is open to the cytosol if one of two things happen. It could, as shown, bind three sodium ions (keeping the calcium site empty) or, it could bind one calcium and keep the sodium sites vacant. This second option does not often happen, since sodium is at high concentration in the extracellular medium and one or more of the sodium sites is usually occupied. Therefore nearly all the switches from open-to-outside to open-to-inside are of the type shown. Once the channel has opened to the low sodium environment of the cytosol, the sodium ions tend to leave.

Once the carrier is open to the cytosol, it can switch back to the open-to-outside form by binding either one calcium or three sodium ions. Since sodium is scarce in the cytosol, the latter event is unlikely. More frequently a calcium ion will bind and will be carried out. The carrier is now ready to bind sodium again and perform another cycle.

The overall effect of one cycle is to carry three sodium ions into the cell down their electrochemical gradient, and one calcium ion out of the cell up its electrochemical gradient. A very simple rule about when the carrier can switch shape has produced a machine that uses the energy currency of the sodium gradient to do work in pushing calcium ions out of the cell. Figure 17.4 gives a simple illustration of what is happening. The circle represents one cycle of operation, from open to extracellular medium back to open to extracellular medium again.

Carriers with an Enzymic Action: The Calcium ATPase

The electron transport chain, ATP synthase and the Na^+/K^+ ATPase are carriers with an additional level of complexity in that they carry out an enzymic action as well as their carrier function. The operation of these carriers is too complicated to discuss here, but the principles of their operation are illustrated by the Ca^{++} ATPase (Fig. 17.5), which like the sodium/calcium exchanger is found in the plasmalemma. The transmembrane part of this carrier forms a tube that can be open to the cytosol or to the extracellular medium. Another part lies in the cytosol and can hydrolyze ATP. Small changes in the shape of the cytosolic region are transmitted to the transmembrane region and force it between the open-to-inside and open-to-outside shapes. The figure shows in simplified form how this might happen. (1) is the relaxed shape of the protein. (2) When a calcium ion moving in from the cytosol pushes the tube open, the distortion is transmitted to the cytosolic region and allows a kinase catalytic site to operate on an aspartate residue. (3) The kinase transfers the γ phosphoryl group from ATP to the aspartate residue (page 218). (4) The addition of two negative charges to the aspartate alters the balance of electrical forces within the protein, which changes shape to bring the negatively charged phosphoryl group closer to positively charged residues. This shape change is transmitted to the transmembrane region, which is forced into a wide open-to-outside shape. The calcium ion, which can no longer bind to the sites on both sides of the tube, is held only weakly and tends to escape into the extracellular medium. Meanwhile the phosphoryl group is cleaved from the aspartate residue by a phos-

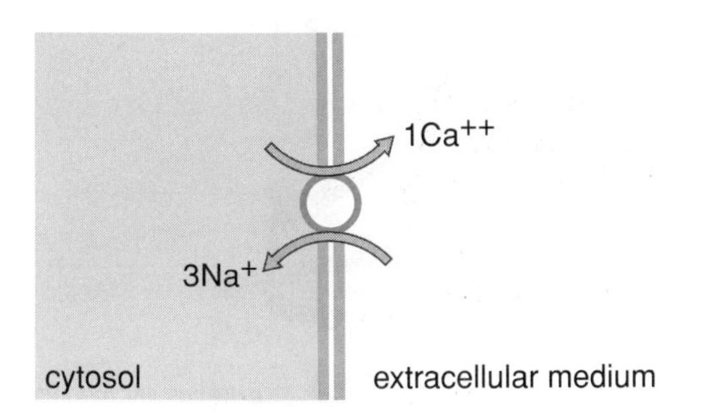

Figure 17.4
The action of the sodium/calcium exchanger.

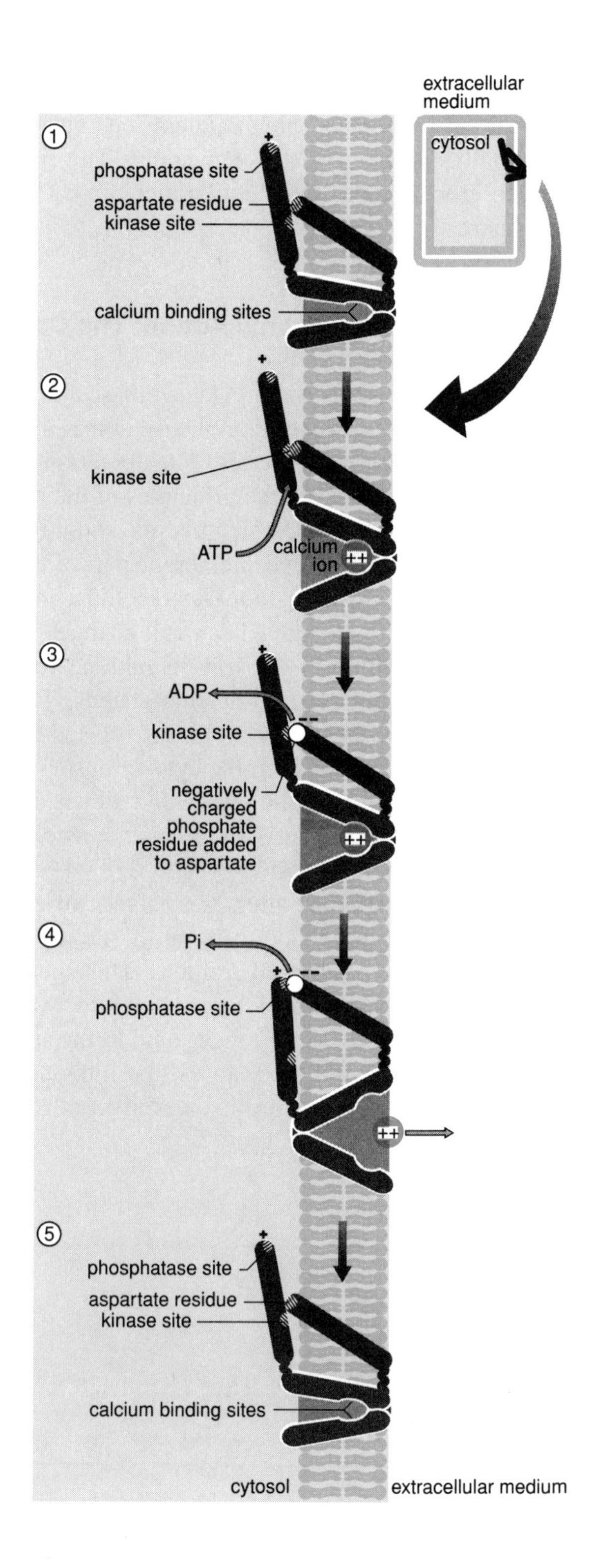

Figure 17.5

The calcium ATPase is sometimes open to the cytosol, and sometimes open to the extracellular medium. A cycle of phosphorylation and dephosphorylation drives the shape changes.

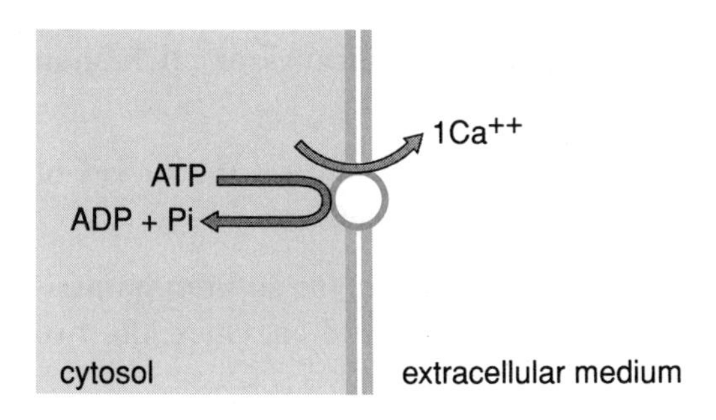

Figure 17.6
The action of the calcium ATPase.

phatase catalytic site. (5) Once phosphate leaves, the protein is no longer held in the open-to-outside shape by electrical forces and relaxes into shape (1). The carrier is now ready to accept another calcium ion from the cytosol and repeat the process.

Figure 17.6 summarizes what is happening. The circle represents one cycle of operation. One molecule of ATP is hydrolyzed to ADP and inorganic phosphate, and one calcium ion moves out of the cell. The energy released by ATP hydrolysis has been used to drive one calcium ion up its electrochemical gradient out of the cell. Box 17.3 lists all the carriers described in this book. Again, it represents only a few of the total known.

SUMMARY

1. Channels are membrane proteins with a central water-filled hole through which hydrophilic solutes, including ions, can pass from one side of the membrane to the other. Changes in protein structure may act to gate the channel but are not required for movement of the solute from one side to the other.

2. The presence of potassium channels in the plasmalemma, and the resting voltage that they set up, means that almost all the energy of ATP hydrolysis by the Na^+/K^+ ATPase is stored in the sodium gradient, while potassium ions are close to equilibrium.

3. The resting voltage repels chloride ions from the cell interior.

4. Carriers, like channels, form a tube across the membrane, but the tube is always closed at one end. Solutes can move into the tube through the open end. When the carrier changes shape, so that the end that was closed is open, the solute can leave to the solution on the other side of the membrane.

5. The glucose carrier is present in the membranes of all human cells.

6. The sodium/calcium exchanger is a carrier that uses the energy of the sodium gradient to push calcium ions out of the cell.

7. The electron transport chain, ATP synthetase, the sodium/potassium ATPase and calcium ATPase are both carriers and enzymes, the two actions being tightly linked.

BOX 17.3

Channels and Carriers

This box is intended as a reference aid.

CHANNELS

Large Channels

Name	Gap junction channel
Chapter	2
Location	Plasmalemma of many cells
Selective for	any solute of MW < 1,000
Opened by	Contact with second gap junction channel
Comments	Allows solutes to pass from cytosol to cytosol.

Name	Porin
Chapter	15
Location	Outer mitochondrial membrane
Selective for	any solute of MW < 10,000
Opened by	Always open
Comments	Allows easy passage of ions and nucleotides across the outer mitochondrial membrane

Potassium Channels

Name	Potassium channels
Chapter	17
Location	Plasmalemma of all cells
Selective for	Potassium ions
Opened by	Some are open all the time, others are opened by depolarization of the plasmalemma
Comments	Responsible for the resting voltage.

Calcium Channels

Name	Voltage-gated calcium channel
Chapter	18
Location	Plasmalemma of sea urchin egg, axon terminal, muscle, and other cells
Selective for	Calcium ions
Opened by	Depolarization of plasmalemma
Comments	Changes the transmembrane voltage and raises the calcium concentration in the cytosol by the inward movement of calcium ions.

Name	Inositol trisphosphate gated calcium channel
Chapter	19
Location	Membrane of endoplasmic reticulum
Selective for	Calcium ions
Opened by	Binding of cytosolic inositol trisphosphate
Comments	Part of the system allows an extracellular solute to raise cytosolic calcium concentration.

Name	Mitochondrial calcium channel
Chapter	16
Location	Inner mitochondrial membrane
Selective for	Calcium ions
Opened by	Always open
Comments	Allows a rise of cytosolic calcium to activate mitochondria.

Sodium and Unselective Cation Channels

Name	Voltage-gated sodium channel
Chapter	18
Location	Plasmalemma of nerve and muscle cells
Selective for	Sodium ions
Opened by	Depolarization of plasmalemma
Comments	Produces brief action potentials.

Name	Nicotinic acetylcholine receptor channel
Chapter	20
Location	Nerve and muscle cells
Selective for	Sodium and potassium
Opened by	Extracellular acetylcholine
Comments	An ionotropic cell surface receptor that causes the plasmalemma to depolarize.

Name	Glutamate receptor channel
Chapter	20
Location	Nerve cells

Selective for	Sodium, potassium and calcium
Opened by	Extracellular glutamate
Comments	An ionotropic cell surface receptor that causes the plasmalemma to depolarize.

Name	Cyclic AMP-gated channel
Chapter	19
Location	scent-sensitive nerve cells
Selective for	Sodium and potassium
Opened by	Cyclic AMP in the cytosol
Comments	Depolarizes the plasmalemma so that the cell produces action potentials.

Chloride Channels

Name	GABA receptor channel
Chapter	20
Location	Nerve cells
Selective for	Chloride ions
Opened by	Extracellular GABA
Comments	More difficult to depolarize the cell to threshold when channel is open.

CARRIERS

Carriers with No Enzymic Action

Name: Glucose carrier
Chapter: 17
Location: Plasmalemma of all cells
Mode of action:

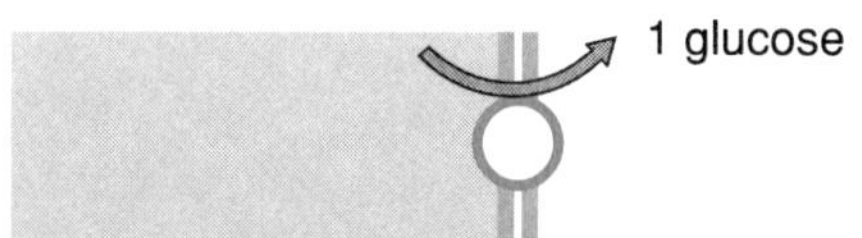

Comments: Required by all cells.

Name: ADP/ATP exchanger
Chapter: 15
Location: Inner mitochondrial membrane
Mode of action:

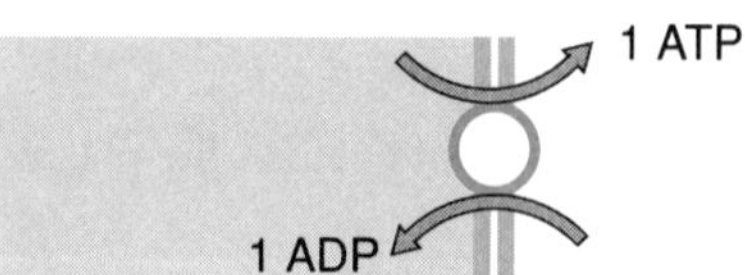

Comments: Gets ATP and ADP across inner mitochondrial membrane.

Name Na$^+$/Ca^{++} exchanger
Chapter 17
Location Plasmalemma of many cells
Mode of action

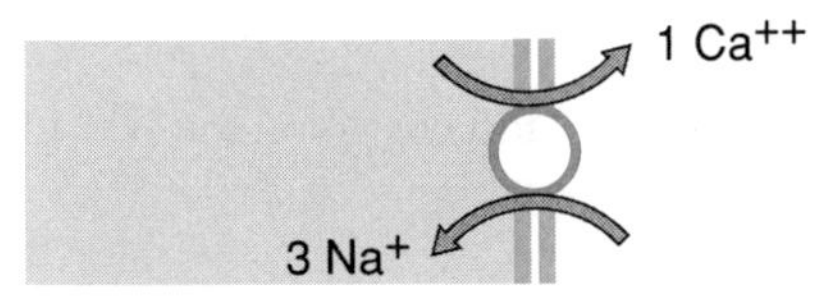

Comments Pushes calcium ions out of the cytosol.

Name β-galactoside permease
Chapter 10
Location Bacterial plasmalemma
Mode of action Linked inward movement of H$^+$ and sugar.
Comments One of the products of the *lac* operon.

Carriers with Both Enzymic and Carrier Action

Name ATP synthase
Chapter 15
Location Inner mitochondrial membrane
Mode of action

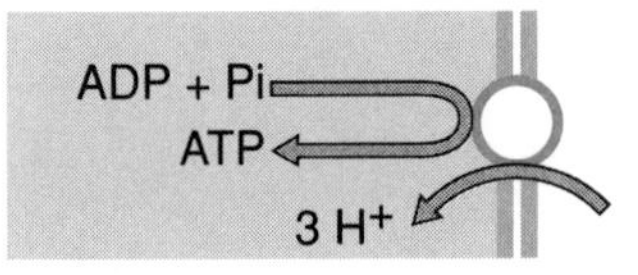

Comments Converts between energy in the H$^+$ gradient and energy as ATP.

Name Ca^{++} ATPase
Chapter 17
Location Plasmalemma of many cells
Mode of action

1 Ca++
ATP
ADP + Pi

Comments Pushes calcium ions out of the cytosol. All cells have either this or the Na$^+$/Ca^{++} exchanger; many have both.

Name Na$^+$/K$^+$ ATPase
Chapter 15
Location Plasmalemma of all cells

Mode of action	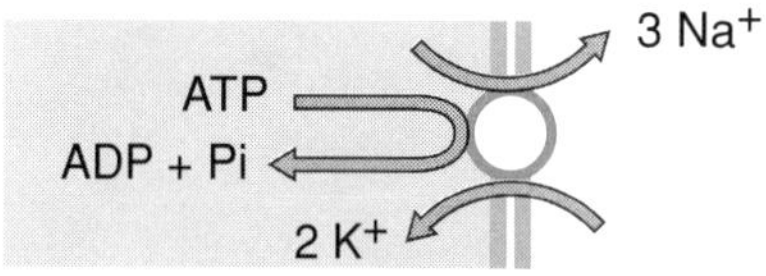
Comments	Converts between energy as ATP and energy in the Na^+ gradient.
Name	Electron transport chain
Chapter	15
Location	Inner mitochondrial membrane
Mode of action	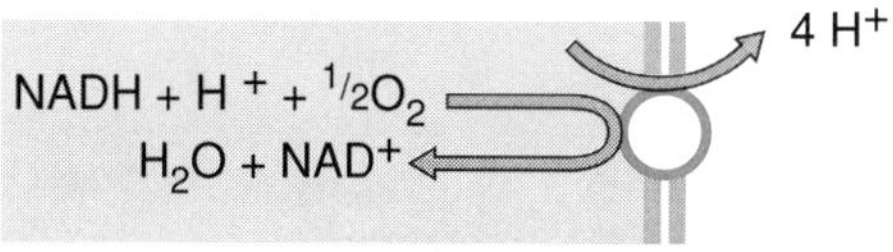
Comments	Converts between energy as NADH and energy in the H^+ gradient.

CHAPTER
18

THE ACTION POTENTIAL

An action potential is an explosive change in the voltage across the plasmalemma. The most sophisticated electrically excitable cells—cells that can produce action potentials—are the nerve cells, which allow our brains to carry out the complex electrical data processing called thought. However, we will begin with a much simpler electrically excitable cell: the sea urchin egg.

THE CALCIUM ACTION POTENTIAL IN SEA URCHIN EGGS

In most animals, eggs that are fertilized by more than one sperm fail to develop. A number of mechanisms exist to prevent this, one of which is based on action potentials.

Effect of Egg Membrane Voltage on Sperm Fusion

An experiment first carried in 1976 out by Jaffe is illustrated in Fig. 18.1a. She impaled a sea urchin egg with a micropipette (Box 17.2). Instead of just measuring the natural resting transmembrane voltage (around -70 mV), she *depolarized* the plasmalemma by passing positive current through the pipette. We use the word depolarization to mean any positive shift in the transmembrane voltage, whatever its size or cause. In this case an electronic feedback circuit was used to alter the value of the current passed through the electrode so that

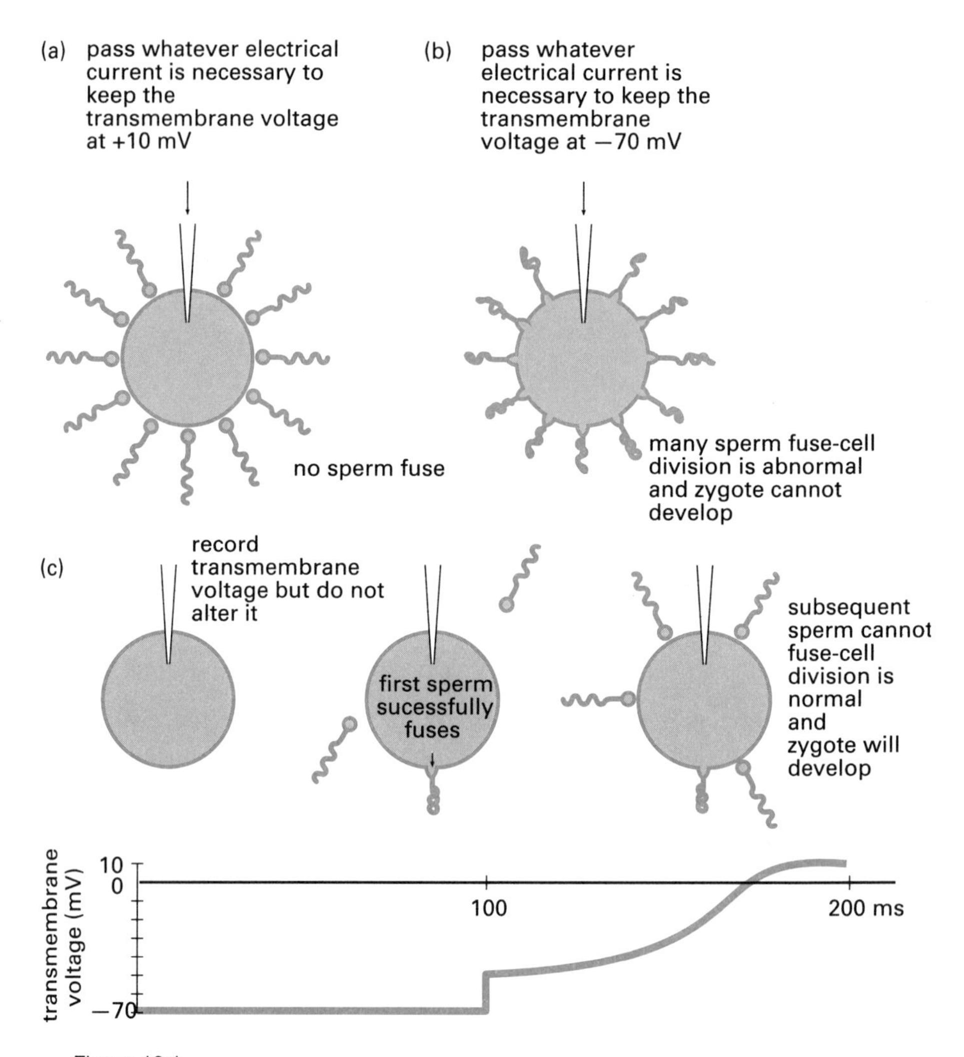

Figure 18.1

In sea urchin eggs transmembrane voltage acts as a switch to control sperm fusion.

the transmembrane voltage was depolarized to +10 mV and held at that value. This technique, in which the experimenter, not the cell, determines the value of the transmembrane voltage, is called voltage clamp.

Jaffe observed that sperm would not fuse with eggs in which the transmembrane voltage was clamped at +10 mV, and the eggs remained unfertilized (Fig. 18.1a). However, an egg clamped at -70 mV (Fig. 18.1*b*) was fertilized by multiple sperm, leading to a defective embryo. The transmembrane voltage therefore acts like a switch that controls the fusion of sperm and egg.

When no current is passed through the micropipette the transmembrane voltage is set by the channels in the egg plasmalemma (Fig. 18.1*c*). The resting voltage of the unfertilized egg is set by potassium movement through potassium channels. The first sperm that arrives can fuse because the transmembrane voltage is -70 mV. But within 100 ms, the transmembrane voltage depolarizes to +10 mV and further sperm fusion is blocked. Thus transmembrane voltage changes act to prevent multiple sperm fusion. The sea urchin egg, which normally when unfertilized has a transmembrane voltage of -70 mV and is receptive to fertilization, depolarizes its membrane following successful fertilization to protect itself from subsequent fertilization. The result is shown on the right of Fig. 18.1c. Only one sperm has fused, so the zygote has the correct number of chromosomes and can go on to develop into an embryo and then into a mature sea urchin.

The Voltage-Gated Calcium Channel

The rapid depolarization that occurs upon fertilization of the sea urchin egg is an example of an action potential. Like many cells urchin eggs can generate action potentials because their plasmalemmae contain, together with potassium channels, an ion-conducting channel called the voltage-gated calcium channel.

Figure 18.2 shows how the voltage-gated calcium channel works. On the left is the shape of the protein when the transmembrane voltage is -70 mV—as in an unfertilized egg. The protein forms a tube, but it is not open all the way through the membrane. If the membrane depolarizes so that the cytosol is less negative or even becomes positive, positively charged amino acid residues on the protein are repelled, popping the tube into an open state. Calcium ions can

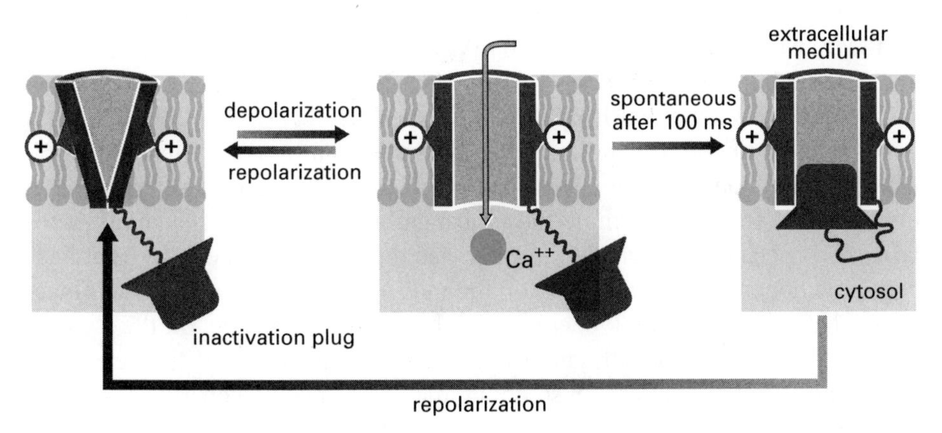

Figure 18.2
The voltage-gated calcium channel.

now pass through the channel. If the membrane is then repolarized, then the channel quickly recloses because the positive charges are pulled back toward the inside of the cell.

A protein domain in the cyosol called the inactivation plug is constantly jiggling about at the end of a flexible link and can bind to the inside of the open channel. The resulting blockage is called *inactivation*, and it occurs after about 100 ms. As long as the plasmalemma remains depolarized, the voltage-gated calcium channel will remain inactivated. When the plasmalemma is repolarized, the positive charges are attracted back toward the inside of the cell, squeezing the plug out. In summary:

1. When the transmembrane voltage is -70 mV, the voltage-gated calcium channel is gated shut.
2. When the plasmalemma is depolarized, the channel opens rapidly and after about 100 ms inactivates.
3. After the channel has gone through this cycle, it must spend at least 100 ms with the transmembrane voltage at the resting voltage before it can be opened by a second depolarization.

The voltage-gated calcium channel goes through this process whether or not calcium ions are present. To understand the action potential, we must now think about the effect that movements of calcium ions through this channel have on the transmembrane voltage.

The Calcium Action Potential

At *A* in Fig. 18.3 the egg is unfertilized. The potassium channels are open and set the transmembrane voltage to -70 mV. This negative voltage closes the voltage-gated calcium channels.

At *B* we show what happens if an experimenter passes enough current down a micropipette to depolarize the plasmalemma by ten millivolts, to -60 mV, and then stops passing current and once again allows the ion channels in the plasmalemma to determine the transmembrane voltage. This small depolarization opens only a very few voltage-gated calcium channels, so what happens next depends almost entirely on the movement of ions through the potassium channels. The slight decrease in the negative voltage of the cytosol reduces its attraction to potassium ions, and they move out of the cell down their concentration gradient. In so doing, they carry positive charge away from the cytosol, and the transmembrane voltage returns to its resting value of -70 mV.

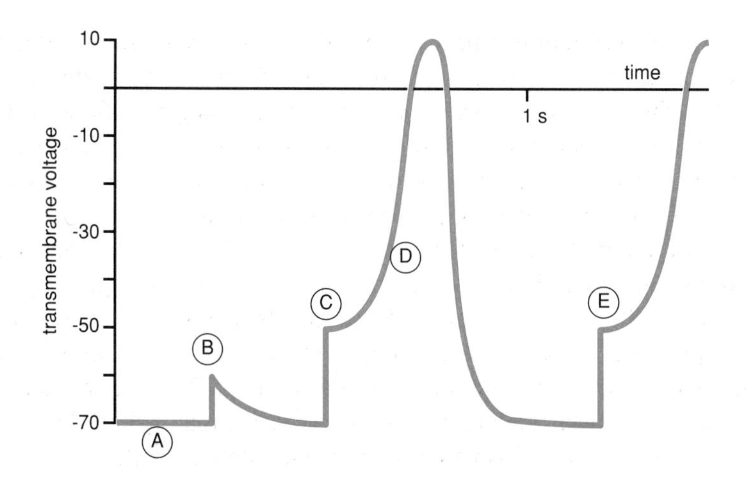

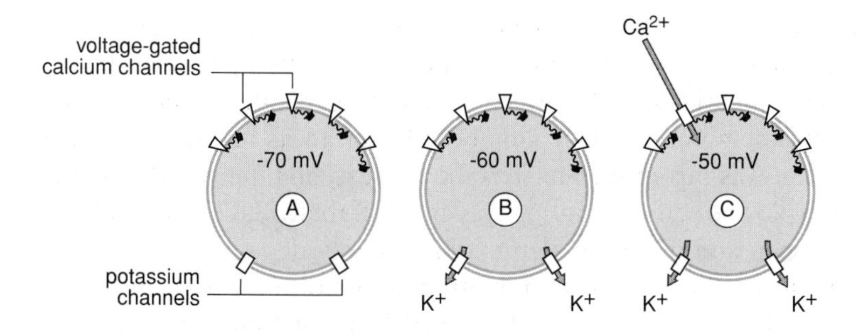

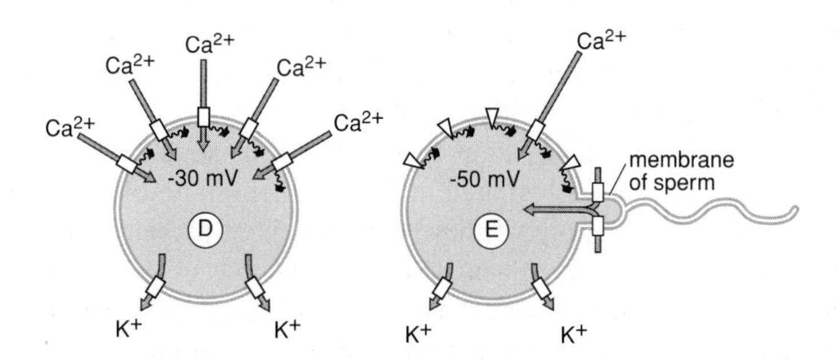

Figure 18.3

The calcium action potential in a sea urchin egg.

At *C* the experiment is repeated but this time the membrane is fleetingly depolarized by 20 mV to -50 mV. Again most of the voltage-gated calcium channels stay shut, but some open, and calcium ions begin to flood in to the cell down their electrochemical gradient. The flow of positive charge inward on the calcium ions outweighs the outward flow of positive charge on potassium ions, so the cytosol is gaining positive charge and its voltage is moving in the positive direction. As a result more voltage-gated calcium channels open, and this allows more calcium ions to flood in. This is an example of positive feedback (page 280) because every calcium ion that moves in makes the cytosol more positive, causing more calcium channels to open, and therefore making it easier for the next calcium ion to move in. At *D* all the calcium channels are open, the inward calcium current is much greater than the outward potassium current, and the plasmalemma is rapidly depolarizing.

However after 100 ms the voltage-gated calcium channels begin to inactivate, and the inward flow of calcium ions stops. The plasmalemma stops depolarizing and returns to -70 mV as potassium leaves through its channels.

This experiment shows the critical feature of an action potential: It is all or nothing. The depolarization at *B* was too small to generate an action potential, and the transmembrane voltage simply fell back to its resting state. At *C* the depolarization was big enough to start the process, and the action potential took off in an explosive self-amplifying way until all the voltage-gated calcium channels were open and the plasmalemma had greatly depolarized. In all excitable cells there is a *threshold* for initiating an action potential; in the sea urchin egg it lies between -60 mV and -50 mV. Depolarizations to below the threshold elicit nothing; depolarizations to voltages more positive than the threshold elicit the complete action potential. At the heart of the action potential is a "vicious circle." Depolarization causes voltage-gated calcium channels to open, and open calcium channels cause depolarization. The scientific name for such a cycle of interactions is positive feedback.

In nature the initial depolarization that sets off an action potential is produced by the sperm. This is illustrated at *E*. The membrane of the sperm contains open channels. The nature of these channels is at present unknown, but their effect is to depolarize the plasmalemma of the egg by 20 mV to -50 mV, enough to cause the critical number of voltage-gated calcium channels to open. The mechanism of the action potential takes over, and the plasmalemma depolarizes rapidly to +10 mV. The graph has been cut off at the right-hand edge (where the voltage gated calcium channels inactivate) where other mechanisms, not described here, take over and maintain a positive transmembrane voltage. The action potential has served its purpose in preventing second and subsequent sperm fusions in the 100 ms following the first fertilization.

THE VOLTAGE-GATED SODIUM CHANNEL IN NERVE CELLS

The voltage-gated calcium channel appeared early in evolution and is found in all eukaryotes and in a wide variety of cell types. Multicellular animals have a second, related channel that is selective for sodium instead of calcium ions. It operates like the voltage-gated calcium channel, but both opening and inactivation are faster (Fig. 18.4). In summary:

1. When the transmembrane voltage is -70 mV, the sodium channel is gated shut.
2. When the plasmalemma is depolarized, the channel opens rapidly and then, after about 1 ms, inactivates.
3. After the channel has gone through this cycle, it must spend at least 1 ms with the transmembrane voltage at the resting voltage before it can be opened by a second depolarization.

Like calcium, sodium is at a much higher concentration outside the cell than inside. In a mammal, the sodium concentration in the blood is 140 mmol liter^{-1}, whereas in the cytosol it is 10 mmol liter^{-1}. When sodium channels open, sodium ions rush into the cell carrying positive charge and depolarizing the plasmalemma. This then favors the opening of more sodium channels, and so on. This positive feedback produces a depolarization to +30 mV—the sodium action potential. Because the sodium channels inactivate so quickly, the sodium-based action potential lasts only for 1 ms. Not surprisingly, the sodium channel

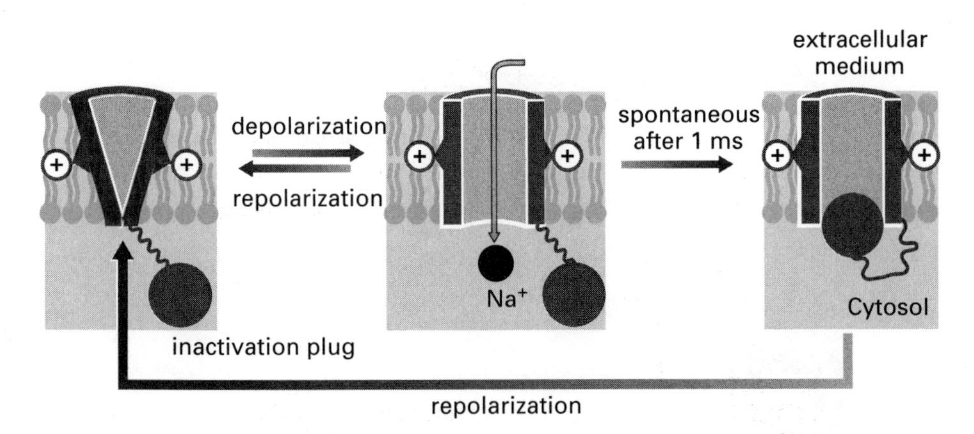

Figure 18.4
The sodium channel.

is found in cells that are specialized for rapid signaling like nerve and muscle cells.

Figure 18.5 shows how one nerve cell, a pain receptor, uses sodium-based action potentials to signal to the central nervous system. The cell body is close to the spinal cord and extends an *axon* that branches to the spinal cord and out to the body. The particular pain receptor illustrated in Fig. 18.5 sends its axon almost a meter to the tip of a finger, an extraordinary distance for a single cell. Potentially damaging events are detected at the finger and the message is passed on to another nerve cell (a pain relay cell) in the spinal cord. We will now explain how this function is performed.

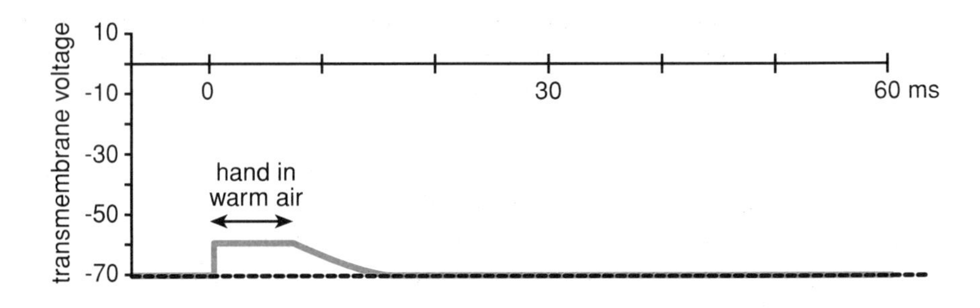

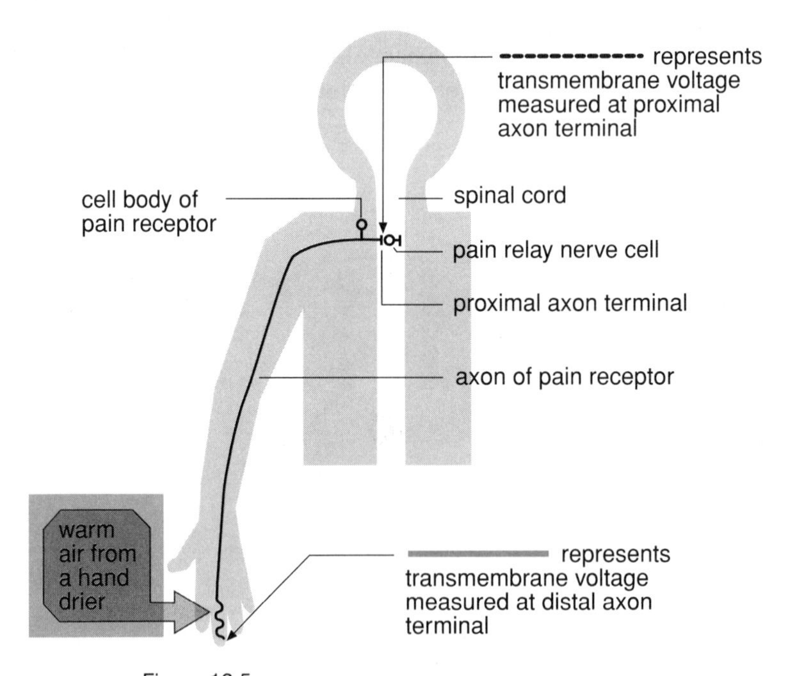

Figure 18.5*a*

Sodium action potentials carry sensations from the skin.

Figure 18.5*a* shows the transmembrane voltage of the pain receptor at the two axon terminals, one near the tip of the finger (blue line) and one at the opposite end of the cell, in the spinal cord (dashed black line). The axon terminal in the skin is the distal, or far away, terminal and that in the spinal cord is the proximal one. When the finger is passed through the warm air from the hand dryer, the heat causes the plasmalemma of the distal terminal to depolarize by 10 mV. This is not enough to open more than a very few sodium channels, since the transmembrane voltage is below threshold. As soon as the finger leaves the warm air jet, the transmembrane voltage sags back to -70 mV under the influence of the potassium channels. The axon is so long that the small

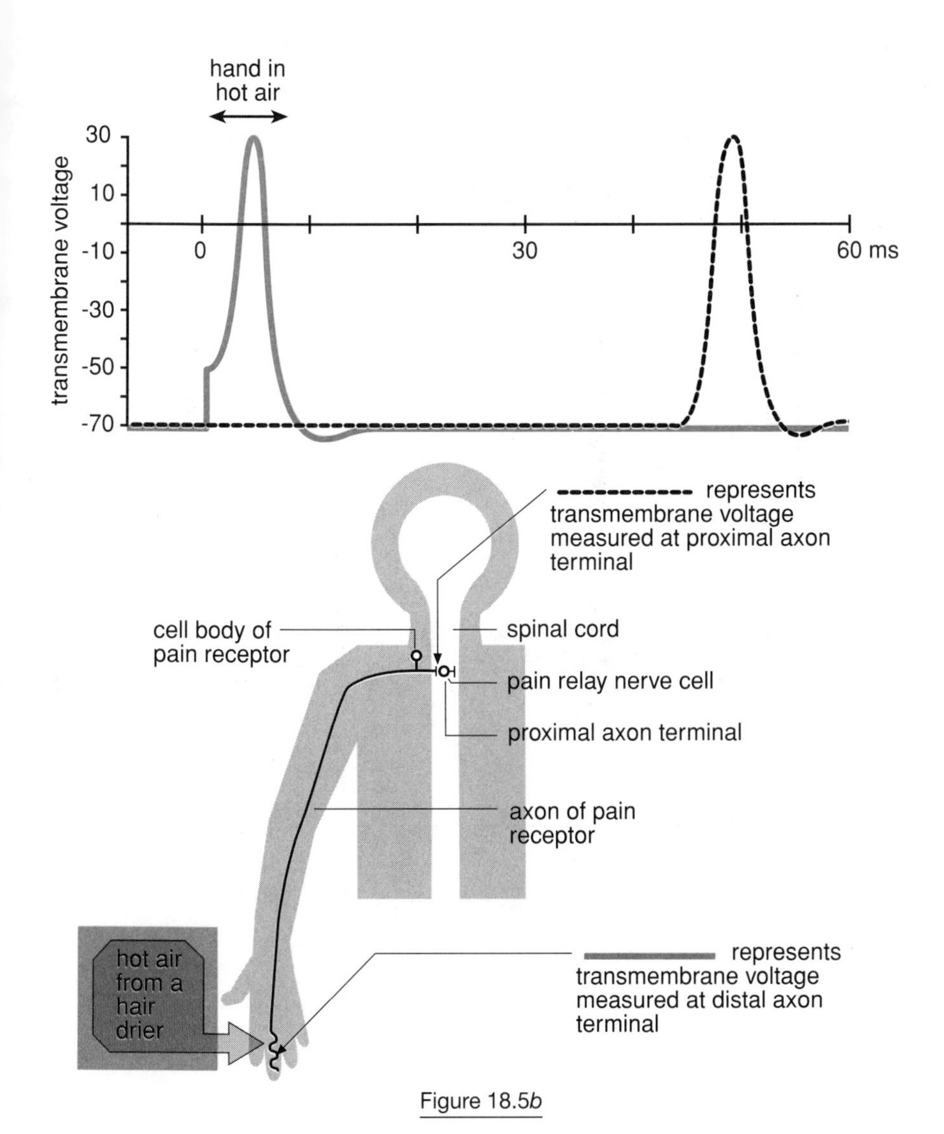

Figure 18.5*b*

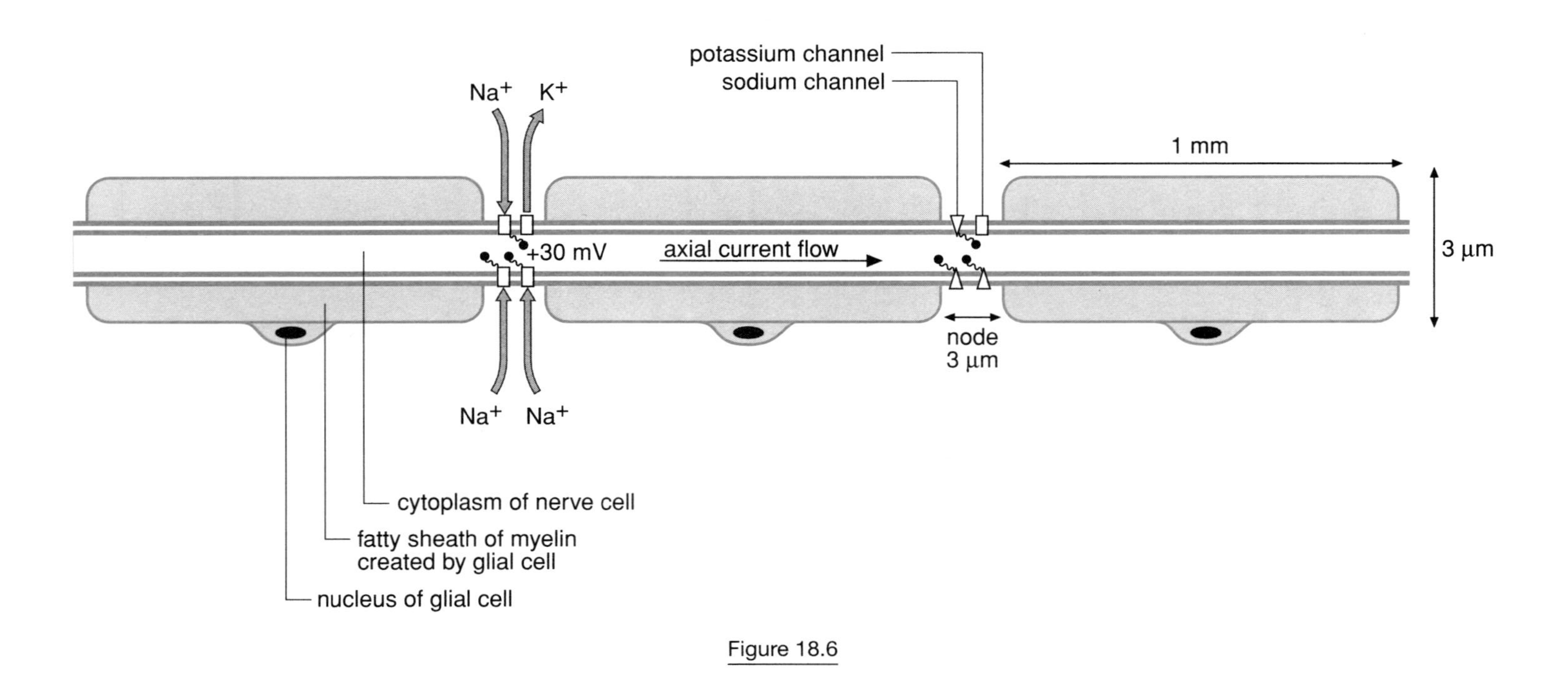
potassium channel
sodium channel
Na+
K+
+30 mV
axial current flow
1 mm
3 µm
node
3 µm
Na+
Na+
cytoplasm of nerve cell
fatty sheath of myelin
created by glial cell
nucleus of glial cell

Figure 18.6

voltage change at the distal terminal does not affect the voltage at the proximal terminal at all. No signal passes to the pain relay cell, and the subject feels no sensation of pain.

Figure 18.5*b* shows what happens if we pass our finger through hotter air. The depolarization caused by the heat is greater and more sodium channels open. Just as in the sea urchin egg at fertilization, the plasmalemma then depolarizes rapidly (in this case as more and more sodium enters) until the transmembrane voltage is +30 mV. After only 1 ms, the sodium channels inactivate and the transmembrane voltage returns to -70 mV, thanks to the action of the potassium channels. At the proximal terminal (dashed black line), a meter away, there is also an action potential after a short delay. We describe later (page 318) how the signal passes on to the brain and the person notices the hot air in time to pull the hand away before too much damage is caused. The pain receptor is but one example of how we are quickly made aware of changes in our environment by sodium action potentials travelling rapidly along nerve cell axons.

Myelination and Rapid Action Potential Transmission

The axon transmits the action potential so rapidly because for most of its length it is insulated by a fatty sheath called myelin made by glial cells. Figure 18.6 shows part of the axon of the pain receptor and its associated glial cells, each of which wraps around the axon to form an electrically insulating sheath. Only at short gaps called *nodes* is the membrane of the nerve cell exposed to extracellular fluid. Note that the vertical and horizontal scales of this diagram are completely different. The axon together with its myelin sheath is only 3 μm across, but the section of axon between nodes is 1mm long, a large distance by normal cell standards. Between nodes the axon is an insulated electrical cable: just as voltage changes at one end of an insulated metal wire have an almost instantaneous effect on the voltage at the other end, so a change in transmembrane voltage at one node has an almost instantaneous effect at the next node along. The plasmalemma at the nodes has potassium and sodium channels and can therefore generate action potentials.

Figure 18.6 represents a snapshot of one brief instant in the operation of the nerve cell. The node on the left is generating an action potential. Sodium channels are open, sodium ions are flooding in, and the transmembrane voltage is +30 mV. Some of the charge flowing in is carried out again by potassium ions flowing through the potassium channels. However, some positive charge flows axially up and down the cytosol of the nerve cell. Because myelin is an electrical insulator, this charge cannot leave the cytosol until the next node. At the node on the right, the incoming charge depolarizes the plasmalemma to threshold, and sodium channels at this node open as well. Sodium rushes in, and this node generates an action potential. The action potential has jumped from node to node in the extremely short time of 50 μs. Now sodium ions will rush into the node on the right, current will pass axially up the interior of the axon to the next node up, and in a further 50 μs the action potential will have jumped to the next node, and so on, all the way up to the proximal terminal at an overall speed of 20 meters per second. This process, in which the action potential jumps from node to node, is named after the Latin for "to jump," *saltere*, so it is saltatory conduction. The fastest axons in our bodies conduct at 60 meters per second.

Nerve conduction is exploited when a dentist uses local anaesthetic. A patient feels pain from the drill because pain receptors in the tooth are depolarized and transmit action potentials to the brain. The site at which they are being depolarized is inside the tooth and therefore inaccessible to drugs until the drill has made a hole. However, the axons of the pain receptors run through the gum. Local anaesthetics injected into the gum close to the axon bind to the nerve cell membrane at the node and prevent the opening of sodium channels. Drilling into the tooth still depolarizes the pain receptor membrane, and action potentials begin their journey toward the brain, but they cannot pass the injection site because the nodes there cannot generate an action potential. The patient therefore feels no pain.

SUMMARY

1. The voltage-gated calcium channel is shut at the resting voltage but opens upon depolarization. After about 100 ms the channel inactivates.

2. While the calcium channel is open, calcium ions pour into the cell down their electrochemical gradient. The mutual effect of current through the channel on transmembrane voltage, and of transmembrane voltage on current through the channel, constitutes a positive feedback system. If the membrane is initially depolarized to threshold, the positive feedback of the calcium current system ensures that depolarization continues in an all-or-nothing fashion. Repolarization occurs when the calcium channel inactivates. The entire cycle of depolarization and repolarization is called an action potential.

3. The same process occurs in cells expressing voltage-gated sodium channels. But, since these channels inactivate within about 1 ms, the action potential lasts only this long. Long nerve cell processes called axons transmit action potentials at speeds up to 60 meters per second. They can transmit the signal at such high rates because myelin, a fatty sheath, insulates the 1 mm distances between nodes.

BOX 18.1

Frequency Coding in the Nervous System

At the distal axon terminal of the pain receptor, the intensity of the stimulus is represented by the amount of depolarization produced. The hotter the air blowing on the skin, the more positive the transmembrane voltage of the terminal becomes. Other sensory cells behave in the same way. For instance, in Chapter 19 we will meet the scent-sensitive nerve cells in the nose that detect smell chemicals. The higher the concentration of the smell chemical, the more these cells depolarize.

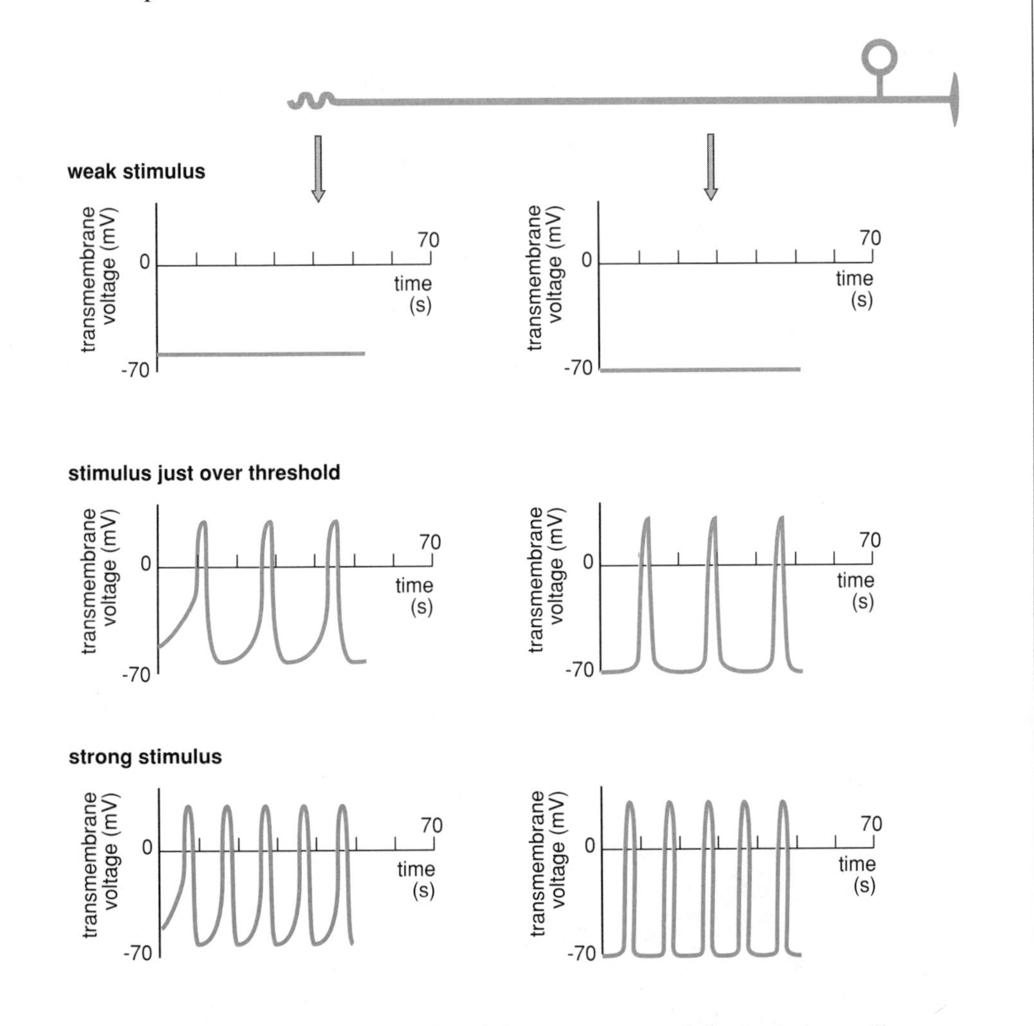

However, we have seen that the signal that passes toward the brain is an all-or-none action potential. Either the stimulus produces a depolarization that is below threshold, in which case no signal passes to the brain, or the membrane depolarizes beyond threshold, in which case an action potential is generated. How then is the brain told about the intensity of the stimulus?

Action potentials continue to be produced and travel up to the brain as long as the stimulus is maintained. The diagrams above represent what happens when the hand is being held in a constant air stream. If the stimulus is strong, the membrane of the pain receptor depolarizes rapidly, so that after each action potential threshold is soon reached once more. Thus the time between each action potential is short. The stronger the stimulus, the higher the frequency of action potentials generated.

AM and FM are familiar terms from radio. Here we are seeing exactly the same two coding strategies in the nervous system. AM, amplitude modulation, is used in the distal terminals of the pain receptor. The strength of stimulus is coded for by the amplitude of the depolarization. FM, frequency modulation, is used in almost all axons, including those of the pain receptor. The strength of stimulus is coded for by the frequency of action potentials, each of which has the same amplitude.

Man-made systems use exactly the same strategy. In a computer, for instance, the joystick, microphone, and mouse code the stimulus in terms of the amplitude of a voltage change. This is then recoded in all-or-none digital pulses for transmission to the processor.

CHAPTER

19

INTRACELLULAR MESSENGERS

The behavior of cells is not constant. Cells change their behavior in response to internal changes or in response to external events, and these responses involve proteins and organelles at different locations in the cell. The signals that tell proteins and organelles to change their activity are carried by special cytosolic solutes called intracellular messengers. We will begin by continuing the story begun in Chapter 18 and explaining how calcium ions carry a signal from the plasmalemma of nerve cells to vesicles deep within the cytosol. The rest of the chapter will describe other cells where calcium is an intracellular messenger and will introduce two more intracellular messengers: cyclic adenosine monophosphate and cyclic guanosine monophosphate.

CALCIUM

Calcium ions are present at a very low concentration (about 100 nmol $liter^{-1}$) in the cytosol of a resting cell. An enormous number of processes in many types of cells are activated when the concentration of calcium rises. Calcium can move into the cytosol from two sources: the extracellular medium or the endoplasmic reticulum.

Calcium Can Enter from the Extracellular Medium: Exocytosis at the Axon Terminal

In Chapter 18 we saw how heating of the hand causes action potentials to travel from the hand to the spinal cord along the axons of pain receptors. When the action potential reaches the proximal axon terminal, the amino acid glutamate is released onto the surface of another nerve cell, the pain relay cell. Glutamate stimulates the pain relay nerve cell so that the message that a finger is being damaged is passed toward the brain.

Before going any further, we need to define two terms. An *agonist* is a solute that stimulates a cell. An agonist that is released by one cell to affect another cell is a *transmitter*. Glutamate is both an agonist and a transmitter because it is released by the pain receptor and stimulates the pain relay nerve cell. Glutamate in the form of monosodium glutamate (MSG) is traditionally added to Chinese food, and in excess can produce a condition of headaches and thirst known as Chinese restaurant syndrome.

The proximal axon terminal of the nerve cell is unmyelinated, so the plasmalemma is exposed to the extracellular medium (Fig. 19.1). The membrane contains not only potassium channels and voltage-gated sodium channels but also voltage-gated calcium channels. These are closed in a resting cell, but when an action potential travels in from the skin and depolarizes the plasmalemma of the proximal axon terminal, the voltage-gated calcium channels open. Calcium ions pour in, increasing their concentration in the cytosol by ten times, from the normal concentration of 100 nmol liter^{-1} to 1 μmol liter^{-1}. At the proximal axon terminal the cytosol of the nerve cell contains regulated exocytotic vesicles (page 205). In the case of the pain receptor, these vesicles are filled with sodium glutamate. The special property of regulated exocytotic vesicles is that when the concentration of cytosolic calcium shoots up, they move to the plasmalemma and fuse with it, releasing their contents into the extracellular medium. The glutamate then diffuses across the gap to the pain relay cell, stimulating it (in the manner described on page 341). As the action potential in the axon terminal of the pain receptor cell is over in 1 ms, the voltage-gated calcium channels do not have time to inactivate. They simply return to the ready-to-open state, where they can be re-opened immediately by the next action potential that arrives.

In the process that we have described, calcium ions act as a link between the depolarization of the plasmalemma and the regulated exocytotic vesicles within the cytosol. Calcium ions are intracellular messengers. Regulated exocytotic vesicles that are triggered by a rise of cytosolic calcium concentration were first discovered in nerve cells but are now known to be a feature of almost all cells.

Before we continue with the topic of intracellular messengers, we will

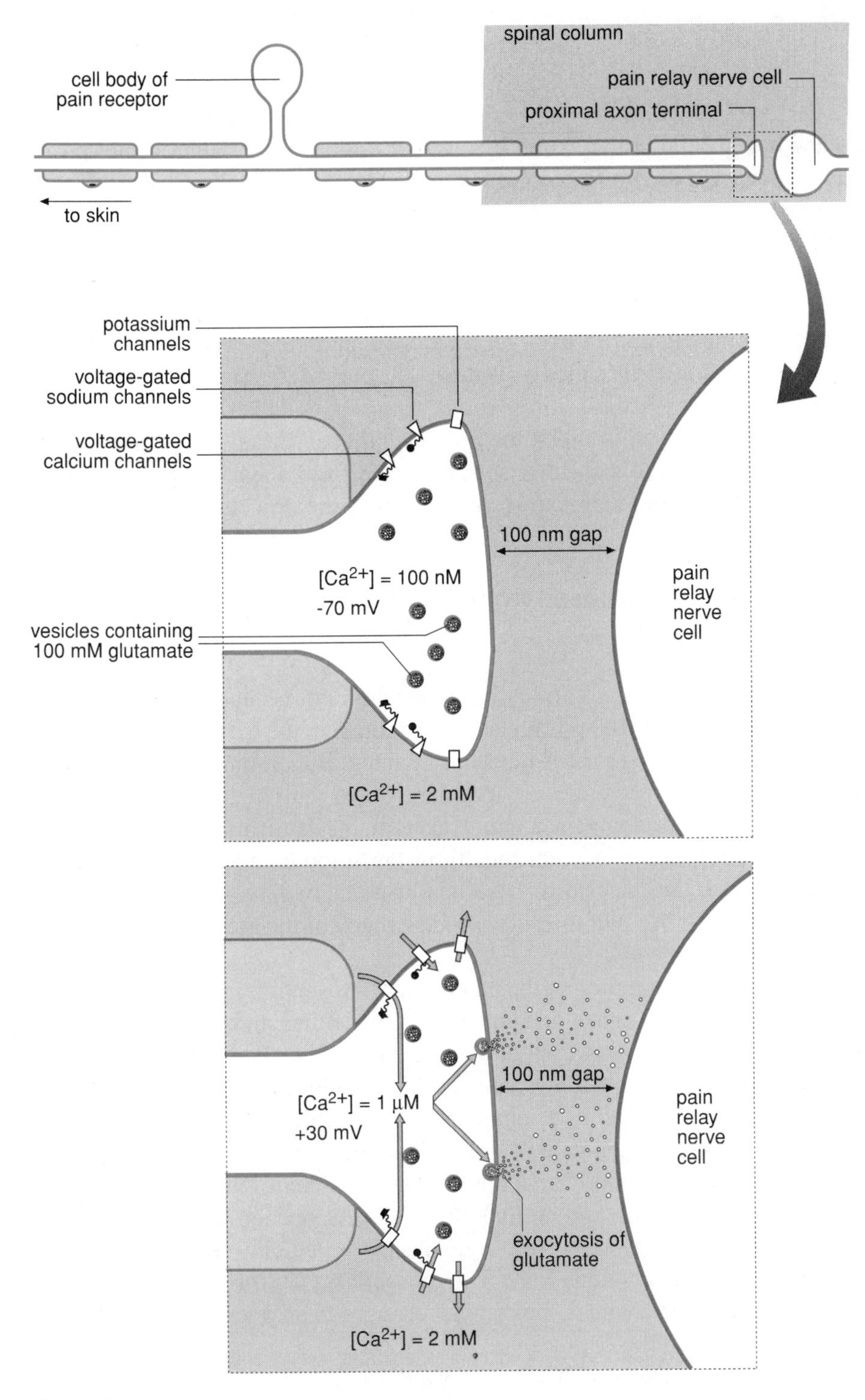

Figure 19.1

Calcium ions entering the cytosol from the extracellular fluid activate exocytosis in the proximal axon terminal of the pain receptor cell.

review two general points raised by our description of the pain receptor proximal axon terminal. The first point is that different nerve cells release different transmitters. Glutamate is in the exocytotic vesicles of many nerve cells, but other nerve cells release other transmitters (Chapter 20). The second point concerns nomenclature. Many nerve cells, like the pain receptor, have their axon terminals close to a second cell and release their transmitter onto it. In such cases the complete unit of axon terminal, gap, and the part of the cell that receives the transmitter is called a *synapse*. The part of the axon terminal that releases transmitter is called a *presynaptic terminal*, and the cell upon which the transmitter is released is called the *postsynaptic cell*. Many nerve cells do not come close enough to a specific second cell to form a synapse but simply release transmitter from their axon terminals into the extracellular medium.

Calcium Can Be Released from the Endoplasmic Reticulum: Platelet Activation

Cells may show an increase of cytosolic calcium not because of an action potential but because of the appearance of an agonist in the extracellular medium. The presence of the agonist is detected by integral membrane proteins—each one a receptor that recognizes a particular agonist with high specificity. These receptors then participate in a more general mechanism, the end result of which is the release of calcium ions from the smooth endoplasmic reticulum into the cytosol. This mechanism will be illustrated with the particular example of blood platelets. We will then discuss how much of the mechanism is general to a wider range of cells.

Platelets are common in the blood. They are small fragments of cells and contain no nucleus, but they do contain a plasmalemma and some endoplasmic reticulum. Blood platelets use the release of calcium from the endoplasmic reticulum as one step in the mechanism of blood clotting (Fig. 19.2). Two new mechanisms are involved in calcium release from the endoplasmic reticulum. The inositol trisphosphate-gated calcium channel (Fig. 19.2, on page 322), like the voltage-gated calcium channel, allows only calcium ions to pass. Most of the time its gate is shut, and no ions flow. It is not opened by a change in the transmembrane voltage across the endoplasmic reticulum membrane. Instead, when the intracellular solute *inositol trisphosphate* (*IP_3* for short) binds to the cytosolic face of the channel, the channel changes to an open shape. This channel is usually called the *IP_3 receptor*.

The second new mechanism that we must describe is the one that makes IP_3.

If a blood vessel is cut, cytosol from damaged cells at the edge of the cut can leak into the blood. The appearance of solutes that normally are found only inside cells is a sure sign that damage has occurred. Adenosine diphosphate (ADP) is one such solute, and it acts as an agonist on platelets, causing them to begin a blood clot to help plug the damaged vessel. The plasmalemma of the platelet contains a protein receptor that binds ADP, that is, ADP is its ligand. When the ADP has bound, the receptor, which is free to move in the plasmalemma, associates with and activates an enzyme called phosphatidylinositol phospholipase C, which we will call *phospholipase C* or just *PLC*. PLC hydrolyzes *phosphatidylinositol bisphosphate* (*PIP_2* for short), a phospholipid in the plas-

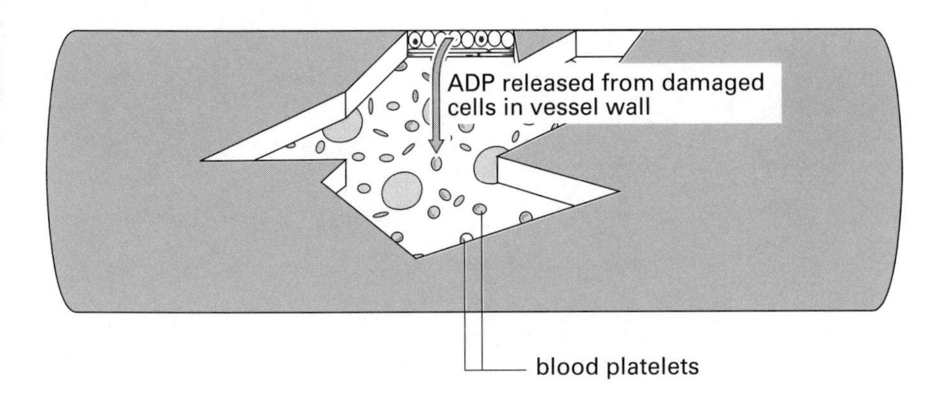

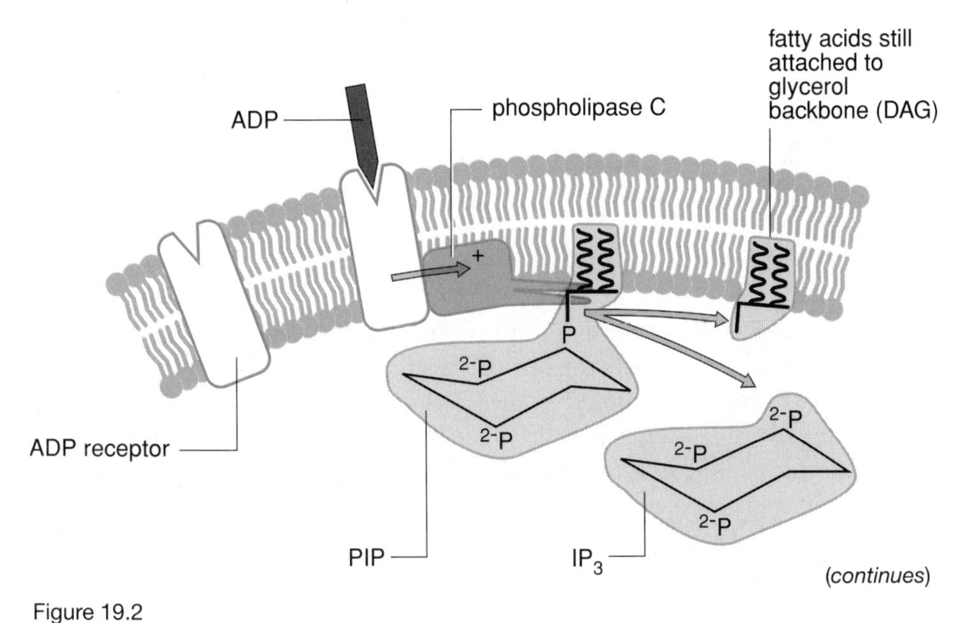

(continues)

Figure 19.2

In the initial stages of blood clotting, inositol trisphosphate is cleaved from a plasmalemmal lipid and causes release of calcium ions from the endoplasmic reticulum.

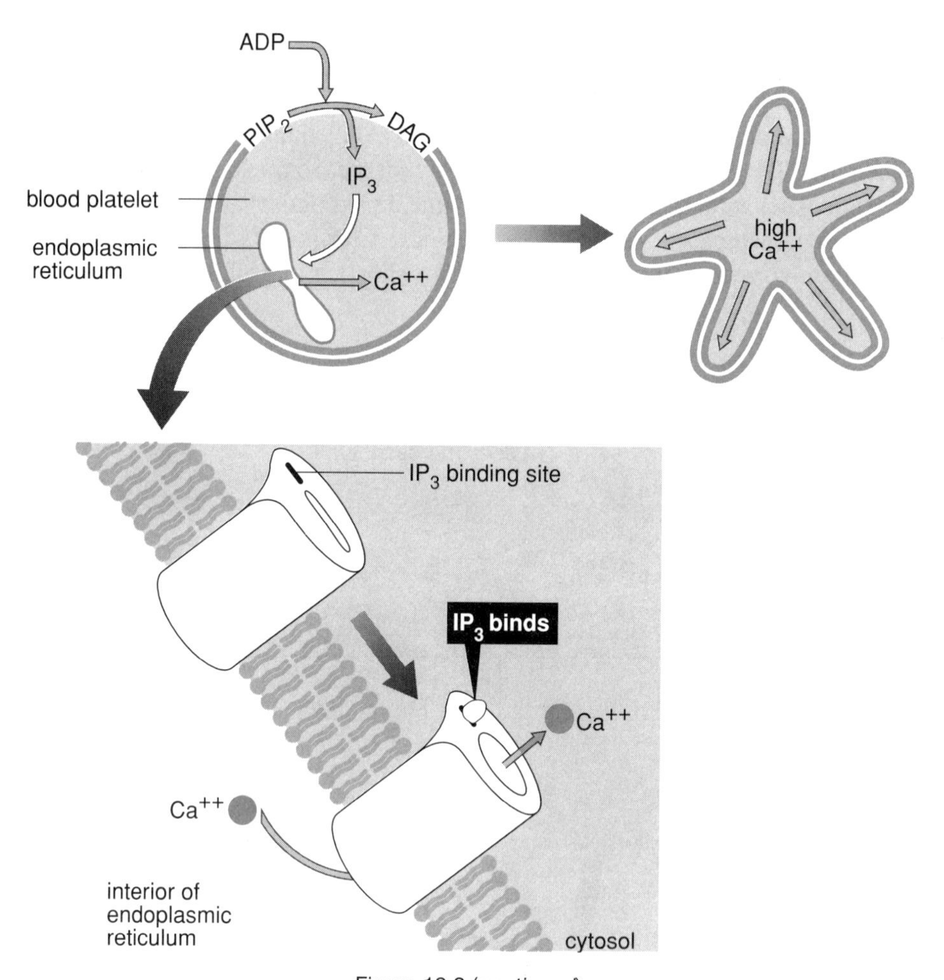

Figure 19.2 (*continued*)

malemma that has IP_3 as its polar head group. Hydrolysis releases the IP_3 to diffuse freely in the cytosol, leaving behind the glycerol backbone with its attached fatty acids (This residual lipid is called diacyglycerol or DAG). As it diffuses through the cytosol, IP_3 reaches the endoplasmic reticulum and binds to the inositol trisphosphate-gated calcium channels. The inositol trisphosphate-gated calcium channels open, and calcium ions pour out of the endoplasmic reticulum into the cytosol. The rise of calcium concentration causes the platelet to change shape and to become very sticky, so platelets begin to clump together in a clot.

Phosphatidylinositol bisphosphate, PLC, and the inositol trisphosphate-gated calcium channel are found in almost all eukaryotic cells, but only platelets are sensitive to extracellular ADP, because only platelets have the ADP receptor in their plasmalemmae. Other cells respond to other agonists. Each produces a receptor specific for that agonist which then activates PLC. Over a hundred such receptors are known. We will meet two more in the next chapter.

The Processes Activated by Cytosolic Calcium are Extremely Diverse

The targets activated by a rise of cytosolic calcium differ between different cells. A rise of cytosolic calcium is a crude signal that says "do it" but contains no information about what the cell should do. This depends on what the cell was designed to do. Cells and cell regions designed for exocytosis (e.g., salivary gland cells and axon terminals) exocytose when cytosolic calcium rises, cells designed to contract (e.g., muscle cells) contract when calcium rises, cells waiting in the G1 phase of the cell cycle begin DNA synthesis, and so on. In each case the calcium ions bind to a calcium-binding protein and the calcium ion: calcium-binding protein complex activates the target process.

In skeletal muscle fibers (page 13) a transmitter released from nerve axon terminals leads to the escape of calcium from the endoplasmic reticulum (Fig. 19.3). This activates several processes. First, calcium ions bind to a protein called troponin that is attached to the cytoskeleton. This causes the cytoskeleton to contract, using the energy released by ATP hydrolysis to do mechanical work. Second, the calcium binds to the protein calmodulin, which in turn activates phosphorylase kinase and hence glycogen breakdown as part of the feedforward control of energy metabolism (page 281).

Mitochondria are also affected by calcium, which passes down its electrochemical gradient from the cytosol into the mitochondrial matrix through a calcium channel. Once there, calcium stimulates the mitochondria to increase the production of NADH and ATP (page 281).

The same simple intracellular messenger, calcium, has many different actions inside the skeletal muscle cell. Under its influence the cytoskeleton begins to use ATP, glycogen phosphorylase releases more glucose, and the mitochondria produce more ATP. The skeletal muscle cell is an exquisite example of how diverse mechanisms inside a cell can be integrated by the action of intracellular messengers.

Return of Cytosolic Calcium to Resting Levels

As soon as a stimulus, be it depolarization or extracellular agonist, disappears, cytosolic calcium concentration falls again. Calcium ions are pumped up their electrochemical gradients from the cytosol into the extracellular medium or

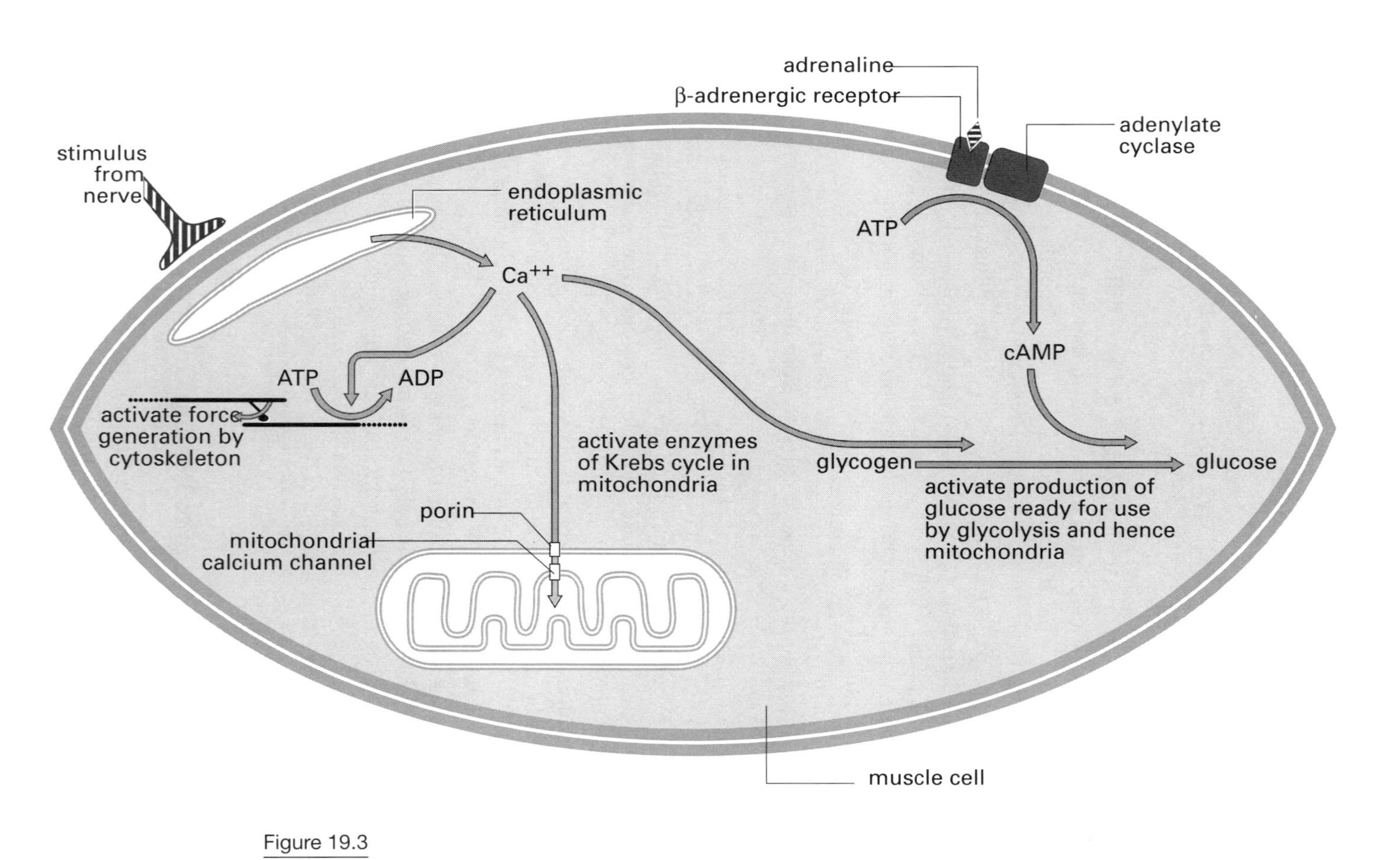

Figure 19.3
Calcium and cyclic AMP activate distinct but overlapping sets of target processes in skeletal

into the endoplasmic reticulum by two carriers. The Ca^{++} ATPase uses the energy released by ATP hydrolysis to move calcium ions up their electrochemical gradient out of the cytosol, while the Na^{+}/Ca^{++} exchanger uses the energy released by sodium ion movement into the cytosol to do the same job.

CYCLIC ADENOSINE MONOPHOSPHATE (cAMP)

We have already met the nucleotide cyclic adenosine monophosphate or cAMP in the context of the regulation of the lac operon (page 155). In eukaryotes, cAMP is also important and acts as an intracellular messenger in a great many cells, including the scent-sensitive nerve cells in our nose (Fig. 19.4). These cells have their cell bodies in the skin of the air passages in the nose. Each cell sends an axon into the brain and shorter processes (dendrites) into the mucus lining the air passages. Scent-sensitive nerve cells are stimulated by scents in the air. Particular chemicals in the air act as agonists on these cells because the cells have protein receptors that specifically bind the scent. When the scent binds, the receptor activates an enzyme called *adenylate cyclase* that converts ATP to cAMP. The next stage in the detection of scents is a channel in the plasmalemma called the cAMP-gated channel. Like the inositol trisphosphate-gated calcium channel, the cAMP gated channel is opened by a cytosolic solute, in this case cAMP. When the channel is open, it allows sodium and potassium ions to pass through. The electrochemical gradient pushing sodium ions into the cell is much greater than the electrochemical gradient pushing potassium ions out of the cell. Thus, when the cAMP-gated channel opens, sodium ions pour in, carrying their positive charge and depolarizing the plasmalemma. The plasmalemma also contains voltage-gated sodium channels. Thus, when enough cAMP-gated channels open, the transmembrane voltage reaches threshold for generation of sodium action potentials. These then propagate along the axon to the brain, and the person becomes aware of the particular scent that was an agonist for that scent-sensitive nerve cell.

Adenylate cyclase is found in many cells, but only scent sensitive nerve cells are sensitive to scents, because only they have specific scent receptors in their plasmalemmae. Other cells that use cAMP as an intracellular messenger are sensitive to other specific agonists because each makes a receptor that binds an agonist, whatever it may be, and then activates adenylate cyclase.

Most of the symptoms of the deadly disease cholera are caused by a toxin released by the bacterium *Vibrio cholera*. The toxin is an enzyme that enters the cytosol of many cells and turns adenylate cyclase fully and permanently on. The cAMP concentration in the cytosol then shoots up. This does not much matter in scent-sensitive nerve cells. Unfortunately, ion channels in the cells of the gut are also opened by a rise of cAMP, allowing ions to leak from the cells

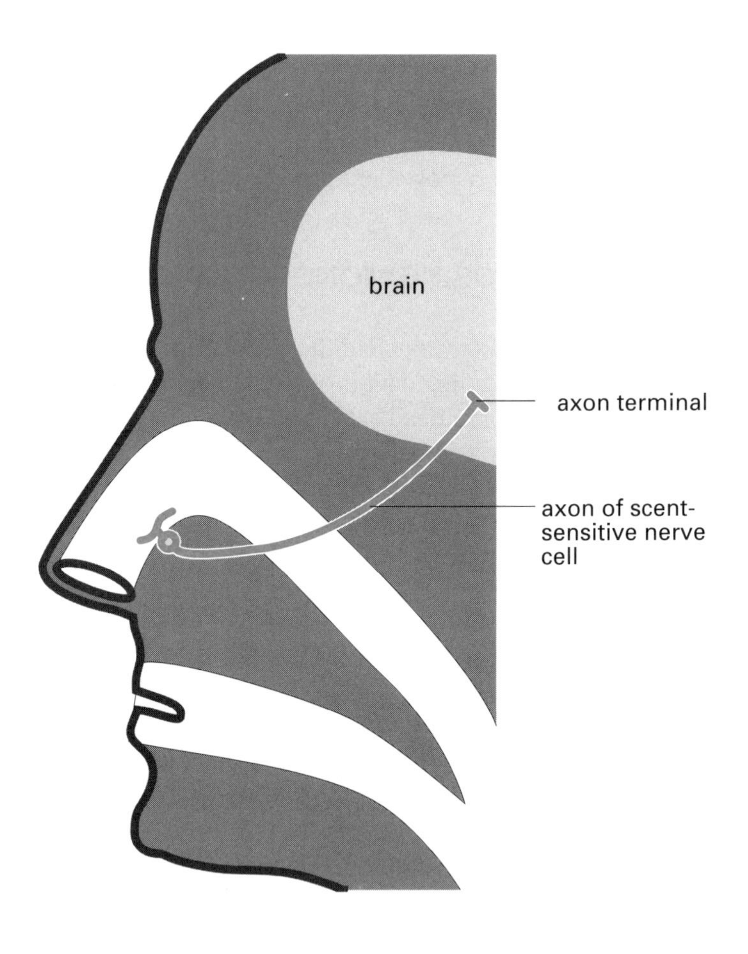

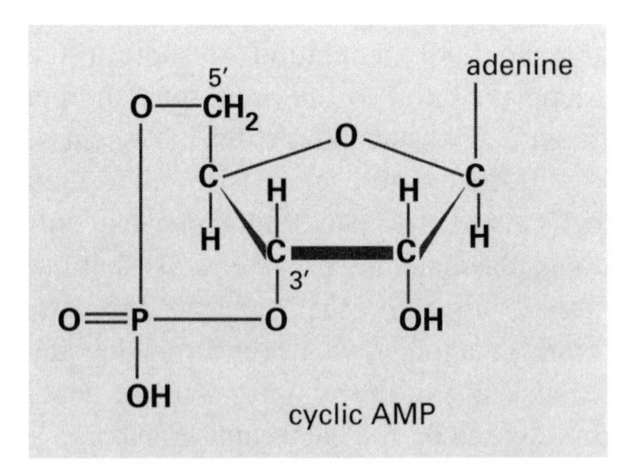

Figure 19.4

Cyclic AMP is the intracellular messenger produced in response to scent chemicals by scent-sensitive nerve cells.

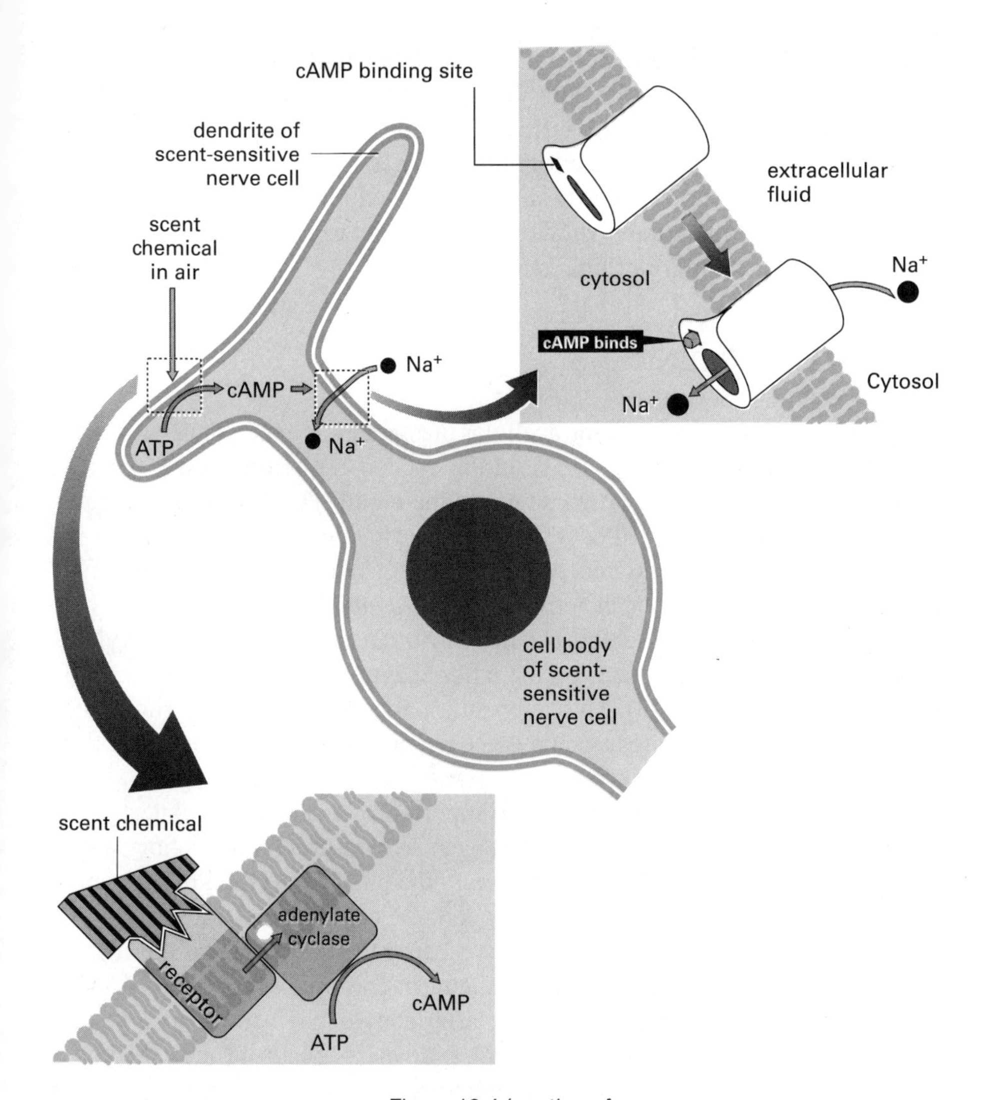

Figure 19.4 (*continued*)

into the gut contents. If untreated this loss of ions and of the water that accompanies them leads to death from dehydration.

CYCLIC GUANOSINE MONOPHOSPHATE (cGMP)

Another nucleotide that acts as a second messenger is cGMP. In light-sensitive nerve cells (photoreceptors), cGMP plays a role like that of cAMP in scent-sensitive nerve cells. In the dark the enzyme guanylate cyclase makes cGMP from GTP. cGMP binds to and opens a channel that allows sodium and potassium ions to pass, so that in the dark the photoreceptor is depolarized to about –40 mV because of the constant influx of sodium ions through the cGMP-gated channel. In the light the concentration of cGMP in the cytosol falls, and the transmembrane voltage changes to the more typical resting voltage of -70 mV. The changing transmembrane voltage is transmitted to other nerve cells, making us aware of the pattern of light and dark.

MULTIPLE MESSENGERS

Many cells use more than one intracellular messenger at once. Skeletal muscle cells are a good example (Fig. 19.3). In the excitement before a race, the runner's adrenal glands release adrenaline into the blood. This binds to a receptor on skeletal muscle cells called the β-adrenergic receptor. The complex of adrenaline plus β-adrenergic receptor can now activate adenylate cyclase. This in turn generates cAMP, which activates *cAMP-dependent protein kinase*. This enzyme, which is given the short name of *protein kinase A*, phosphorylates phosphorylase kinase and turns on the latter enzyme even when cytosolic calcium is low (see Fig. 16.13). The end result is that even before the runner begins to run, the muscles break down glycogen and make the glucose that they will need once the race begins.

The last example illustrates two important general points. The first is that many eukaryotic intracellular messengers act via protein phosphorylation. Many of the actions of cAMP occur through the phosphorylation of other proteins by protein kinase A. In particular, cAMP can activate transcription of many eukaryote genes, but it does so by causing the phosphorylation of transcription factors, and hence activating them, in sharp contrast to the cAMP-CAP system in *E. coli* (page 155). Similarly many of the actions of cytosolic calcium—for instance, activation of DNA synthesis—occur through the phosphorylation of other proteins by calcium-activated kinases. The second general point is that second messenger systems show crosstalk: one messenger can produce some or all of the effects of the other. In skeletal muscle cells cAMP mimics the effect of cytosolic calcium in causing glycogen breakdown but (unlike calcium) does not cause contraction.

SUMMARY

1. An agonist is any extracellular solute that stimulates a cell.

2. An intracellular messenger is an intracellular solute whose concentration changes in response to cell stimulation; it activates or modu-

lates a variety of cellular processes. The most important intracellular messengers are calcium ions, cyclic adenosine monophosphate (cAMP), and cyclic guanosine monophosphate (cGMP).

3. The increased cytosolic calcium may be derived from two sources: the extracellular fluid, or the smooth endoplasmic reticulum.

4. A complex cascade of reactions causes release of calcium from the endoplasmic reticulum in response to stimulation of the cell. Binding of agonist to a cell surface receptor activates the enzyme phospholipase C, which cleaves the hydrophilic head group inositol trisphosphate (IP_3) from the lipid phosphatidylinositol bisphosphate. Inositol trisphosphate binds to and opens a calcium channel in the membrane of the endoplasmic reticulum called the IP_3 receptor.

5. Many of the actions of calcium are mediated through the calcium-binding protein calmodulin, and a large subset of these are mediated by calmodulin-dependent protein kinases.

6. A second enzyme activated by the binding of ligand to a cell surface receptor is adenylate cyclase, which makes cyclic AMP. Many of the actions of cAMP are mediated by cAMP-dependent protein kinase (protein kinase A).

7. The majority of transmitters are released from cells by exocytosis induced by a rise of cytoplasmic calcium concentration.

8. Many nerve cells have their axon terminals close to a second cell, and they release their transmitter onto it. The complete unit of axon terminal, gap, and the part of the cell that receives the transmitter is called a synapse. The part of the axon terminal that releases transmitter is called a presynaptic terminal, and the cell upon which the transmitter is released is called the postsynaptic cell.

BOX 19.1

Caffeine and Ryanodine Receptors

A article in the *New York Times* once proposed that coffee, brought from Arabia to Europe, caused the Renaissance. Perhaps there is some truth to the joke, for it is certainly hard to imagine modern life without caffeine.

The inositol trisphosphate receptor is found in all eukaryotes. Vertebrates also possess a second, related calcium channel in the membrane of the endoplasmic reticulum. Two plant toxins bind to the channel: ryanodine, which has given the protein its name, the ryanodine receptor, and caffeine. Caffeine sensitizes this channel, and makes it more likely to open.

Ryanodine receptors are caused to open by a rise in the calcium concentration in the cytoplasm. When they open, calcium ions pour out of the endoplasmic reticulum into the cytoplasm, amplifying the original change in a process called calcium-induced calcium release. The released calcium can then activate the many calcium-dependent processes within the cell. By sensitizing the ryanodine receptors, caffeine makes our cells—and therefore our bodies—more responsive to stimuli.

CHAPTER

20

INTERCELLULAR COMMUNICATION

The millions of cells that make up a multicellular organism can work together only because they continually exchange chemical messages. Transmitters released from one cell act as agonists on others, changing their behavior. Here we describe how systems of cells use transmitters to co-operate for the good of the organism. Most of the many transmitter chemicals known are found in all animals and probably evolved with ancestral multicellular organisms more than a billion years ago. We will introduce the general principles of intercellular communication and then illustrate how transmitters operate in a single tissue, the gastrocnemius muscle.

CLASSIFYING TRANSMITTERS AND RECEPTORS

Transmitter mechanisms can be classified in two ways. The first depends on their longevity in the extracellular fluid. A transmitter that is rapidly broken down or taken up into cells acts only near its release site. One that is broken down slowly can diffuse a long way and may act on cells a long way away. The shortest-lived transmitters of all are those released at synapses (page 320), where the distance from release site to receptor is only 100 nm.

At the other extreme are transmitters that last many minutes or, sometimes, even longer. *Hormones* are long-lived transmitters that are released into the blood and travel around the body before being broken down. Most are released by specialized groups of secretory cells that form an endocrine gland. *Paracrine transmitters* also last many minutes before being broken down, but they are released into specific tissues rather than into the blood and only diffuse within the tissue before they are destroyed.

The second way of classifying transmitter mechanisms depends on the location and action of the receptor on the target cell. There are three types of receptors upon which agonists can act: ionotropic cell surface receptors, metabotropic cell surface receptors, and intracellular receptors.

Ionotropic Cell Surface Receptors

Ionotropic cell surface receptors are channels that open when a specific agonist binds to the extracellular face of the channel protein. The nicotinic acetylcholine receptor (Fig. 20.1), so named because the drug nicotine binds to it, is one example. In the absence of acetylcholine, the channel is closed. When acetylcholine binds, the channel opens and allows sodium and potassium ions to pass through. The electrochemical gradient pushing sodium ions into the cell is much greater than that pushing potassium ions out of the cell, so when the channel opens, sodium ions pour in carrying positive charge and depolarizing the plasmalemma. This mechanism is similar to that of two channels we met in Chapter 19, the inositol trisphosphate-gated calcium channel and the cAMP-gated channel. There is however a major difference: those two channels were opened by cytosolic solutes, while ionotropic cell surface receptors are opened by extracellular solutes.

Metabotropic Cell Surface Receptors

Metabotropic cell surface receptors are linked to enzymes—as we have already seen for the ADP receptor, the receptor for smell chemicals, and the β-adrenergic receptor. When the ADP receptor binds extracellular ADP, it causes a rise of cytosolic calcium concentration. The other two receptors are linked to adenylate cyclase, so an agonist binding to them increases cytosolic cAMP. The α-adrenergic receptor (Fig. 20.2) is another receptor that causes cytosolic calcium concentration to rise. The α- and β-adrenergic receptors are distinct proteins that bind the same ligands, adrenaline and noradrenaline. To simplify the issue somewhat, we can say that noradrenaline acts mainly on α receptors and adrenaline acts mainly on β receptors. Because the α and β receptors are distinct proteins, it is possible to design drugs (α and β blockers) that interfere with one or the other.

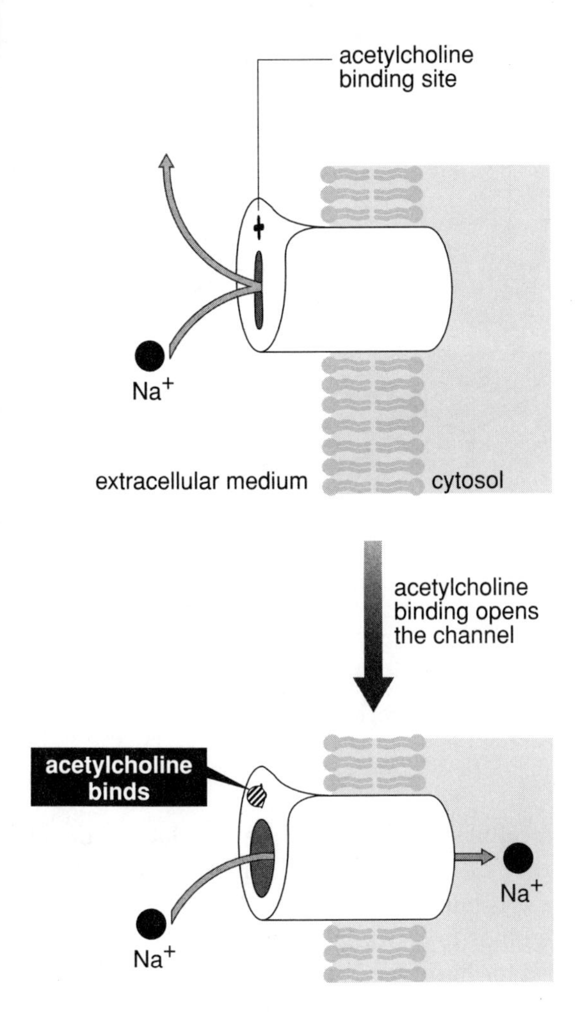

Figure 20.1
The nicotinic acetylcholine receptor is an ionotropic cell surface receptor that opens when acetylcholine in the extracellular fluid binds.

Intracellular Receptors

Intracellular receptors lie within the cell (in the cytosol or in the nucleus) and bind transmitters that diffuse through the plasmalemma. They always exert their effects by activating enzymes. The receptors for nitric oxide and steroid hormones are two examples.

Nitric oxide or NO is a transmitter in many tissues. It is not stored ready to be released but is made from arginine at the time it is needed. NO diffuses easily through the plasmalemma and binds to various cytosolic proteins that are NO receptors. One particularly important NO receptor is the enzyme guanylate cyclase which in the presence of NO converts the nucleotide guanosine triphosphate (GTP) to the intracellular messenger cyclic guanosine monophosphate or cGMP.

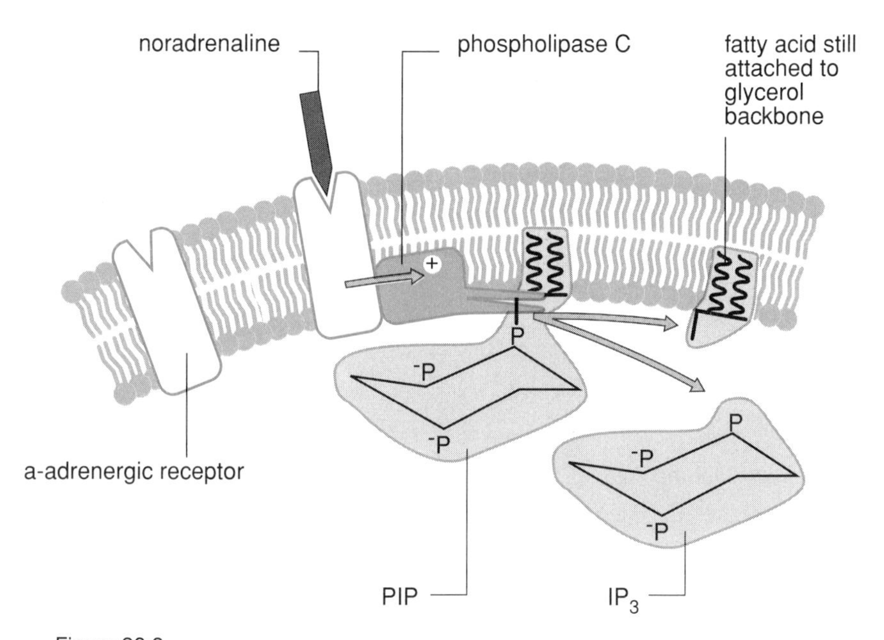

Figure 20.2

The α-adrenergic receptor is a metabotropic cell surface receptor that activates phospholipase C when noradrenaline in the extracellular fluid binds.

Steroid hormones have intracellular receptors such as the glucocorticoid receptor (Fig. 10.14). In the absence of hormone this receptor remains in the cytoplasm and is inactive because it is bound to an inhibitor protein. However, when the glucocorticoid hormone binds to its receptor, the inhibitor protein is displaced. The glucocorticoid receptor hormone complex now moves into the nucleus. Here two molecules of the complex bind to a 15-bp sequence known as the hormone response element (HRE), which lies upstream of the TATA box (page 160). The HRE is an enhancer sequence. The binding of the glucocorticoid hormone receptor to the HRE stimulates transcription (Figure 10.14).

INTERCELLULAR COMMUNICATION IN ACTION: THE GASTROCNEMIUS MUSCLE

The gastrocnemius muscle illustrates how transmitters and receptors operate in a single tissue. This is the calf muscle at the back of the lower leg. When it contracts, it pulls on the Achilles tendon so that the toes push down on the ground. Most of the bulk of the muscle is made up of one type of cell, skeletal muscle fibers.

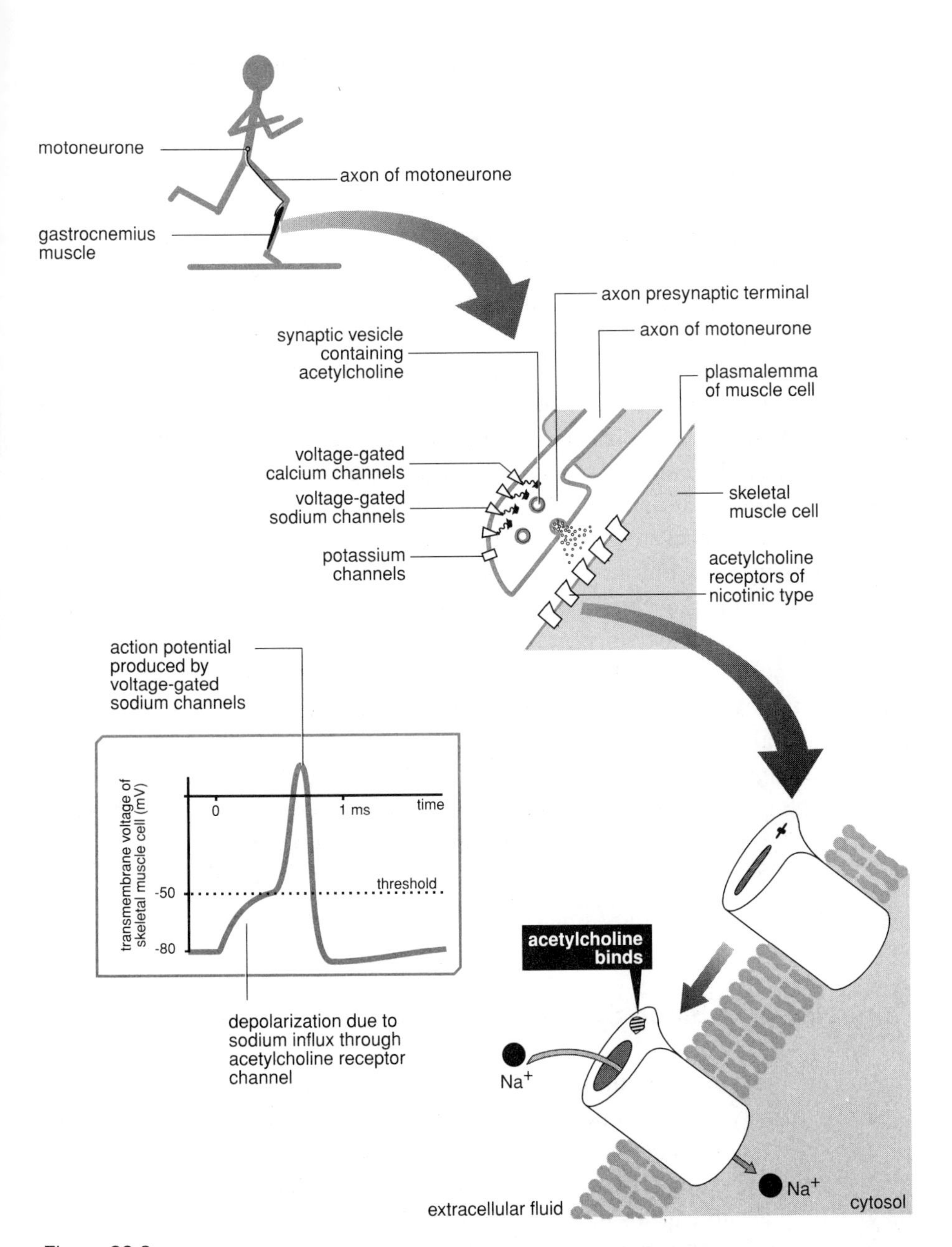

Figure 20.3

Motoneurones release the transmitter acetylcholine that binds to nicotinic receptors on the muscle cells. The plasmalemma of the muscle cell is depolarized to threshold and fires an action potential.

Telling the Muscle to Contract: The Action of Motoneurones

All skeletal muscles are relaxed until they receive a command to contract from nerve cells called motoneurones. Motoneurone cell bodies are in the spinal cord, while the myelinated axons run to the muscle. The axon terminal of the motoneurone that controls the gastrocnemius muscle contains vesicles of the transmitter acetylcholine (Fig. 20.3). To press the foot down, action potentials travel from the spinal cord down the motoneurone axon to its terminal, opening voltage-gated calcium channels in the plasmalemma. Calcium ions pour in, raising their cytosolic concentration from 100 nmol liter^{-1} to 1 μmol liter^{-1}. The calcium torrent causes the synaptic vesicles containing acetylcholine to fuse with the plasmalemma, releasing the transmitter into the extracellular fluid.

Acetylcholine survives only 0.2 ms in the extracellular fluid, since it is quickly hydrolyzed into choline and acetate by the enzyme acetylcholinesterase. However, its target—the skeletal muscle cell—is only 100 nm away. It hits its mark before it is broken down and binds to the nicotinic acetylcholine receptors present in the plasmalemma of the skeletal muscle cell. These open, allowing both potassium and sodium ions to pass. There is no large electrochemical gradient for potassium, but sodium is at a high concentration in the extracellular fluid and rushes in, depolarizing the plasmalemma of the skeletal muscle cell. The plasmalemma of the skeletal muscle cell also contains voltage-gated sodium channels. When the flow of sodium ions through the nicotinic acetylcholine receptors depolarizes the plasmalemma to threshold, the skeletal muscle cell fires an action potential, which in turn, causes release of calcium from the endoplasmic reticulum. The resulting rise of calcium concentration in the cytosol of the muscle cell causes it to contract.

Controlling the Blood Supply: Paracrine Transmitters

Figure 20.4 is a cutaway drawing of a blood vessel in the muscle. Lining the tube is a thin layer of endothelial cells. Wrapped around these are muscle cells of a different type, smooth muscle cells. Both endothelial cells and smooth muscle cells are much smaller than skeletal muscle cells. A small blood vessel may be only as large as a single skeletal muscle cell. Smooth muscle cells have no input from motoneurones. Instead, like many other internal organs, they are supplied by a separate system called autonomic nerves. These usually use one of two transmitters: acetylcholine or noradrenaline. Two other transmitters, adrenaline and nitric oxide, also help to adjust the blood flow on a time scale of seconds to minutes. The nerve cells that release noradrenaline onto the smooth muscle of blood vessels are called vasoconstrictors because they cause blood vessels to contract. The smooth muscle cells have α-adrenergic receptors in

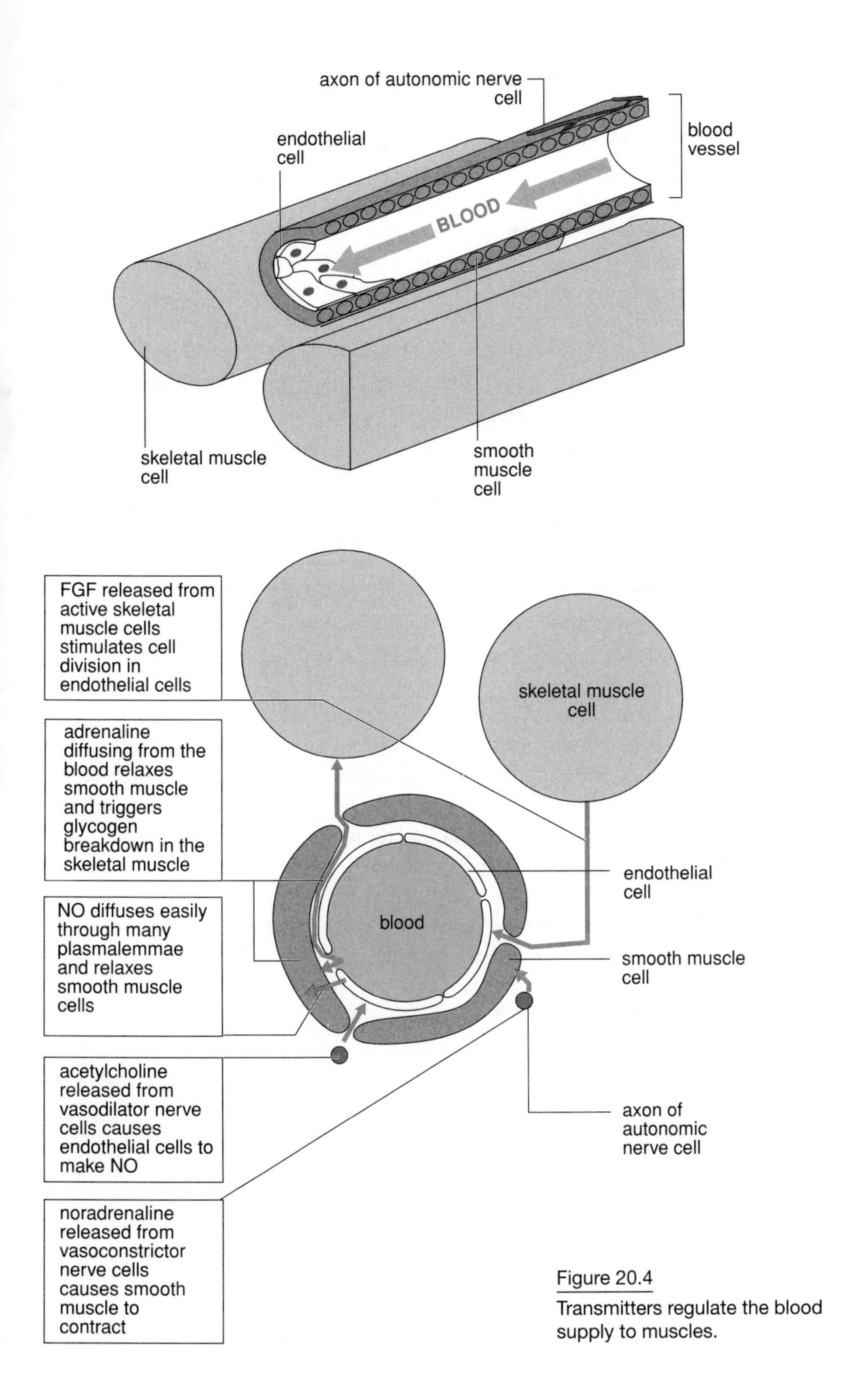

Figure 20.4

Transmitters regulate the blood supply to muscles.

their plasmalemmae. Binding of noradrenaline to α-adrenergic receptors releases calcium from the endoplasmic reticulum into the cytosol. This increase causes the smooth muscle cells to contract, constricting the blood vessel and reducing the flow. Vasoconstrictor nerves are used to restrict the flow of blood to muscles and other organs that are not in heavy use.

The hormone adrenaline is chemically related to noradrenaline but is more stable, lasting a minute or so in the extracellular fluid before being broken down. It is released from an endocrine gland (the adrenal gland) during times of stress and spreads around the body in the blood. Adrenaline that diffuses to the skeletal muscle cells stimulates them to begin breaking down glycogen to glucose (page 281). The smooth muscle cells of blood vessels within skeletal muscles also have β-adrenergic receptors connected to adenylate cyclase. However, they do not contain glycogen, and cAMP has another effect in these cells. When cAMP rises, it activates cAMP-dependent protein kinase. This in turn phosphorylates proteins that relax the smooth muscle cell. The action of adrenaline is therefore to increase the blood supply to all the muscles of the body in preparation for flight or fight. If we are very frightened and too much adrenaline is released, so much blood is diverted from the brain to the muscles that we faint.

Just as autonomic vasoconstrictor nerve cells are used to shut down blood flow to particular regions, autonomic vasodilator nerves are used to dilate blood vessels in muscles that are about to be used. They release acetylcholine, but neither smooth muscle cells nor endothelial cells have nicotinic acetylcholine receptors. Endothelial cells have, instead, a different receptor called the muscarinic acetylcholine receptor. This is named for a chemical that binds it: muscarine, from the poisonous fly agaric mushroom *Amanita muscaria* (Box 20.1). Like the α-adrenergic receptor, the muscarinic acetylcholine receptor is linked to phospholipase C, so that binding of acetylcholine causes the concentration of calcium to rise. Endothelial cells do not contract in response to a such a calcium rise. However, they have an enzyme that makes nitric oxide, and this enzyme is activated by calcium and calmodulin. When endothelial cells are stimulated with acetylcholine, they make nitric oxide. This intercellular transmitter easily passes through the plasmalemma of both endothelial and smooth muscle cells and reaches its receptor within the smooth muscle cells. Here it activates guanylate cyclase, causing a rise of cGMP which in turn acts to cause a rise of cAMP and hence relaxes the smooth muscle cells.

Nitric oxide lasts for only about four seconds before being broken down. It is therefore a paracrine transmitter, able to diffuse through and relax all the smooth muscle cells surrounding the endothelial cells but without lasting long enough to pass into more remote tissues. The discovery in 1987 that nitric oxide was a transmitter explained why nitroglycerine (more familiar as an explosive than as a medicine) relieved angina pectoris. Angina is a pain in the heart

BOX 20.1

Cigarettes, Mushrooms, and Insecticides

Nicotine is one of the most addictive substances known. The reason for this is still rather unclear. The muscle weakness that new smokers experience is due to nicotine blocking receptors on the muscle cells, but the site in the brain that nicotine acts upon to cause a pleasurable experience is still unknown.

The fly agaric mushroom *Amanita muscaria* is seldom taken recreationally. Its popular use is more prosaic, in killing flies, hence its English and Latin names. However, Robert Graves has argued that it was used by the classical Greeks and by the Vikings to produce a state of berserk energy with hallucinations. Any such use would be highly dangerous: muscarine acts by stimulating the muscarinic acetylcholine receptor and the difference between a dose of muscarine large enough to cause pharmacological effects and a lethal one is very small.

Amanita muscaria has an even more dangerous relative, *Amanita phalloides*, the death cap mushroom. The toxin it produces, phalloidin, does not affect acetylcholine signaling. Instead, it interferes with the actin cytoskeleton.

Farmers and others who suffer insecticide toxicity show similar symptoms to those of muscarine poisoning. This is because these insecticides work by blocking the enzyme acetylcholinesterase which breaks down acetylcholine. With the enzyme blocked, acetylcholine hangs around in the extracellular fluid, stimulating the receptor for longer than it should. Nerve gases developed to kill people work in the same way. Paradoxically, sufferers are treated with another toxin, atropine, from the deadly nightshade plant. Atropine turns off the muscarinic acetylcholine receptor.

caused by poor blood flow. Nitroglycerine spreads throughout the body via the blood stream and slowly breaks down, releasing nitric oxide that then dilates the blood vessels of the heart.

New Blood Vessels in Growing Muscle: Growth Factors

All the phenomena we have discussed so far occur on a time scale of seconds. However, when a muscle is repeatedly exercised over many days, it becomes stronger: the individual skeletal muscle cells enlarge. This is because high cytosolic calcium stimulates the transcription of genes coding for structural proteins. Furthermore new blood vessels sprout and grow into the enlarging muscle. A paracrine transmitter called FGF is released by stimulated muscle and binds to metabotropic cell surface receptors on endothelial cells. Activated FGF receptors then activates a series of signaling steps that eventually trigger cell division and hence the growth of new blood vessels (Box 2.2). FGF is an example of a growth factor. These are paracrine transmitters with lifetimes of approximately ten minutes that can therefore diffuse quite long distances in tissues, for instance, from the skeletal muscle cells to the blood vessel. Growth factors bind to specific metabotropic cell surface receptors and promote cell division. Many of the receptors for growth factors are tyrosine kinases and work in the same way as the FGF receptor does.

In the gastrocnemius muscle and its blood supply, there are examples of all types of transmitter mechanism. Acetylcholine acts as a synaptic transmitter at the axon terminal of the motoneurone. Adrenaline is a hormone. The other transmitters are paracrine. The nicotinic acetylcholine receptor is an ionotropic cell surface receptor. The nitric oxide receptor is intracellular. The other receptors are metabotropic cell surface receptors. There is a wide variety of time scales of action. The acetylcholine released from the axon terminal of the motoneurone causes a contraction of the skeletal muscle cell within 5 ms, by which time the acetylcholine has already been destroyed. Adrenaline lasts one minute and dilates blood vessels for all this time. FGF lasts ten minutes, but its effects are much longer lasting. FGF triggers the synthesis of proteins, which then act to cause cell proliferation that lasts for days. A similar pattern of intercellular communication, using some of the same transmitters and receptors but also many others, is found in every tissue of the body.

SUMMARY

1. Transmitter mechanisms can be classified in two ways. One depends on their longevity in the extracellular medium, one on the location and action of the receptor on the target cell.

2. In increasing order of longevity, transmitters can be divided into synaptic transmitters, paracrine transmitters, and hormones.

3. Receptors can be divided into ionotropic cell surface receptors, metabotropic cell surface receptors, and intracellular receptors. Ionotropic cell surface receptors are ion channels that open in response to ligand binding. The effect upon the target cell is electrical. Metabotropic cell surface receptors are linked to enzymes. The effect upon the target cell is metabolic. Intracellular receptors lie within the target cell and bind transmitters that are able to cross the cell membrane by simple diffusion.

4. The gastrocnemius muscle provides examples of all types of intercellular signaling operating in concert to fit the operation of the tissue to the requirements of the organism.

BOX 20.2

Interactions Mediated by Ionotropic Cell Surface Receptors

The synapse between a motoneurone and a skeletal muscle cell is unusual in that a single action potential in the presynaptic cell releases enough transmitter to depolarize the postsynaptic cell to threshold and hence produce an action potential in the postsynaptic cell. The events at a more typical synapse, that between a pain receptor and a pain relay nerve cell, are shown below. Time is on the horizontal axis. An action potential in the axon terminal of the pain receptor raises cytosolic calcium and releases the transmitter glutamate. This diffuses to the pain relay cell where it binds to glutamate receptors, which are ionotropic and allow both potassium and sodium ions to pass. The postsynaptic nerve cell depolarizes as sodium ions move in through the glutamate receptor channels. However, the depolarization is not enough to take the transmembrane voltage to threshold. As soon as glutamate is removed from the extracellular fluid, the transmembrane voltage returns to the resting level. In the axon of the relay cell, some distance from the synapse, the transmembrane voltage does not change, and no message passes on to the brain. The subject does not feel pain.

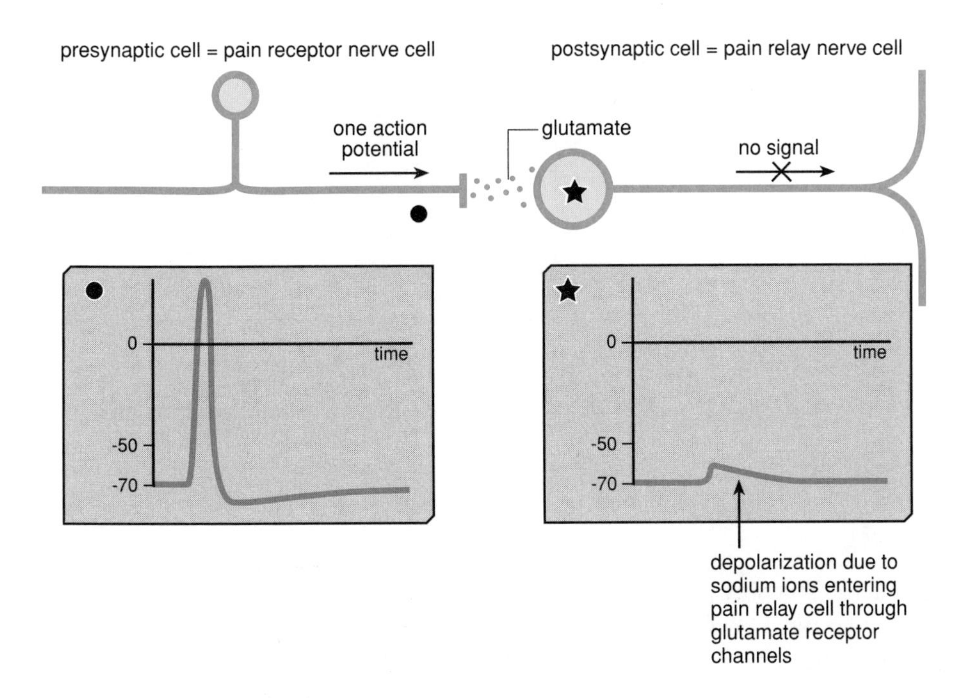

A subject *does* feel pain when he or she moves a finger into the hot air from a hair dryer. This is because the hot air heats a large area and causes many pain receptors to fire action potentials. Many of the pain receptors synapse onto one relay cell, which therefore receives many doses of glutamate. Enough glutamate receptor channels open to depolarize the relay cell to threshold, and an action potential travels along the axon of the relay cell to the brain. The subject feels pain.

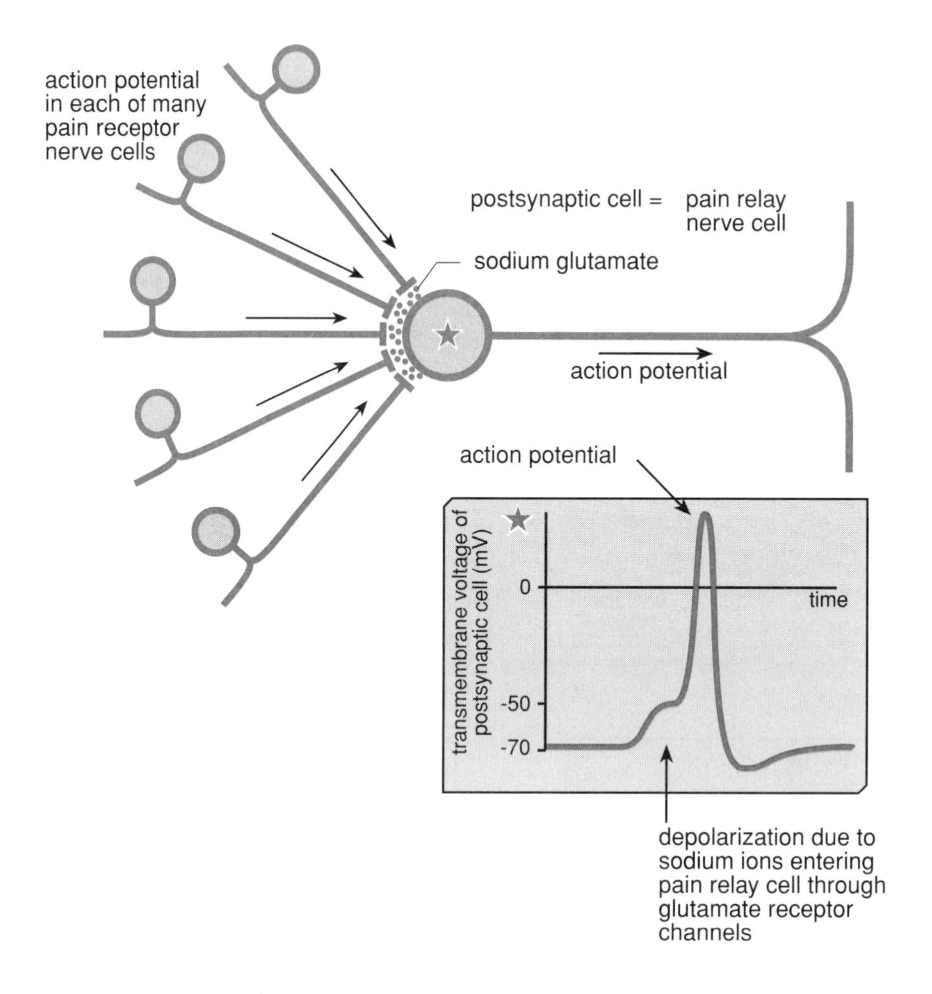

An intense stimulus to a small area is also painful. When the subject stabs himself or herself with a needle, only one pain receptor is activated, but that receptor is intensely stimulated and fires a rapid barrage of action potentials, each of which causes the release of sodium glutamate and an extra depolarization of the pain relay cell. Soon the transmembrane voltage of the relay cell reaches threshold, and an action potential travels along its axon toward the brain and the subject feels pain. Such summation only occurs if the presynaptic action potentials are frequent enough to ensure that the depolarizations produced in the postsynaptic cell add.

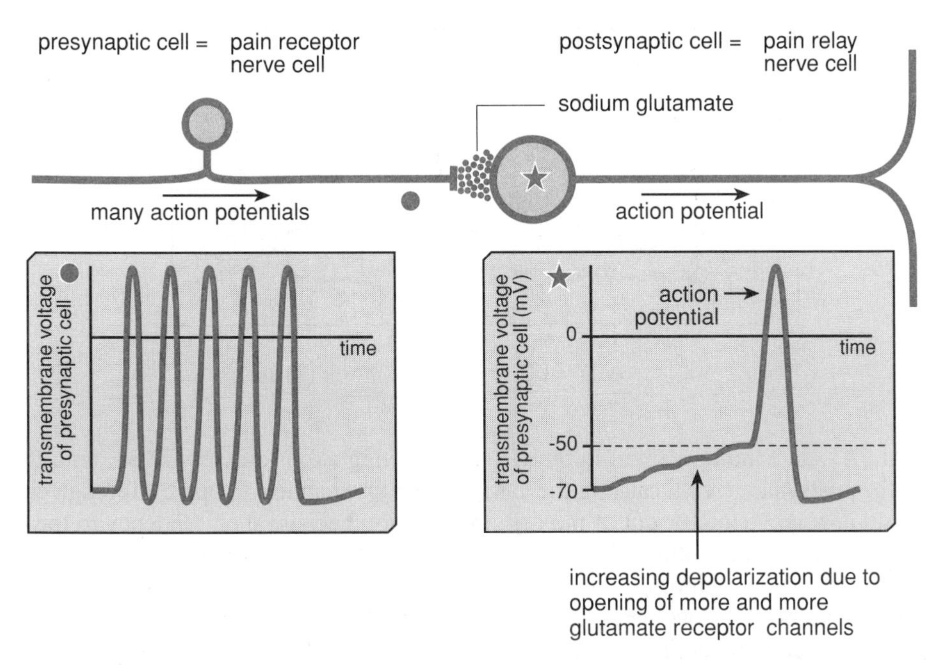

Another amino acid, γ amino butyrate (GABA) (Figure 13.1), is also used as a synaptic transmitter. Its receptor is an ionotropic cell surface receptor that forms a channel selective for chloride ions. The diagram below illustrates a single nerve cell bearing both glutamate and γ amino butyrate receptors.

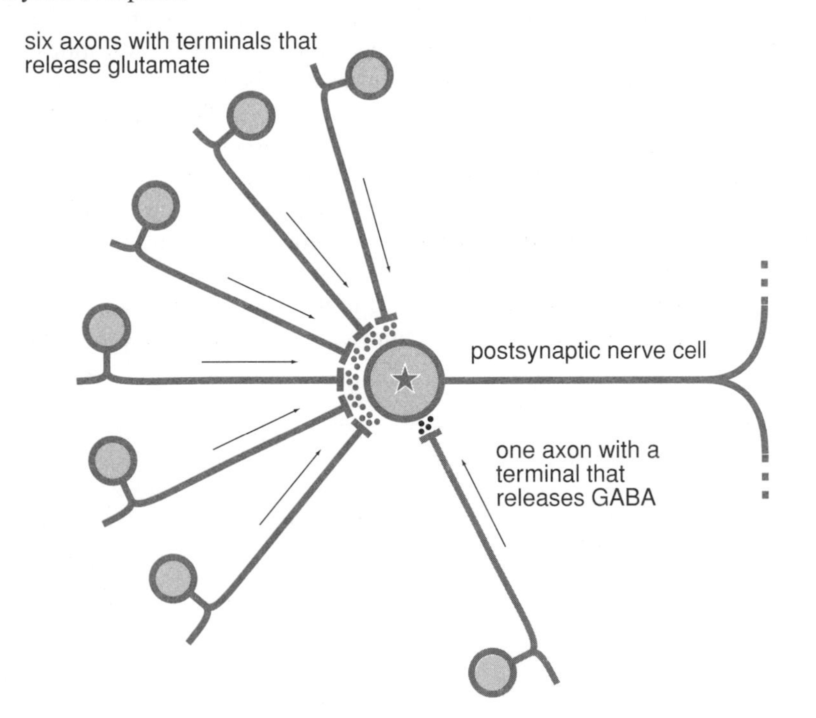

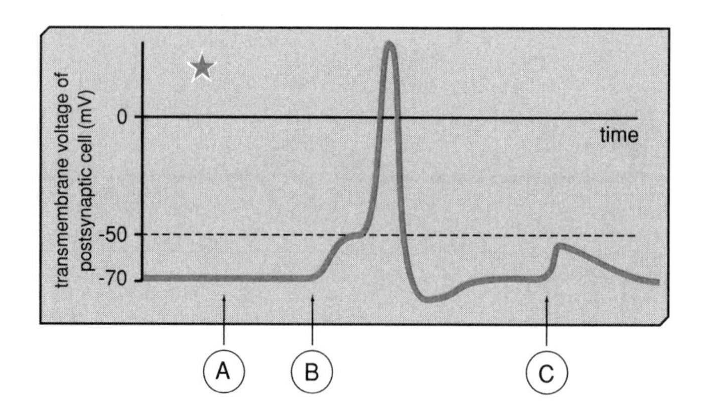

At {A}, an action potential in the GABA-secreting axon releases GABA onto the surface of the postsynaptic cell, causing the GABA receptor channels to open. Although chloride ions could now move into or out of the cell, they do not, because their tendency to travel into the cell down their concentration gradient is balanced by the repulsive effect of the negative voltage of the cytosol: chloride ions are at equilibrium (Fig. 17.2). The opening of GABA channels causes no ion movements, and therefore does not alter the transmembrane voltage.

At {B}, action potentials occur simultaneously in six glutamate-secreting axons. In this example, this activity provides enough glutamate to depolarize the postsynaptic cell to threshold, and the postsynaptic nerve cell fires an action potential.

At {C}, action potentials occur simultaneously in six glutamate-secreting axons and also in the GABA-secreting axon. The same number of glutamate receptor channels open as before, and the same number of sodium ions flow into the postsynaptic nerve cell, tending to depolarize it. However, as soon as the transmembrane voltage of the postsynaptic cell deviates from the resting value, chloride ions start to enter through the GABA receptor channels because the cytosol is no longer negative enough to prevent them from entering the cell down their concentration gradient. The inward movement of negatively charged chloride ions neutralizes some of the positive charge carried in by sodium ions moving in through the glutamate receptor channels. The postsynaptic nerve cell therefore does not depolarize as much, and does not reach threshold. No action potential is generated in the postsynaptic nerve cell.

Anti-anxiety drugs such as Valium act on the γ amino butyrate receptor and increase the chance of its channel opening. Nerve cells exposed to the drug are less likely to depolarize to threshold. Valium therefore reduces action potential activity in the brain, calming the patient.

CHAPTER

21

CYSTIC FIBROSIS

Genetics and molecular biology often seem oddly detached from the real world of health and disease. Twenty years ago medical students often complained about having to study genetics. What, they asked, was its relevance (the most overused word in the undergraduate vocabulary)? Genes seemed to have no real bearing on the problem of making sick people better. Why not get on to brain surgery straight away?

Now everything has changed. The average medic is desperate to learn more about inherited disorders: Carrier screening and gene therapy are the stuff of first-year lectures. But still, the cynic might ask, what is the relevance of all this to treating disease? The commonest single-gene inborn error in the world is sickle-cell anemia. In West Africa one child in 400 is born with the condition. Migration means that the disease is also a problem in North and South America and, increasingly, in Europe. Sickle cell anemia was the first disease to be understood at the molecular level. The error in the protein chain was worked out more than thirty years ago, and the details of the DNA change are now known from many different human populations.

None of this has been of any help in treatment. The care of sickle-cell homozygotes still depends on the classical traditions of empirical medicine. The splendid science that revealed the biology of sickle cell anemia remains just that, splendid science. The best it can offer—and it is certainly valuable—is carrier testing and prenatal diagnosis. Those at risk of passing on the gene can be told, and parents whose unborn child has two copies of the damaged gene can be offered the option of terminating the pregnancy. Even this is of little relevance to the millions of parents at risk in Africa where population screening is not feasible for economic reasons.

Scientists—particularly medical scientists—carry out research for reasons that go beyond the disinterested search for knowledge. The impetus behind the human genome project is the hope that understanding the nature of inherited illness is the first step toward treating it. When the idea was first mooted, there was great confidence: Work out what had gone wrong, and soon there would be drugs designed to correct the defect and even a chance of cutting out damaged genes and replacing them with a working copy.

Almost none of this has yet happened. The biggest disappointment has been in gene therapy. This was just around the corner ten years ago, and that is where most of it still is. However, the promise of modern genetics is still very much there, and the subject is being treated with more realism than when it began its explosive growth.

Nowhere have the prospects been more discussed than in the study of cystic fibrosis. This, the commonest single-locus inborn lethal disease in the western world, has been tracked down to its precise location using the methods of molecular biology. The task involved a race between several competing groups. It was as much a model of the sociology of science as it was of science itself.

Remarkably soon after CF was mapped there was an explosion of information about how it can go wrong. What seemed to be a simple genetic change turned out to be a complex series of independent mutations. The cystic fibrosis story is a *tour de force* of what molecular genetics can do, and—for the first time—there is the prospect of using the information it provides to understand, and perhaps treat, a common inherited disease.

CYSTIC FIBROSIS IS A SEVERE GENETIC DISEASE

Among white-skinned people about one child in 2,500 is born with cystic fibrosis. Inheritance is simple: when both parents are heterozygotes, there is a 1 in 4 chance that a child will have the disease. The disease is a distressing one. Most of its symptoms arise from faults in the way the body moves liquids, leading in particular to a buildup of inadequately hydrated, sticky mucus in various parts of the body. In the lungs this leads to difficulty in breathing, a persistent cough, and a much increased risk of infection. The pancreas—which provides a digestive secretion that flows to the intestine—is also affected and may be badly damaged (which explains the disease's full name; cystic fibrosis of the pancreas). Often there are digestive problems because the damaged pancreas is not producing enzymes. The reproductive system is harmed and most males who survive to adolescence are infertile. Even though there are so many varied symptoms, the simple Mendelian pattern of inheritance led researchers to believe that the cause was an abnormality or absence of a single protein.

Until recently babies with CF did not survive to their first birthday. Today the life expectancy is 22 years. This remarkable improvement is wholly due to old-fashioned medicine that intervenes to reverse individual symptoms. Digestive enzymes are taken by mouth to replace those the pancreas fails to produce. Physiotherapy—helping patients to cough up the mucus in their lungs by slapping their backs—reduces the severity of the lung disease. Yet, however remarkable the increase in survival, each life that extends only to early adulthood is still a tragedy.

THE FUNDAMENTAL LESION LIES IN CHLORIDE TRANSPORT

A Swiss children's song says that a baby whose brow tastes unusually salty will die young. However, the link between salty sweat and a disease known for its effects on the pancreas and lungs was not made until 1953 and gave the first clue as to the fundamental defect in CF. Sweat glands have two regions that perform different jobs (Fig. 21.1). The secretory region deep in the skin produces a fluid that has a ionic composition similar to that of extracellular medium; that is, it is rich in sodium and chloride. If the sweat glands were simply to pour this liquid onto the surface of the skin, they would do a good job of cooling, but the body would lose large amounts of sodium and chloride. A reabsorbtive region closer to the surface removes ions from the sweat, leaving the water (plus a small amount of sodium chloride) to flow out

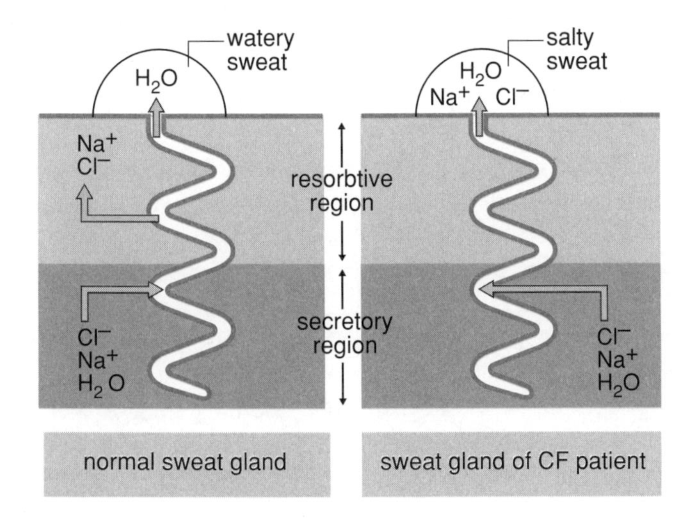

Figure 21.1
Sodium, chloride, and water transport in the sweat gland.

of the end of the gland. CF patients produce normal volumes of sweat, but this contains lots of sodium and chloride, implying that the secretory region is working fine but that the reabsorbtive region has failed. The pathways by which sodium and chloride ions are removed are distinct. Which one has failed in CF? A simple electrical test gave the answer. A normal sweat gland has a small negative transepithelial voltage (Fig. 21.2). In CF patients the inside of the gland is much more negative than usual. This result tells us at once that it is chloride transport (and not sodium transport) that has failed. In CF sweat glands the reabsorbtive region has the transport systems to move sodium, but the chloride ions remain behind in the sweat, giving it a negative voltage. The sodium transport system cannot continue to move sodium ions out of the sweat in the face of this larger electrical force pulling them in, so sodium movement stops too, and sodium chloride is lost in the sweat, which then tastes salty.

Every one of the symptoms suffered by CF patients is caused by a failure of chloride transport.

HOMING IN ON THE GENE USING CLASSICAL GENETICS

The first family studies of the disease were carried out more than fifty years ago, and pedigrees showing CF as a classic case of recessive autosomal inheritance were published in 1946 (Fig. 6.4). Although different families differed in

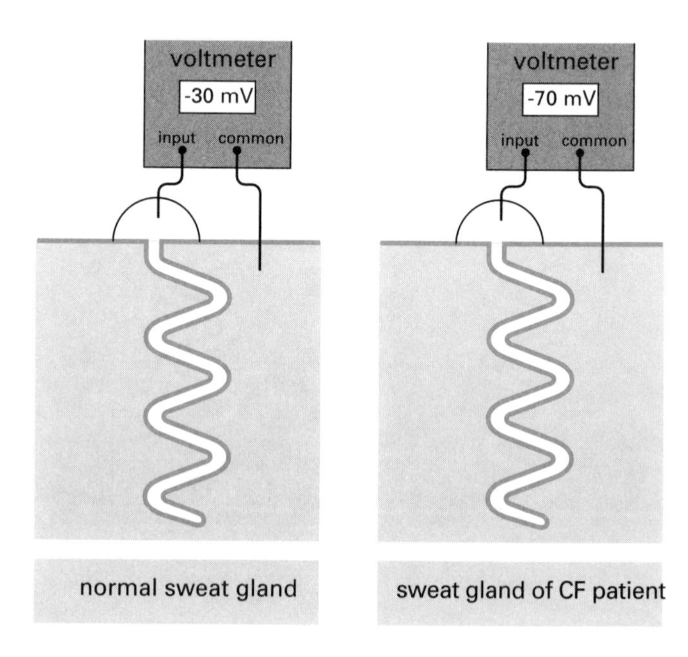

Figure 21.2
Measuring the transepithelial voltage of human sweat glands *in situ*. The voltmeter measures the voltage of the input terminal with respect to the common terminal. In this case a reading of -30 mV indicates that the droplet of sweat has a voltage 30 mV negative to the extracellular medium.

the severity of their symptoms, the disease seemed to be a rather simple entity, following Mendel's rules quite precisely.

The cystic fibrosis allele seemed at first to be recessive, however it was looked at. Unlike, say, sickle-cell anemia (in which the discovery of the defect led to an immediate test for an abundant protein—mutant hemoglobin—that identified heterozygotes, most of whom knew nothing of their condition) there was no obvious way to identify the millions of people who carry a single copy of the CF gene. It was also impossible to check whether a developing fetus of a mother known to be at risk of the disease had two copies and would hence be affected.

There were many attempts to find a test. For a time it seemed that a simple stain might do the job, but this idea was abandoned. Other tests were more eccentric. For example, there was a claim that an extract from the blood of cystic fibrosis patients, and perhaps even from that of their parents, who must be heterozygotes, slowed the beating of the cilia of oysters.

Most other claims that particular gene products were peculiar to heterozygotes also failed to stand up or proved to be symptoms rather than cause. For instance, trypsinogen, a precursor of the digestive enzyme trypsin, is found in fluid bathing fetuses homozygous for CF, and not in others—but this proves to be because the pancreas is already dying at this time. In the 1980s there was hope that this result could be used to diagnose affected pregnancies, but the approach was quickly overtaken by developments in studying the gene itself. As for most inborn diseases, the big problem was that nobody had any real idea what the faulty gene did; looking for a defect in an unknown protein seemed a hopeless task.

The English biologist J.B.S. Haldane suggested more than fifty years ago that one way of identifying carriers of genetic disease would be to see if those who had two copies of a damaged gene were also similar in other, more easily detected, genes that might be linked to the one that had gone wrong. If every sufferer from such a disease had, for example, two copies of a particular blood group allele that was rare in unaffected individuals, then that allele might be a test both for carriers and (if it was expressed in the developing embryo) for damaged fetuses. His idea was the seed of much of modern clinical genetics.

In the 1960s and 1970s many attempts were made to find an association between the cystic fibrosis gene and others, such as blood groups. Without exception they failed. Given that only a few variable systems could be tested for and that the cystic fibrosis gene is just one among tens of thousands, it is not surprising that no linked loci were found.

One simple technical trick altered everything. This was the use of restriction endonuclease enzymes (page 170) to cut DNA at specific points. Each of these enzymes—and there are dozens of them—searches out and attacks a particular DNA sequence as a target. If two people differ in where the target is sited, the lengths of the cut pieces will differ. Thousands of these "restriction fragment length polymorphisms" are now known. One guess is that there may be a hundred thousand such sites on each chromosome. There is hence a good chance that one may be linked to (or perhaps even part of) a damaged functional gene and can be used to flag its presence.

Because cystic fibrosis is so common, this gave immediate hope of mapping the disease. After all, although each family might be small and involve only a single generation, there are many families. Optimists hoped that every patient might share a common DNA polymorphism that could act as a diagnostic test and give a hint as to where the gene was located. Although the initial optimism was overstated, the approach of looking for joint patterns of inheritance of the CF gene and others was the key to finding the gene itself.

For much of the 1980s the task was a thankless one. Dozens of polymorphisms (either in the DNA itself or in proteins) were screened. None seemed to share a common inheritance with cystic fibrosis. Nevertheless, since many of the markers used had already been mapped to particular chromosomes, the negative results served at least to exclude great sections of the genome as a possible home for the gene.

MOLECULAR GENETICS CARRIED THE SEARCH FORWARD

Family studies published in 1986 showed that the damaged gene was inherited with a particular DNA variant. Now there was something that could be used to find the chromosome on which the CF gene was located. The next step—and nearly all the subsequent steps—went beyond Mendelian genetics. The crucial linked DNA variant was discovered to be on the long arm of chromosome 7, so the CF gene had to be on that chromosome. Several more DNA markers closely linked to the CF gene soon turned up and confirmed its location to one small part of the chromosome. Despite these advances there was still no indication of how the CF gene might be recognized if it were to be found. The next step involved finding more linked markers and mapping them in relation to each other and to the gene itself. More than 20 were found, spanning the gene. The area of the chromosome into which it must be confined grew smaller. Two markers were close to each other. From patterns of recombination, it seemed certain that the CF locus must be between them—somewhere in a length of one and a half million nucleotides.

Eventually a genomic clone was isolated that was thought to contain part of the CF gene. This clone was used as a gene probe to screen a cDNA library isolated from the sweat glands of a normal individual. Success was then close at hand. The sweat gland cDNA was the tool needed to isolate the entire CF gene. After a great deal of work and much screening of genomic libraries, the CF gene was isolated. It is 250 kB in length and has 24 exons. We now know that the gene encodes a large protein of 1,480 amino acids. Once the sequence of the protein from a normal individual was known, it was compared to the sequences derived from cDNAs isolated from the sweat glands of CF patients. In 70% of patients, one triplet codon was missing. In the normal protein this encodes the amino acid phenylalanine at position 508 in the protein. However, this is not the only mutation to cause CF, and to date 300 CF gene mutations have been discovered. Some cause a complete failure to express the protein, and produce severe symptoms. Other mutations produce a protein that works, but not as well as the normal protein. Individuals with these mutations have relatively mild symptoms.

The protein encoded by the CF gene had no sequence similarity with any known ion channel gene. Could it be that the product of the gene was not itself a chloride channel but was in some way necessary for a pre-existing chloride channel to operate? Hedging their bets, scientists named the protein product of the gene the CF transmembrane regulator, or CFTR. In fact the CFTR really is a chloride channel (Fig. 13.12). We know this because cells without any chloride channels can be injected with CFTR cDNA so that they will make the CFTR protein. When such cells are then studied by the patch clamp technique (Box 17.2), they are found to have chloride channels.

A CURE FOR CF—OR A PREMATURE HOPE?

In 1993 there were reports of a trial of a genetic treatment for CF. Introducing plasmids carrying the normal CFTR gene into the lungs of CF sufferers might result in the epithelial cells making some normal CFTR. The cells would then be able to transport chloride ions. The trial had two components (Fig. 21.3). First, the group used transgenic animal technology (Box 11.2, page 184) to create a mouse lacking a CFTR gene and hence suffering from CF. In the second phase of the trial, the mice inhaled droplets containing the CFTR plasmid. The results were spectacular: The lung symptoms disappeared for many days. Sadly, clinical trials on humans have been much less successful, mainly because human lungs are so much larger. Only the bits of the lung close to the throat get a dose of the plasmid and hence a little of the normal CFTR. At present, it is too early to say whether this problem will be overcome by adjusting the method of plasmid delivery.

A cure for CF? No, not even if or when the lung disease can be treated effectively. Other organs where the CFTR acts will fail in adult life: Already, as CF patients survive longer, they are increasingly affected by diabetes and cirrhosis. Reduction of CF as a medical problem will result not from the genetic

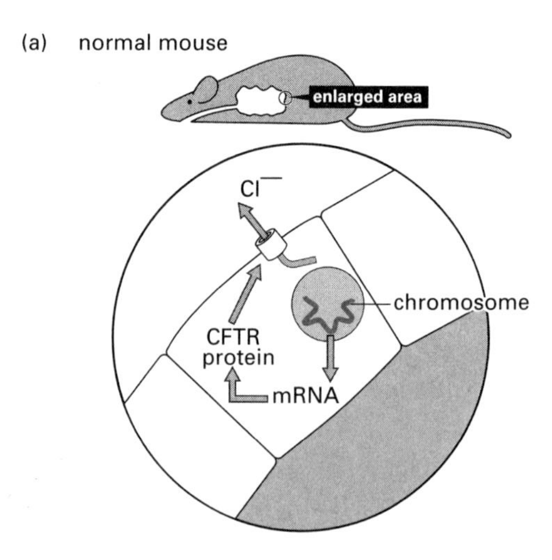

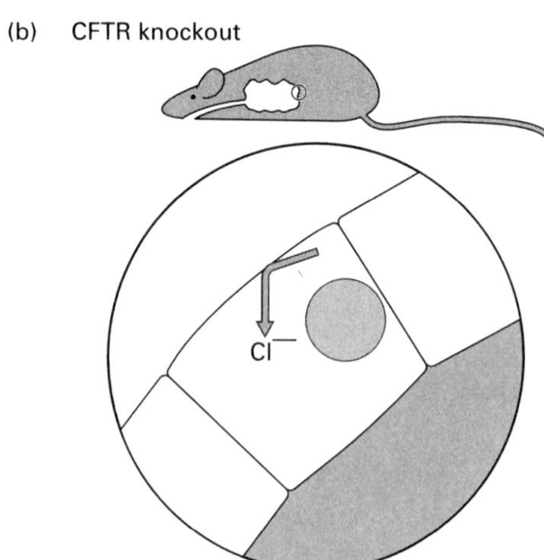

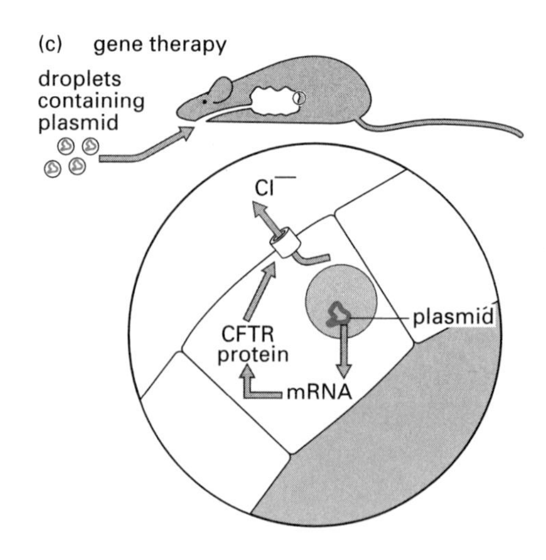

Figure 21.3

The CFTR knockout mouse is a model that can be used to test gene therapy strategies.

modification of individual organs in homozygotes but from reducing the frequency with which homozygotes are produced. At present, this occurs through genetic counseling and subsequent termination of pregnancy. In the distant future it may perhaps be possible to directly modify the DNA in the germ cells of carriers but as yet there is no sign of success in so doing, and in most countries such experiments are regarded as unethical.

The investigation of CF demonstrates the enormous power of modern cell biology to explain. Where previous generations of biologists could only describe what the genes were producing, now at last there is the hope of understanding, and perhaps changing, the genes themselves.

SUMMARY

1. Cystic fibrosis is the most common serious single-locus illness in the western world. Many organs are affected. Sticky mucus builds up in the reproductive tract and in the lungs. The pancreas is always affected and usually fails completely.

2. Electrical measurements show that the basic problem is in chloride transport.

3. CF is a classical autosomal recessive. Genetic counseling needed a test that would reveal both heterozygote carriers and homozygote fetuses, but this did not arrive until the use of restriction enzymes uncovered thousands of variants, each detectable in heterozygotes.

4. After a long search one variant of this kind was found to travel down the generations in consort with CF. This linked locus was chased down to chromosome 7. Soon other markers were found, and by using the techniques of physical mapping, a segment of the gene was found. A cDNA library from the sweat gland of a normal individual contained this sequence.

5. From this result it was possible to identify the gene and infer its amino acid sequence in both normal and mutated form. The gene codes for a chloride channel. Hundreds of different CF mutations have now been found.

6. Gene therapy of affected individuals has so far yielded disappointing results. Even the optimists accept that for the time being gene therapy only holds out the hope of alleviating the lung problems, leaving other less accessible organs untreated.

SUGGESTED FURTHER READING

This book is intended as a first step in university level study of Genetics and Cell Biology. These topics are covered in more detail in a number of excellent textbooks.

Alberts, B., Bray, D., Lewis, J., Raff, M., Roberts, K., and Watson, J. D. 19. *Molecular Biology of the Cell,* 3rd edition. New York: Garland. Especially for Chapters 1–5 and 9–21.

Klug, W. S., and Cummings, M. R. 1993. *Essentials of Genetics.* New York: Macmillan. Especially for Chapters 6–8.

Lewin, B. 1994. *Genes.* 5th edition. New York: Oxford University Press. Especially for Chapters 5 and 9–11.

Lodish, H., Baltimore, D., Berk, A., Zipursky, S. L., Matsudaira, P., and Darnell, J. 1995. *Molecular Cell Biology*, 3rd edition. New York: Scientific American Books, W. H. Freeman. Especially for Chapters 5, 8–12, and 21.

Stryer, L. 1995. *Biochemistry*, 4th edition. New York; W. H. Freeman. Especially for Chapters 5, 9, 10 and 12–16.

Voet, D., and Voet, J. G. 1995. *Biochemistry*, 2nd edition. New York: Wiley. Especially for Chapters 5, 9, 10, and 12–16. Provides more detail than other standard biochemistry texts.

More detail on the content of particular chapters can be found in the following publications:

Chapter 1. Cells and Tissues

Campbell, N. A. 1996. 4th edition. *Biology*, Menlo Park, CA: Benjamin Cummins. An excellent overall textbook.

Chapter 2. Membranes and Organelles

McCormick, F. 1993. *Signal Transduction: How receptors turn Ras on. Nature* 363:15–16. A discussion of how, in the space of a few weeks, the mechanism by which growth factors activate cell division became clear.

Margulis, L. 1981. *Symbiosis in Cell Evolution.* New York: W. H. Freeman. Where organelles came from.

Novikoff, A. N., and Holtzman, E. 1984. *Cells and Organelles*, New York: Holt, Reinert and Winston. A standard textbook.

Chapter 3. The Cytoskeleton and Cell Movement

Amos, L. A. and Amos, W. B. 1991. *Molecules of the Cytoskeleton*. New York: Macmillan.

Bray, D. 1992. *Cell Movements*. New York: Garland. Concentrates on how cells behave.

Kreis, T., and Vale, R. 1993. *Guidebook to the Cytoskeletal and Motor Proteins*. Oxford: Oxford University Press. A comprehensive text on protein machinery.

Preston, T. M., King, C. A., and Hyams, J. S. 1990. *The Cytoskeleton and Cell Motility*. Glasgow: Blackie. A simple textbook.

Murray, A., and Hunt, T. 1993. *The Cell Cycle*. New York: W. H. Freeman. A simple textbook.

Hutchinson, C., and Glover, D. M. 1995. *Cell Cycle Control*. Frontiers in Molecular Biology. Oxford: IRL Press. A more comprehensive text.

Chapter 5. DNA Structure and the Genetic Code

Adams, R. L. P., Knowler, J. T., and Leader, D. P. 1992. *The Biochemistry of the Nucleic Acids*, 11th edition. London: Chapman & Hall. A standard textbook.

Watson, J. D., and Crick, F. H. C. 1953. A structure for deoxyribose nucleic acid. *Nature* 171:737–738. The paper in which they proposed the double-helix structure of DNA.

Watson, J. D. 1968. *The Double Helix*. New York: Atheneum. An easy-to-read history of the discovery.

Chapter 6. Patterns of Inheritance

Iltis, H. 1932. *Life of Mendel*, translated by E. Paul and C. Paul. New York: W. W. Norton. A biography of the father of genetics.

Jones, Steve. 1993. *The Language of the Genes: Biology, History and the Evolutionary Future*. London: Harper-Collins. Genetics for the general reader.

Chapter 7. Mapping the Genes

Davies, K. E., and Read, A. P. 1992. *Molecular Basis of Inherited Disease*. Oxford: IRL Press. A small, readable book.

Primrose, S. B. 1995. *Principles of Genome Analysis*. Oxford: Blackwell. A comprehensive book

Chapter 8. Gene Mutation

Auerbach, C. 1976. *Mutation research: Problems, Results and Perspectives*. London: Chapman and Hall.

Cooper, D. N., and Krawczak, M. 1993. *Human Gene Mutation*. Oxford: Bios.

McNamara, J. O., and Fridovich, I. 1993. Human Genetics: Did radicals strike Lou Gehrig? *Nature* 362:20–21. A clear, but out-of-date, discussion of superoxide dismutase and amyotrophic lateral sclerosis.

Chapter 9. DNA Replication and DNA Repair

Radman, M., and Wagner, R. 1988. The high fidelity of DNA duplication. *Scientific American* 259:40–46.

Chapter 10. Transcription and the Control of Gene Expression

Jacob, F., and Monod, J. 1961. Genetic regulatory mechanisms in the synthesis of proteins. *Journal of Molecular Biology* 3:318–356. The description of the lac operon.

Tjian, R. 1995. Molecular machines that control genes. *Scientific American* 272:38–45. A recent review.

Chapter 11. Recombinant DNA and Genetic Engineering

Gasser, C. S., and Fraley, R. T. 1992. Transgenic crops. *Scientific American* 266:62–69.

Mullis, K. B. 1990. The unusual origin of the polymerase chain reaction. *Scientific American* 262:56–65. An account of how he devised the technique.

Watson, J. D., Witkowski, J., Gilman, M., and Zoller, M. 1992. *Recombinant DNA*, 2nd edition. New York: Scientific American Books. W. H. Freeman. A standard textbook.

Chapter 12. Translation and Protein Targeting

Austen, B. M., and Westwood, O. M. R. 1991. *Protein Targeting*. In focus series. Oxford: IRL Press. A useful short book.

Chapter 13. Protein Structure

Branden, C., and J. Tooze, 1991. *Introduction to Protein Structure*. London: Garland. A well-illustrated introduction to patterns in protein structures.

Creighton, T. E. 1993. *Proteins*, 2nd edition. New York: W. H. Freeman. An approachable advanced text.

McGee, H. 1986. *On Food and Cooking*. London: George Allen and Unwin. An entertaining introduction to the chemistry of foods and food preparation.

Perutz, M. 1992. *Protein Structure: New Approaches to Disease and Therapy*. New York: W. H. Freeman. An excellent discussion of how protein structure is related to protein function and how this may apply to the treatment of disease.

Chapter 14. How Proteins Work

Cornish-Bowden, A. 1995. *Fundamentals of Enzyme Kinetics*, 2nd edition. London: Portland Press. An excellent and comprehensive treatment of enzyme kinetics.

Fersht, A. 1985. *Enzyme Structure and Mechanism*, 2nd edition. New York: W. H. Freeman. The standard textbook of enzymology covering both kinetics and enzyme mechanisms.

Palmer, T. 1995. *Understanding Enzymes*, 4th edition. London: Prentice Hall/Ellis Horwood. A good textbook.

Price, N. C., and Stevens, L. 1989. *Fundamentals of Enzymology*, 2nd edition. Oxford: Oxford University Press. A good textbook with a better explanation of the relation to cell biology than many others.

Chapter 15. Energy Trading within the Cell

Kapit, W., Macey, R., and Meisami, E. 1987. *The Physiology Coloring Book*. New York: Harper & Row. A highly recommended book that simply explains difficult concepts of ions and voltages.

Matthews, G. G. 1991. *Cellular Physiology of Nerve and Muscle*, 2nd edition. Oxford: Blackwell. A small book on electrical phenomena in cell biology.

Chapter 16. Metabolism

Devlin, T. M., ed. 1992. *Textbook of Biochemistry with Clinical Correlations*, 3rd edition. New York: Wiley. A general textbook with an emphasis on human disease.

Duchen, M. R. 1992. Ca^{2+}-dependent changes in the mitochondrial energetics in single dissociated mouse sensory neurons. *Biochemical Journal*. 283:41–50. Shows how calcium controls mitochondrial metabolism on the minute time scale.

Lea, P. J., Leegood, R. C., eds. 1993. *Plant Biochemistry and Molecular Biology*. New York: Wiley. A good introduction to the biochemistry and molecular biology of plants.

Lehninger, A. L., Nelson, D. L., and Cox, M. M. 1993. *Principles of Biochemistry*, 2nd edition. New York: Worth. A standard and popular textbook.

Chapter 17. Ions and Voltages

Jan, L. Y., and Jan, Y. N. 1994. Potassium channels and their evolving gates. *Nature* 371:119–122. A discussion of how gated ion channels evolved.

Kapit, W., Macey, R., and Meisami, E. 1987. *The Physiology Coloring Book*. New York: Harper & Row. A highly recommended book that simply explains difficult concepts of ions and voltages.

Matthews, G. G. 1991. *Cellular Physiology of Nerve and Muscle*. 2nd edition. Oxford: Blackwell. A small book on electrical phenomena in cell biology.

Chapter 18. The Action Potential

Matthews, G. G. 1991. *Cellular Physiology of Nerve and Muscle*, 2nd edition. Oxford: Blackwell. A small book on electrical phenomena in cell biology.

Chapter 19. Intracellular Messengers

Berridge, M. J. 1993. Inositol trisphosphate and calcium signalling. *Nature* 361:315–325. A review article from one of the giants of modern cell physiology.

Chapter 20. Intercellular Communication

Hall, Z. W. 1992. *An Introduction to Molecular Neurobiology*. Sunderland, MA: Sinauer. An inclusive text that concentrates on the nervous system.

Chapter 21. Cystic Fibrosis

Collins, F. S., and Wilson, J. M. 1992. Cystic fibrosis: a welcome animal model. *Nature* 358:708–709. Explains why the development of the mouse with its CF gene deleted is such an advance for research into this disease.

Armstrong, J. 1992. Cystic fibrosis: another protein out in the cold. *Nature* 358:709–710. The commonest mutation in CF patients causes a failure to target the channel to the plasmalemma.

And Also . . .

Card, O. S. 1986. *Speaker for the Dead*. New York: Tom Doherty Associates. A novel that gives one person's view of where science might be taking us—and a wonderful, humane view of life and civilization.

GLOSSARY

9 + 2 axoneme: Structure of cilium or flagellum that describes the number and arrangement of microtubules—a ring of 9 doublets around a central 2.

acid: An acid is a molecule that will give an H^+ to water. Acid solutions are those with a pH less than 7.

acrosome: A structure that helps the sperm to penetrate the egg at fertilization.

actin: The protein that makes up microfilaments.

actin-binding proteins: Proteins that bind to and modulate the function of G-actin or F-actin.

action potential: An explosive depolarization of the plasmalemma.

activation energy: The height of the free energy barrier between reactants and products of a chemical reaction.

adenine (A): One of the four bases present in DNA and RNA; adenine is a purine.

adenosine: Adenine linked to the sugar ribose.

adenosine diphosphate (ADP): Adenosine with two phosphoryl groups attached to the 5′ carbon of ribose.

adenosine monophosphate (AMP): Adenosine with one phosphoryl group attached to the 5′ carbon of ribose.

adenosine triphosphate (ATP): Adenosine with three phosphoryl groups attached to the 5′ carbon of ribose. ATP is one of the cell's energy currencies.

adenylate cyclase: An enzyme that converts ATP to the intracellular messenger cyclic AMP (cAMP).

adhering junction: A junction that strongly attaches cells together.

adipocyte: Cells that store fats (triacylglycerols).

adipose tissue: A type of fatty connective tissue.

ADP (adenosine diphosphate): An adenosine with two phosphoryl groups attached to the 5′ carbon of ribose.

ADP/ATP exchanger: A carrier in the inner mitochondrial membrane. ADP is moved in one direction and ATP in the other.

adrenaline: A hormone released into the blood when an individual is under stress.

aerobic: In the presence of air. Used to describe reactions that require oxygen.

agonist: Any extracellular solute that stimulates a cell. An agonist released from another cell is a transmitter, but agonists can also arrive from the outside world or be produced by chemical processes in the extracellular medium.

alkaline: An alkaline solution is one with a pH more than 7.

allele: One of the alternative forms of a gene, each ultimately reflecting a difference in DNA sequence.

allolactose: A disaccharide made of glucose and galactose. Lactose and allolactose are interconverted by the enzyme β galactosidase. An inducer of lac operon transcription.

allosteric; allostery: When the binding of a ligand to one site on a protein affects the binding at another site. These interactions can be between sites for the same ligand or different ligands. Molecules that show allosteric behavior almost always have a quaternary structure.

aminoacyl site (A-site): The site on ribosome occupied by an incoming tRNA and its linked amino acid.

aminoacyl tRNA: tRNA attached to an amino acid by an ester bond.

aminoacyl tRNA synthetase: An enzyme that attaches an amino acid to a tRNA.

amino terminal: The end of a peptide or polypeptide that has a free α-amino group. This end is made first at the ribosome.

AMP (adenosine monophosphate): An adenosine with one phosphoryl group attached to the 5′ carbon of ribose.

anabolism: Those metabolic reactions that build up molecules: biosynthesis.

anaerobic: Without air.

anaerobic respiration: The partial breakdown of sugars and other cellular fuels that can be accomplished in the absence of oxygen.

anaphase: The period of mitosis or meiosis during which sister chromatids or homologous chromosome pairs separate; consists of anaphase A and anaphase B.

anaphase A: Part of the anaphase in which the chromosomes move to the spindle poles.

anaphase B: Part of the anaphase in which the spindle poles are separated.

anaphase I: Anaphase of the first meiotic division (meiosis I).

anaphase II: Anaphase of the second meiotic division (meiosis II).

anemia: A disease characterized by a lack of red blood cells or the hemoglobin they should contain.

aneuploid: Having an incorrect number of chromosomes; often an excess or a deficiency of a single one as a result of nondisjunction.

anterograde: Forwards. When applied to axonal transport it means away from the cell body.

antibiotic: A chemical that kills organisms. The most useful antibiotics to humans are those that are selective for prokaryotes.

antibody: A protein made by the immune system that binds to and helps remove foreign molecules in the body. Because they must not bind to the bodies' own molecules, antibodies are very specific about their targets, called antigens.

anticipation: (in genetics) The tendency for the symptoms of particular inherited diseases to become more severe, or to show their effects at an earlier age, in succeeding generations. Often due to the accumulation of extra copies of short sequences of DNA within the gene.

anticodon: Three bases on a tRNA molecule that hydrogen-bond to the codon on a mRNA molecule.

antigen: A foreign molecule for which the immune system makes an antibody.

AP endonuclease: A DNA repair enzyme that cleaves phosphodiester bonds on either side of a depurinated or depyrimidinated sugar residue. AP stands for *ap*urinic/*ap*yriminic.

apoptosis (programmed cell death): A process in which a cell actively promotes its own destruction, as distinct from necrosis. Apoptosis is important in vertebrate development, where tissues and organ are shaped by the death of certain cell lineages.

aporepressor: A protein that when it is complexed with another molecule binds to an operator region of DNA and represses transcription.

A-site (aminoacyl site): A site on ribosome occupied by an incoming tRNA and its linked amino acid.

assay: A term for a single experiment, for instance one in which the activity of an enzyme reaction is measured.

ATP (adenosine triphosphate): An adenosine with three phosphoryl groups attached to the 5′ carbon of ribose.

ATP synthase: A carrier of the inner mitochondrial membrane. Three H^+ enter the mitochondrial matrix, and a single ATP is made.

autonomic nerves: The nerves that control processes of which we are usually not consciously aware, such as blood vessel size.

axoneme: The microtubule-based structure within a cilium or flagellum.

axon: The long process of a nerve cell, specialized for the rapid conduction of action potentials.

backcross: A technique in genetics in which the offspring of a cross are then crossed with one of the parental strains.

bacteriophage (or **phage**): A virus that infects bacterial cells.

basal: At the base.

basal body: The structure from which cilia and flagella arise; it has the same structure as the centriole.

base: Either *i* a molecule that will accept an H^+ from water or *ii* a type of nitrogen-rich organic molecule, used to make nucleotides.

basement membrane: A thin layer of modified extracellular matrix that supports epithelial cells.

B-DNA: A right-handed DNA double helix.

β-galactosidase: An enzyme that cleaves lactose to produce glucose and galactose, and also catalyzes the interconversion of lactose and allolactose.

binding site: A region of a protein that specifically binds a ligand. A property of the protein's tertiary structure.

bivalents: Paired sister chromosomes seen during the zygotene stage of meiotic prophase.

blue-green algae: Photosynthetic prokaryotes (now known as cyanobacteria).

blunt ends: Ends of a DNA molecule produced by an enzyme that cuts the two DNA strands at sites directly opposite one another.

bright field microscopy: A basic form of light microscopy in which the specimen appears against a bright background.

caffeine: A plant toxin that sensitizes ryanodine receptors.

calcineurin: A phosphatase that is activated by calcium: calmodulin.

calcium ATPase: A carrier that uses the energy released by ATP hydrolysis to move calcium ions up their concentration gradient out of the cytosol. Located both at the plasmalemma and in the membrane of the endoplasmic reticulum.

calcium-binding protein: Any protein that binds calcium. Calmodulin is a calcium-binding protein found in many cells, while troponin is a calcium-binding protein found only in muscle cells.

calcium-induced calcium release: A process in which a rise of calcium concentration in the cytosol triggers the release of more calcium from the endoplasmic reticulum.

calmodulin: A calcium-binding protein found in many cells. When calmodulin binds calcium, it can then activate other proteins such as the enzymes calcineurin and phosphorylase kinase and the carrier calcium ATPase.

cAMP (cyclic adenosine monophosphate): A nucleotide produced from ATP by the action of the enzyme adenylate cyclase. cAMP is an intracellular messenger in many cells.

cAMP-gated channel: A channel found in the plasmalemmae of scent-sensitive nerve cells. The channel opens when cAMP binds to its cytoplasmic face and allows sodium and potassium ions to pass.

cap: A methylated guanine added to the 5′ end of a eukaryotic mRNA.

CAP: Catabolite activator protein—a protein that binds to cAMP—then the CAP:cAMP complex binds within the promoter region of some bacterial operons and helps RNA polymerase to bind to the promoter.

carboxylation: The addition of a carboxyl group (–COOH).

carboxyl terminal: The end of a peptide or polypeptide which has a free α-carboxyl group. This end is made last at the ribosome.

carcinogenesis: The creation of cancer. Carcinogenic chemicals usually work either by causing somatic mutation of cell cycle control proteins, or by selecting for cells that have already so mutated.

cardiac muscle: The type of muscle that makes up the heart.

carrier: An integral membrane protein that forms a tube through the membrane and is never open all the way through. Solutes can move into the tube through the open end. When the channel changes shape so that the end that was closed is open, the solute can leave on the other side of the membrane.

catabolism: Those metabolic reactions that break down molecules to derive chemical energy.

catabolite activator protein (CAP): A protein that binds to cAMP—then the CAP:cAMP complex binds within the promoter region of some bacterial operons and helps RNA polymerase to bind to the promoter.

catalytic rate constant: The proportionality constant which relates the maximal initial velocity (V_m) of an enzyme reaction to the enzyme concentration, $k_{cat}=V_m/[E]$. The units are reciprocal time. See also **turnover number** and **maximal velocity**.

cdc mutants (cell division cycle mutants): Mutants with abnormalities in the cell division cycle. Temperature sensitive cdc mutants of the yeasts *Saccharomyces cerevisiae* and *Schizosaccharomyces pombe* have been particularly useful in determining the proteins that control cell division.

cDNA (complementary DNA): A DNA copy of an mRNA molecule.

cDNA library: A collection of bacterial cells, each of which contains a different foreign cDNA molecule.

cell cycle or **cell division cycle:** An ordered sequence of events that must occur for successful cell division; consists of G1, S, G2, and M phases.

cell division cycle (cdc) mutants: Mutants with abnormalities in the cell division cycle. Temperature sensitive cdc mutants of the yeasts *Saccharomyces cerevisiae* and *Schizosaccharomyces pombe* have been particularly useful in determining the proteins that control cell division.

cell center: The region of the cell where the centrosome and Golgi apparatus are found.

cell junctions: Points of cell-cell interaction in tissues; included are tight junctions, adhering junctions, and gap junctions.

cell membrane: The membrane that surrounds the cell, also known as the plasmalemma.

cell motility: The movement of cells or cell constituents.

cellulose: The major structural polysaccharide of the plant cell wall.

cell wall: A rigid coat that surrounds a cell. The cell wall of plants lies outside the plasmalemma and is composed mainly of cellulose fibers.

centriole: The structure found at the centrosome of animal cells and composed of microtubules.

centromere: The region of the chromosome at which the kinetochore is formed.

centrosome (microtubule organizing center): A structure from which cytoplasmic microtubules arise.

channel: An integral membrane protein that forms a continuous water-filled hole through the membrane.

charge: An excess or deficit of electrons giving a negative or positive charge, respectively.

charged tRNA: A tRNA to which an amino acid is attached.

chiasmata (singular **chiasma**)**:** The structures formed during crossing over between the chromatids of homologous chromosomes during meiosis. The physical manifestation of the genetic recombination of linked genes.

chloroplast: The photosynthetic organelle of plant cells.

chromatin: A complex of DNA and certain DNA-binding proteins such as histones.

chromatophore: The pigment cell of fish and amphibia.

chromosome: A single, enormously long, linear molecule of DNA, together with its accessory proteins. Chromosomes are the unit of organization of the nuclear chromatin and carry many genes.

ciliates: A group of single-celled organisms notable for their large size and large numbers of cilia. *Paramecium* is a ciliate.

cilium (plural **cilia**)**:** The locomotory appendage of some epithelial cells and single celled organisms.

cis-face: The surface of the Golgi apparatus that receives proteins from the endoplasmic reticulum.

cisternae: A flattened membrane-bound sac. The term is especially used of the sacs that make up the Golgi apparatus.

Citric acid cycle (Krebs cycle): The series of reactions in the mitochondrial matrix in which acetate is completely oxidised to CO_2 and water, with the attendant reduction of NAD^+ to NADH and FAD to $FADH_2$.

clathrin: The three-legged, 180kD protein that forms the "coat" of coated vesicles.

cleavage furrow: In animal cells, the structure that constricts the middle of the cell during cytokinesis.

clone: A number of genetically identical individuals.

clone library: A collection of bacterial cells, each of which contains a different foreign DNA molecule.

cloning: Strictly the creation of a number of genetically identical organisms. In molecular genetics, the term is used to mean the multiplication of particular sequences of DNA by an asexual process such as bacterial cell division.

closed promoter complex: A structure formed when RNA polymerase binds to a promoter sequence prior to the start of transcription.

coated pit: The region of the cell membrane that invaginates and is coated with clathrin on the cytoplasmic side. Coated pits pinch off into the cytosol, forming coated vesicles.

coated vesicle: A vesicle enclosed within a basket of the protein clathrin. The endocytotic vesicles produced by receptor-mediated endocytosis are initially coated.

codon: A sequence of three bases in a mRNA molecule that specifies a particular amino acid.

coenzyme: A molecule that acts as a second substrate for a group of enzymes. ATP (and ADP), NADH (and NAD^+), and acyl-coenzyme A (and coenzyme A) are all good examples.

cofactor: A nonprotein molecule or an ion necessary for the activity of a protein. It is closely associated with the protein but can be removed. Examples are pyridoxal phosphate in aminotransferases and zinc in zinc finger proteins.

colchicine: The plant toxin from the autumn crocus, *Colchicum autumnale*, that binds to tubulin.

collagen: The major structural protein of the extracellular matrix found in tissues such as connective tissue.

columnar: A cell that is taller than it is broad, used as a description of some types of epithelial cell.

complementary (of DNA and RNA): Having the appropriate sequence to allow base pair formation.

complementary DNA (cDNA): A DNA copy of a mRNA molecule.

condenser lens: The lens of light and electron microscopes that focuses light (or electrons) onto the specimen.

conditional mutations: Mutations that manifest their effect only under certain conditions, such as in the presence of an antibiotic or high temperature. (See permissive and restrictive conditions.)

connective tissue: A loosely defined term for the material that fills the space between the cells in animal tissues; includes the extracellular matrix.

cortex: The outer part of any organ or structure. For instance, the tissue outside the vascular tissue of plants and the tissue that forms the outer region of the brain are called cortex.

covalent: A covalent bond is a strong attachment between atoms represented as a solid line, e.g. H-H.

covalent modification: A molecule is covalently modified when a group becomes attached to it by a covalent bond.

c-ras: The cellular form of the ras-oncogene.

cristae: A name given to the folds of the inner membrane of mitochondria.

crossing over: The physical exchange of DNA that takes place between homologous chromosomes during recombination and is manifest in the formation of chiasmata.

crosstalk: Two messenger systems show crosstalk when one messenger can produce some or all of the effects of the other.

C terminal (carboxyl terminal): The end of a peptide or polypeptide that has a free α-carboxyl group. This end is made last at the ribosome.

cyanobacteria: The photosynthetic prokaryotes, formerly known as blue-green algae.

cyclic adenosine monophosphate (cAMP): A nucleotide produced from ATP by the action of the enzyme adenylate cyclase. cAMP is an intracellular messenger in many cells.

cyclic guanosine monophosphate (cGMP): A nucleotide produced from GTP by the action of the enzyme guanylate cyclase. cGMP is an intracellular messenger in light sensitive nerve cells (photoreceptors) and vascular smooth muscle cells.

cyclin B: One of the two proteins that make up maturation promoting factor (MPF); one of a family of proteins whose level oscillates (cycles) through the cell division cycle.

cyclin D: A protein that binds to $p33^{cdk2}$ to form a complex that is active at the G1 control point of the cell division cycle.

cyclin E: A protein that binds to $p33^{cdk2}$ to form a complex that is active at the G1 control point of the cell division cycle.

cystine: A double amino acid formed by two cysteine molecules joined by a disulphide bond.

cytochemistry: The use of chemical compounds to stain specific cell structures and organelles.

cytokinesis: A process by which a cell physically divides in two; part of the M-phase of the cell division cycle.

cytology: The study of cell structure by microscopy.

cytoplasm: The insides of a cell; everything inside the plasmalemma except the nucleus. Cytoplasm consists of cytosol plus organelles.

cytoplasmic dynein: A motor protein that moves organelles along microtubules in a retrograde direction.

cytoplasmic streaming: Movement of cytoplasm commonly seen in plant cells and in amoebae, generated by actin and myosin.

cytosine: One of the four bases present in DNA and RNA. Cytosine is a pyrimidine.

cytoskeleton: A cytoplasmic filament system consisting of microtubules, microfilaments, and intermediate filaments.

cytosol: The aqueous medium inside a cell in which organelles are suspended.

deamination: Removal of an amino group from a molecule. Deamination of cytosine to form uracil is a form of DNA damage.

deletion: Loss of part of a chromosome.

dendrite: A branching cell process. The term is commonly used to name those processes of nerve cells that are too short to be called axons.

deoxyribonucleic acid (DNA): A polymer of deoxyribonucleotides. DNA specifies the inherited instructions of a cell.

deoxyribonucleotide: A monomeric unit of DNA that is made up of a nitrogenous base and the sugar deoxyribose to which a phosphoryl group is attached.

deoxyribose: The sugar residue found in DNA.

dephosphorylation: The removal of a phosphoryl group from a molecule.

depolarization: Any positive shift in the transmembrane voltage, whatever its size or cause.

depurination: The removal of either of the purine bases, adenine and guanine, from a molecule. Depurination of DNA is a common form of DNA damage.

desmin: A protein that makes up the intermediate filaments in muscle cells.

desmosomes: A type of adhering junction common in epithelial cells such as skin.

diakinesis: In meiosis, prophase I is divided into five stages: leptotene, zygotene, pachytene, diplotene, and diakinesis.

dideoxyribonucleotide: A man-made molecule similar to a deoxyribonucleotide but lacking a 3′-hydroxyl group on its sugar residue. Used in DNA sequencing.

dihybrid cross: A cross involving parents differing in genetic constitution at two loci.

differentiation: The process whereby a cell becomes specialized for a particular function.

diploid: Containing two sets of chromosomes.

diplotene: In meiosis, prophase I is divided into five stages: leptotene, zygotene, pachytene, diplotene, and diakinesis.

distal: Far away, usually as in far away from the cell body, or the center, or the start.

disulphide bridge (bond): A covalent bond between two sulfur atoms. In proteins it forms by oxidation of two thiol (–SH) groups of cysteine residues. Expecially common in extracellular proteins.

DNA (deoxyribonucleic acid): A polymer of deoxyribonucleotides. DNA specifies the inherited instructions of a cell.

DNA fingerprint: The individual pattern of DNA fragments determined by the number and position of specific repeated sequences.

DNA ligase: An enzyme that joins two DNA molecules by catalyzing the formation of a phosphodiester bond.

DNA polymerase: An enzyme that synthesizes DNA by catalyzing the formation of a phosphodiester bond. DNA is always synthesized in the 5′ to 3′ direction.

DNA repair enzymes: Detect and repair altered DNA.

DNA replication: A process in which the two strands of the double helix unwind and each acts as a template for the synthesis of a new strand of DNA.

docking protein (signal recognition particle receptor): A receptor on the endoplasmic reticulum to which the signal recognition particle binds.

domain: A separately folded segment of the polypeptide chain of a protein.

dominant: An allele that manifests its phenotype even when in heterozygous condition with a recessive allele.

downstream: A general term meaning the direction in which things move. When applied to the DNA within and adjacent to a gene, it means lying on the side of the transcription start site that is transcribed, that is, towards the 'S' end of the DNA molecule.

double helix: The structure formed when two filaments wind about each other, most commonly applied to DNA.

duplication: (as a form of mutation) The doubling-up of a particular sequence of the genetic material.

dynamic instability: A term that describes the behavior of microtubules in cells, where growing microtubles tend to keep growing, and those that are shortening tend to keep shortening.

dynein: A motor protein of the microtubule cytoskeleton. There are two types of dynein: Dynein arms and cytoplasmic dynein. Cytoplasmic dynein moves organelles along microtubules in a retrograde direction.

dynein arms: Structures that power ciliary and flagellar beating by generating sliding between adjacent outer doublet microtubules.

ectoplasm: A viscous, gel-like outer layer of cytosol.

effective stroke: The part of the beat cycle of a cilium that moves the extracellular medium, in contrast to the recovery stroke.

electrically excitable: Able to produce action potentials.

electrochemical gradient: The free energy gradient for an ion in solution. The arithmetical sum of the gradients due to concentration and voltage.

electron gun: The source of electrons in electron microscopes.

electron transport chain: The series of electron acceptor/donator molecules found in the inner mitochondrial membrane that transport electrons from the reduced coenzymes NADH and $FADH_2$ (thus reoxidizing them) to oxygen. The entire complex forms a carrier that uses the energy of NADH and $FADH_2$ oxidation to transport hydrogen ions up their electrochemical gradient out of the mitochondrion.

electrophoresis: A method of separating charged molecules by drawing them through a filtering gel material using an electrical field.

endocytosis: The inward budding of vesicles off the plasmalemma—a process by which cells take up material from their surroundings.

endocytotic vesicle: An endocytosed coated vesicle loses its clathrin coat to form an endocytotic vesicle.

endonuclease: An enzyme that splits nucleic acids by breaking internal phosphodiester bonds. See **exonuclease**.

endoplasm: A fluid, inner layer of cytoplasm that moves during cytoplasmic streaming.

endoplasmic reticulum (abbr. **ER**)**:** A network (reticulum) of membrane-delimited tubes and sacs that extends from the outer membrane of the nuclear envelope almost to the cell membrane. There are two types of ER, rough endoplasmic reticulum with a surface coating of ribosomes and smooth endoplasmic reticulum.

endosymbiotic theory: A proposal that the organelles of eukaryotic cells originated as free living prokaryotes.

enhancer: A specific DNA sequence to which a protein binds and in turn increases the rate of transcription of a gene.

endothelium: A layer of cells that lines blood vessels and other body cavities that do not open to the outside.

endothelial cells: Cells that line blood vessels and other body cavities that do not open to the outside.

energy currency: A source of energy for many cellular processes. The four energy currencies are NADH, ATP, the H^+ gradient across the mitochondrial membrane, and the Na^+ gradient across the plasmalemma.

enzyme: A protein that acts as a catalyst, reducing the activation energy for a particular reaction.

epidermis: The protective outer cell layer of an organism.

epithelial cells: The cells that make up an epithelium.

epithelium: The sheet of cells covering the surface of the human body and those internal cavities such as the lungs and gut that open to the outside.

equilibrium: A process or object is in equilibrium if the tendency to go in one direction is exactly equal to the tendency to go in the other direction. For ion movement this condition is equivalent to saying that the electrochemical gradient for that ion is zero.

equilibrium voltage: The transmembrane voltage that will exactly balance the concentration gradient of a particular ion. The equilibrium voltage can be calculated (using the Nernst equation) for each ion at a membrane; all the values will usually be different.

ER (endoplasmic reticulum): The network (reticulum) of membrane channels that extends from the outer membrane of the nuclear envelope almost to the cell membrane. There are two types of ER, rough endoplasmic reticulum with a surface coating of ribosomes and smooth endoplasmic reticulum.

erythrocyte: A red blood cell.

ester bond: The bond formed between an alcohol group and a carboxyl group. Water is formed when this bond is made.

euchromatin: That portion of the nuclear chromatin that contains genes that code for proteins; see also heterochromatin.

eukaryotic: A cell containing a distinct nucleus and other organelles. Almost all cells except bacteria and cyanobacteria (blue-green algae) are eukaryotic.

exon: Sequences within a gene that encode protein. Exons are separated by introns.

exocytosis: The release of material from a cell by a process in which a vesicle moves to the plasmalemma and the two membranes then fuse, releasing the vesicle contents to the extracellular medium

exonuclease: An enzyme that digests nucleic acids by cleaving phosphodiester bonds successively from one end of the molecule.

expression (of a gene): The appearance of the protein for which the gene codes.

expression vector: A cloning vector containing a promoter sequence. This enables a foreign DNA insert to be transcribed into mRNA.

extracellular matrix: The solid material outside the cells of a multicellular organism. The major structural elements of the matrix are proteins.

F-actin: Actin in its polymerized, microfilaments form.

fat cell (adipocyte): Cells that store fats (triacylglycerols).

fibroblast: A cell that secretes components of connective tissue.

fibroblast growth factor (FGF): A paracrine transmitter that causes many cells, including fibroblasts, to divide.

fibroblast growth factor receptor (FGFR): A metabotropic cell surface receptor that activates cell division upon binding its ligand, fibroblast growth factor.

First Law: Mendel's first law states that characters are controlled by pairs of alleles that segregate during gamete formation and are restored at fertilization.

fluid mosaic model: A suggested structure for the cell membrane in which integral membrane proteins float in a sea of lipid.

focal contact: The points at which a locomoting cell makes contact with its substrate.

formyl methionine (fmet): A methionine modified by the attachment of a formyl group. fmet is the first amino acid in all newly made bacterial polypeptides.

frameshift mutation: A mutation that changes the mRNA reading frame, caused by the insertion or deletion of bases.

G0: Describes the quiescent state of cells that have left the cell division cycle.

G1 (gap 1): The period of the cell division cycle that separates mitosis from the following S phase.

G2 (gap2): The period of the cell division cycle between the completion of S phase and the start of cell division or M phase.

G-actin: A globular subunit form of the cytoskeletal protein actin.

gamete: A haploid sex cell (sperm or egg), carrying just a single copy of the genetic message.

gap 1 (G1): The period of the cell division cycle that separates mitosis from the following S phase.

gap2 (G2): The period of the cell division cycle between the completion of S phase and the start of cell division or M phase.

gap junctions: A type of cell junction that allows solutes to be shared between the cytosol of neighbouring cells.

gap junction channel: A channel that runs all the way through the plasmalemma of one cell, across a small extracellular gap, and through the plasmalemma of a second cell. Solutes of MW < 1000 can pass from the cytosol of one cell to the cytosol of the other.

gastrocnemius muscle: The muscle at the back of the shin. When it contracts, the toes move down.

gated, gating: A channel is gated if it can switch to a shape in which the tube through the membrane is closed.

gelsolin: A type of actin-binding protein that binds to and fragments actin filaments.

gene: The fundamental unit of heredity. It often contains the information needed to code for a single polypeptide.

gene family: A group of genes that share sequence similarity and usually code for proteins with a similar function.

gene probe: A radiolabeled cDNA or genomic DNA fragment used to detect a specific DNA sequence to which it is complementary in sequence.

gene therapy: A correction or alleviation of a genetic disorder by the introduction of a normal gene copy into an affected individual.

genetic code: The relationship between the sequence of the four bases in DNA and the amino acid sequence of proteins.

genome: The complete set of genes in an organism.

genomic DNA library: A collection of bacterial cells, each of which contains a different fragment of foreign genomic DNA.

genotype: The allelic constitution of a particular individual; the term can be used either for a single locus or to refer to the entire genome.

germ cells: Cells that give rise to the eggs and sperm.

glial cells: An electrically inexcitable cell found in the nervous system.

glucocorticoid: Steroid hormones produced by the adrenal cortex that form part of the system controlling blood sugar levels.

glycerol:

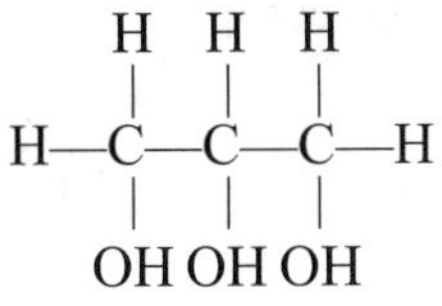

```
     H   H   H
     |   |   |
 H—C—C—C—H
     |   |   |
    OH OH OH
```

Glycerol forms the backbone of many lipids.

glucocorticoid receptor: The intracellular receptor to which glucocorticoid hormones bind.

gluconeogenesis: The synthesis of glucose from noncarbohydrate precursors such as amino acids and lactate.

glucose carrier: A plasmalemmal protein that carries glucose into or out of cells.

glycogen: A glucose polymer that can be quickly broken down to yield glucose 1-phosphate.

glycogen phosphorylase: An enzyme that acts on glycogen to release glucose 1-phosphate for use in respiration.

glycosidic bond: A bond joining two sugar molecules through a C-O-C linkage. As a glycosidic bond forms, a water molecule is lost.

glycosylation: The addition of sugar residues to a molecule. Both proteins and lipids can be glycosylated.

glyoxisome: A type of peroxisome found in plant cells and concerned with the conversion of fatty acids to carbohydrate.

Golgi apparatus: A system of flattened cisternae concerned with glycosylation and other modifications of proteins destined for secretion.

grana: Distinctive structures within the chloroplast formed by the stacking of the thylakoid membranes.

growth factor: A paracrine transmitter that modifies the developmental pathway of the target cell, often by causing cell division.

guanine: One of the four bases found in DNA and RNA; guanine is a purine.

hairpin loop: A loop in which a linear object folds back on itself. Used of the loop formed in a RNA molecule due to complementary base pairing.

haploid: Containing a single copy of each chromosome.

head group: The hydrophilic portion of a phospholipid or other membrane lipid.

helicase: An enzyme that helps unwind the DNA double helix during replication.

hemicellulose: A polysaccharide component of the plant cell wall; it links cellulose fibrils together.

hemizygous: Having only one copy of a gene because the gene is on a sex chromosome.

hemoglobin: The oxygen-carrying, iron-containing protein of the blood.

heritability: The proportion of the total phenotypic variation in a population that is due to genetic variation.

hermaphrodite: An individual that is simultaneously male and female.

heterochromatin: That portion of the nuclear chromatin that consists of highly repetitive DNA with no coding function.

heterozygote: A diploid genotype involving two different alleles at a particular locus.

histone: A positively charged protein that binds to negatively charged DNA and helps to fold DNA into chromatin.

histone octamer: Two molecules each of histones H2A, H2B, H3, and H4, the whole combining with DNA to form a nucleosome.

homophilic: Likes to bind to similar molecules.

homozygote: A diploid genotype involving two copies of the same allele at a particular locus.

hormones: Long-lived transmitters that are released into the blood and travel around the body before being broken down.

hormone response element (HRE): A specific DNA sequence to which a steroid hormone receptor binds.

housekeeping gene: A gene that is transcribed into mRNA nearly all the time. In a multicellular eukaryote a housekeeping gene is one that is transcribed in almost all the cells; in bacteria a housekeeping gene is one that is always being transcribed.

HRE (hormone response element): A specific DNA sequence to which a steroid hormone receptor binds.

hybridization: The association of unlike things. In molecular genetics, the association of two nucleic acid strands (either RNA or DNA) by complementary base pairing.

hydrogen bond: A relatively weak bond formed between a hydrogen atom and two electron-grabbing (e.g., nitrogen or oxygen) where the hydrogen is shared between the other atoms.

hydrogen ion gradient: An energy currency. Hydrogen ions are more concentrated outside the mitochondrion than inside, and this chemical gradient is supplemented by a voltage gradient pulling hydrogen ions in. If hydrogen ions are allowed to rush in down their electrochemical gradient, they release 20,000 $Jmol^{-1}$.

hydrolyse, hydrolysis: To hydrolyse is to break a covalent bond by adding water. A simple example is the hydrolysis of pyrophosphate to phosphate:

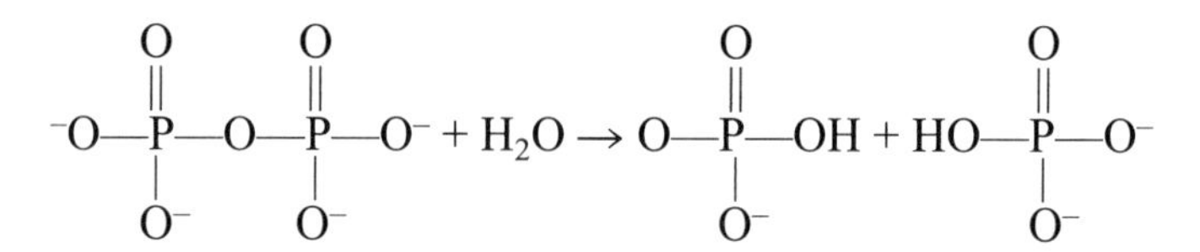

hydrophilic: Describes a molecule or part of a molecule that can interact with water.

hydrophobic: A molecule or part of a molecule that is unable to interact with water and so tends to associate with other hydrophobic molecules.

ionizing radiations: High energy radiation such as X rays and gamma rays.

inactivation (of voltage-gated channels): Blockage of the open channel with a plug that is attached to the cytosolic face of the protein.

Inclusion cell disease (I cell disease): A human disease in which affected individuals synthesize their lysosomal proteins normally, but these proteins are not targeted to the lysosomes.

indirect immunofluorescence microscopy: A form of light microscopy that uses a fluorescent antibody to localize specific subcellular structures.

inducible operon: An operon that is transcribed only when a specific substance is present.

initial velocity (of a reaction): The rate at which an enzyme converts substrate to product in the absence of product.

initiation complex: A complex of transcription factors bound to the TATA box that enables RNA polymerase to bind at the beginning of a gene to be transcribed. Also known as the preinitiation complex.

inositol trisphosphate: A small (MW = 420) phosphorylated molecule that is released into the cytosol by the action of phospholipase C on the membrane lipid phosphatidylinositol bisphosphate and that acts to cause release of calcium ions from the endoplasmic reticulum.

inositol trisphosphate-gated calcium channel: A channel found in the endoplasmic reticulum of many cells. The channel opens when inositol trisphosphate binds to its cytosolic aspect. It allows only calcium ions to pass.

inositol trisphosphate receptor: Another name for the inositol trisphosphate-gated calcium channel.

in situ hybridization: The binding of a particular labeled sequence of RNA (or DNA) to its matching sequence in the genome as a way of searching for the location of that sequence.

integral protein (of a membrane): A class of protein that is tightly associated with a membrane (cf. peripheral protein).

intermediate filament: One of the filaments that makes up the cytoskeleton; it is composed of various subunit proteins.

intermembrane space: In organelles such as mitochondria, chloroplasts, and nucleii that are bound by two membranes the aqueous space between the inner and outer membranes is called the intermembrane space. The intermembrane space of nucleii is continuous with the lumen of the ER. The intermembrane space of mitochondria has the ionic composition of cytosol because porin in the outer mitochondrial membrane allows solutes of MW $<10{,}000$ to pass.

interphase: A period of synthesis and growth that separates one cell division from the next; it consists of the G1, S, G2, phases of the cell division cycle.

intracellular messenger: A cytosolic solute that changes in concentration in response to external stimuli or internal events, and that acts on intracellular targets to change their behavior. Calcium ions, cyclic AMP, and cyclic GMP are the three common intracellular messengers.

intracellular receptors: Receptors that are not on the plasmalemma but which lie within the cell and bind transmitters that diffuse through the plasmalemma.

intron: Segments of a gene that do not code for protein. Introns are removed during RNA splicing.

inversion: A re-ordering of the chromosomal material involving the reversal of the order of the genetic material within a chromosome.

ion: An atom or molecule that has gained or lost electrons and is therefore charged.

ionotropic cell surface receptors: Channels that open when a specific agonist binds to the extracellular face of the channel protein.

IP_3 (Inositol trisphosphate): A small (MW = 420) phosphorylated molecule that is released into the cytosol by the action of phospholipase C on the membrane lipid phosphatidylinositol bisphosphate and that acts to cause release of calcium ions from the endoplasmic reticulum.

IP_3-gated calcium channel: A channel found in the endoplasmic reticulum of many cells. The channel opens when inositol trisphosphate binds to its cytosolic aspect. It allows only calcium ions to pass.

IP_3 receptor: Another name for the IP_3-gated calcium channel.

isomers: Molecules composed of the same numbers of atoms arranged differently.

keratin: Protein that makes up the intermediate filaments in epithelial cells.

kinesin: The molecular motor protein responsible for movement along microtubules in the anterograde direction.

kinetochore: The point of attachment of the chromosome to the spindle. The kinetochore forms around the centromere.

Krebs cycle (tricarboxylic acid cycle): The series of reactions in the mitochondrial matrix in which acetate is completely oxidized to CO_2 and water with the attendant reduction of NAD^+ to NADH and FAD to $FADH_2$.

lactose: a disaccharide made of glucose and galactose.

lac (lactose) operon: The cluster of three bacterial genes that encodes enzymes and carriers involved in metabolism of lactose.

lagging strand: The strand of DNA that grows discontinuously during replication.

lamins: Proteins that make up the nuclear lamina. Lamins are chemically related to intermediate filaments.

leading strand: A strand of DNA that grows continuously in the 5′ to 3′ direction by the addition of deoxyribonucleotides.

leptotene: In meiosis, prophase I is divided into five stages: leptotene, zygotene, pachytene, diplotene, and diakinesis.

ligand: When two molecules bind together, one (often the smaller one) is called the ligand, and the other (often the bigger one) is called the receptor.

lignin: A component of the plant cell wall. It is a member of the class of chemical compounds called polyphenols.

linkage, linked (of genes): The physical association of genes on the same chromosome. Linked genes tend to be inherited together.

linkage map: A map of the order of genes based on crossing experiments and on deviations from free recombination.

linker DNA: A stretch of DNA that separates two nucleosomes.

lipid: An ester of glycerol and fatty acid.

lipid bilayer: Two layers of lipid molecules that form a membrane.

locus: The physical position of a gene on the chromosome.

lumen: The inside of any bag- or tube-shaped structure.

lysosome: A membrane-bound organelle containing digestive enzymes.

M-phase: Period of the cell division cycle during which the cell divides; consists of mitosis and cytokinesis.

macrophage: A phagocytic housekeeping cell of the blood that engulfs and digest bacteria and dead cells.

major groove (of DNA): The larger of the two grooves along the surface of the DNA double helix.

malignant: A term that describes the ability of some tumors to spread around the body.

maturation-promoting factor (MPF): A complex of $p34^{cdc2}$ and cyclin B that regulates the G2/M phase transition of the cell division cycle.

maximal velocity (of a chemical reaction): The limiting value of the initial velocity of an enzyme reaction as the substrate concentration is increased. Occurs when the enzyme is saturated with substrate.

meiosis: The form of cell division that produces gametes, each with half the genetic material of the cells that produce them.

meiosis I and II: The first and second meiotic divisions.

meiotic spindle: A bipolar, microtubule-based structure on which chromosome segregation occurs during meiosis I and II.

melanin: A dark brown insoluble pigment that is a complex polymer made chiefly from dihydroxyphenylalanine (DOPA), itself derived from tyrosine.

membrane voltage: The voltage difference between one side of a membrane and the other. It is usually stated as the voltage inside with respect to outside. Also called the transmembrane voltage or membrane potential.

messenger RNA (mRNA): The coding RNA molecule that is made on DNA then leaves the nucleus and specifies the amino acid sequence of a polypeptide chain.

metabolism: All of the chemical reactions in a living cell.

metabotropic cell surface receptors: Receptors in the plasmalemma that are linked to, and activate, enzymes.

metaphase: The period of mitosis or meiosis at which the chromosomes align prior to separation at anaphase.

metaphase plate: The equator of the spindle; point at which the chromosomes congregate at metaphase of mitosis or meiosis.

Michaelis constant: The substrate concentration at which one measures an initial velocity that is half as fast as the maximal velocity (V_m) of an enzyme reaction.

microfilament: One of the important filaments of the cytoskeleton; also known as actin filament or F-actin. The thin filament of striated muscle is now known to be a microfilament.

microtubule: A component of the cytoskeleton. Microtubules are polymers of tubulin subunits.

microtubular molecular motor: A protein that moves organelles along microtubules. Examples are cytoplasmic dynein and kinesin.

microtubule organizing center (abbr. **MTOC**): The structure from which cytoplasmic microtubules arise; synonymous with the centrosome.

microvilli (singular **microvillus**): Projections from the surface of epithelial cells that increase the absorptive surface, and contain actin filaments.

minor groove (of **DNA**): The smaller of the two grooves along the surface of the DNA double helix.

missense mutation: A base change in a mRNA molecule that changes the codon sense from one amino acid to that for another.

mitochondrial inner membrane: The inner membrane of mitochondria that is elaborated into christae.

mitochondrial matrix: The aqueous space inside the mitochondrial inner membrane where the enzymes of the Krebs cycle are located.

mitochondrial outer membrane: The outer membrane of mitochondria that is permeable to solutes of MW <10,000 because of the presence of the channel porin.

mitochondrion (plural, mitochondria): The cell organelle concerned with aerobic respiration. In mammals, the vast majority of ATP is made in mitochondria.

mitosis: The type of cell division found somatic cells, in which each daughter cell receives the full complement of genetic material present in the original cell.

monomer, monomeric: On its own, not attached to another similar molecule.

mononucleate: Having only one nucleus. This is the normal cellular condition.

motoneurone: The nerve cell that carries action potentials from the spine to the muscles. It releases the transmitter called acetylcholine onto the muscle cells, causing them to depolarize and hence contract.

MPF (maturation promoting factor): A complex of $p34^{cdc2}$ and cyclin B that regulates the G2/M phase transition of the cell division cycle.

M phase: The period of the cell division cycle during which the cell divides; it consists of mitosis and cytokinesis.

mRNA (messenger RNA): The RNA molecule that carries the genetic information from DNA to the ribosome. The order of bases on mRNA specifies the amino acid sequence of a polypeptide chain.

MTOC (Microtubule organizing center): The structure from which cytoplasmic microtubules arise; it is synonymous with the centrosome.

multinucleate: Having more than one nucleus. Skeletal muscle cells are multinucleate.

muscle fibers: The large, multinucleate cells of which skeletal muscles are made.

mutagen: A physical or chemical agent that gives rise to mutations.

mutation: An inherited change in the structure of a gene or chromosome.

myelin: A fatty substance that is wrapped around nerve cell axons by glial cells.

N terminal (amino terminal): The end of a peptide or polypeptide that has a free α-amino group. This end is made first at the ribosome.

necrosis: Cell death that is due to damage so severe that the cell cannot maintain the level of its energy currencies and therefore falls apart, as distinct from apoptosis.

negative feedback: Any system in which a change in some parameter activates a mechanism that reverses the change in that parameter.

negative regulation (of transcription): The inhibition of transcription due to the presence of a particular substance that is often the end product of a metabolic pathway.

Nernst equation: This allows the equilibrium voltage of an ion across a membrane to be calculated (Box 17.1).

nerve cell: Electrically excitable cells with long axons specialized for transmission of (usually sodium) action potentials.

nervous tissue: A tissue formed of nerve and glial cells that carries out electrical data processing.

neurofilaments: The type of intermediate filaments found in nerve cells.

neuron: A nerve cell.

neuronal transport: The ATP-dependent movement of material along a nerve cell process; it can be outward (anterograde) or inward (retrograde).

nicotinamide adenine dinucleotide (NADH, NAD^+): A combination of two nucleotides. The reduced form, NADH, is a strong reducing agent and one of the cell's energy currencies.

nitric oxide (NO): A paracrine transmitter that acts on intracellular receptors, the most important of which is guanylate cyclase.

node (of a myelinated axon): The gap between adjacent glial cells where the nerve cell plasmalemma contacts the extracellular medium.

nondisjunction: A failure of separation of chromosomes into daughter cells during meiosis or mitosis, leading to aneuploidy.

nonsense mutation: A base change that causes an amino acid codon to become a stop codon.

nuclear envelope: The double membrane system enclosing the nucleus; it contains nuclear pores and is contiguous with the endoplasmic reticulum.

nuclear lamina: The meshwork of lamin fibers lining the inner face of the nuclear envelope.

nuclear pores: Holes running through the nuclear envelope that allow and regulate the traffic of proteins and nucleic acids between the nucleus and the cytoplasm.

nuclease: An enzyme that degrades nucleic acids.

nucleic acid: A polymer of nucleotides joined together by phosphodiester bonds. DNA and RNA are nucleic acids.

nucleolar organizer regions: A region of one or more chromosomes at which the nucleolus is formed.

nucleolus (plural, nucleoli): Region(s) of the nucleus concerned with the production of ribosomes.

nucleoside: A purine or pyrimidine base attached to either ribose or deoxyribose.

nucleosome: A beadlike structure formed by a stretch of DNA wrapped around a histone octamer.

nucleotide: A purine or pyrimidine base attached to either ribose or deoxyribose that has a phosphoryl group on its 5′-carbon atom.

nucleus: The cell organelle housing the chromosomes and enclosed within a nuclear envelope.

objective lens: The lens of light or electron microscope that forms a magnified image of a specimen.

Okazaki fragment: A series of short fragments that are joined together to form the lagging strand during DNA replication.

olfactory neurone: The nerve cell found in the nose, whose dendrites are sensitive to smell chemicals.

oligonucleotide: A short DNA fragment.

oligosaccharide: Short chain of sugar residues.

oligosaccharide transferase: The enzyme that adds an oligosaccharide group onto a protein.

oncogene: A gene that, upon mutation, causes cancer. Oncogenes code for proteins with important functions in the control of the cell division cycle.

oocyte: A cell that undergoes meiosis to give rise to an egg.

open promoter complex: A structure formed when the two strands of the double helix separate so that transcription can commence.

operator: The DNA sequence to which a repressor protein binds to prevent transcription from an adjacent promoter.

operon: A cluster of genes encoding proteins involved in the same metabolic pathway.

origin of replication: The site on a chromosome at which DNA replication can commences.

organelle: A membrane-bound, intracellular structure such as a mitochondrion or chloroplast.

osmosis: The movement of water across a semipermeable membrane.

outer doublet microtubules: The paired microtubules that make the "9" of the 9 + 2 axoneme.

oxidize, oxidation: Molecules are oxidized when they lose electrons in a reaction, for instance, when oxygen atoms are added to them or hydrogen atoms are removed from them.

P-site (peptidyl site): Site on ribosome occupied by the growing polypeptide chain.

pachytene: In meiosis, prophase I is divided into five stages: leptotene, zygotene, pachytene, diplotene, and diakinesis.

pain receptor: A nerve cell whose distal axon terminal is depolarized by potentially damaging events such as heat or stretching.

pain relay cell: The nerve cell upon which a pain receptor synapses and which carries the message on toward the brain.

paracrine transmitters: Agonists released by cells into the extracellular fluid that can last many minutes and can therefore diffuse widely within the tissue before they are destroyed.

patch clamp: A technique in which a glass micropipette is sealed to the surface of a cell to allow electrical recording of cell properties.

PCR (polymerase chain reaction): A method for making many copies of a DNA sequence whose sequence, at least at each end, is known.

pectin: The polysaccharide component of the plant cell wall that determines its thickness.

peptide: A short linear polymer of amino acids.

peptide bond: The bond between amino acids. The bond is formed between the carboxyl group of one amino acid and the amino group of the next.

Peptidyl site (P-site): Site on ribosome occupied by the growing polypeptide chain.

peptidyl transferase: An enzyme that catalyzes the formation of a peptide bond between two amino acids.

peripheral protein: A class of protein that is easily detached from a cell membrane (see integral protein).

permissive conditions: The environmental conditions that allow a conditional lethal mutation to survive (see restrictive conditions).

peroxisome: A class of cell organelles of diverse function that frequently contains the enzyme catalase.

phage: Short for bacteriophage, a virus that infects bacteria.

phase contrast microscopy: A type of light microscopy in which differences in refractive index of a specimen are converted into differences in brightness.

phenotype: The physical appearance or behavior of an organism due to an interaction between its genotype and the environment in which it is placed.

phosphate: Properly, a name for the ions $H_2PO_4^-$, HPO_4^{2-}.and PO_4^{3-}. The word is also very commonly used to mean the group

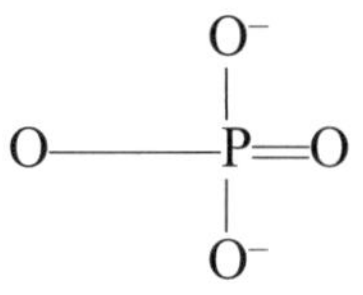

```
        O⁻
        |
O ──────P═O
        |
        O⁻
```

which is properly called a phosphoryl group.

phosphatidylinositol bisphosphate (PIP_2): A membrane lipid that releases inositol trisphosphate into the cytosol upon hydrolysis by phospholipase C.

phosphodiester bond: The bond X—O—P

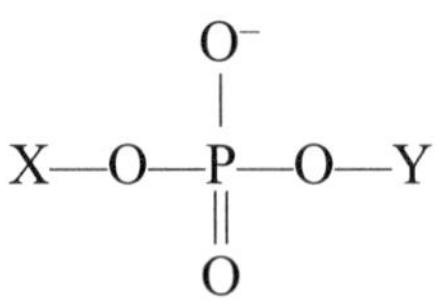

```
          O⁻
          |
X—O—P—O—Y
          ‖
          O
```

where X and Y are any two molecules. In nucleic acids, phosphodiester bonds join the adjacent nucleotides.

phospholipase C: An enzyme that hydrolyzes the membrane lipid phosphatidylinositol bisphosphate (PIP_2) to release inositol trisphosphate into the cytosol.

phospholipid: A glycerol molecule linked by ester bonds to two fatty acids and a hydrophilic head group. Membranes are made of phospholipid.

phosphoryl group: The group

$$O^{-}\text{—}\overset{O^{-}}{\underset{O^{-}}{|}}\!\!P{=}O$$

which is commonly, but incorrectly, called a phosphate group.

phosphorylase kinase: A kinase that is activated by the calcium: calmodulin complex and phosphorylates glycogen phosphorylase, activating the latter enzyme.

phosphorylation: The addition of the phosphoryl group to a molecule. Usually added to an hydroxyl group to form a phosphoester bond. Sometimes it is added to an acid to form a phosphoanhydride.

phragmoplast: The structure associated with the formation of a new cell wall during cytokinesis in plant cells.

physical map: A map of the order of the genes based on the physical examination of chromosomes or DNA rather than the interpretation of crossing experiments (cf. linkage map).

Pi: Inorganic phosphate, that is, the ions $H_2PO_4^-$, HPO_4^{2-} and PO_4^{3-}.

PIP_2 (Phosphatidylinositol bisphosphate): A membrane lipid that releases inositol trisphosphate into the cytosol upon hydrolysis by phospholipase C.

PKA (Protein kinase A): An enzyme that transfers a phosphoryl group from ATP to a protein. It is activated by the intracellular messenger cAMP.

plaque: An area of dead bacteria in a lawn of live bacteria that is caused by infection of bacterial cells by a bacteriophage.

plasmalemma: The membrane that surrounds the cell. Often called the cell membrane.

plasma membrane: The plasmalemma.

plasmid: A circular DNA molecule that replicates independently of the host chromosome in bacterial cells.

plasmodesmata (singular **plasmodesma**)**:** A type of cell junction unique to plant cells that provides a much bigger hole for passage of substances between the cytoplasm of the two cells than do gap junctions.

PLC: Phospholipase C: An enzyme that hydrolyzes the membrane lipid phosphatidylinositol bisphosphate (PIP_2) to release inositol trisphosphate into the cytosol.

polyadenylation: The process whereby a poly A tail is added to the 3′ end of a eukaryotic mRNA.

poly A tail: A string of adenine residues added to the 3′ end of a eukaryotic mRNA.

polycistronic mRNA: An mRNA that, when translated, yields more than one polypeptide.

polymerase: An enzyme that makes polymers, that is, long chains of identical or very similar subunits. DNA and RNA polymerase respectively, are involved in making DNA and RNA.

polymerase chain reaction (PCR): A method for making many copies of a DNA sequence whose sequence, at least at each end, is known.

polymorphism: The existence of alternative allelic forms at a particular locus within a population.

polypeptide: A polymer of more than fifty amino acids joined by peptide bonds.

polyploid: An individual with three or more sets of chromosomes.

polyribosome: A chains of ribosomes attached to an mRNA molecule—also known as a polysome.

polysome (polyribosome): A chain of ribosomes attached to an mRNA molecule.

porin: A channel found in the mitochondrial outer membrane. It is always open and allows all solutes of MW <10,000 to pass.

positive feedback: A process in which the consequences of a change act to increase the magnitude of that change so that a small initial change tends to get bigger and bigger.

positive regulation (of transcription): A process whereby transcription is activated in the presence of a particular substance.

postsynaptic cell: The cell upon which a nerve cell releases its transmitter at a synapse.

preinitiation complex: A complex of transcription factors bound to the TATA box that enables RNA polymerase to bind at the beginning of a gene to be transcribed. Also known as the initiation complex.

presynaptic terminal: An axon terminal that is specialized for exocytosis of transmitter.

primary cell wall: The first layer of the plant cell wall.

primary endosome: An acidic cell compartment with which endocytotic vesicles fuse.

primary structure (of a protein): The sequence of amino acids held together by peptide bonds making up a polypeptide.

primase: The enzyme that synthesizes the RNA primers needed for the initiation of synthesis of the leading and lagging DNA strands.

primer: A short sequence of nucleic acid (RNA or DNA) that acts as the start point at which a polymerase can initiate synthesis of a longer nucleic acid chain.

primosome: The protein complex, including the enzyme primase, involved in the synthesis of RNA primers for DNA replication.

profilin: A type of actin-binding protein that regulates the assembly of actin filaments.

programmed cell death (apoptosis): A process in which a cell actively promotes its own destruction, as distinct from necrosis. Apoptosis is important in vertebrate development, where tissues and organ are shaped by the death of certain cell lineages.

projector lens: The lens of light or electron microscope that carries image to the eye, more commonly known as the "eyepiece."

prokaryotic: A type of cellular organization found in bacteria in which the cells lack a distinct nucleus and other organelles.

prometaphase: A period of mitosis or meiosis that sees the breakdown of the nuclear envelope and the attachment of the chromosomes to the mitotic spindle.

promoter: The region of DNA to which RNA polymerase binds to initiate transcription.

pronuclei: The nuclei of the egg and sperm prior to fusion.

prophase: The period of mitosis or meiosis in which the chromosomes condense.

prophase I: The first meiotic prophase, subdivided into five stages: leptotene, zygotene, pachytene, diplotene, and diakinesis.

prophase II: The prophase of the second meiotic division (meiosis II).

prosthetic group: A nonprotein molecule necessary for the activity of a protein. The concept overlaps with the concept of cofactor. The difference is just how tightly they are bound: A prosthetic group is very tightly bound and cannot be removed without at least partial unfolding of the protein. Examples are the heme groups of myoglobin, haemoglobin, and the cytochromes.

protein: A polypeptide with a complex three-dimensional shape.

protein kinase: An enzyme that transfers a phosphoryl group from ATP to a protein.

protein kinase A (PKA): An enzyme that transfers a phosphoryl group from ATP to a protein. It is activated by the intracellular messenger cAMP.

protein phosphorylation: The addition of a phosphoryl group to a protein. The addition of the charge on the phosphoryl group can markedly alter the tertiary structure and therefore function of a protein.

protein targeting: Delivery of proteins to their correct cellular location.

protofilament: The chains of subunits that make up the wall of a microtubule.

proto-oncogene: A normal cellular gene that can give rise to a cancer forming oncogene by mutation or translocation.

protozoa: Simple single-celled animals.

proximal: Close to, usually as in close to the cell body, or the start, or the center.

pseudogene: A gene that has mutated such that it no longer codes for a protein.

pseudopodium: A projection extended by an amoeba or other crawling cell in the direction of movement.

P-site (peptidyl site): The site on ribosome occupied by the growing polypeptide chain.

purine: A nitrogenous base found in nucleotides; adenine and guanine are purines.

pyrenoid: A starch-synthesizing structure unique to the chloroplasts of algae.

pyrimidine: A nitrogenous base found in nucleotides; cytosine, thymine, and uracil are pyrimidines.

pyrophosphate: The ion

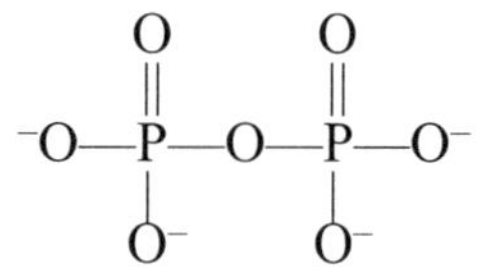

$$
\begin{array}{ccccccccc}
 & & O & & & & O & & \\
 & & \| & & & & \| & & \\
^{-}O & - & P & - & O & - & P & - & O^{-} \\
 & & | & & & & | & & \\
 & & O^{-} & & & & O^{-} & &
\end{array}
$$

In the cytosol pyrophosphate is rapidly hydrolyzed to two phosphate ions.

quaternary structure: The structure in which subunits of a protein, each of which has a tertiary structure, associate to form a more complex molecule. The subunits associate tightly but not covalently, and may be the same or different.

ras oncogene: The gene coding for the ras protein. Mutations in ras are very common in cancer cells.

ras protein: The product of thc ras oncogene; a protein that forms part of the pathway signaling a cell to enter the cell division cycle.

Rb: A type of tumor suppressor gene; mutations lead to the formation of retinoblastomas in the eye.

reading frame: A reading of the genetic code in blocks of three bases—there are three possible reading frames for each mRNA only one of which will produce the correct protein.

receptor: A protein that specifically binds a particular solute, called its ligand. Receptors can be transmembrane or cytosolic proteins. Particular receptor proteins perform other functions (ion channel, enzyme, activator of endocytosis, transcription factor, etc.) that may be activated by the binding of the solute. Proteins are often named for the ligand they bind, even when binding that ligand is not something the protein is designed to do—an example being the ryanodine receptor.

receptor-mediated endocytosis: A process by which ligands bind to specific receptors in the cell membrane for internalization.

recessive: An allele that does not manifest its phenotype when it is present in heterozygous condition with a dominant allele.

recombinant, recombination: The production of new phenotypes during sexual reproduction, often used especially of the production of new combinations of alleles at different loci through crossover at meiosis.

recombinant plasmid: A plasmid into which a foreign DNA sequence has been inserted.

recovery stroke: The part of the beat cycle of a cilium in which extracellular fluid is not moved, in contras to the effective stroke.

reduce, reduction: Molecules are reduced when they gain electrons in a reaction, for instance when hydrogen atoms are added to them or oxygen atoms are removed from them.

reducing agent: A molecule, such as NADH, that donates electrons to other molecules (and in doing so becomes oxidized).

refractive index: A measure of the speed of light in a material. The higher the refractive index, the slower does light move.

repetitious DNA: A DNA sequence that is repeated many times within the genome.

replication fork: The Y-shaped structure formed when the two strands of the double helix separate during replication.

repressible operon: An operon whose transcription is repressed in the presence of a particular substance—often the final product of the metabolic pathway.

residue: A monomer that is part of a polymeric structure; for example, an amino acid linked to other amino acids in a polypeptide is said to be an amino acid residue.

resolving power: Defines the smallest object that can be distinguished in a microscope.

resting voltage: The voltage across the plasmalemma of an unstimulated cell, typically -70 to -90 mV.

restriction enzyme endonuclease: An enzyme that cleaves phosphodiester bonds within a specific sequence in a DNA molecule.

restrictive: The environmental conditions, such as temperature, that are fatal to a conditional lethal mutation.

reticulocyte: The cells that synthesize the blood protein hemoglobin and give rise to red blood cells or erythrocytes.

retrograde: Backwards. When applied to axonal transport it means towards the cell body.

retrovirus: A virus whose genetic information is stored in RNA.

reverse transcriptase: An enzyme of some viruses that copies RNA into DNA.

reverse transcription: The process whereby RNA is copied into DNA.

ribonuclease: An enzyme that cleaves phosphodiester bonds in RNA.

ribonuclease H: An enzyme that cleaves phosphodiester bonds in an RNA molecule that is joined to a DNA molecule by complementary base pairing.

ribonucleic acid (RNA): A polymer of ribonucleoside monophosphates; see **mRNA, tRNA**, and **rRNA**.

ribonucleoside monophosphate: A nitrogenous base attached to the sugar ribose which has one phosphoryl group on its 5′-carbon atom; also known as a ribonucleotide.

ribose: A sugar used to make RNA.

ribosomal RNA (rRNA): Both the large and small subunits of a ribosome contain their specific types of RNA.

ribosome: The particle, made of both RNA and protein, where proteins are made.

ribozyme: An RNA molecule with enzymelike catalytic activity.

RNA (ribonucleic acid): A polymer of ribonucleoside monophosphates; see mRNA, tRNA, and rRNA.

RNA polymerase: An enzyme that synthesizes RNA.

RNA splicing: The removal of introns from an RNA molecule and the joining together of exons to form the final RNA product.

rough endoplasmic reticulum: Endoplasmic reticulum with ribosomes attached, concerned with the synthesis of proteins destined for various organelles or secretion.

rRNA (ribosomal RNA): Both the large and small subunits of a ribosome contain their specific types of RNA.

ryanodine: A plant toxin that binds to the ryanodine receptor, with complex effects on gating of the channel.

ryanodine receptor: A calcium channel found in the membrane of the endoplasmic reticulum. It opens in response to a rise of calcium concentration in the cytoplasm and therefore tends to cause calcium-induced calcium release.

σ factor (sigma factor): A subunit of bacterial RNA polymerase that recognizes the promoter sequence.

S phase: The phase of the cell division cycle during which new DNA is synthesized.

S value (sedimentation coefficient): A value that describes how fast particles sediment in a centrifuge.

saltatory conduction: The jumping of an action potential from node to node down a myelinated axon.

salt bridge: An interaction in protein structure between a positively charged amino acid residue (e.g., arginine) and a negatively charged residue (e.g., aspartate).

sarcomere: A contractile unit of striated muscle.

sarcoplasmic reticulum: A type of smooth endoplasmic reticulum found in muscle and cells and highly specialized for the storage and rapid release of calcium ions.

satellite DNA: A DNA sequence that is tandemly repeated many times.

saturated (of **fatty acids**)**:** Containing no carbon-carbon double bonds.

saturated (of an enzyme or carrier)**:** Working as fast as it can - see saturation kinetics.

saturation kinetics: A condition whereby the rate of a reaction does not continue to increase as the concentration of reactants increases. The rate becomes limited by the availability of binding sites on the catalyst for the reactant.

Schwann cell: A type of glial cell.

secondary cell wall: The layers of a plant cell wall formed external to the primary cell wall.

secondary structure: A regular, repeated folding of the backbone of a polypeptide. The side chains of the amino acids have an influence but are not directly involved.

Second Law: Mendel's second law states that patterns of allele segregation at one locus are independent of those at other loci.

secretory vesicles: The vesicles derived from the trans Golgi network that transport secreted proteins to the cell membrane with which they fuse.

sedimentation coefficient (S value): A value that describes how fast particles sediment in a centrifuge.

segregation: The separation of alleles as the gametes are formed; heterozygotes produce two different classes of gamete, and homozygotes only one.

self-compatible: A description of those organisms that allow the eggs of a hermaphrodite to be fertilized by sperm (or pollen) from the same individual if gametes from another individual are not available.

selfing: The formation of a zygote from the gametes of a single individual organism.

semiconservative replication: Both strands of the double helix serve as templates for the synthesis of new daughter strands.

sex chromosomes: Chromosomes whose constitution differs in males and females. In mammals, one of them, the Y chromosome, is very short and carries only a few genes.

sex linkage, sex linked inheritance: Association of particular loci with the sex chromosomes. For instance, in mammals the Y carries a testis-determining factor.

Shine-Dalgarno sequence: A sequence on a bacterial mRNA molecule to which the ribosome binds.

side chain (of polypeptides): The group attached to the a carbon of an amino acid.

sigma factor (σ factor): A subunit of bacterial RNA polymerase that recognizes the promoter sequence.

signal peptidase: The enzyme that cleaves the signal sequence from a polypeptide as it enters the lumen of the endoplasmic reticulum.

signal sequence: A short stretch of amino acids within polypeptides which targets them to the endoplasmic reticulum.

signal recognition particle: The ribonucleoprotein particle that recognizes and binds to the signal sequence of a polypeptide.

signal recognition particle receptor: The receptor on the endoplasmic reticulum to which the signal recognition particle binds.

simple diffusion: The process by which uncharged solutes can enter cells by passing through the lipid component of the membrane.

single-stranded binding protein: The protein that binds to the separated DNA strands to keep them in an extended form during replication and thus preventing the double helix from reforming.

sister chromatids: At mitosis or meiosis, the paired, replicated chromosomes attached at their centromeres.

skeletal muscle fibers: Large multinucleate muscle cells that are attached at one end to bone. Most cuts of meat are mainly skeletal muscle.

small nuclear RNAs: Small RNA molecules found in the nucleus that play a role in RNA splicing.

smooth endoplasmic reticulum: A portion of the endoplasmic reticulum that stores calcium ions to be released on cell stimulation. Synthesis of lipids and steroids also occurs here.

smooth muscle: The muscle that forms part of blood vessels and visceral organs such as the uterus.

smooth muscle cells: Small mononucleate muscle cells that make up smooth muscle.

sodium action potential: An action potential driven by the opening of sodium channels and the resulting sodium influx.

sodium-calcium exchanger: A carrier in the plasmalemma. Three sodium ions move into the cell, and one calcium is moved out up its concentration gradient.

sodium gradient: An energy currency. Sodium ions are more concentrated outside the cell than inside, and this chemical gradient is usually supplemented by a voltage gradient pulling sodium ions in. If sodium ions are allowed to rush in down their electrochemical gradient, they release 14 $Jmol^{-1}$.

sodium/potassium ATPase: A plasmalemmal carrier. For every ATP hydrolyzed, three Na^+ ions are moved out of the cytosol and two K^+ ions are moved in.

somatic cells: Cells that make up all the tissues of the human body except the germ cells.

sorting signal: A length of peptide within a protein that causes the cell to direct the protein to a specific cell compartment such as the nucleus or mitochondrion. The first sorting signal discovered was the signal sequence, which targets a protein to the endoplasmic reticulum.

sorting vesicle: A vesicle in which proteins destined for distinct membrane compartments are separated.

specific activity: The ratio between activity and amount of material. For an enzyme it is the ratio between catalytic activity and mass of protein, an indicator of the purity of an enzyme preparation.

specificity constant: The ratio between the catalytic rate constant (k_{cat}) and the Michaelis constant (K_M) for an enzyme. It is a rate constant with dimensions of liter $mol^{-1}s^{-1}$ and is used to compare different substrates for the same enzyme or to compare the effectiveness of one enzyme with another.

spermatid: A cell formed by meiosis and which differentiates to form a to spermatozoon.

spermatozoon: The motile male gamete.

S phase: A period of the cell division cycle during which DNA replication occurs.

spindle: The microtubule—based structure responsible for moving the chromosomes during anaphase.

splicesome: A complex of proteins and small RNA molecules involved in RNA splicing.

squamous: Flat, especially used to describe epithelial cells.

START: The point in G1 where cells can leave the cell division cycle.

stereocilia: Sensory projections on the surface of the hair cells of the vertebrate cochlea. Stereocilia contain microfilaments.

steroid hormone: These agonists act on intracellular receptors to activate transcription of particular genes.

sticky ends (of DNA): The short single-stranded ends produced by cleavage of the two strands of a DNA molecule at sites that are not opposite to one another.

stratified epithelium: A type of epithelium consisting of several layers, such as is found in the skin.

stress fiber: A bundle of actin filaments commonly seen in cultured, non-motile animal cells.

striated muscle: Muscles whose cytoskeleton is organized into obvious sarcomeres, giving them a striped appearance in a microscope. Skeletal and cardiac muscles are striated.

summation (at synapses): The additive effects of more than one presynaptic action potential upon the postsynaptic voltage.

superoxide dismutase: The enzyme that catalyses the formation of hydrogen peroxide from the highly reactive superoxide ion.

S value (sedimentation coefficient): A value that describes how fast ribosomal subunits and rRNA molecules sediment in a centrifuge.

synapse: The structure formed from the axon terminal of a nerve cell and the adjacent region of the postsynaptic cell at which transmitter released by the axon terminal diffuses to and acts upon the postsynaptic cell.

synapsis: The alignment of chromosomes during the zygotene stage of meiotic prophase I.

synaptonemal complex: The structure essential for recombination that is present between bivalents during the pachytene stage of meiosis I.

tandem repeats: Many copies of the same DNA sequence that lie side by side on the chromosome.

TATA box: A sequence found about 20 bases upstream of the beginning of many eukaryotic genes that forms part of the promoter sequence and is involved in positioning RNA polymerase for correct initiation of transcription.

taxol: A compound obtained from the bark of the Pacific yew *Taxus brevifolia* that binds to tubulin; it is a powerful anticancer drug.

telophase: The final period of mitosis or meiosis in which the chromosomes decondense and the nuclear envelope reforms.

telophase I: The telophase of the first meiotic division (meiosis I).

telophase II: The telophase of the second meiotic division (meiosis II).

temperature-sensitive (of a mutation): A type of mutation (e.g., a cdc mutant) in which the gene product is functional at one temperature (the permissive temperature) and inactive at another (the restrictive temperature).

tetraploid: Having four copies of each chromosome, rather than the two copies in a normal diploid.

terminally differentiated: A term that describes a cell that cannot return to the cell division cycle.

tertiary structure: The three-dimensional folding of a polypeptide chain into a biologically active protein molecule. It usually includes regions of secondary structure. Interactions of the amino acid side chains are central in its formation.

thermogenin: An H^+ channel found in the inner mitochondrial membrane of brown fat cells. It acts as an uncoupler.

threshold (voltage): The plasmalemmal transmembrane voltage at which enough calcium or sodium channels open to create an action potential.

thick filament: One of the two filaments that generate the contraction of striated muscle; composed of the protein myosin.

thin filament: One of the two filaments that generate the contraction of striated muscle; composed of actin.

thylakoid: The folded inner membrane of the chloroplast and site of the light reaction of photosynthesis.

thymine: One of the four bases found in DNA; thymine is a pyrimidine.

tight junctions: A type of cell junction in which a tight seal is formed between adjacent cells occluding the extracellular space.

Ti plasmid: A naturally occurring plasmid used in plant genetic engineering experiments.

tissue: A group of cells in the body that have a common function.

topoisomerase: An enzyme that cuts and reforms DNA in response to strain. Topoisomerase I relieves rotational strain in one chromosome by cutting and reforming one DNA strand. Topoisomerase I is essential during DNA replication since, unless the helix can rotate, the two strands cannot be separated. Topoisomerase II relieves the stress when two chromosomes are tangled by cutting an entire chromosome, passing another chromosome through the gap, then joining the cut ends.

transcription: The synthesis of a RNA molecule from a DNA template.

transcription complex: A complex of RNA polymerase and various transcription factors.

transcription factor: A protein (other than RNA polymerase) that is required for gene transcription.

trans-face (of the **Golgi apparatus**)**:** The face distant from the nucleus from which vesicles leave for the trans Golgi network and eventual secretion.

transfection: The introduction of foreign DNA into a eukaryotic cell.

transfer RNA (tRNA): The RNA molecule that carries an amino acid to a mRNA template.

transferrin: An iron-containing protein of the blood, actively taken up into reticulocytes.

transformation: In molecular genetics, the introduction of foreign DNA into a cell.

transgenic animal: An animal carrying a gene from another organism, the foreign gene is usually injected into the nucleus of a fertilized egg.

translation: The synthesis of a protein molecule from a mRNA template.

translocation: Movement. When used of the ribosome, it means the movement, three nucleotides at a time, of the ribosome on the mRNA molecule. When used of chromosomes it means the transfer of one chromosomal segment to another, non-homologous, chromosome.

transmembrane proteins: A class of proteins that spans the cell membrane. Transmembrane proteins are always integral membrane proteins.

transmembrane voltage: The voltage difference between one side of a membrane and the other. It is usually stated as the voltage inside with respect to outside. Also called the membrane voltage or membrane potential.

transmitter: A chemical that is released by one cell and which acts as an agonist on another cell.

transport vesicle: A membrane vesicle that transports proteins from one membrane compartment to another.

transposase: An enzyme involved in the transposition of certain DNA segments from one site to another within the genome.

triacylglycerol: Three fatty acids joined to glycerol. Triacylglycerols are completely hydrophobic and are familiar to us as fats and oils.

tricarboxylic acid cycle (Krebs cycle): The series of reactions in the mitochondrial matrix in which acetate is completely oxidized to CO_2 and water, with the attendant reduction of NAD^+ to NADH and FAD to $FADH_2$.

tRNA (transfer RNA): The RNA molecule that carries an amino acid to a mRNA template.

troponin: A calcium-binding protein found in muscle cells.

tryptophan (trp) operon: A cluster of five bacterial genes involved in the synthesis of the amino acid tryptophan.

tubulin: A subunit protein of microtubules that exists as α-, β-, and γ-forms.

tumor: The proliferative cell mass associated with many cancers.

tumor suppressor gene: A gene concerned with the detection of DNA damage; if mutated, it can cause the cell to become cancerous.

turgid: Stiffened by internal pressure acting against a flexible, inextensible, outer wall.

turnover number: The number of moles of substrate converted to product per mole of enzyme per unit time under conditions where the enzyme is saturated with substrate. Another term for the catalytic rate constant, k_{cat}.

tyrosine kinase: An enzyme that transfers a phosphoryl group from ATP to a tyrosine residue on a protein. Many growth factor receptors are tyrosine kinases.

ultramicrotome: A machine for cutting thin sections (<100 nm) for electron microscopy.

ultrastructure: The fine structure of the cell and its organelles revealed by electron microscopy.

ultraviolet (UV) absorbance: The degree to which ultraviolet light (below 400 nm in wavelength) is absorbed by a molecule. Molecules absorb light of different wavelength because of differences in their structures.

uncouples, uncoupler, uncoupled (of mitochondria): A mitochondrion is uncoupled when the tight link between the electron transport chain and ATP synthesis is broken. Since the link is the gradient of H^+ across the inner mitochondrial membrane, any chemical that allows H^+ ions to cross the inner mitochondrial membrane is an uncoupler. If the chemical is present in more than a small fraction of the cells of the body, it will kill, but expression of the uncoupler thermogenin in brown fat allows that tissue to generate heat.

unsaturated (of fatty acids): Containing carbon-carbon double bonds.

untranslated sequence: The sequence of bases in a mRNA molecule that does not code for protein and is found at the 5′ and 3′ ends of an mRNA.

upstream: A general term meaning the direction from which things have come. When applied to the DNA within the transcription start site that is not transcribed, that is, towards the 3′ end of the DNA molecule.

uracil: One of the four bases found in RNA; uracil is a pyrimidine.

uracil-DNA glycosidase: A DNA repair enzyme that recognizes and removes uracil from DNA molecules.

vacuole: A membrane-bound compartment of plant cells containing sugars and pigments.

valium: An anti-anxiety drug that increases the chance that the GABA receptor channel will open.

variable number tandem repeats (VNTRs): DNA sequences that are repeated many times within the human genome. Each person carries a different number of these repeats.

vascular tissue: Blood vessels. The term is used by analogy to describe the water transporting and support tissue of plants that is composed of xylem and phloem.

vasoconstrictor: Anything that constricts blood vessels.

vector: Something that carries something else. The term is often used of a plasmid or bacteriophage that carries a foreign DNA molecule and is capable of independent replication within a bacterial cell.

vesicle: A closed, more or less spherical membranous bag within a cell.

villin: A type of actin-binding protein that cross-links actin filaments.

vimentin: Protein that makes up the intermediate filaments in fibroblasts and other cells.

villus (plural **villi**)**:** A finger-like fold of epithelial surface.

viral oncogene: An altered copy of an oncogene carried by a virus.

virus: A packaged fragment of DNA or RNA that uses the synthetic machinery of a host cell to replicate its component parts.

VNTRs (Variable number tandem repeats): DNA sequences that are repeated many times within the human genome. Each person carries a different number of these repeats.

voltage clamp: A technique in which a current is passed to one side of a membrane to artificially maintain the value of the transmembrane voltage at a desired level.

voltage-gated calcium channel: A channel found in the plasmalemma of many cells that is selective for calcium ions and that opens upon depolarization.

voltage-gated sodium channel: A channel found in the plasmalemma of nerve and muscle cells that is selective for sodium ions and that opens upon depolarization.

v-ras: A viral form of the ras-oncogene that codes for a ras protein which is permanently switched on so that cells infected by the virus are constantly triggered to divide.

wobble (in tRNA binding): Flexibility in the base pairing between the 5′ position of the anticodon and the 3′ position of the codon.

Xeroderma pigmentosum: An inherited human disease caused by defective DNA repair enzymes. Affected individuals are sensitive to ultraviolet light and contract skin cancer when exposed to sunlight.

Z-disc: Structures that delineate the borders of the sarcomere in striated muscle cells.

Z-DNA: The left-handed helical form of DNA.

Zellweger's syndrome: An inherited human disease resulting from aberrant targeting of proteins to the peroxisome.

zinc fingers: A structural motif in some families of DNA binding proteins in which a zinc ion coordinated by cysteines and histidines stabilizes protruding regions that "finger" the edges of the base pairs exposed in the major groove of DNA.

zygote: The fertilized egg.

zygotene: In meiosis, prophase I is divided into five stages: leptotene, zygotene, pachytene, diplotene, and diakinesis.

INDEX

The suffix G indicates a glossary entry. This is usually a good place to begin reading about a subject, so glossary entries, if present, are listed first. F, T and B indicate figures, tables, and boxes, respectively.

–10 box, 148–149
–35 box, 148–149
2-oxoglutarate, 265F
  conversion to glutamate, 247F
  in Krebs cycle, 265
30S subunit, of ribosome, 194
3-dimensional structure of proteins, 219–230
  specification in primary structure, 230
3-hydroxybutyrate, 272B
3-phosphoglycerate, 267F
  in carbon fixation, 279
  in gluconeogenesis, 273F
  in glycolytic pathway, 267
50S subunit, of ribosome, 194
5-bromo uracil, as mutagen, 125
70S initiation complex, 194–195
9+2 axoneme, 361G, 50
α-adrenergic receptor, 332, 334F, 336
α-amino acid, 210
α-blocker, 332
ABO blood groups, 97B, 99, 99B
Acanthamoeba, 41
acceptor (in hydrogen bond), 19B
acetoacetate, 272B
acetoacetyl ACP, 278
acetone, on breath, 272B
acetylcholine,
  at motoneurone:muscle synapse, 335–336
  in control of blood supply, 337–338
  receptor
    muscarinic, 337–338
    nicotinic, 332–333, 335–336
acetylcholinesterase,
  at synapse, 336
  inhibition, 339B
  specificity constant, 244T
acetyl CoA, 264
Achilles tendon, 334
achondroplastic dwarfism, 98, 120
acid, 361G, 17B
  as amino acid side group, 211, 213
  organic, naming the atoms, 210
ACP (acyl carrier protein), 275
acridine mustard, 125
acridine orange, 125
acrosome, 361G, 53
actetate, as breakdown product of acetylcholine, 336
actin, 361G, 53–56
  at cell junctions, 29, 30B
  -binding proteins, 361G, 53
  in cytoplasmic streaming, 56

actin (*continued*)
in cell locomotion, 56
in muscle, 54–55
target for phalloidin, 339B
action potential, 361G, 303–316
activation energy, 361G, 245
acyl carrier protein, 275
address labels, on proteins, 200, 202F
adenine, 361G, 73–74
bonding with thymine, 76–77
adenosine, 361G
adenosine diphosphate, 361G
receptor, 321, 332
sign of cell damage, 321
adenosine monophosphate, 361G
adenosine triphosphate, 361G, 34, 254F
as an energy currency, 253
adenylate cyclase, 361G, 325–326
adhering junction, 361G, 29, 30B
adipocyte, 361G
adipose tissue, 362G, 11
ADP, 362G
receptor, 321, 332
sign of cell damage, 321
ADP/ATP exchanger, 362G, 260, 300B
ADP-ribose, and diptheria toxin, 197
adrenal gland, 338
adrenaline, 362G, 338
acts mainly at β receptor, 332
promotion of glycogen breakdown, 281–282, 324F, 328
action on blood vessels, 337–338
adrenergic receptor, 332
α, 332, 334F, 336
β, 324F, 328, 332, 338
aequorin, 208B
aerobic, 362G
aerotaxis, 47
agarose gel, 182F
agonist, 362G, 318
α helix , 222–223
as chiral structure, 224B
alanine, 215
aminotransferase, 247F
conversion to pyruvate, 247F
genetic code, 84F
albino, 100
Alexander the Great, 87
algae, 10
alkaline, 362G, 18B
alkaline phosphatase, use in DNA cloning, 176F
allele, 362G, 91
alligator, sex determination, 117B
allolactose, 362G, 152–153
galactosides and the lac operon, 152–156
all or nothing, of action potentials, 308
allostery, 362G, 236
alpha-adrenergic receptor, 332, 334F, 337–338
alpha-blocker, 332
alpha helix, 222–223
as chiral structure, 224B
ALS (amyotrophic lateral sclerosis), 131B
Amanita muscaria, 338, 339B
Amanita phalloides, 339B
Ames, Bruce, 125
test, 125
amino acid, 210F
attachment to tRNA, 191F
breakdown to feed Krebs cycle, 269–271
synthesis, 278
aminoacyl site, 362G, 192
aminoacyl tRNA, 362G
synthetase, 362G, 191
amino terminal, 362G
aminotransferase
mechanism, 246–247
in amino acid synthesis, 278
amoeba, 1, 56
AMP, 362G
amplitude modulation, 315B
amyotrophic lateral sclerosis, 131B
anabolism, 362G
anaerobic, 362G
direction of energy flow, 260
glycolysis, 268
anaphase, 362G, 67–68
anemia, 363G
sickle-cell, 99B
nature of mutation, 86, 127
treatment, 345
thalassemia, 128
aneuploidy, 363G, 127
angina pectoris, 338
animal cell, summary, 9F
animalcules, 41
anterograde, anterograde transport, 363G, 51–52
anthocyanin, 10
anti-anxiety drugs, 344B
antibiotic, 363G, 200
resistance, 80, 120
in DNA cloning, 170
antibody, 363G, 42–43
manufacture in bacteria, 183
anticipation, in genetics, 363G, 130
anticodon, 363G, 190
antigen, 363G, 42
anti-mutagens, 125
antiparallel
β sheet, 222, 225F
DNA structure, 76–77

anxiety, 344B
AP endonuclease, 363G, 143
apoptosis, 363G, 59
  role of p53 in response to DNA damage, 70
aporepressor, 363G, 156–157
arginine, 84F, 212
arum lily, 262B
A-site (aminoacyl site), 363G, 192
asparagine, 84F, 214
aspartate, 211
  in genetic code, 84F
  phosphorylation, 218–219, 295–296
assay, 364G, 241
ATP, 364G, 34, 252–254
ATPase, calcium, 295–297, 325
ATPase, sodium/potassium, 259
ATP synthase, 364G, 258, 301B
atropine, 339B
autonomic nerves, 364G, 336– 337
autophagic vacuole, 38
autumn crocus, 45
Avery, Oswald, 78
axon, 364G
  axonal transport, 51–52
  electrical transmission, 310–314
axoneme, 364G, 50

backcross, 364G, 109–110
backward reaction, 241
bacteria, 2B
  flagellum and motility, 47–48
  bacterial lawn, in DNA cloning, 175
bacteriophage, 364G, 82
  in DNA cloning, 170, 175
β-adrenergic receptor, 324F, 328, 332, 338
Bam H1, 171T, 172F
basal, basal body, 364G, 44, 49F, 50F
base, 364G
  amino acid side chain, 213
  analogues, as mutagens, 125
  as opposed to acid, 18B
  in DNA, 73, 76–77
  pair, 76–77
basement membrane, 364G, 11–12
β barrel structure, of proteins, 225
β-blocker, 332
B blood group, 97B, 99, 99B
B-DNA, 364G, 78
beam scanner, 5F
belt desmosomes, 29
berserk, 339B
beta-adrenergic receptor, 324F, 328, 332, 338
beta barrel structure, of proteins, 225
beta-blocker, 332
beta galactosides and galactosidase, 364G
  β-glycosidic bond, 27
  in lac operon, 152–156
  reactions catalyzed, 153F, 240
beta-galactosidase permease, 301B
beta oxidation of fatty acids, 269, 271F
beta pleated sheet, 222, 225F
beta sheet, 222, 225F
β galactosides and galactosidase, 364G
  β-glycosidic bond, 27
  in lac operon, 152–156
  reactions catalyzed, 153F, 240
β-galactoside permease, 301B
binding, binding site, 364G
  of molecules to proteins, 235
  of mRNA on ribosome, 193
  of ions on proteins, 217
bivalents, 364G, 67F, 69
blood
  clotting, 320–321
  flow regulation, 336–339
  groups, 97B, 99, 99B
  vessels, 13, 339
Bloom's syndrome, 144B
blotting, of DNA, 181–182
blue-green algae, 364G, 10
blunt ends, 364G, 170
bond
  ester, 373G, 22
  glycosidic, 376G, 27
  hydrogen, 377G, 19B
    between bases in DNA, 76–77
    in α helix, 222–223
    in β sheet, 222, 225F, 226F
    in protein structure, 222, 227B
  peptide, 210
  phosphodiester, 386G, 76, 147F
  van der Walls, 227B
borage, 40B
β_0-thallassemia, 159B
β oxidation of fatty acids, 269, 271F
β pleated sheet, 222, 225F
breakdown of food chemicals, 263–283
breathing, of proteins, 235
bright field microscopy, 364G, 3, 6F
brown fat, 262B
brush border, 12F
β sheet, 222, 225F
budding yeast, 63
building blocks, of proteins, 210–217

cadherin, 30B
caffeine, 364G, 329B
calcineurin, 364G, 164B, 248B

calcium
ATPase, 365G, 295–297, 301B, 325
action potential, 306–308
activation of NO synthase, 338
binding protein, 365G
binding sites, 217, 228F
cell locomotion and motility, 56
channel
IP_3 gated, 299B, 320, 322F
mitochondrial, 299B
ryanodine receptor, 329B
voltage gated, 299B, 305
chloride, in DNA cloning, 171
control of glycogen breakdown, 281–282
control of mitochondria, 281
detection with aequorin, 208B
induced calcium release, 365G
influx from extracellular medium, 318–319
intracellular messenger, 317–325
release from endoplasmic reticulum, 36, 320, 322F
removal from cytosol, 323
stimulation of gene transcription, 339
trigger for exocytosis, 205
calf muscle, 334–335
calming, by Valium, 344B
calmodulin, 365G, 228F
activation of NO synthase, 338
calcineurin activation, 164B
regulator of phosphorylase kinase, 281–282
structure, 225, 228F
study by stopped-flow, 248B
camouflage, 51–52
cAMP, 365G, 325–327
dependent protein kinase, 281–282
gated channel, 365G, 300B, 325–327
deriviation during evolution, 228
gene expression triggered by, 152, 155F
intracellular messenger, 325–327
skeletal muscle cells, role in, 324F, 328
smooth muscle cells, role in, 337–338
structure, 326F
cancer, 70–71, 101, 125
cap (on mRNA), 365G, 158
CAP (catabolite activating protein), 365G
structure, 226, 229F
function, 152, 155F
capping, of mRNA, 365G, 158
carbon fixation, 278–279
carboxylation, 365G
carboxyl group, 17B
carboxyl terminal, 365G
carcinogenesis, 365G, 125
cardiac muscle, 365G, 13
carrier, 365G, 292–297, 300B
enzymic action, 295–297
exchange mechanisms for energy currencies, 256–261
glucose, 293
catabolism, 365G
catabolite activating protein, 365G
structure, 226, 229F
function, 152, 155F
catalase
function, 126
k_{cat}, 241
location, 36
specificity constant, 244T
catalyst, 240
catalytic rate constant, 366G, 240
catenin, 30B
Cat's Cradle, 233B
cdc2, 62, 65F
cdc (cell division cycle) mutants, 366G, 62–64
cDNA, 366G, 168
cell center, 366G, 4F, 37, 44
cell cycle, 366G, 59–65
control points, 60
cell division, 59–71
cdc mutants, 366G, 62–64
promotion by growth factors, 339, 31B
spindle structure, 44
cell, 1
energy currencies, 251–256
junctions, 366G, 29
locomotion, 54–56
membrane, 366G, 9F, 21
motility, 366G, 46–53, 54–56
mitosis, 66
surface receptor, 332
theory, 1
cellulose, 366G, 24, 28B, 45
cell wall, 366G, 9, 24, 29
central dogma, 79
centriole, 366G, 44
location in cell, 4F
absence in plants, 10
centromere, 366G, 66–67
centrosome, 366G, 44
CF (Cystic Fibrosis), 345–353
gene product, 229–230
inheritance, 95, 106
CFTR, 351
cGMP (cyclic GMP), G369, 327–328
elevation in response to NO, 333, 337–338
gated channel, 328
photoreceptors, role in, 327
smooth muscle cells, role in, 337–338

chain initiation
  at bacterial ribosome, 193
  at eukaryotic ribosome, 199
chain release factors, 197, 199F
channel, 366G, 291, 298B
  chloride-selective, GABA, 343B
  potassium, 286
  solute-gated, 320, 325, 332
chaotropic agents, 231
Chargaff, Erwin, 76
charge, 367G
  on amino acid side group, 217
charged tRNA, 367G
chemotaxis, 47–48
Chernobyl, 122
chiasma, chiasmata, 367G, 69, 110
chimpanzee, 114
Chinese restaurant syndrome, 318
chiral structures, 224B
chloramphenicol, 200
chloride channel
  GABA, 343–344
  product of cystic fibrosis gene, 229–230, 351
chloride ion gradient, 291
chloroplast, 367G, 34–35, 10
  appearance, 9F, 35F
  carbon fixation, 278
  cytoplasmic streaming, 56
  energy flux, 260
  genetic entities, 100
  origin as cyanobacteria, 10
cholera, 325
choline
  breakdown product of actetylcholine, 336
  in phospholipid structure, 22F
chromatid, 66–67
chromatin, 367G, 9F, 33, 80
chromatophore, 367G, 51–52
chromosome, 367G, 33
  site of genes, 105
  changes at mitosis, 33, 62, 67F
  packaging of DNA, 80–81
cigarettes, 125, 339B
cilia, 367G, 46, 49F, 50F
  on epithelia, 11
ciliates, 367G, 49
cis-face, 367G
cisternae, 367G, 37
citrate, in Krebs cycle, 265
citric acid cycle, 367G, 264–266
classification of transmitters, 331–334
clathrin, 367G
  in endocytosis, 37
  in Golgi apparatus, 205
  in sorting vesicle, 206F
cleavage furrow, 367G, 67–68
cleavage sites, of restriction enzymes, 171T
clone, 367G
clone library, 367G, 171
cloning, 367G
  cDNA, 168–175
  gene, 175–178
  vector, 169–170
closed promoter complex, 367G, 148
clotting of blood, 320–321
coated pit, 367G, 37
coated vesicle, 368G, 37
code, genetic, 84F
coding, by action potential frequency, 315B
codon, 368G, 82
coenzyme, 368G
  coenzyme A, 264
cofactor, 368G, 246–247
colchicine, 368G, 45–46
collagen, 368G, 11
columnar, 368G
communication between cells, 29, 331–344
complementarity of DNA chains, 77–78
complementary DNA, 368G, 168
condenser lens, 368G, 3, 5F
conditional mutation, 368G, 120
connective tissue, 368G, 11
constitutive exocytotic vesicles, 205
continuous characters, 100
contraction of muscle, 55F
  activation by calcium, 323–324
control of blood flow, 336–339
cooking as chemistry, 232B
copying DNA by PCR, 186
corepressor, 156, 157F
cortex, 368G, 13F, 14
Corynebacterium diphtheriae, 197
cottage cheese, 209
covalent, 368G
c-ras, 368G, 70
crawling (of amoeboid cells), 56
Creutzfeldt–Jacob disease, 233B
Crick, Francis, 76, 82
Cri du chat syndrome, 113, 127
cristae, 369G, 34–35
crossing-over, 369G, 69, 110
crosstalk, 369G, 328
C-terminal, 369G
CTP, as an energy currency, 253
currency, of cellular energy, 251–256
cuticle, 13F, 14
cyanide, 259B
cyanobacteria, 369G, 10

cyclic adenosine monophosphate, 369G, 325–327
  dependent protein kinase, 281–282
  gated channel, 365G, 300B, 325–327
    deriviation during evolution, 228
  gene expression triggered by, 152, 155F
  intracellular messenger, 325–327
  skeletal muscle cells, role in, 324F, 328
  smooth muscle cells, role in, 337–338
  structure, 326F
cyclic guanosine monophosphate (cyclic GMP), 369G, 327–328
  elevation in response to NO, 333, 337–338
  gated channel, 328
  photoreceptors, role in, 327
  smooth muscle cells, role in, 337–338
cyclin, 63, 65F
cyclin B, 369G, 63, 65
cyclin D, 369G, 65
cyclin E, 369G, 65
cyclosporin A, 164B
cysteine
  disulfide bridging, 217, 218F
  genetic code, 84F
  protonation, 213
  structure, 212
cystic fibrosis, 345–353
  gene product, 229–230
  inheritance, 95, 106
  transmembrane regulator, 351
cystine, 369G, 217
cytidine triphosphate, as an energy currency, 253
cytochemistry, 369G, 3
cytochrome oxidase, 258
cytochromes, 258
cytokinesis, 369G, 68
cytology, 369G
cytoplasm, 369G
cytoplasmic
  dynein, 369G, 51
  inheritance, 100
  streaming, 369G, 56
cytosine, 370G, 73–74
  bonding with guanine, 76–77
cytoskeleton, 370G, 41–57
  changes at mitosis, 62
cytosol, 370G

DAG (diacylglycerol), 322
D-amino acids, 224B
DAPI, 43
dATP (deoxyadenosine triphosphate), 75
dCTP (deoxycytidine triphosphate), 75
DDT, 125
deadly nightshade, 339B
deamination, 370G, 140, 141F
death cap mushroom, 339B
degeneracy, of genetic code, 83
deletion, 370G, 111, 113, 127
Delgarno, 194
denaturation of proteins, 230
dendrite, 370G
deoxyadenosine triphosphate, 74–75
deoxycytidine triphosphate, 75
deoxyguanosine triphosphate, 75
deoxyribonucleic acid (DNA), 370G, 73–88
  binding protein, 80
    helix-turn-helix type, 229F
  blotting, 181
  chemical structure, 75F
  classes, 115–116
  cloning, 167–178
  content
    of *E. coli,* 80
    of human cell, 79
  damage, 140, 141F, 142F
  fingerprint, 371G, 181
  helicase, 136
  ligase, 371G, 168, 169F
  minisatellite, 115
  mutations, 127
  packaging in eukaryotes, 80–81
  packaging in prokaryotes, 80
  polymerase, 371G
    I, 138
    III, 136F, 137, 139
    thermostable, 186
  proof that it is the genetic material, 78
  purification, 167–178
  recombinant, 167–187
  repair, 371G, 140–144
  repetitious, 115
  replication, 371G, 133–140
  satellite, 115
  sequencing, 178–181
  staining for fluorescence microscopy, 43
  structure, 73–78
  synthesis
    control during cell cycle, 65, 70–71
    from RNA, 169F
    position in cell cycle, 59
deoxyribonucleotide, 370G, 73–74
deoxyribose, 370G
  in DNA, 73
  structure, 26B
deoxythymidine triphosphate, 75
dephosphorylation, 370G
  driving shape change of calcium ATPase, 295
  role in control of MPF and cell cycle, 65
depolarization, 370G, 303
  of postsynaptic cell, 336

deprotonation, 18B
depurination, 370G, 140, 141F
designer probes, 208B
desmin, 370G, 57
desmosomes, 370G, 57
detergents, as denaturing agents, 230
dGTP (deoxyguanosine triphosphate), 75
diabetes mellitus, 272B
diacylglycerol, 322
diakinesis, 370G, 69
dideoxy chain termination technique, 178–181
dideoxyribonucleotide, 370G
diet
  metabolism of food chemicals, 263–283
  requirement for essential amino acids, 278
differentiation, 370G
diffusion limit to enzyme-catalyzed reactions, 244
diffusion, simple, 395G, 24
  of steroid hormones, 162
digitalis, 259B
dihybrid cross, 370G, 92, 93F
dihydroxyacetone phosphate, in glycolytic pathway, 266
dimerization, of cysteine, 217, 218F
diphosphoglycerate, in glycolytic pathway, 266
diphtheria toxin, 197
Diplococcus pneumoniae, transformation by DNA, 78, 79F
diploid, 370G, 66
diplotene, 370G, 69
dipoles, in protein structure, 227B
dissolve, 16B
distal, 371G
disulfide bond (bridge), 371G
  formation, 217, 218F
  function, 227B
DNA, 371G, 73–88
  binding protein, 80
    helix-turn-helix type, 229F
  blotting, 181
  chemical structure, 75F
  classes, 115–116
  cloning, 167–178
  content
    of *E. coli,* 80
    of human cell, 79
  damage, 140, 141F, 142F
  fingerprint, 371G, 181
  helicase, 136
  ligase, 371G, 168, 169F
  minisatellite, 115
  mutations, 127
  packaging in eukaryotes, 80–81
  packaging in prokaryotes, 80
  polymerase, 371G
    I, 138
    III, 136F, 137, 139
    thermostable, 186
  proof that it is the genetic material, 78
  purification, 167–178
  recombinant, 167–187
  repair, 371G, 140–144
  repetitious, 115
  replication, 371G, 133–140
  satellite, 115
  sequencing, 178–181
  staining for fluorescence microscopy, 43
  structure, 73–78
  synthesis
    control during cell cycle, 65, 70–71
    from RNA, 169F
    position in cell cycle, 59
docking protein, 371G, 201, 203F
domain, 371G, 30, 226
  relation to intron in the gene, 115
dominant, dominance, 371G, 91
  deviations from Mendel's laws, 99B
  in human genetics, 95
donor (in hydrogen bond), 19B
double bonds in fatty acids, 39B
double helix, 371G, 76
Down's syndrome, 127
downstream, direction of DNA molecule, 371G, 146, 148F
driving of energetically unfavourable reactions, 252
Drosophila
  eye color, 97B, 106–107, 109F
  linkage map, 111, 112F
  notch mutant, 127
  vestigial wing, 107
drugs, manufacture in bacteria, 183
dTTP (deoxythymidine triphosphate), 75
duplication, 371G, 127
dynamic instability, 371G, 45, 46F
dynein, 371G, 50–52
dystrophy, muscular, 123

economics, of cell currencies, 251–262
Eco R1, 171T
ectoplasm, 372G, 54, 56, 56F
effective stroke, 372G, 48
egg
  dimensions, 2B
  white, 232B
electrical force, on ions at a membrane, 287
electrically excitable, 372G, 303–316
electrical signalling, 11, 309–314

electrochemical gradient, 372G
  chloride, 291, 292F
  H^+, 255
  potassium, 287
electron gun, 372G
electron microscope, 5F, 6–7
electron transport chain, 372G, 258
electrophoresis, 372G
electrostatic interactions, in protein structure, 227B
Elodea, 56
elongation factors, 196–197
elongation of protein chain at ribosome, 194–197
embedding for microscopy, 7B
endocrine gland, 338
endocytosis, 372G, 37, 202F
endocytotic vesicle, 372G, 37
endonuclease, 372G
  restriction endonucleases, 392G, 170
    in search for CF gene, 349
endoplasm, 372G, 54, 56,
endoplasmic reticulum, 372G, 36
  appearance, 4F
  changes at mitosis, 62
  rough, 9F, 36, 200–204
  site of fat synthesis, 278
  source of calcium ions, 320, 322F
endosymbiotic theory, 372G, 10
endothelium, endothelial cells, 372G, 336–337
energy
  currency, 372G, 251–256
  flow
    in aerobic cell, 257F
    in anaerobic cell, 261F
  trading within the cell, 251–262
engineering
  genetic, 167–187
  of proteins, 184, 208B
enhancer, 372G, 161
environment-gene interaction, 101–102
enzyme, 373G, 240–247
  assay, 241
  carrier action, 295–297
  concentration, effect on reaction rate, 244
  molecular switches, 246
  purification, 244
  regulation, 246
  substrate:enzyme complex, 240
epidermis, 373G, 13F, 14
epithelia, epithelium, 373G, 11
equilibrium, 373G
  chemical reaction, 241
  voltage, 373G, 287, 292F
    calculation, 288B
ER, 373G, 36
  appearance, 4F
  changes at mitosis, 62
  rough, 9F, 36, 200–204
  site of fat synthesis, 278
  source of calcium ions, 320, 322F
erythrocyte, 373G
Escherichia coli, 47
essential amino acids, 278
esterase, acetylcholine-specific, 336
ester bond, 373G, 22
ethyl methyl sulphonate, 125
euchromatin, 373G, 33
eukaryotic, eukaryotic cell, 373G
  distinguishing features, 8, 8T, 9F
  evolutionary origin, 10
evening primrose oil, 40B
evolution
  constraint applied by palette of amino acids, 218
  of cAMP gated channel, 228
  of eukaryotic cell, 10
exchange mechanisms, for energy currencies, 256–261
excision repair, 126, 143, 143F
  failure in Bloom's syndrome and Xeroderma pigmentosum, 144B
exocytosis, 373G, 202F, 205
  at axon terminal, 318–319
exon, 373G, 114–115
  correspondance with domain within protein, 228
exonuclease, 373G
  activity, of DNA polymerase III, 139f, 140
explosive, nitroglycerine, 338
expression (of a gene), 373G, 145, 149, 160
expression vector, 374G, 175
extracellular matrix, 374G, 11, 24, 29
eye
  color in *Drosophila*, 97B, 106–107, 109F
  light detection, 327
eyepiece, 3, 5F

F-actin, 374G, 53
Factor VIII
  manufacture in bacteria, 183
  manufacture in sheep, 184B
fat, fat cell, 374G
  adipose tissue, 11
  brown fat, 262B
  chemical stucture, 22–23
  as energy store, 272B
fatty acid
  breakdown, 269, 271F
  structure, 22
  synthesis, 274–278

feedback, 279
feedforward, 280–281
FGF, 31B, 339
FGFR, 31B
fibroblast, 374G, 11
  growth factor and receptor, 374G, 31B, 339
First Law of genetics, 374G, 92
fission yeast, 63
fitness, 98
fixation
  of carbon, 278, 279F
  of nitrogen, 278
  of tissue for microscopy, 7B
flagella, 44F, 46, 49F, 50F
flagellin, 48
flavin adenine nucleotide, 264
fluid mosaic, 374G, 23F, 24
fluorescence microscopy, 42–43
fly agaric mushroom, 338, 339B
fmet, 193
focal contact, 374G, 54, 56F
formyl methionine, 374G, 193
forward reaction, 241
foxgloves, 259B
fragile-X syndrome, 129
frameshift mutation, 374G, 85
Franklin, Rosalind, 76
free energy, 251
  of solute at concentration [I], 288B
  change during reaction, 251–253
frequency coding, 315B
frequency modulation, 315B
fructose, 26B
  6-phosphate and 1,6-bisphosphate
    in gluconeogenesis, 274
    in glycolytic pathway, 266
fumarase, 244T
fumarate, 266

G0, 374G, 60
G1, 374G, 60, 65
G2, 374G, 60
GABA, 210, 218
  at synapse, 343B
  receptor channel, 300B, 343B
G-actin, 374G, 53
galactose, 26B
  lac operon, 152–156
gamete, 374G
  formation, 68–69, 92
γ-amino butyrate, 210, 218
  at synapse, 343B
  receptor channel, 300B, 343B
gap 1, 374G, 60, 65
gap 2, 374G, 60
gap junction, 375G, 29, 32
  channel, 291, 298B
gastrocnemius muscle, 375G, 334–340
gated, gating (of channels), 375G, 32, 292
gelsolin, 375G, 56
gene, 375G, 91, 105
  environment-gene interaction, 101–102
  expression
    regulation in eukaryotes, 160–163
    regulation in prokaryotes, 149–157
  family, 375G, 115
  interaction, 99–100
  knockout, 187B, 352F
  probe, 375G, 177, 350
  therapy, 375G, 351, 352F
genetic
  code, 375G, 82–86
  counseling, 96
  engineering, 167–187
  fingerprinting, 183F
genetics
  classical Mendelian, 89–103
  somatic cell, 89
genome, 375G, 32, 105–118
  mapping, 107–114
genomic DNA clones, library, 375G, 175–178
genotype, 375G, 91
Gerard, 290B
germ cells, 375G, 66
Gherig, Lou, 131B
Gibbs free energy, 251
  of solute at concentration [I], 288B
  change during reaction, 251–253
gland
  adrenal, 338
  sweat, 347, 348F
glial cell, 375G, 11
glucocorticioid and receptor, 375G, 376G,
    162–163, 333
gluconeogenesis, 376G, 272–274
glucose, 25
  breakdown, 266–268
    in absence of oxygen, 268
  carrier, 376G, 235, 293, 300B
  energetics of phosphorylation, 252
  glycogen as source, 270F
  6-phosphate
    energetics of dephosphorylation, 252
    in gluconeogenesis, 274
    in glycolytic pathway, 266
  repression of lac operon, 155F
  storage as glycogen, 274
  structure, 25
  synthesis, 272–274

glutamate, 211
  charge on side chain, 217
  conversion to 2-oxoglutarate, 247F
  genetic code, 84F
  phosphorylation of side group, 218, 219F
  receptor channel, 173B, 299B
  structure, 211
  transmitter action, 318–319, 341B
glutamine, 84F, 214
gluteraldehyde, 7B
glyceraldehyde 3-phosphate, 266
glycerol, 375G, 22
glycine, 84F, 214
glycogen, 376G, 27–28
  breakdown, 269–270
    regulation by calcium and cyclic AMP, 281, 282F
  glucose source, 269–270
  phosphorylase, 376G
    action, 269
    regulation by phosphorylation, 281, 282F
  phosphorylase kinase
    narrow specificity, 240
    action, 281, 282F
  structure, 27–28
  synthesis, 274, 276F
glycolysis, 266–268
  in absence of oxygen, 268
  negative feedback control, 280
glycosidic bond, 376G, 27
glycosylation, 376G
  in ER, 202–203
  in Golgi apparatus, 204
glyoxisome, 376G
Golgi apparatus, 376G, 37
  changes at mitosis, 62
  function, 202F, 204
  location in cell, 4F
Gordian knot, 87
gout, 45
gramicidin A, 224B
grana, 376G, 34–35
gratuitous inducers, 155
Graves, Robert, 339B
Greeks, use of fly agaric, 339B
Griffith, Fred, 78
ground substance, 11
growth factor, 376G, 31B, 340
growth hormone, expression in transgenic mice, 184B
GTP
  as an energy currency, 253
  in protein synthesis, 194
guanine, 376G, 73–74
  bonding with cytosine, 76–77
guanosine triphosphate
  as an energy currency, 253
  in protein synthesis, 194
guanylate cyclase, 333, 337F, 338

H^+ gradient, 255
H^+ transport out of mitochondria, 258
hair, 209–210
hairpin loop, 376G, 148, 150F
haploid, 376G, 66–67
Hartridge, 248B
head group, 376G, 22F, 23
heart, 13
  disease, genetic basis, 101
heat
  as denaturing agent, 230, 232B
  generation in brown fat, 262B
helicase, 376G, 136
helix
  α-helix, 222–223
  DNA helix, 76–78
  helix destabilizing proteins, 136
  helix-turn-helix protein, 228, 229F
hemicellulose, 376G, 29
hemizygote, 376G, 106
hemoglobin, 376G, 231F
  allosteric effects, 236
  gene family, 115
  oxygen loading curve, 237F
  sickle cell anemia, 86, 127
hemophilia, 97B, 107–108
herbicide resistance, by transgenic technology, 187B
heritability, 377G, 101
hermaphrodite, 377G, 90
heterochromatin, 377G, 33
heterozygote, 377G, 91
Hiroshima, 121
histidine, 212
  genetic code, 84F
  essential amino acid, 278
  pH sensitivity, 213, 238
  protonation, 213, 238
  structure, 212
histone, 377G, 80–81
homophilic, 377G, 30B
homozygote, 377G, 91
Hooke, Robert, 1
hooves, 57
hormone, 377G, 331
  response element, 377G, 162
horns, 57
housekeeping gene, 377G, 161
HRE (hormone response element), 377G, 162
hsp90, 162F

human
DNA, introduction into cloning vector, 172F
genetics, 94–97
proteins, manufacture in bacteria, 183
Huntingdon's disease, 94, 130
hybridization, 377G
hydration shell, 17B
hydrogen bond, 377G, 19B
between bases in DNA, 76–77
in α helix, 222–223
in β sheet, 222, 225F, 226F
in protein structure, 222, 227B
hydrogen ion
gradient, 377G, 255
transport out of mitochondria, 258
hydrogen peroxide, 36
hydrolysis, 377G
hydrophilic, 378G, 23
amino acid side chains, 211
hydrophobic, 378G, 23
amino acid side chains, 211
interactions, in protein structure, 227B
3-hydroxybutyrate, 272B

ice-9, 233B
I cell disease, 206B
identical twins, detection by Southern blotting, 183F
imino acid, 217
immunofluorescence microscopy, 42–43
immunoglobulin gene family, 115
immunosuppressants, 164B
inactivation, 378G, 305F, 306, 309F
inbred lines, 101
Inclusion cell disease, 378G, 206B
independent assortment, 93
and linked genes, 109–110
indirect immunofluorescence, 378G, 42–43
induced fit model of enzyme action, 246
inducible operon, 378G, 152
infertility, 51
inheritance, 89–103
cytoplasmic, 100
multifactorial, 100–101
sex-linked, 106, 107F, 108F
inhibition
of acetylcholinesterase, 339B
of muscarinic acetylcholine receptor, 339B
of phosphofructokinase by ATP, 280
initial velocity, 378G, 241
initiation complex, 378G, 194
initiation factors, 194
initiation of protein synthesis
at bacterial ribosome, 193
at eukaryotic ribosome, 199
inosine, in transfer RNA, 191
inositol trisphosphate, 378G, 320, 321F
gated calcium channel, 378G, 299B, 320, 322F
input–output relations at a synapse, 341B
insecticides, 339B
in situ hybridization, 378G, 181
insulation, of axons, 312–313
insulin, 272B
manufacture in bacteria, 183
integral protein (of membrane), 378G, 24, 229
export to plasmalemma, 205
integration, of cell processes by intracellular messengers, 323
interaction of genes, 99–100
intercellular messengers, 331–344
interleukin 2, 164B
intermediate filament, 379G, 56–57
intermembrane space, 379G
mitochondrial, 34
nuclear, 32
interphase, 379G, 33, 59
intestine, 12F, 29
intracellular
messenger, 379G, 317–330
receptor, 379G, 333
transport, 51–53
intron, 379G, 114–115
inversion, 379G, 127
ion, 379G
binding site on protein, 217
force due to voltage, 287
hydration in solution, 17B
membrane phenomena, 285–298
ionizing radiations, 378G, 121, 124
ionotropic cell surface receptor, 379G, 332
IP_3, 379G, 320, 321F
gated calcium channel, 379G, 299B, 320, 322F
IPTG (isopropylthio-β-D-galactoside), 155, 175
isocitrate, 265
isoleucine, 215
essential amino acid, 278
genetic code, 84F
isomers, 379G, 25B
isopropylthio-β-D-galactoside, 155, 175

Jaffe, 303
jellyfish, 208B

Kartegener's syndrome, 51
k_{cat}, 240
keratin, 379G, 57, 210
ketone bodies, 272B
Khorana, Gobind, 83

kinesin, 380G, 51–52
kinetics, of chemical reaction, 242–243
kinetochore, 380G, 66
K_M, 242–243
knockout, of gene, 187B, 351
Koshland, 245
Krebs cycle, 380G, 264–266
Krebs, Hans, 264

lac operon, 380G, 152–156
  in DNA cloning, 175
lac repressor protein, 152, 154F, 236
lactate, 17B, 268
lactic acid, 17B
lactose, 380G, 27
  operon, 152–156
lagging strand, 380G, 136F, 138
lambda (bacteriophage), 82, 175
lamins, 380G, 57
law of independent assortment, 93
  fails for linked genes, 109F, 110
law of segregation, 92
lead citrate, 7B
leader sequence, 193
leading strand, 380G, 136F, 138
lemon juice, as denaturing agent, 232B
lens, of the eye, 209
leptotene, 380G, 69
lethal alleles, 97–98
lethal genes, 98
leucine, 215
  essential amino acid, 278
  genetic code, 84F
  structure, 210F, 215
library, of DNA clones, 171
ligand, 380G
  protein:ligand binding, 209
light
  detection, 327
  microscope, 3–5
  -sensitive nerve cells, 327
lignin, 380G, 29
limit, to speed of enzyme-catalyzed reactions, 244
Ling, 290B
linkage, 380G, 111
  analysis, in search for CF gene, 350
  map, 380G, 111–112
linker DNA, 380G, 80
linoleic acid, 39B
linolenic acid, 40B
lipid, 380G, 22–23
  bilayer, 380G, 23
  breakdown, 269
lipolysis, 269
liver cell
  regulation of glycogen breakdown, 282F
  lactate oxidation, 268
local anaesthetic, 314
locus, loci, 380G, 91
longevity, of transmitters in extracellular fluid, 331
lumen, 380G
lung cancer, genetic component, 102
lymphocyte, 164B
lysine, 212
  essential amino acid, 278
  charge on side chain, 217
  genetic code, 84F
  maize enrichment, 187B
  structure, 212
lysosome, 380G, 38, 202F, 205, 206F
lysozyme, 222
  specificity constant, 244T
  structure, 220–222

μ, 3B
M13 (bacteriophage), 82
MacLeod, Colin, 78
macrophage, 380G, 11
mad cow disease, 233B
maize, improvement by transgenic technology, 187B
major groove, 381G, 76–77
malaria and the sickle-cell gene, 99B
malate, 266
  dehydrogenase, regulation by calcium, 281
maleness determination, 114, 117B
malignant, 381G, 70
malonyl CoA, 275
mannose, 26B
mannose-6-phosphate and lysosome targeting, 205–206
Manx cat, 98
mapping the genome, 111–114
Marfan's syndrome, 11
matrix, 34–35
maturation promoting factor, 381G
maximal velocity, 381G, 242, 243F
McCarty, Maclyn, 78
measurement of membrane voltage, 290B
medieval saints, 272B
meiosis, 381G, 67–69
melanin, 381G, 217
membrane
  barrier function, 285
  fluidity, 39B
  structure, 23F
  voltage, 381G, 285–298
    effect on sperm fusion, 304
    measurement, 290B

Mendel, Gregor, 89–103
  laws, 89–103
    first law, 92
    second law, 93
      fails for linked genes, 109F, 110
Menten, Maud, 243
Meselson, Matthew, 134
Meselson-Stahl experiment, 134–135
messenger RNA, 381G, 145
  processing after transcription, 158–160
  release from template DNA, 150F
  splicing, 158F, 159, 159B
messengers, intercellular, 331–344
metabolism, 381G, 263–283
metabotropic cell surface receptor, 381G, 332
metallothionein gene expression, 184B
metaphase, 381G, 67–68
  plate, 381G, 68
methionine, 216
  essential amino acid, 278
  formyl, 193
  genetic code, 84F
  in protein synthesis, 193, 199
  structure, 216
methylation of mRNA, 159
Michaelis, Leonor, 243
  constant, 381G, 242–243
Michaelis–Menten equation, 243
microfilament, 381G, 53–56
micron, 3B
micropipette, 290B
microscope, 2–7
Microspora, 10
microtubule, 381G, 41–53
  location in cell, 4F
  molecular motor, 382G, 50–53
  organizing center, 382G, 44
  spindle, 68
microvilli, 382G, 11–12, 53
millimeter, 3B
minisatellite DNA, 115, 181
minor groove, 382G, 76–77
mirror images, 224B
missense mutation, 382G, 85F, 86
mitochondria,mitochondrial, 382G
  appearance, 34–35
  calcium channel, 281, 299B
  function, 251–262
  genetic entities, 100
  inner membrane, 258, 260, 382G
  Krebs cycle, 264–266
  matrix, 382G
  origin as bacteria, 10
  outer membrane, 382G
mitosis, 382G, 66–68
  nuclear reorganization, 33
  spindle structure, 44, 45
  stimulation by growth factors, 31B
μm, 3B
mm, 3B
mobile DNA, 129
molecular motor, 51
monomer, 382G, 73
mononucleate, 382G
monosodium glutamate (MSG), 318
Morgan, Thomas Hunt, 106–107
motility, 46–56, 66
motoneurone, 382G, 336
  disease, 131B
MPF (M-phase promoting factor), 382G, 62, 65F
M phase, 380G, 60
  promoting factor (MPF), 62, 65F
mRNA, 382G, 145
  processing after trancription, 158–160
  release from template DNA, 150F
  splicing, 158F, 159, 159B
MSG (monosodium glutamate), 318
MTOC (microtubule organizing center), 382G, 44
multifactorial inheritance, 100–101
multinucleate, 383G
multiple alleles, 97B
muscarine, 339B
muscarinic acetylcholine receptor, 338, 339B
  inhibition by atropine, 339B
muscle, 383G, 13
  activation, 323, 335–336
  calf, 334
  fibers, 14
  gastrocnemius, 334–340
  glycogen breakdown, 282F
  mechanism of contraction, 54
  skeletal, 13, 334
  smooth, 13, 336
muscular dystrophy, 123
mushroom
  fly agaric, 338, 339B
  death cap, 339B
mutagenesis, site-directed, 184
mutagen, mutagenesis, 383G, 123–125
mutation, 383G, 119–131
  CF, 351
  frameshift, 85
  missense, 85F, 86
  nonsense, 85F, 86
  rate, 123T, 119–123
  repair, 125
  types, 85F, 126–128, 140
mutator gene, 128
myelin, 383G, 313

myoglobin, 236
myosin, 54–56

Na^+/Ca^+ exchanger, 294–295, 301B
Na^+/K^+ ATPase, 259, 259B, 295, 301B
Na^+ gradient, as an energy currency, 256
NAD^+, NADH, 253–254
NADH dehydrogenase, 258
NADH oxidation, 258
nanometre, 3B
nature or nurture, 101–102
necrosis, 383G
negative feedback, 383G, 279
  in trp operon of *E. coli*, 156
negative regulation of transcription, 152, 383G
Neher, 290B
Nernst equation, 383G, 288B
nerves, nerve cell, 383G, 11
  autonomic, 336
  axon, 310–314
    terminal, 318
  electrical signalling, 310–314
  intermediate filaments, 57
  light sensitive, 327
  motoneurone, 336
  scent-sensitive, 325
  voltage gated sodium channel, 309
nerve gases, 339B
neurofilaments, 383G, 57
neuron, 383G
neuron (nerve cell), 11
  autonomic, 336
  axon, 310–314
    terminal, 318
  electrical signaling, 310–314
  intermediate filaments, 57
  light sensitive, 327
  motoneurone, 336
  scent-sensitive, 325
  voltage gated sodium channel, 309
neutral pH, 17B
NFAT, 164B
nicotinamide adenine dinucleotide, 383G, 253–254
nicotine, 18B, 339B
nicotinic acetylcholine receptor, 299B, 332, 336
nightshade, 339B
Nirenberg, Marshall, 83
Nitella, 56
nitric oxide (NO), 383G, 333, 337F, 338–339
  receptor, 333
nitrogen fixation, 278
nitroglycerine, 338
nm, 3B
NO (nitric oxide), 383G, 333, 337F, 338–339
  receptor, 333
node, 383G, 313
non-disjunction, 383G, 127
non-muscle myosin, 54
non-polar, 15B
non-secretor allele, 99
nonsense mutation, 384G, 85F, 86
noradrenaline, 332, 336
NO receptor, 333
nose, scent receptors, 325
notch mutant in Drosophila, 127
N-terminal, 383G
nuclear bombs, accidents and mutation, 121–122
nuclear
  calcium, detection with aequorin, 208B
  envelope, 384G, 4F, 32
    changes at mitosis, 62
  lamina, 384G, 32, 57
  pore, 384G, 4F, 32
nuclease, 384G
nucleic acid, 384G, 73–88, 133–149
nucleolar organizer region, 384G, 33
nucleolus, 384G, 33
  appearance, 4F
  changes at mitosis, 62
nucleoside, 384G, 75
  triphosphates, as energy currencies, 253
nucleosome, 384G, 80–81
nucleotide, 384G
  in DNA, 73
  in RNA, 145
nucleus, nuclear, 384G, 32–34
  appearance, 4F, 33F
  calcium, detection with aequorin, 208B
  envelope, 384G, 4F, 32
    changes at mitosis, 62
  lamina, 384G, 32, 57
  pore, 384G, 4F, 32
numbering of bases on DNA, 146, 148F

objective lens, 384G, 3, 5F
O blood group, 97B, 99, 99B
octane, 15B
odor of sanctity, 272B
Okazaki, Reiji, 138
  fragment, 384G, 138, 136F
oleic acid, 39B
olfactory neurone (scent-sensitive nerve cell), 384G, 325–326
oligonucleotide, 384G
oligosaccharide, 384G, 27B
  transferase, 384G, 204
olive oil, 22
oncogene, 384G, 70

oocyte, 385G
formation, 69
in cloning of a receptor protein, 173B
open promoter complex, 385G, 148
operator, 385G, 152
operon, 385G, 151–157
optical isomer, 25B
organelle, 385G, 8, 21, 32–38
detection of calcium concentration, 208B
origin of replication, 385G, 134
osmium tetroxide, 7B
osmosis, 385G
outer doublet microtubules, 385G, 50
oxaloacetate, 265F
de novo generation from pyruvate, 268
in gluconeogenesis, 274
in Krebs cycle, 266
oxidize, 385G
2-oxoglutarate, 265F
conversion to glutamate, 247
in Krebs cycle, 265
oxoglutarate dehydrogenase, regulation by calcium, 281
oxygen loading curve of hemoglobin and myoglobin, 237F

$p34^{cdc2}$, 62, 65F
as regulated enzyme, 246
p53 tumor supressor protein, 70
mutant forms and susceptibility to cancer, 101
pachytene, 385G, 69
pain receptor, 385G, 310–314
synapse, 319F, 341–344
pain relay cell, 385G, 310
palette of amino acids, constraints on evolution, 218
palmitic acid, 278
paracrine transmitters, 385G, 332
parallel β sheet, 222, 225F
paramecium, 49, 50F
parenchyma, 13F, 14
patch clamp, 385G, 290B
in cloning of glutamate receptor, 173B
PCBs as mutagens, 125
PCR, 385G, 185–186
peas, in Mendel's experiments, 90–94
pectin, 385G, 29
pedigrees, 94
peptide, 385G
peptide, 210
bond, 385G, 210
generation on ribosome, 194
peptidyl site (P site), 386G, 192
peptidyl transferase, 386G, 194, 196F, 210
pericentriolar material, 44
peripheral protein, 386G, 24
permissive, 386G, 64, 121
peroxisome, 386G, 36, 202F
pH, 17B
cytoplasmic, 19B
denaturing effect, 230, 232B
gradient, as an energy currency, 255
protein shape affected by, 238, 239F
receptor/ligand disassociation, 206F
phage, 386G
phalloidin, 339B
phase contrast microscopy, 386G, 5, 6F
phenotype, 386G, 91
phenylalanine, 215
essential amino acid, 278
genetic code, 84F
hydroxylase, 271
toxicity in phenylketonuria , 271
UV absorbance, 217
phenylketonuria, 271
phenylpyruvate, 271
phloem, 13F, 14
phosphate, 386G, 252
phosphatidylcholine, 22F
phosphatidylinositol bisphosphate, 386G, 321
phosphatidylinositol phospholipase C, 321
phosphodiester bond, 386G, 76, 147F
phosphoenolpyruvate, 267
phosphofructokinase, 266
regulation, 280
3-phosphoglycerate, 267F
in carbon fixation, 279
in gluconeogenesis, 273F
in glycolytic pathway, 267
phospholipase C, 386G, 321
in activation of endothelial cells, 338
phospholipid, 386G, 23
phosphorylase kinase, 387G, 240, 281, 282F
phosphorylation, 387G, 218
amino acid side group, 218, 219F
control of gene expression, 31B, 164B
control of MPF and cell cycle, 63, 65F
effect on protein shape, 238–239
of calcium ATPase, 295–296
energy cost, 252
generator of motile force, 240
mediator of intracellular messenger effects, 328
protein targeting, 205–206
phosphoryl group, 387G, 218
photoreceptors, 327
phototaxis, 47
phragmoplast, 387G, 68
physical map, 387G, 111–114
Pi, 387G
pigment cells, 51

PIP_2 , 387G, 321
pipette, for electrical measurements on cells, 290B
PKA (Protein kinase A), 387G, 328
  regulation of glycogen breakdown, 281
pK_a, 18B
  of amino acid side chain , 213
plant, plant cells, 13F
  unique features, 8
  carbon fixation, 278
  cytoplasmic streaming, 56
plaque, 387G, 175
plasmalemma, 387G, 21
  sodium gradient, 256
  endocytosis, 37
plasma membrane, 387G
plasmid, 387G, 80
  in DNA cloning, 170
  in gene therapy for CF, 352F
plasmodesmata, 387G, 9F
platelets, 320
PLC, 387G, 321
plug, inactivation, 305F, 306
pneumonia, 78
polar body, 69
polarity, of DNA chain, 76
polar molecule, 15B
polyadenylation, 387G, 159
poly A tail, 388G, 159
polychlorinated biphenyls as mutagens, 125
polycistronic mRNA, 388G, 151
polymerase, 388G
  I, 138
  III, 136F, 137, 139
  chain reaction, 388G, 185–186
  thermostable, 186
  RNA, 146, 148
polymorphism, 388G, 128
polypeptide, 388G, 210
  synthesis, 189–208
    in bacteria, 193–199
    in eukaryotes, 199
polyploid, 388G, 126
polyribosome, 388G, 197
polysaccharide, 27B
polysome, 388G, 197
porin, 388G, 292, 298B
positive feedback, 388G, 280
  in action potential, 308
positive regulation of transcription, 388G, 155
postsynaptic cell, 388G, 320
potassium
  channel, 286, 298B
  concentration gradient, 286
  movement into cytosol, 259
preferred states, of proteins, 235
pregnancy test kit, 209
preinitiation complex, 388G, 160, 161F
presynaptic terminal, 388G, 320
Pribnow box, 148
primary
  antibody, 42
  cell wall, 388G
  endosome, 388G, 37
  structure, 388G
    of lysozyme, 220F
    of proteins, 222
primase, 389G, 136F, 138
primer, 389G, 136F, 138
primosome, 389G
processed pseudogenes, 116
processing, of mRNA, 158–160
product, of reaction, 240
profilin, 389G, 53
programmed cell death, 389G, 59
projector lens, 389G, 3, 5F
prokaryote, prokaryotic, 389G, 8, 9F, 10
proline, 216
  genetic code, 84F
prometaphase, 389G, 66–67
promoter, 389G, 148
pronuclei, 389G
proof reading, during DNA synthesis, 139
prophase, 389G, 66–67
prophase I, 389G
prophase II, 389G
prosthetic group, 389G, 230, 246
protein, 389G, 209–249
  basic structure, 211
  binding to protein, 30B-31B
  catalysts, 240
  domains, relation to introns in the gene, 115
  energy source, 269–271
  engineering, 184
    in production of calcium probes, 208B
  kinase, 389G
    in cell cycle control, 63, 65F
  kinase A, 390G, 328
    in regulation of glycogen breakdown, 281
  phosphorylation, 390G, 218
    control of gene expression, 164B
    control of MPF and cell cycle, 63, 65F
    effect on shape, 238–239
      of calcium ATPase, 295–296
    generator of motile force, 240
    mediator of intracellular messenger effects, 328
  primary structure, 222
  quaternary structure, 230
  secondary structure, 222
  size and molecular weight, 210

structure
dynamic nature, 235
primary, 222
quaternary, 230
secondary, 222
stabilization, 227B
tertiary, 225
synthesis, 189–208
in bacteria, 193–199
in eukaryotes, 199
targeting, 390G, 200, 202F
error in inclusion cell disease, 206B
to lysosome, 206F
tertiary structure, 225
proteoglycan, 24
protofilament, 390G, 45
proton, 17B
protonation, 18B
of cysteine, 213
of histidine, 213
proto-oncogene, 390G
protozoa, 390G
proximal, 390G
PrP, in mad cow disease, 233B
pseudogene, 390G, 116
pseudopodium, 390G, 54, 56F
P-site, 385G, 192
Pst 1, 171T
pure line, 90
purine, 390G, 73
bonding with pyrimidine, 76–77
puromycin, 200
purple eye gene of *Drosophila,* 107, 109F
pyrenoid, 390G
pyridoxamine, pyridoxal phosphate, 246–247
pyrimidine, 390G, 73
bonding with purine, 76–77
pyrophosphate, 390G
pyruvate, 267F
conversion to alanine, 247F
dehydrogenase, 268
regulation by calcium, 281
generation from glucose, 266–268
in gluconeogenesis, 274
uses in cell, 268

quaternary structure, 390G, 230
Queen Victoria, 108F

radiation as mutagen, 121, 124
rapid reaction techniques, 248B
ras, ras oncogene, 390G, 31B, 70
rate of mutation, 119–123
Rb (retinoblastoma), 391G, 71
R bacteria, 78
reading frame, 391G, 82F, 85
receptor, 391G
α-adrenergic, 332, 334F, 336
ADP, 332
β-adrenergic, 324F, 328, 332, 338
glucocorticoid, 333
intracellular, 333
ionotropic cell surface, 332
-mediated endocytosis, 391G, 37
metabotropic cell surface, 332
muscarinic acetylcholine, 338, 339B
nicotinic acetylcholine, 332, 336
nitric oxide, 333
steroid, 333
tyrosine kinases, 340
recessive, 391G, 91, 95
lethal gene, 98
reciprocal cross, 92, 106
recognition sites, of restriction enzymes, 171T
recombinant, recombination, 391G, 92, 111
DNA, 167–187
plasmid, 391G, 170–171, 172F
proteins, 183
recovery stroke (of cilium), 391G, 48
recycling, of ribosome subunits, 198
red eye gene, of *Drosophila*, 106
reduce, reduction, 391G
reducing agent, 391G, 253
refractive index, 391G, 5
regulated exocytosis, 205, 318
regulation
of blood flow, 336–339
of enyme activity, 246
of gene expression, 149–157, 160–163
release factors, 197, 199F
Renaissance, 329B
repair of DNA after mutation, 125, 140–143
repeats, tandem, 115
repetitious DNA, 391G, 115
replication
fork, 391G, 134, 136F
of DNA, 133–140
repressible operon, 391G, 156
repressor, repressor protein, 152, 156
RER, 9F, 36, 200–204
residual bodies, 38
residue, 392G
resistance to antibiotics, in bacteria, 120
resolving power, 392G, 2– 3
resting voltage, 392G, 287
restriction enzymes (endonucleases), 392G, 170, 171T
in search for CF gene, 349
restriction fragment length polymorphisms, 350

restrictive, 392G, 64, 121
reticulocyte, 392G
retinoblastoma, 71
retrograde, 392G
retrograde transport, 51
retrovirus, 392G
  and cancer, 70–71
  exception to central dogma, 79
reverse
  transcription, 392G, 79
  transcriptase, 392G, 168
Rhesus blood group, 95
ribonuclease, 392G
  H, 392G, 138, 168, 169F
  specificity constant, 244T
ribonucleic acid (RNA), 392G
  as catalytic molecule, 240
  polymerase, 146–149
    in eukaryotes, 157
  primer, in DNA synthesis, 136F, 138
  splicing, 159, 159B
  structure, 145, 147F
  synthesis, 145–165
    in eukaryotes, 157
  transfer, 190F
  types, 145
ribonucleoside monophosphate, 392G
ribose, 392G, 26B
  in RNA, 145
ribosome, ribosomal , 392G, 33, 37, 192
  -binding site, on mRNA, 193
  location in cell, 4F
  initiation complex, 195F
  recycling, 198
  ribosomal RNA, 145
ribozyme, 392G, 240
ribulose bisphosphate, 278
  carboxylase, 278
RNA, 392G
  catalytic molecules, 240
  polymerase146–149
    in eukaryotes, 157
  primer, in DNA synthesis, 136F, 138
  splicing, 393G, 159, 159B
  structure, 145, 147F
  synthesis, 145–165
    in eukaryotes, 157
  transfer, 190F
  types, 145
root, root hair, 14
rotary motion, 48
rough endoplasmic reticulum, 393G, 9F, 36, 200–204
Roughton, 248B
rRNA, 393G, 33, 145
Rsa 1, 171T
ryanodine, 393G, 329B
  receptor, 393G, 329B

Saccharomyces, 63–64
saints, odor of sanctity, 272B
Sakmann, 290B
Salmonella typhimurium, 47
saltatory conduction, 393G, 314
salt bridge, 393G, 217, 227B
salty sweat, in cystic fibrosis, 347
sanctity, the odor, 272B
Sanger, Frederick, 178
sarcomere, 393G, 54
sarcoplasmic reticulum, 393G
satellite DNA, 393G, 115
saturated, 393G
saturation
  kinetics, 393G, 242
  of fatty acids, 393G
S bacteria, 78
scanning electron microscope, 5F, 7
scent-sensitive nerve cells, 325–326
schizophrenia, 113B
Schizosaccharomyces, 63–64
Schleiden, Matthias, 1
Schwann, Theodor, 1
  Schwann cell, 394G
scrapie, 233B
sea cucumber, 53
sea urchin egg, 303
secondary
  antibody, 42
  cell wall, 394G
  structure, 394G
    of lysozyme, 221F
    of protein, 222
Second law, 394G, 93
  fails for linked genes, 110
secretion granules, 204
secretor allele, 99
secretory vesicles, 394G
sedimentation coefficient (S value), 394G, 192
segregation, 394G, 92
selection
  of cDNA clones, 176F
  of genomic DNA clones , 177F
self-compatible, 394G, 90
self-correcting enzyme, 140
selfing, 394G, 90
semi-conservative replication, 394G, 133, 135F
sequencing
  gel, 178F
  of DNA, 178–181

serine, 213
  genetic code, 84F
  phosphorylation of side group, 218, 219F
sex
  chromosomes, 394G
  determination, 114
  linkage (in inheritance), 394G, 106
σ factor, 393G, 148
SH2 domain, 31B
Shine–Dalgarno sequence, 394G, 194
Siamese cat, 101
sickle-cell anemia, 86
  complications with dominance, 99B
  nature of mutation, 86, 127
  treatment, 345
side chain, 394G, 210
sigma factor, 394G, 148
signal peptidase, 394G, 201
signal recognition particle and receptor, 395G, 201, 203F
signal sequence, 394G, 200, 203F
silk, 209
simple diffusion, 395G, 24
  of steroid hormones, 162
single-copy genes, 115
single stranded binding protein, 395G, 136
sister chromatids, 395G, 66
site-directed mutagenesis, 184
skeletal muscle (fibers), 395G, 13
  activation, 336
  control of energy production, 281, 328
  mechanism of contraction, 54
Sma H1, 171T
small nuclear RNAs, 395G, 159
smell detection, 325
smooth endoplasmic reticulum, 395G, 36
  site of fat synthesis, 278
  source of calcium ions, 320
smooth muscle (cells), 395G, 13
  in blood vessels, 336
snRNAs (small nuclear RNAs), 159
sodium
  action potential, 395G, 309
  /calcium exchanger, 395G, 294, 325
  channel, voltage gated, 309
  chloride, dissolving, 16B
  gradient, 395G, 256
  /potassium ATPase, 395G, 259, 289
solubility, of polypeptides, 210–217
solute-gated channel, 320, 325, 332
solvent, 16B
somatic cell, 395G, 66
  genetics, 89
sorting signal, 396G, 164B, 200
sorting vesicle, 396G, 205
Southern blotting, 181
specific activity, 396G, 244
specificity constant, 396G, 244, 244T
sperm, spermatid, 396G, 69
  actin in acrosome, 53
  formation, 69
  fusion, effect of membrane voltage, 304
  motility, 51
SPF (S-phase promoting factor), 65
S phase, 393G, 59
S-phase promoting factor (SPF), 65
spindle, 396G
  function, 66
  microtubule structure, 44, 45
  trigger to formation, 62
spliceosome, 396G, 159
splicing, of mRNA, 159, 159B
spongiform encephalopathy, 233B
spot desmosomes, 29
squamous, 396G
SRY (sex determining region, Y chromosome), 117B
SSB (single stranded binding protein), 136
stabilization of protein structure, 227B
Stahl, Franklin, 134
START (control point in cell cycle), 396G, 60F, 62
start codon, 85, 193, 199
starvation, 272B
stearic acid, 39B
stereocilia, 396G
sterile male technique, 98
steroid hormone, 396G
  receptor and mechanism of action, 162, 333
  simple diffusion, 24
  synthesis, 36
sticky ends, 396G, 170
stoma, stomata, 14
stop codon, 85, 197, 199F
stopped flow, 248B
stratified epithelium, 396G, 11
streptomycin
  action, 200
  resistance, 120
stress and fear, 338
stress fiber, 396G, 53
striated muscle, 396G, 13
  activation, 336
  control of energy production, 281, 328
  mechanism of contraction, 54
stroma, 35F, 278
substrate, of enzyme, 240
succinate, 265
sugar–phosphate backbone, of DNA, 76
sugar, 25B-28B
summation at synapse, 397G, 341B

superoxide dismutase, 397G, 126
  motoneurone disease, 131B
S value, 393G, 192
sweat gland, 347
swimming
  bacterial, 47–48
  cilia and flagellae, 49F
synapse, 397G, 320
  calcium and exocytosis, 320
  effect on postsynaptic cell, 336, 341B
  transmitter, 331
synapsis, 397G, 69
synaptonemal complex, 397G, 69
synthesis, 272–278
  amino acids, 278
  DNA, 133–140
  fatty acids, 274–278
  glucose, 272–274
  glycogen, 274, 276F
  proteins, 189–208
  RNA, 145–151

tactic response, 47–48
tail (of phospholipid), 23
tandem repeats, 397G, 115
tangling, of chromosomes, 87–88
targeting, of proteins, 200
  to lysosome, 205
targets of raised cytosolic calcium, 323
TATA box, 397G, 160
taxol, 397G, 45
Tay–Sachs disease, 97B
telophase, 397G, 67–68
telophase I, 397G
telophase II, 397G
temperature-sensitive mutation, 397G, 121
  in cell cycle research, 63–64
  reduction of protein stability, 232
  Siamese cat, 101
template, for DNA synthesis, 134
tendon, Achilles, 334
terminal differentiation, 397G, 60
termination
  of protein synthesis, 197
  of RNA synthesis (transcription), 148
tertiary structure, 397G
  of lysozyme, 220F
  of proteins, 225
testis determining factor, 114, 117B
tetracycline, 200
tetraploid, 397G
TFIID (transcription factor IID), 160
TGN (trans Golgi network), 204
thalassemia, 128, 159B
thermogenin, 398G, 262B
thermostable DNA polymerase, 186
thick filament, 398G, 54
thin filament, 398G, 54
three dimensional structure, of proteins, 219–230
  specification in primary structure, 230
three point cross, 111
threonine, 213
  genetic code, 84F
  essential amino acid, 278
  phosphorylation of side group, 218, 219F
threshold voltage, 398G, 308
  requirement for many presynaptic action potentials, 341B
thylakoid, 398G, 35
thymidine triphosphate, as an energy currency, 253
thymine, 398G, 74
  bonding with adenine, 76–77
  dimer, 126, 140
  in DNA helix, 73
Thyone, 53
tight junctions, 398G, 29
Ti plasmid, 398G
tissue, 398G, 1
T lymphocyte, 164B
tobacco smoke, as mutagen, 125
tomato, transgenic, 187B
topoisomerase, 398G
  I, 136
  II, 87–88
transcription, 398G, 145–165
  bubble, 148
  complex, 398G
  factor, 398G, 160
transepithelial voltage, 348F
trans-face, 398G
transfection, 398G
transferrin, 398G
transfer RNA, 398G, 190
transfer vesicles, 204
transformation, 399G
  by heat killed bacteria, 78
  by recombinant plasmid, 171
transgenic animal, 399G, 184B
  as model for CF, 351
transgenic plants, 187B
trans Golgi network, 202F, 204
transition state, of chemical reaction, 245F
translation, 399G, 189–208
translocation, 399G
  of parts of chromosomes, 113, 127
  of ribosome along mRNA, 194
transmembrane proteins, 399G
transmembrane voltage, 399G, 285–298
  measurement, 290B

sperm fusion, effect on, 304
voltage clamp, 401G, 304
voltage-gated calcium channel, 401G, 305
voltage-gated sodium channel, 401G, 309
transmission electron microscope, 5F, 6
transmitter, 399G, 318, 331–334
transpiration, 14
transport vesicle, 399G, 205
transposable elements, 128
transposase, 399G, 129
triacylglycerol, 399G, 22
tricarboxylic acid cycle, 399G, 264–266
triglyceride, 22
trioleoylglycerol, 22
trisomy-21, 127
tRNA, 399G, 190
troponin, 399G, 323
trp operon, 156
trypsinogen, indicator of pancreatic disease, 244
tryptophan, 216
essential amino acid, 278
genetic code, 84F
operon, 399G, 156, 280
synthesis, 156, 280
UV absorbance, 217
ts factor, 197
TTP, as an energy currency, 253
tubulin, 399G, 45
tu factor, in protein synthesis, 197
tumor, 399G, 70–71
suppressor gene, 399G, 70–71
turgid, turgidity, 399G, 9
Turner's syndrome, 114
turnover number, 400G, 240
two-dimensional electrophoresis, 122
types of RNA, 145
tyrosine, 214
genetic code, 84F
kinase, 400G, 31B
receptors, 340
phosphorylation of side chain, 218, 219F
UV absorbance, 217

UDP glucose synthesis, 275F
ultramicrotome, 400G, 7B
ultrastructure, 400G
ultraviolet light (UV)
absorbance, 400G, 217
as mutagen, 124
production of thymine dimers in DNA, 140
supersensitivity in Xeroderma pigmentosum, 144B
unambiguity, of genetic code, 83
uncoupler, uncoupling, 400G, 262B
uniformity, increase with time, 251
unique properties of individual amino acids, 217
unsaturated fatty acid, 400G
untangling, of chromosomes, 87–88
untranslated sequence, 400G, 193
unwinding, of DNA double helix, 136
upstream, 400G, 146, 148F
uracil, 400G, 146F
genetic code, 83
in RNA, 145
uracil-DNA glycosidase, 400G, 141
uranyl acetate, 7B
urea, as denaturing agent, 231
uridine diphosphate glucose synthesis, 275F
uridine triphosphate (UTP), as an energy currency, 253
UV light
absorbance, 217
as mutagen, 124
production of thymine dimers in DNA, 140
supersensitivity in Xeroderma pigmentosum, 144B

vaccines, manufacture in bacteria, 183
vacuole, 400G, 9
valine, 215
genetic code, 84F
essential amino acid, 278
valinomycin, 224B
Valium, 400G, 344B
van der Waals interactions, 227B
van Leeuwenhoek, Anton, 1, 41
variable number tandem repeats, 400G, 181
variation in populations, 101
variegation in leaves, 100
vascular tissue, 400G
in plants, 14
vasoconstrictor, 401G, 336
vector, 401G
expression, 175
for DNA cloning, 169, 170F
velocardiofacial syndrome, 113B
vesicle, 401G
vestigial wing gene, of *Drosophila*, 107
Vibrio cholera, 325
Victoria (British Queen), 108F
vikings, use of fly agaric, 339B
villin, 401G, 53
villus, villi, 401G, 11
vimentin, 401G, 57
vinegar, as denaturing agent, 232B
viral oncogene, 401G
virus, 401G, 82
viscosity (of cytoplasm), 53
vitamin B_6, 246
V_m (maximal velocity), 242

VNTRs, 401G, 181
v_0, 241
voltage, 285–298
  clamp, 401G, 304
  -gated calcium channel, 401G, 305
  -gated sodium channel, 401G, 309
  measurement, 290B
  sperm fusion, effect on, 304
  transepithelial, 348F
v-ras, 401G, 70

water, 15B
Watson–Crick model of DNA structure, 76
Watson, James, 76
wheat, polyploid genome, 126
white eye gene, of *Drosophila*, 106
Wilkins, Maurice, 76
wobble, 401G, 190

Xenopus oocyte, as expression system, 173B
Xeroderma pigmentosum, 401G, 126, 144B
X-rays
  mutagen, 124
  solving DNA structure, 76
xylem, 14

yeast, 1, 2B
  energy flows when anaerobic, 260
  in cell cycle research, 63–64
yew, 45

Z-disc, 401G, 54
Z-DNA, 401G, 78
Zellweger's syndrome, 402G, 36
zinc finger, 402G, 162
zygote, 402G
zygotene, 402G, 69